Progress in Pesticide Biochemistry and Toxicology

Volume 5

Insecticides

Edited by

D. H. Hutson

and

T. R. Roberts

Shell Research Limited, Sittingbourne

A Wiley–Interscience Publication

JOHN WILEY & SONS

CHICHESTER · NEW YORK · BRISBANE · TORONTO
SINGAPORE

The Library of Congress has cataloged this serial publication as follows:

Progress in pesticide biochemistry and toxicology.
  —Vol. 3
  (1983)- — Chichester [West Sussex] ; New York : Wiley, 1983–
     v. : ill. ; 23 cm.
  Editors: 1983-   D. H. Hutson and T. R. Roberts.
  Running title: Progress in pesticide biochemistry
  Continues: Progress in pesticide biochemistry.

  1. Pesticides—Physiological effect—Congresses.
  2. Pesticides—Metabolism—Congresses.
  3. Pesticides—Toxicolgy—Periodicals.
  4. Biological chemistry—Periodicals.
  I.   I. Hutson, D. H. (David Herd), 1935–   II. Roberts, T. R. (Terence Roberts), 1943–
  III. Title: Progress in pesticide biochemistry.

QP801.P38P76   632′.95—dc19   83-647760   AACR 2   MARC-S

British Library Cataloguing in Publication Data:

Insecticides.—(Progress in pesticide biochemistry
  and toxicology; v. 5)
  1. Insecticides
  I. Hutson, D. H.   II. Roberts, T. R.   III. Series
  668′.651   SB951.5

ISBN 0 471 90758 8

Printed and bound in Great Britain

Contributors to Volume 5

M. Cool — *Department de Chemie, Université de Moncton, Moncton, New Brunswick, Canada, E1A 3E9.*

J. Drabek — *Agricultural Division, Ciba-Geigy Limited, CH-4002, Basle, Switzerland.*

R. Edwards — *Department of Biochemistry, Royal Holloway College, University of London, Egham, Surrey TW20 0EX, UK.*

A. J. Gray — *Searle Research and Development, P.O. Box 53, Lane End Road, High Wycombe, Bucks HP12 4HL.*

N. P. Hajjar — *Dynamac Corporation, Dynamac Building, 11140 Rockville Pike, Rockville, Maryland 20852, USA.*

D. H. Hutson — *Shell Research Limited, Sittingbourne, Kent ME9 8AG, UK.*

C. K. Jankowski — *Department de Chemie, Université de Moncton, Moncton, New Brunswick, Canada, E1A 3E9.*

O. T. Jones — *Leiner Gelatins Limited, Treforest, Mid-Glamorgan, UK.*

P. Millburn — *Department of Biochemistry, St Mary's Hospital Medical School, Norfolk Place, London W2 1PG, UK.*

R. Neumann — *Agricultural Division, Ciba-Geigy Limited, CH-4002, Basle, Switzerland.*

T. R. Roberts — *Shell Research Limited, Sittingbourne, Kent, ME9 8AG UK.*

R. M. Sawicki — *Crop Protection Division, Rothamsted Experimental Station, Harpenden, Herts. AL5 2JQ, UK.*

D. M. Soderlund — *Department of Entomology, New York State Agricultural Experimental Station, Cornell University, Geneva, New York 14456, USA.*

Contents

Preface

This volume of the series differs from its predecessors in that it deals exclusively with a single theme, namely insecticides. Future volumes will similarly specialize on a specific class of pesticides or on a theme. The last thirty years have seen the development of four major chemical classes of insecticide ranging from the effective, persistent organochlorines, followed by the more toxic but less persistent organophosphorus compounds and carbamates to the highly effective, relatively non-toxic, non-persistent pyrethroids. Remarkably safe examples of each have been developed and economically important examples of each class are in wide use. Of the pesticides generally, the insecticides have taken the brunt of the concern over environmental impact. This is because all members of the major classes are neuroactive via mechanisms that have their counterpart in man, animals, fish and birds. Thus, the animal toxicology of the insecticides is somewhat more interesting and pertinent than that of the other pesticide classes. Small groups of insecticides or insect control agents acting via alternative mechanisms have been developed and have important selective uses.

This and future volumes will maintain the scope of the earlier volumes which must be wide because of the nature of the subject. Thus, the biochemistry and toxicology of insecticides is deemed to cover the following broad areas: (i) mode of action (i.e. biocidal action in target species); (ii) biotransformation in target species (which may, of course, be related to mode of action); (iii) biotransformation in non-target species (which may include soils, bacteria, insects, plants, fish, birds and mammals, including man); (iv) environmental effects (e.g. effects on the ecology of treated areas); (v) environmental chemistry (distribution and fate in the environment); (vi) biochemical toxicology in mammals, including man. All of these cannot, of course, be covered in one volume but we have endeavoured to deal with many important aspects. The insecticides have also received considerable coverage in early volumes of the series, the contents of which are listed at the end of this volume.

We have included an introductory chapter that describes the current status of the various classes of insecticides in terms of use and prospects. This chapter is also used to complement the content of the following chapters to provide a balanced treatment. Chapter 2 addresses the concept of the pro-insecticide, i.e. a chemical which is bioactivated to an insecticide in the target organism (or in the host). Several of the older insecticides have been shown to act via bioactivation but, as our knowledge of mode of action and metabolism improves, successful attempts at the rational design of pro-insecticides are on the increase. The carbamate insecticides are still very important and widely used. Their metabolic fate is reviewed in Chapter 3. Several aspects of the pyrethroid insecticides have been described in earlier volumes of the series. Their increasing importance warrants continuing coverage. Two chapters review in some detail, respectively, resistance to pyrethroids, and the mammalian toxicology of these compounds. In the past, these two aspects have provided much of the impetus for change, both within a chemical class and to new classes.

Continuing concern about the environmental effects of pesticides is reflected by the inclusion (Chapter 6) of a discussion on the metabolism and toxicity of insecticides to fish. The pyrethroids are again prominent in the authors' treatment of this topic. The chitin synthesis inhibitors are reviewed in Chapter 7. This mode of action is specific to insects and is, therefore, an attractive prospect for development. The final chapter highlights the vast range of natural chemicals which control the behaviour of insects. This is a fascinating area of biochemistry which is proving difficult to exploit in practice. However, the future of insect control must lie partly in this direction.

The aim of this volume is not to cover all aspects of the biochemistry and toxicology of insecticides comprehensively. Rather, it should serve to provide an update of our knowledge of the major insecticide classes and review related topics of relevance to insecticide biochemistry.

Sittingbourne, 1985   D. H. HUTSON
                                                                          T. R. ROBERTS

Insecticides
Edited by D. H. Hutson and T. R. Roberts
© 1985 John Wiley & Sons Ltd

CHAPTER 1

Insecticides

D. H. Hutson and T. R. Roberts

INTRODUCTION

The commercial use of chemicals for pest control in agriculture developed soon after the Second World War. Recent surveys have shown that the world-wide use of agrochemicals continued to increase in the years up to 1982 and, despite a small downturn in 1983, is likely to continue to grow. The various sectors of the market (principally herbicides, insecticides, fungicides and plant growth regulators) tend to be unevenly distributed. For example, whereas the major herbicide outlets are in North America and Western Europe (developed regions of the World with higher labour cost), insecticides comprise a greater proportion

1

of pesticides used in the tropics and subtropics because of the higher intensity of pests in these climatic conditions.

Taken overall (Anon, 1983; Lewis and Woodburn, 1984), the value of the agrochemical market in 1983 totalled US $12.8 billion with the distribution: herbicides, 38.7%; insecticides, 33.4%; fungicides, 22.0% and other products, 5.9%. This does, in fact, represent a small decline in estimates of the world-wide end-user value of agrochemicals compared with 1982, although this was mainly due to a US Government control programme in which a temporarily reduced acreage of crops was grown in the US in 1983. The growth trend is likely to return from 1984 onwards. (Separate figures published for the US for 1983 (Storck, 1984) give a total value of US $3.9 billion of which insecticides comprised 35%.)

The use of insecticides in any particular year is likely to vary on a regional basis since, in addition to the market and other factors which can influence all sectors of the agrochemical business, insecticide use is highly dependent upon the extent of insect pest infestations. With the introduction of newer insecticides with higher biological activities than their predecessors, lower dose rates have been required to control insect pests. For example, pyrethroids such as FASTAC (Shell Trade Mark, one of the cypermethrin isomers) and deltamethrin are applied at dose rates of 5–30 g/ha.

Most insecticidal products fall within four main classes, the organochlorines, the organophosphates, the carbamates and the pyrethroids. Other compounds include organotin compounds such as fenbutatin oxide, acylureas, formamides, pest control agents and growth regulators including pheromones and juvenile hormones and analogues. Estimates of world-wide market size within these classifications for 1982 and 1983 are summarized in Table 1. The major classes are still the organophosphates and carbamates and a large number of active ingredients fall within these categories. Organophosphates and carbamates gained favour, of course, because they combine high insecticidal activity with

Table 1   Estimated world-wide market for the major groups of insecticides

Insecticide group	Estimated market value (US $m)	
	1982	1983
Organophosphates	1650	1640
Carbamates	1150	1100
Organochlorines	650	560
Pyrethroids	500	600
Others	400	380
Total	4350	4280

After Lewis and Woodburn, 1984.

low persistence when compared with organochlorines although the spectrum of activity of these two newer classes tends to be narrower than that of the organochlorines. Many of the early organophosphates such as parathion are still widely used, although some of the more recently introduced materials (e.g. chlorpyrifos-methyl and tetrachlorvinphos) have much lower acute toxicities (in the 2000–5000 mg/kg range). Carbamates have proved to be particularly useful in cases where organophosphate resistance has occurred and about 25 or so carbamates are currently used, the major ones being carbaryl, aldicarb and methomyl.

Restrictions on the organochlorine insecticides has resulted in a decline in their use in developed countries where the economic situation can accommodate the use of alternatives; however, because of their low cost, they are still routinely used. Regulatory controls will probably result in further decline in organochlorine use on a world-wide basis, as evidenced in Table 1. Recent surveys of organochlorine residues in foodstuffs show a downward trend in the small concentration of certain organochlorines in some food and feed commodities.

Photostable pyrethroids, first introduced in 1976, have rapidly made an impact on the insecticide market. Initially introduced for use in cotton they have

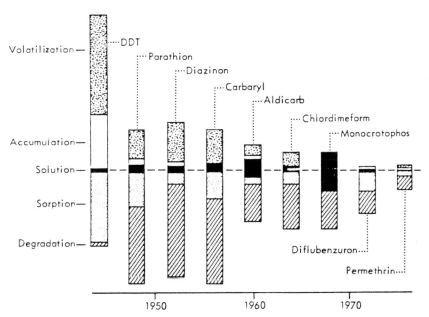

Figure 1 Changes in the potential of insecticides for distribution and accumulation in the environment. Compounds arranged in order of date of introduction. (From Geissbuhler, 1982, with the permission of Birkhauser Verlag, Basel)

been well-received in this outlet because their residual activity results in a smaller number of applications per season compared with other insecticides. Pyrethroids are now also used on a wide range of other crops, including fruit, vegetables, cereals and tobacco. They are unsuitable as soil-applied insecticides owing to their lack of persistence in soil and strong physical adsorption prior to degradation by hydrolysis.

Geissbühler (1982) has pointed out that the potential for insecticides to be distributed and to accumulate in the environment has decreased over a period of years. As illustrated in Figure 1, compounds with less bioaccumulation potential have been steadily introduced; they are generally more readily degraded and applied at lower dose rates than are the earlier generations of insecticides. This is exemplified by the pyrethroids, which have been shown to have limited impact on the aquatic and terrestrial environment due primarily to their physical properties, their ready degradation and to the use of very low dose rates (Shires, 1983; Stephenson, 1983; Shires et al., 1984; Stephenson et al., 1984).

In the future, it is likely that pyrethroids and subsequently introduced novel insecticides will be the centre of interest in the insecticide market (Anon, 1983). It is also likely that broad-spectrum insecticides will continue to be favoured since the development of insecticides for controlling specific insects will be costly and of less interest commercially.

There are surprisingly few books dealing with the chemistry, action, metabolism and toxicology of insecticides generally; however, texts by Hayes (1975), Matsumura (1975) and that edited by Wilkinson (1976) provide good coverage of these areas up to the mid-1970s and Corbett et al. (1984) have recently produced an excellent treatise on the mode of action of pesticides. A recent translation of a book on the chemistry of pesticides (Büchel, 1983) is a valuable source of information on individual insecticides. The proceedings of the 5th International Congress of Pesticide Chemistry held in Kyoto in 1982 have also been published (Miyamoto and Kearney, 1983). Books describing specific classes of insecticides are cited in the appropriate sections on each class which follow below.

ORGANOCHLORINES

There is no doubt that the organochlorine insecticides have played a major role in the history of agrochemicals for pest control. Most organochlorines have high insecticidal activity, low acute mammalian toxicity and residual biological activity. DDT, 1,1-(2,2,2-trichloroethylidene)bis(4-chlorobenzene) (1), has a fascinating history (Müller, 1955). It was first synthesized in 1874 but work at Geigy's laboratory in 1939 led to the discovery of its insecticidal properties. Its original use, for the extensive control of disease-carrying insects during and after World War II, preceded its use in agriculture.

The effective control of mosquitoes, lice and flies had a major impact on the epidemic diseases malaria and typhus resulting in their virtual eradication in many areas. When concern over the bioaccumulation of DDT resulted in its restricted use in some areas, it is notable that the incidence of disease rapidly increased.

In agriculture, DDT and related compounds, methoxychlor and TDE, 1,1-(2,2-dichloroethylidene)*bis*(4-chlorobenzene), have been extensively used as wide-spectrum insecticides for the control of many species with the exception of aphids and mites (Brooks, 1974). Methoxychlor was favoured for use in animal health situations (e.g. spraying dairy barns) as it is less lipophilic and accumulated in fat and milk to a lesser extent than does DDT.

Hexachlorocyclohexane (γ-HCH, lindane) was reported to have insecticidal activity in 1935, this activity residing mostly in the γ-isomer (Slade, 1945). Lindane has a similar spectrum of activity to that of DDT with physical properties more suitable than those of DDT for use as a soil insecticide, namely greater volatility and water solubility than DDT.

Interest in other organochlorine insecticides soon followed the successful introduction and use of DDT. The cyclodienes, notably aldrin, dieldrin, endrin, chlordane and heptachlor were the most important products developed based on Diels–Alder reaction products. Since their commercial introduction in the early 1950s they have been used in a wide range of outlets including soil insect control, malaria eradication campaigns, locust control and in wood preservation. Endrin became a major cotton insecticide as a foliar application. However, the uses of these compounds today are more limited. To date their replacement by compounds of similar biological activity with shorter persistence and bioaccumulation has proved difficult.

Information on the historical development of organochlorine insecticides has been interestingly reviewed by Brooks (1974) and a more recent summary of their chemical and biological properties has been given by Stetter (1983).

Chemistry

DDT (1) is synthesized by the condensation of chlorobenzene and chloral, methoxychlor being made analogously from anisole and chloral. The chemistry of DDT was reported as early as 1945 by Haller *et al.* (1945) and Forrest *et al.* (1946). The major reaction of importance is dehydrochlorination to afford DDE (2) which is no longer insecticidal.

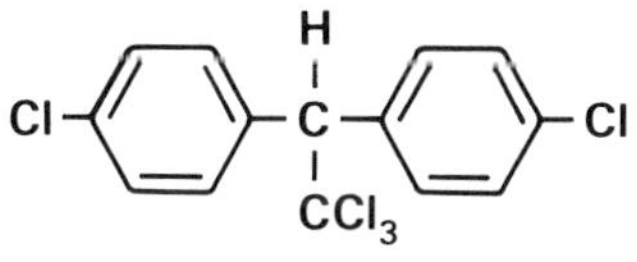

(1, DDT) 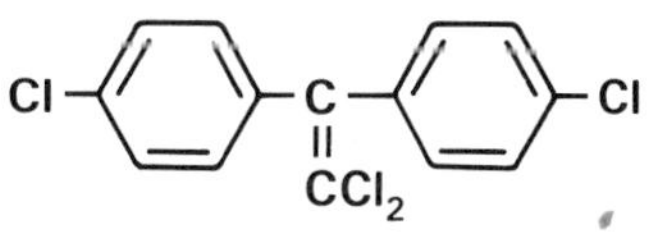   (2, DDE)

As indicated earlier, cyclodienes are prepared using the Diels–Alder reaction with hexachlorocyclopentadiene (HCCP) acting as the diene in most cases. It is reacted with three types of dienophile: acyclic, monocyclic and bicyclic. Endosulfan, for example is formed by condensation of HCCP with *cis*-1,4-diacetoxybut-2-ene followed by cyclization with thionyl chloride. For the synthesis of chlordane, HCCP is combined with the monocyclic cyclopentadiene followed by chlorination. In the case of aldrin (**3**), HCCP is reacted with the bicyclic norbornadiene; dieldrin (**4**) is aldrin epoxide. Endrin (**6**), a stereoisomer of dieldrin, was introduced in 1951. Endrin is prepared by the epoxidation of isodrin (**5**), an isomer of aldrin. The synthesis of isodrin is unusual in that the HCCP is converted first into hexachloronorbornadiene and the Diels–Alder reaction is then carried out with cyclopentadiene. This altered sequence affords the correct stereochemistry (cf. aldrin). These relationships are shown in Figure 2.

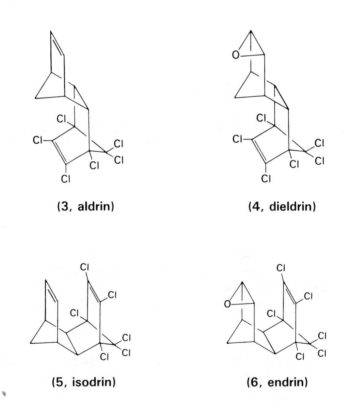

(3, aldrin)   (4, dieldrin)

(5, isodrin)   (6, endrin)

Figure 2 Structures of some cyclodiene insecticides

HCCP can be used in a related mode: dimerization to form the tightly caged structure, mirex (**7**). This compound is an effective insecticide but it is chemically very stable and consequently very resistant to metabolism.

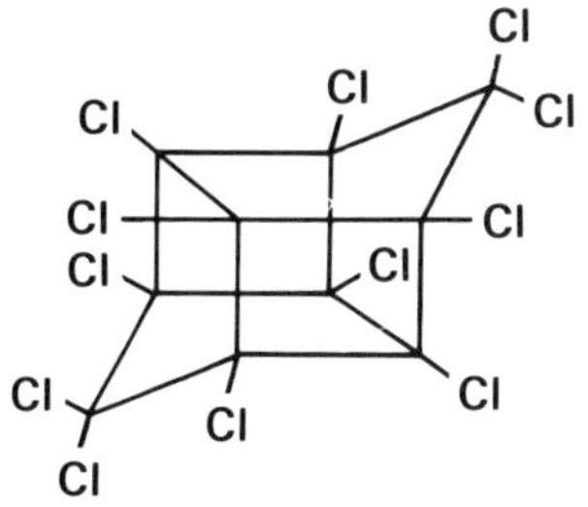

(**7, mirex**)

Mode of action

DDT acts as a nerve poison and models have shown that this insecticide causes a disturbance of the sodium balance of the nerve membranes. The basis for the development of resistance of insects to DDT is the enzyme-catalysed formation of inactive DDE (**2**). Lindane produces tremors, ataxia and convulsions in insects, these signs of toxicity being slightly different to those caused by DDT. Lindane is, however, faster acting.

All cyclodiene insecticides cause similar symptoms of poisoning in insects and have a characteristically slow–acting effect on insects. Soloway (1965) reviewed the structure-activity relationships in the series and the situation is complicated by the formation of active metabolites such as aldrin *trans*-diol from dieldrin (Akkermans *et al.*, 1974). A presynaptic action of dieldrin in the cockroach ganglion has been suggested by Shankland and Schroeder (1973). However, despite the fact that several organochlorines have been intensively studied as models, their precise modes of action at the molecular level remain obscure (Corbett *et al.*, 1984).

Metabolism

The metabolism of organochlorine insecticides in insects, mammals, plants and soils is slow and, consequently, the compounds leave residues in normal use situations. This persistence was seen as an advantage in the early days and indeed is of immense value for industrial use such as termite control. In addition, the efficacy and convenience of one dip per season for the control of ectoparasites of sheep were valuable assets. However, the tendency of residues to partition into animal fat, to be transferred from treated plants to animals and to be transferred from prey to predator resulted in the widespread presence of the organochlorines in the environment and to various poisoning incidents. It must be stressed, however, that the organochlorines form a diverse group of compounds and these very general statements have notable exceptions. As examples, endosulfan and endrin (6) are not persistent whilst at the other extreme, mirex (7) is very resistant to biodegradation. Routes of biotransformation also differ across the range, though various dechlorinations and oxidation predominate.

The metabolism of DDT in man has been reviewed in this series (Hutson, Volume 1) and its metabolism in livestock and other mammals has also been covered (Akhtar, Volume 4). DDT (1) is metabolized to the very persistent DDE (2) by dehydrochlorination. Another important route is initiated by a reductive dechlorination (to DDD), continues via dehydrochlorination (to DDMU) and a series of reductions, hydration and oxidation to 1,1-*bis*-(4-chlorophenyl)acetic acid (DDA).

In contrast, lindane (8, γ-hexachlorocyclohexane) is metabolized by glutathione-dependent dechlorination as well as by simultaneous reduction and oxidation processes of dechlorination (Kurihara *et al.*, 1979).

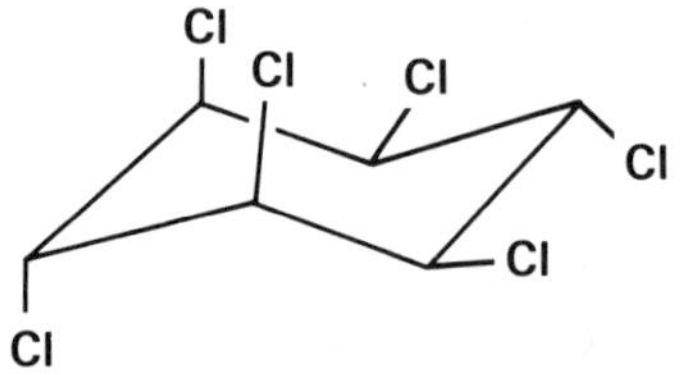

(8, lindane)

Where methylene bridges exist in the cyclodienes, they offer a point of oxidative attack by the cytochrome P450 system of animals. The resulting hydroxylated insecticides are conjugated with sulphate and/or glucose/glucuronic acid prior to excretion. These reactions are very effective detoxifications. The rates of the hydroxylations are markedly affected by the stereochemical environment of the C–H bonds. Such effects may be clearly exemplified by the large difference in the rates of metabolism (hydroxylation) of dieldrin (4), a persistent insecticide, and its isomer endrin (6), which is non-persistent. The reactions are illustrated in Figure 3 in which the thickness of the arrows indicates the approximate rates of metabolism. The major reaction of endrin, hydroxylation *anti* to the epoxide oxygen, is completely blocked in dieldrin, with dramatic consequences on the overall rates of metabolism of the two isomers *in vivo* in mammals (Bedford and Hutson, 1976).

The metabolism of insecticides, including that of organochlorines, in insects and its utilization in the design of selective insecticides has been reviewed by Brooks (1979).

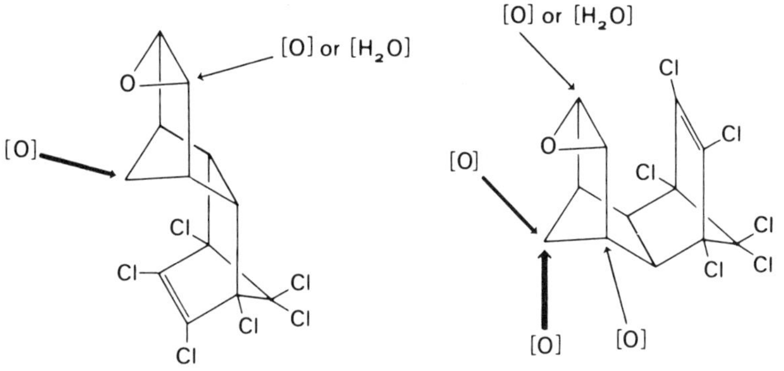

Figure 3 The comparative metabolism of dieldrin and endrin. The thickness of the arrows indicates the rate of hydroxylation

Toxicology

The organochlorines vary quite widely in their acute toxicities to mammals. Some acute oral LD_{50} values (rat) are shown as examples in Table 2. Generally the insect/mammalian toxicity ratios are very favourable and, with the exception of endrin, these insecticides are not particularly toxic to mammals. Their action in mammals is thought to be via the disruption of nerve conduction. As they are relatively slowly metabolized and lipophilic, they partition into body fat. Therefore, high frequent exposures must be avoided. Under conditions of good industrial hygiene and good agricultural practice, damaging exposures are indeed avoided.

Table 2 Acute oral LD_{50} values for some organochlorine insecticides in the rat

Insecticide	LD_{50} (mg/kg)
DDT	115
Endosulfan	20–50
Heptachlor	100–160
Methoxychlor	5000
Lindane (γ-HCH)	90–125
Camphechlor	80–90
Chlordane	280–430
Aldrin	40–60
Dieldrin	45–100
Endrin	10–40

From Gunn, 1975.

The discovery and increasingly wide use of the organochlorines in the 1950s coincided with the development of a new generation of analytical methods (gas-liquid chromatography and the electron-capture detector). Low concentrations of several organochlorines were subsequently found in many sectors of the environment including human diet and human fat. The demonstrated presence of such residues, rather than their toxicological significance, has always been the main problem that the organochlorines have posed to society. Undoubtedly, they have been responsible for the accidental killing of birds, fish and other non-target species and other effects (e.g. possibly egg-shell thinning) have been produced in species near the top of the food chain. However, adverse effects of DDT, and some other organochlorines, in man, when ingesting the insecticide adventitiously in food or via occupational exposures, have been difficult to demonstrate. The ensuing argument, which has generated a huge scientific effort, was the first environmental issue to receive the full publicity possible in a modern industrialized society. The debate took place as much in newspapers, on television and in the courts, as in the primary scientific press. The benefits and risks associated with two organochlorines were reviewed in a thorough, very readable article by Dr D. L. Gunn (1975) entitled 'Uses and abuse of DDT and dieldrin'. The author presents an independent and balanced view but he clearly emphasizes the propaganda and selective information presented in many cases by opponents of pesticides. Though written 10 years ago, the article was prepared just after DDT had been banned in the USA and when dieldrin was under pressure. This may appear to be history from our viewpoint but in fact the organochlorines are still important where cost considerations are critical as in many countries and are important for industrial uses (e.g. termite control) and to public health. Gunn wrote,

'Indiscriminate attacks on pesticides by members of our affluent society should be resisted, especially by the developing nations, by representatives of poorer people in developed nations, and by all persons of good will on behalf of their fellow men and women. Equally, indiscriminate and overlavish use of any pesticide, even by those that can readily afford such waste, should be actively discouraged.'

This remains sound advice.

Two examples, both drawn from the history of dieldrin, serve to illustrate problems with organochlorines in the field of environmental and human toxicology. The first concerns a series of deaths of owls at London Zoo between 1974 and 1976. Post-mortem examinations revealed no evidence of bacterial, viral or parasitic infections. Chemical analyses of livers for heavy metals and organochlorines revealed the presence of dieldrin. A collaborative study between the Zoological Society of London and Shell Research Ltd confirmed the liver residue at 29 µg/g (range 13–46) and concentrations of dieldrin in the brain of 17 µg/g (range 16–19). These concentrations indicated that dieldrin was the sole or major contributory cause of death in most of the owls. The contamination was traced to laboratory mice used to feed the affected owls. The diet of the mice contained only minute traces of the insecticide ($0.007\,\mu g/g$); however, their sawdust bedding was found to contain a wide range around $100\,\mu g/g$. The sawdust was obtained from a builder who was using a timber preservative containing dieldrin to protect against woodworm. The details of this study have been published (Jones et al., 1978).

The second issue is the significance of results of long-term feeding studies of dieldrin in mice. Dieldrin (and some other compounds such as DDT, lindane and phenobarbitone) induce a significant increase in the incidence above background, of hepatocellular carcinomas in certain strians of mice (Thorpe and Walker, 1973). There is, however, no evidence that dieldrin is carcinogenic in humans, and there is much evidence that these compounds are not carcinogenic in other animal species, including humans. This conclusion is supported by a number of expert committees and regulatory authorities. The Joint Expert Committee on Pesticide Residues of the WHO and the FAO concluded in 1977 that 'these new findings again support the view that dieldrin (and aldrin) are not carcinogens on the basis of knowledge available to the meetings'.

In a recent review of the impact of DDT on human health, Spindler (1983) concludes that claims that DDT has had an alarming impact on the environment, including man, are unrealistic when weighed against the enormous benefits resulting from its use in agriculture and more particularly in the control of malaria.

ORGANOPHOSPHATES

The organophosphorus insecticides comprise a very large group of compounds stemming from the observation by Lange and von Krueger (1932) that the simple esters of monofluorophosphoric acid were highly toxic. G. Schrader, who was working on toxic acid fluorides at the time, also became involved in organophosphorus chemistry and was the major force in establishing the potential of the organophosphates for pest control. From the dialkylphosphofluoridates, he developed the phosphorodiamidofluoridates, thence the pyrophosphoramidates and the pyrophosphates. Schrader synthesized O,O-diethyl 4-nitrophenyl phosphorothionate, later known as parathion, in 1944 and recognized its insecticidal activity. However, owing to the confusion in Europe at that time, it could not be patented and its first commercial use was in the United States. The potency and broad spectrum of activity of parathion stimulated an enormous industrial effort on the search for phosphorus-based pesticides. There are currently about 250 compounds registered for use in various parts of the world.

It is possible to effect wide variations in structure around the active phosphorus centre and thus a wide range of physical properties, chemical stabilities, biological stabilities and selectivity may be obtained. This flexibility is reflected in a wide range of uses. The organophosphates are used as contact insecticides on plants and as soil insecticides.

It is also possible to synthesize active compounds with the correct lipophilic–hydrophilic balance for movement within plants, hence the plant systemic insecticides such as dimethoate, mevinphos and monocrotophos. A large range of crops, including the majors such as cotton, rice, corn, wheat, barley, sorghum and soybean, are protected by organophosphates. Their use is also important in top fruit and in vegetables, for both foliage and root protection. Selective toxicity data have been exploited in veterinary uses of the organophosphates; for example, for ectoparasite control on cattle and sheep in the form of ear tags, sprays and dips. Endoparasites may also be controlled; dichlorvos, in a slow-release formulation, is used as an anthelmintic in horses and pigs.

Organophosphates must be considered amongst the most effective insecticides in use over the last 40 years. Not the least important factor in this context is their environmental acceptability. They have a very specific mode of action and their biological activity is generally destroyed by the simplest chemical or biochemical modification. They are also relatively unstable in biological systems. Under correct use conditions, therefore, persistent biological activity away from the site of use is rare. The chemistry and biochemistry of the organosphorus insecticides is described in detail in books by Eto (1974) and Fest and Schmidt (1982).

Chemistry

The essential feature of the organophosphate insecticides may be exemplified using O,O-diethyl 4-nitrophenyl phosphate (**9**, paraoxon). Paraoxon is a phosphoric acid triester. Its phosphorus atom is electrophilic, partly owing to polarization of the P=O bond and partly owing to the electron-withdrawal properties of the 4-nitrophenyl group. The electrophilic nature of the phosphorus atom confers upon it the ability to phosphorylate nucleophiles, including biological nucleophiles. It is aided in this ability by the properties of the 4-nitrophenyl substituent as a good leaving group in the SN_2 phosphorylation reaction. This reaction is central to the mode of action of these compounds. Numerous modifications of this structure alter the intrinsic activity, lipophilicity, biostability and selective toxicity of these molecules. Another important feature is the use of the phosphorothionates (P=S compounds) as opposed to phosphates (P=O compounds). The former would now be classed as propesticides (see Chapter 2) as they do not possess the electrophilic phosphorus atom required for intrinsic reactivity. They must be bioactivated by oxidative desulphurization in the insect, soil, plant or mammal before a toxic interaction can occur (Figure 4). This requirement for bioactivation offers the opportunity for other metabolic processes to operate and confer selective toxicity. Paraoxon (**9**),

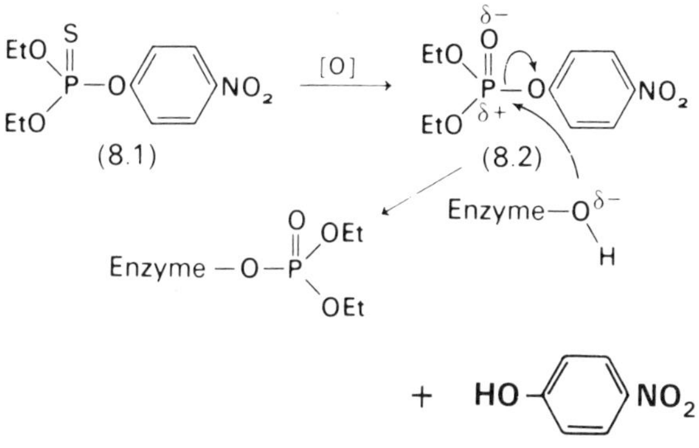

Figure 4 Oxidative bioactivation of parathion and the inhibition of acetylcholinesterase by paraoxon

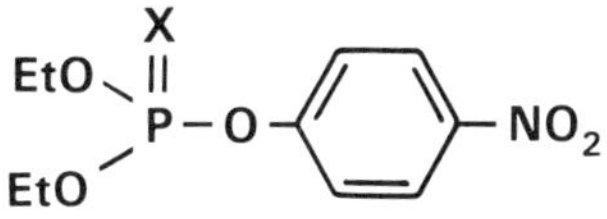

(9, X = O; paraoxon)

(10, X = S; parathion)

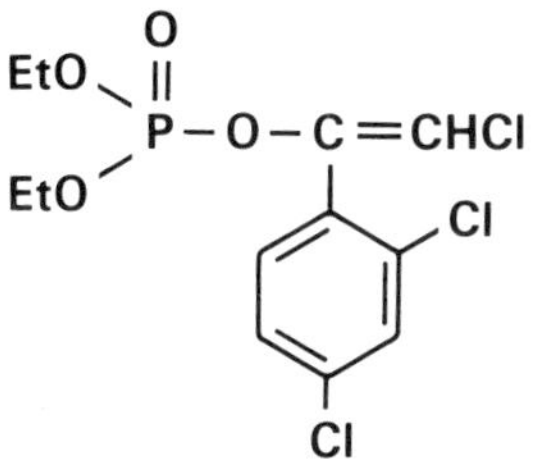

(11, chlorfenvinphos)

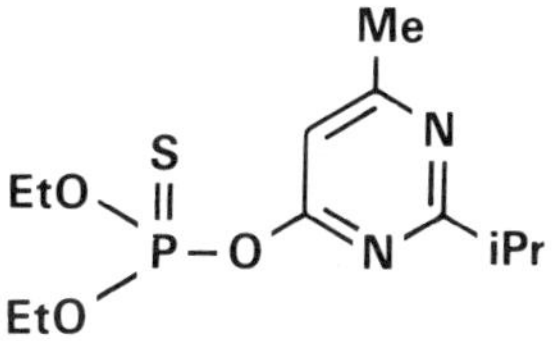

(12, diazinon)

parathion (**10**), chlorfenvinphos (**11**) and diazinon (**12**) are illustrated above as examples of these modifications. The phosphonates (P–C bonds) and the phosphoramidates (P–N bonds) also provide chemical variations around the active centre but essential reactivity, phosphorylation, is retained.

As esters, the organophosphate insecticides are liable to hydrolysis and to other mechanisms of ester cleavage and, in practice, they are biodegradable and are relatively non-persistent in animals, plants and in the environment generally.

Mode of action

Acetylcholine is the chemical transmitter of nerve impulses in cholinergic synapses. It is thought that it is an important neurotransmitter in the insect central nervous system. Following the release of acetylcholine and its diffusion across the synapse, a rapid return to normality is critical to the effective functioning of an organism. This is effected by hydrolysis of the released acetylcholine by an esterase, acetylcholine hydrolase (acetylcholinesterase). This enzyme, as other carboxyesterases, contains a serine hydroxyl group at its active site. The organophosphorus insecticides phosphorylate this hydroxyl group and cause the inhibition of the enzyme. The intrinsic reactivity of such an insecticide can be measured *in vitro* as a concentration for 50% inhibition of an enzyme preparation or, better, as the bimolecular rate constant for the reaction between the inhibitor and the enzyme. The inhibition is an SN_2 reaction in which, for example with paraoxon (**9**), the enzyme is diethylphosphorylated and 4-nitrophenol is released. The rate of the inhibition is affected by electronic, steric and hydrophobic factors. These can be correlated with activity within a structurally related series (Hansch and Deutsch, 1966). Build-up of acetylcholinesterase in the synapses of insects (and of vertebrates) leads to repetitive firing followed by blockage of nerve transmission. Such severe disruption of the nervous system cannot be tolerated for long and the failure of a critical process leads to the death of the organism. This subject has been authoritatively reviewed in detail by Main (1979).

Metabolism

Metabolism is a very important factor in the toxicity of the organophosphorus insecticides. The oxidation of the thionates to the corresponding phosphates is a prerequisite for the action of the former as insecticides. In the case of the phosphates, however, with the exception of certain variations in the leaving group moiety, metabolism affords detoxification. This is because the conversion of a

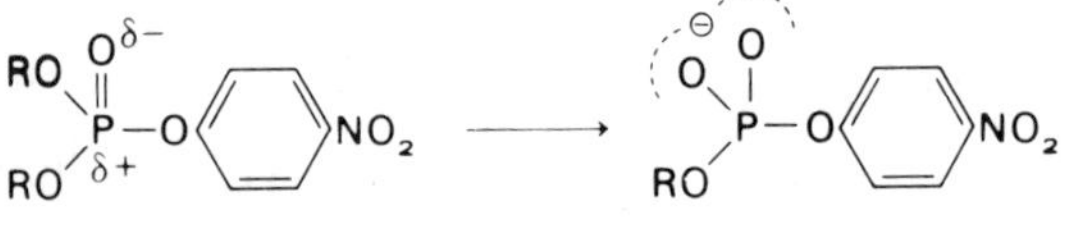

(13)   (14)

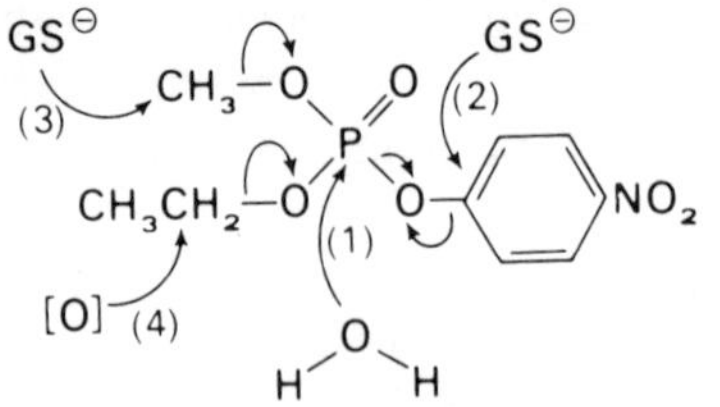

Figure 5 Detoxification mechanisms operating on a phosphoric acid triester

triester (**13**) into a diester (**14**) destroys the electrophilic character of the phosphorus atom and leads to complete detoxification in a single step. Various enzymatic mechanisms for the conversion of a triester into a diester have been discovered. The important ones are illustrated in Figure 2 for ethyl methyl 4-nitrophenyl phosphate; this compound is not an important insecticide but it serves as an example to demonstrate these mechanisms. These processes (Figure 5) are:

(1) Hydrolytic deoxyarylation,
(2) Glutathione-dependent dearylation,
(3) Glutathione-dependent demethylation,
(4) Oxidative de-ethylation.

All or some of these general reactions may occur simultaneously, together with various compound-specific reactions at the periphery of the leaving group (e.g. nitroreduction in the case of parathion (**10**), methyl hydroxylation with diazinon (**12**) and carboxyl ester hydrolysis with malathion). The route of metabolism and its effects on toxicity are both difficult to predict in most cases. A safe prediction, however, is that with the variety of mechanisms and functional groups available for biotransformation, the organophosphorus insecticides will be readily degraded in organisms such as insects, plants and animals. Many studies in experimental models have confirmed rapid metabolism. Beynon *et al.* (1973) have compared the metabolism of a series of vinyl phosphate insecticides in plants, soils and mammals and have shown that similar options for metabolism occur in these life forms. Insects also possess these options but generally a combination of high sensitivity of the acetylcholinesterase and rather slow rate of detoxification affords a lethal action. In some cases, resistance to the organophosphates has been attributable to an increased level of detoxification in the resistant (cf. susceptible) insect strains. For example, Motoyama and Dauterman (1972) have shown that detoxification by glutathione *S*-transferases is a major mechanism of resistance of certain insect species to azinphos-methyl (**15**). Motoyama *et al.* (1983) have described two forms of glutathione *S*-transferase in resistant and susceptible houseflies. Each has different activities towards the model substrates 2,4-dichloro-1-nitrobenzene and 1-chloro-2,4-dinitrobenzene and different specificities in the demethylation and dearylation of diazinon.

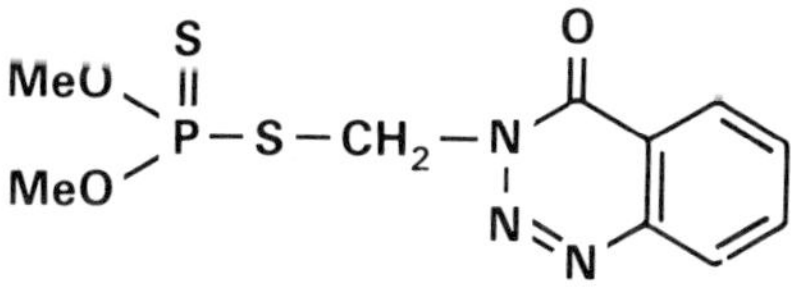

(15, azinphos-methyl)

Metabolism occurring at a peripheral position in the leaving group and which leaves the triester structure intact may detoxify or bioactivate, depending on the chemistry of the molecule. In general, if biotransformation of the leaving group increases its electron-withdrawing properties and/or its efficacy as a leaving group, such biotransformation will increase the intrinsic activity of the insecticide. However, this will also increase its lability towards hydrolysis (nucleophilic attack of OH^- on electrophilic phosphorus) and may result overall in reduced toxicity. Phorate (16) provides an example of such bioactivation. Oxidation to the P=O compound (phorate oxon) provides the active anticholinesterase but this is further bioactivated to the sulphone (17) (Bowman and Casida, 1952). Other sulphur-containing organophosphate insecticides are similarly activated by S-oxidation (Leesch and Fukuto, 1972).

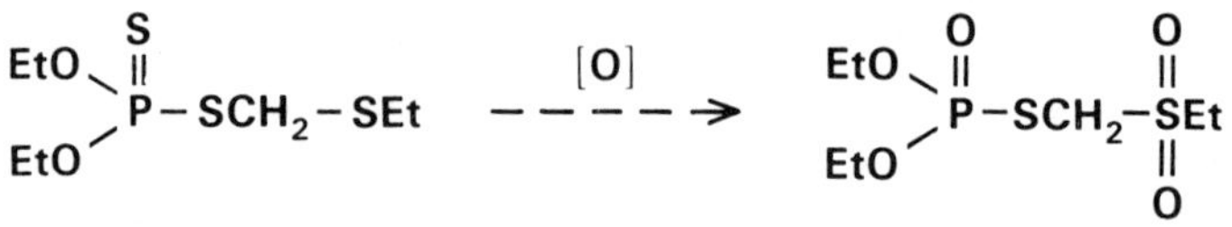

(16, phorate)   (17)

N-Demethylation in the dicrotophos–monocrotophos series is a classical pesticide bioactivation (Figure 6). The biotransformation of dicrotophos (18) to monocrotophos (20) occurs via the N-methylol derivative (19). The reaction occurs again with monocrotophos to afford (21) and (22). Each compound in this series is an anticholinesterase in insects and mammals (Bull and Lindquist, 1964; Menzer and Casida, 1965). The biological activities of this series of products is shown in Table 3. Monocrotophos is still an important insecticide.

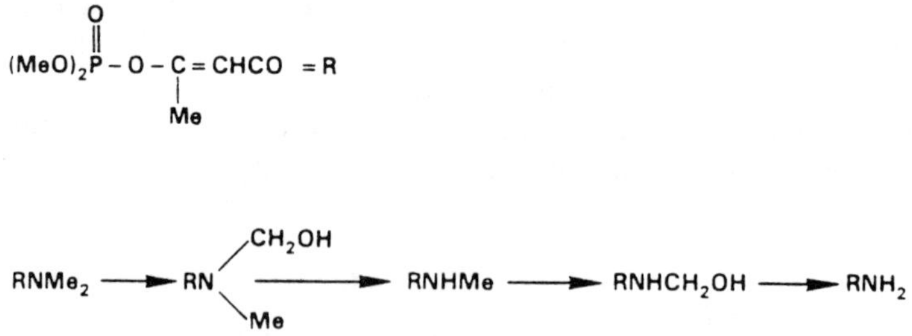

Figure 6 Bioactivation in the dicrotophos (18)–monocrotophos (20) series

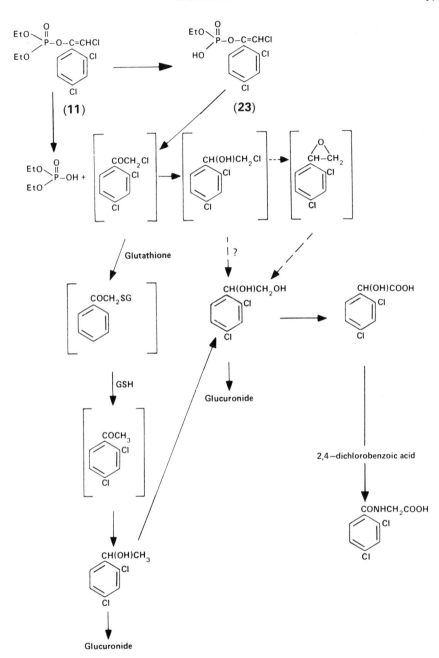

Figure 7 The metabolic fate of chlorfenvinphos in mammals

Table 3   Biological activity in the dicrotophos–monocrotophos series

Property	18	19	20	21	22
pI_{50} (M, fly-head acetylcholinesterase)	7.2	7.0	6.8	6.9	6.5
LD_{50} (mg/kg, fly)	38	14	6.4	30	1
LD_{50} (mg/kg, mouse)	14	18	8	12	3

The reduction of the nitro group of parathion, which occurs in the rumen of cattle, is a detoxification reaction. The 4-amino analogue of paraoxon is a poor inhibitor of cholinesterase as the electron-withdrawing effect of the nitro group has been lost.

Metabolism studies on organophosphorus insecticides are usually carried out with ^{14}C-labelling in the leaving group. ^{32}P-Labelling and ^{14}C-alkyl labelling originally favoured for simplicity do not give enough information on the fate of the bulk of the organic content. Nevertheless, it is important to distinguish between the primary detoxification processes (triester to diester) and secondary processes involving the further metabolism of the leaving group. This may be complex as illustrated in Figure 7 which describes the metabolism of chlorfenvinphos (**11**) in mammals to desethyl chlorfenvinphos (**23**) and on to other terminal products via a variety of intermediates (Hutson, 1981). A similar scheme appears to hold for plants (Beynon *et al.*, 1973) but the fate of the insecticide in insects has not been studied in detail.

Toxicology

The mechanisms of action of organophosphorus insecticides in mammals and insects are similar and the targets in the organisms are also similar. Many neuromuscular functions in the mammal, unlike the insect, are also cholinergic.

Poisoning by organophosphates is actually toxification by endogenous acetylcholine (**24**) which builds up following the inhibition of acetylcholinesterase. The ultimate cause of death in insects is still in dispute but death in mammals is almost certainly due to paralysis of the striated respiratory muscles or to paralysis of the respiratory centre. The most effective antidote to organophosphate poisoning is atropine (**25**) which blocks the acetylcholine receptors of the motor end-plates of the parasympathetic nervous system, those controlling the critical respiratory muscles. Atropine treatment may be followed up with 2-PAM (**26**) which reactivates phosphorylated acetylcholinesterase by nucleophilic attack by the oxime oxygen on phosphorus and displacement of the serinyl oxyanion (Wilson and Ginsburg, 1955). It has always been expected, therefore, that an organophosphorus insecticide will be acutely toxic to experimental animals and to man until the contrary is indicated. The earlier

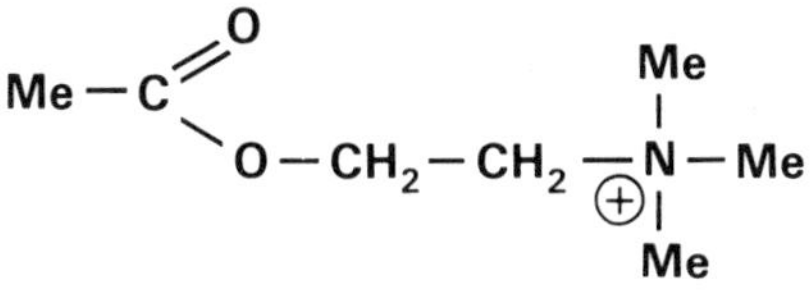

(24)

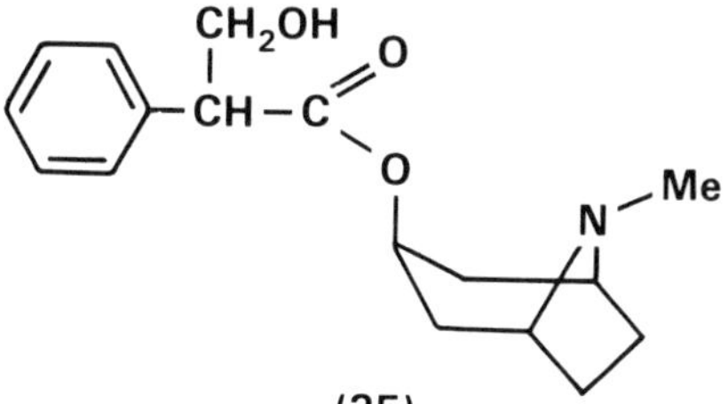

(25)

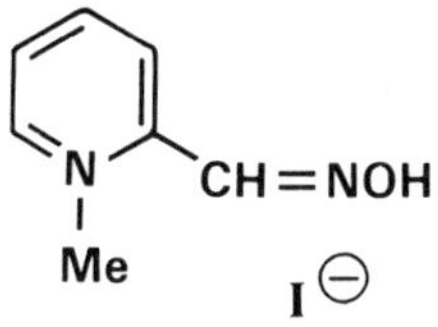

(26)

insecticides, e.g. tetraethyl pyrophosphate (TEPP) and parathion, were very toxic (Table 4) and their accidental or deliberate misuse has caused many fatalities. However, from the several hundred thousand compounds that have now been synthesized and screened there are now a large number of insecticides with very favourable insect/mammal toxicity ratios. Selectivity depends on differential penetration, transport, target sensitivity and metabolism. The interaction of these factors is very complex and, when other factors associated with field use are added, it is not surprising that efficient screening and toxicological programmes are as important as rational design in the discovery of good insecticides. The range of acute mammalian toxicities and an indication of selective toxicity are illustrated with a few examples in Table 4. Relatively small changes in structure within a series can have a remarkable effect on acute toxicity. Similarly, some remarkable species differences in toxicity exist for both insects and mammals. The case of chlorfenvinphos, which exhibits a very large difference in acute toxicity to rats and dogs (Table 4) has been studied in some detail (Hutson, 1981) and found to be due to a number of factors, the most important of which is the initial detoxification by de-ethylation (**11** → **23**; Figure 7).

Delayed neurotoxicity is a special chronic toxicity that is not due to the inhibition of acetylcholinesterase, but to the inhibition of a 'neurotoxic esterase'. This leads to a central-distal axonopathy in sensitive species like man, cat and chicken (Johnson, 1982). Potentially commercial insecticides are tested for this action, usually in the chicken and, if positive, their development is stopped.

A more recent question has been raised concerning the alkylating reactivity of the organophosphate triesters. This is a possibility with all of the *dimethyl* phosphate triesters. The ethylating activity of the ethyl esters is negligible by comparison. It was postulated (Löfroth, 1970) that these compounds may methylate DNA, as does, for example, methyl methanesulphonate, and cause damage to

Table 4 Acute toxicity of some organophosphorus insecticides

Compound	Structure	LD$_{50}$ (mg/kg) Rat	Housefly
Tetraethyl pyrophosphate	$(EtO)_2 \overset{O}{\underset{\|}{P}} - O - \overset{O}{\underset{\|}{P}} (OEt)_2$	1–2	1
Parathion	$\overset{EtO}{\underset{EtO}{}} \searrow \overset{S}{\underset{\|}{P}} - O - \langle \rangle - NO_2$	5	1
Methyl parathion	$\overset{MeO}{\underset{MeO}{}} \searrow \overset{S}{\underset{\|}{P}} - O - \langle \rangle - NO_2$	14	3
Fenitrothion	$\overset{MeO}{\underset{MeO}{}} \searrow \overset{S}{\underset{\|}{P}} - O - \langle \rangle \overset{Me}{-NO_2}$	250	5
Tetrachlorvinphos	$\overset{MeO}{\underset{MeO}{}} \searrow \overset{O}{\underset{\|}{P}} - O - C = CHCl$ Cl, Cl, Cl	5000	4
Chlorfenvinphos	$\overset{EtO}{\underset{EtO}{}} \searrow \overset{O}{\underset{\|}{P}} - O - C = CHCl$ Cl, Cl	20	4

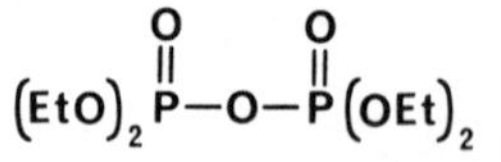

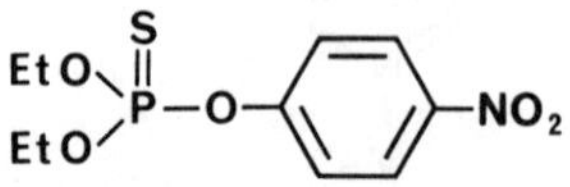

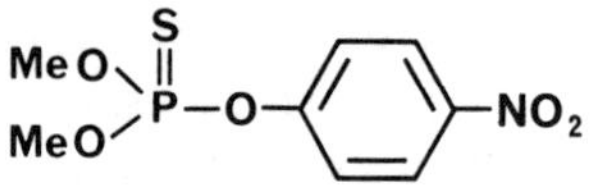

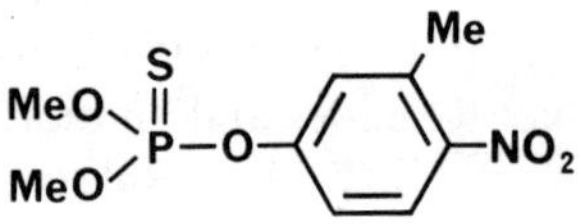

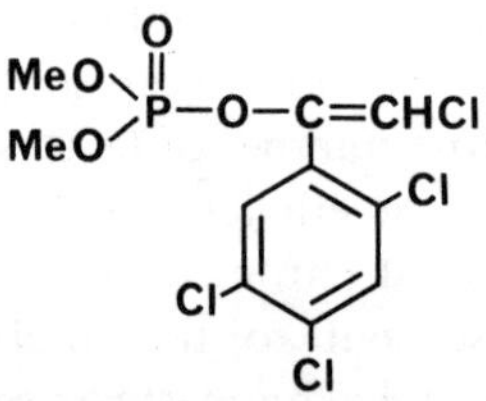

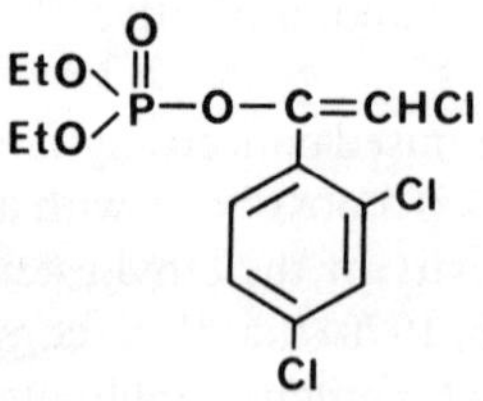

Table 4 Acute toxicity of some organophosphorus insecticides

Compound	Structure	LD$_{50}$ (mg/kg) Rat	Housefly
Tetraethyl pyrophosphate	$(EtO)_2\overset{O}{\underset{\|\|}{P}}-O-\overset{O}{\underset{\|\|}{P}}(OEt)_2$	1–2	1
Parathion	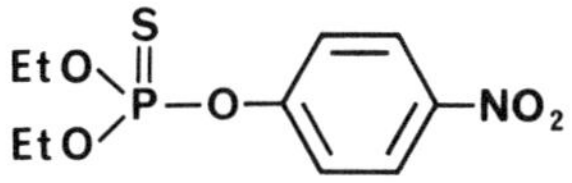 EtO, EtO — P(=S) — O — C$_6$H$_4$ — NO$_2$	5	1
Methyl parathion	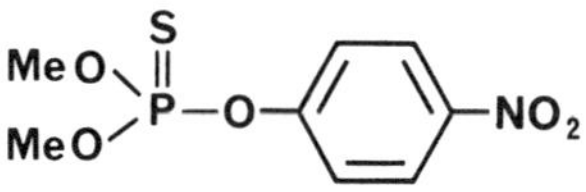 MeO, MeO — P(=S) — O — C$_6$H$_4$ — NO$_2$	14	3
Fenitrothion	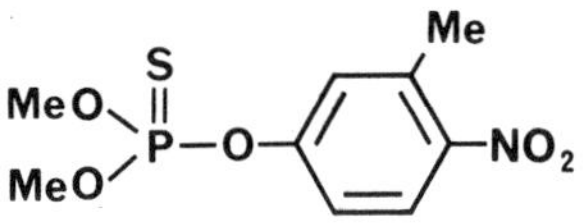 MeO, MeO — P(=S) — O — C$_6$H$_3$(Me) — NO$_2$	250	5
Tetrachlorvinphos	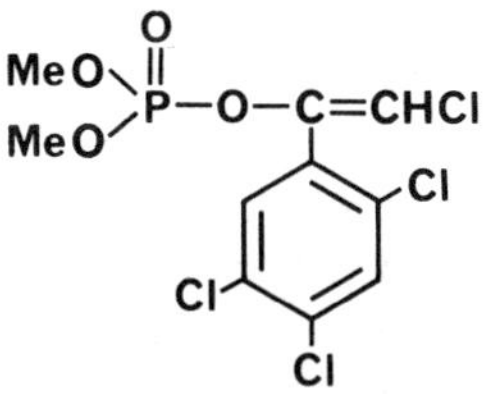 MeO, MeO — P(=O) — O — C=CHCl, C$_6$H$_2$Cl$_3$	5000	4
Chlorfenvinphos	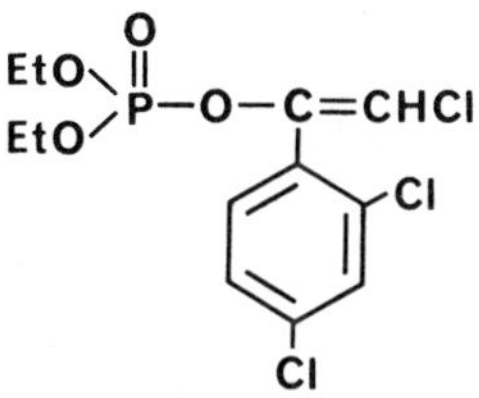 EtO, EtO — P(=O) — O — C=CHCl, C$_6$H$_3$Cl$_2$	20	4

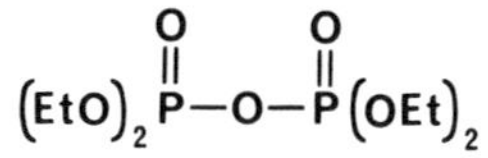

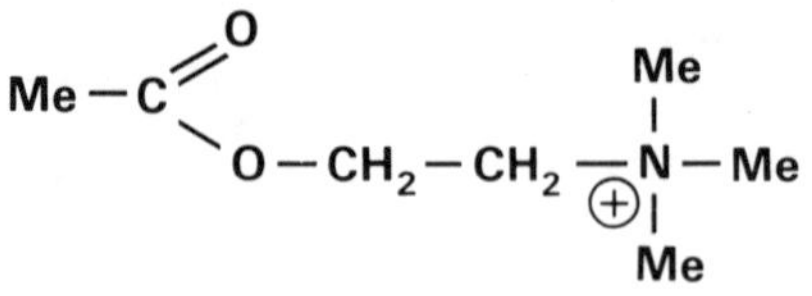

(24)

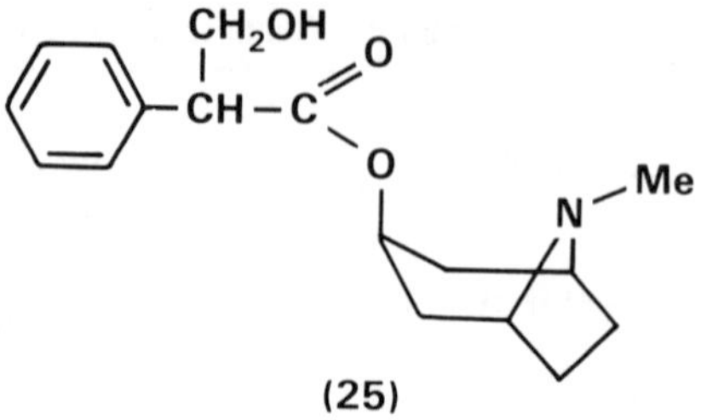

(25)

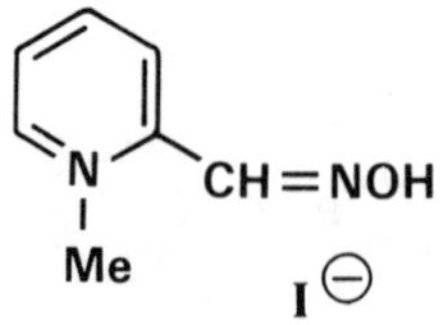

(26)

insecticides, e.g. tetraethyl pyrophosphate (TEPP) and parathion, were very toxic (Table 4) and their accidental or deliberate misuse has caused many fatalities. However, from the several hundred thousand compounds that have now been synthesized and screened there are now a large number of insecticides with very favourable insect/mammal toxicity ratios. Selectivity depends on differential penetration, transport, target sensitivity and metabolism. The interaction of these factors is very complex and, when other factors associated with field use are added, it is not surprising that efficient screening and toxicological programmes are as important as rational design in the discovery of good insecticides. The range of acute mammalian toxicities and an indication of selective toxicity are illustrated with a few examples in Table 4. Relatively small changes in structure within a series can have a remarkable effect on acute toxicity. Similarly, some remarkable species differences in toxicity exist for both insects and mammals. The case of chlorfenvinphos, which exhibits a very large difference in acute toxicity to rats and dogs (Table 4) has been studied in some detail (Hutson, 1981) and found to be due to a number of factors, the most important of which is the initial detoxification by de-ethylation (**11** → **23**; Figure 7).

Delayed neurotoxicity is a special chronic toxicity that is not due to the inhibition of acetylcholinesterase, but to the inhibition of a 'neurotoxic esterase'. This leads to a central-distal axonopathy in sensitive species like man, cat and chicken (Johnson, 1982). Potentially commercial insecticides are tested for this action, usually in the chicken and, if positive, their development is stopped.

A more recent question has been raised concerning the alkylating reactivity of the organophosphate triesters. This is a possibility with all of the *dimethyl* phosphate triesters. The ethylating activity of the ethyl esters is negligible by comparison. It was postulated (Löfroth, 1970) that these compounds may methylate DNA, as does, for example, methyl methanesulphonate, and cause damage to the mammalian gene leading to mutation and/or carcinogenesis. The counter argument, that these insecticides in general and dichlorvos in particular (the compound in question) were effectively and rapidly detoxified by hydrolysis, oxidation and glutathione conjugation, was put forward (Bedford and Robinson, 1972). It was tested by subjecting rats for 12 hours to atmospheres containing the near-use concentration of dichlorvos (0.064 µg/l). The insecticide was labelled with ^{14}C in the methyl groups at maximum specific radioactivity (113 Ci/mol). After exposure, the DNA of soft tissues was isolated, purified and analysed for the presence of 7-methylguanine and 6-O-methylguanine (Wooder et al., 1977). None was found at a detection limit of one methylated base in 6×10^{11} (0.0000005% of the received dose).

The organophosphorus insecticides will remain a valuable resource for crop protection for the foreseeable future. They are the best understood compounds in terms of mode of action, toxicology and metabolism and they still offer a vast range of structures from which active and safe compounds may be derived.

CARBAMATES

Physostigmine is the active principle in the aqueous extract of calibar bean seeds which was used in witchcraft trials in S. E. Calibar Province, Nigeria in the seventeenth and eighteenth centuries. The structure (27) was elucidated in 1925

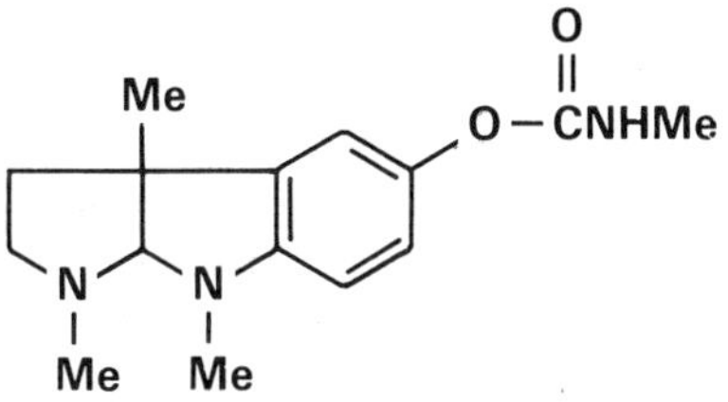

(27)

and, shortly afterwards, one of its activities (that for pupil constriction) was found to be dependent on the N-methylcarbamyl group. Also around that time, pupil constriction was found to be due to inhibition of acetylcholinesterase. Physostigmine itself is inactive as an insecticide and did not serve as a model for the development of the carbamate insecticides. The latter were first discovered and developed from about 1947 by Gysin and coworkers (Gysin, 1954). The initial search was for insect repellents and was concentrated around N,N-dimethylcarbamates. The first monomethylcarbamate insecticide, carbaryl (28), was developed by Union Carbide in the USA in 1953. This proved to be an extremely successful product and has stimulated the search for others of its class. As with the organophosphorus insecticides, toxicity of the carbamates is controlled by a number of factors and the selectivity ratio can be made to vary enormously by alteration of the ROH function carrying the N-methylcarbamyl group.

Since 1958, the carbamates have steadily increased their share of the total insecticide tonnage. Carbaryl is non-residual in comparison with DDT and, moreover, was found to be active against DDT-resistant pests. These factors partly account for its success. Some forty carbamates, mostly phenyl N-methylcarbamates, are in use or in late stages of development. They are mostly used in cotton, fruit and vegetables and fodder crops and some are used in rice. A greater proportion of the carbamates than of the organophosphates are systemically active in plants. This property greatly enhances the value of such compounds, allowing the control of pests on growing shoots and roots. Certain carbamates are of low enough acute toxicity to mammals to be used in veterinary outlets. Butacarb (3,5-di-*tert*-butylphenyl-N-methylcarbamate), for example, with an LD_{50} (rat) greater than 4000 mg/kg, is used as a sheep-dip. Carbaryl (28), with an LD_{50} of about 800 mg/kg, is used on poultry, livestock and buildings and for domestic pets (e.g. in dog shampoos).

Kuhr and Dorough (1976) have produced an excellent monograph dealing with the chemistry, biochemistry and toxicology of the carbamate insecticides.

Chemistry

Three classes cover most of the carbamate types. These are exemplified by carbaryl (28), an aryl N-methylcarbamate, pirimicarb (29), a dimethylcarbamyl ester of a heterocyclic hydroxy compound, and methomyl (30), an N-methylcarbamyl ester of an oxime (the so-called oxime carbamates). The rationale for the synthesis of the latter class is given below in the Mode of action section. All are carbamyl esters of very weak acids, e.g. phenols and oximes. This ester linkage introduces chemical instability and biodegradability into the molecules. Their chemistry is relatively simple and allows for ready synthesis of large numbers of analogues, though experience has shown that the N-methylcarbamyl group is the most successful of the esterifying groups. These insecticides are direct-acting, i.e. bioactivation is not a generally important phenomenon. However, in

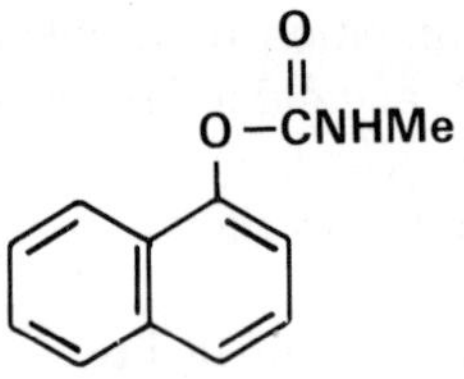

(28)

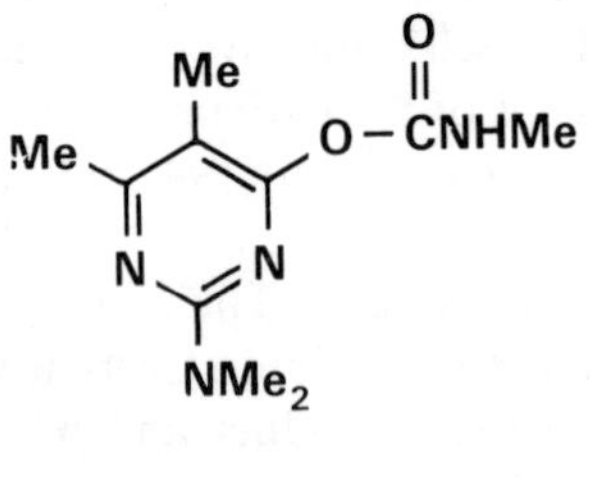

(29)

1962, the Boots Company found that *N*-acetylation of the monomethylcarbamates afforded reduced mammalian toxicity without much loss of insecticidal activity (Fraser *et al.*, 1968). The metabolic basis of this selectivity, a bioactivation, is described below in the section on Metabolism.

Mode of action

N-Methylcarbamates inhibit acetylcholinesterase by *N*-methylcarbamylation of the serine hydroxyl group of the enzyme in an exactly analogous way to the dialkylphosphorylation by the organophosphate triesters. The carbamate is, of course, destroyed in the process, with the liberation of a phenol or an oxime. A significant difference, however, is that the carbamylated enzyme is much less stable than the phosphorylated enzyme. The inhibited enzyme is reactivated in the presence of water to yield free, active enzyme and methylcarbamic acid. This is an acid–base-catalysed hydrolysis. Much work has been carried out on the kinetics of enzyme inhibition *in vitro* with both insect and mammalian enzymes. Such studies are a very reliable aid in the search for maximum anticholinesterase activity but when compounds of high intrinsic activity are applied to whole insects the search may be confounded by adverse penetration and metabolism. The situation in the field is even more uncertain. Intrinsic activity in the phenyl carbamates requires that the 2- or 3-positions contain electron-rich substituents; this may facilitate binding of the inhibitor to the enzyme. Lipophilicity is also important and correlates with inhibitory potency (Hansch and Deutsch, 1966).

High intrinsic activity has been obtained with the oxime carbamates via a consideration of the structure of the active site of the enzyme. This has been defined in relation to the natural substrate acetylcholine (**24**) and some structural analogues. The enzyme is perceived to have an anionic binding site and an esteratic binding site. The oxime carbamates, e.g. methomyl (**30**), present to the enzyme a carbamylating reagent of correct size and reasonable rigidity. These oxime carbamates and the relatively simple phenyl *N*-methylcarbamates together provide most of the currently successful carbamate insecticides.

Metabolism

The metabolism of the carbamates is dominated in mammals and plants by hydrolysis to the phenol, oxime (or other hydroxy compound) and methyl carbamic acid (which is further metabolized to carbon dioxide and ammonia). The phenyl groups are often extensively hydroxylated before and after carbamate hydrolysis. The phenols and the oximes are conjugated with sugars and sulphate which effects their removal or further detoxification. These reactions are summarized in simplified form for carbaryl in Figure 8. The enzyme 'carbamyl

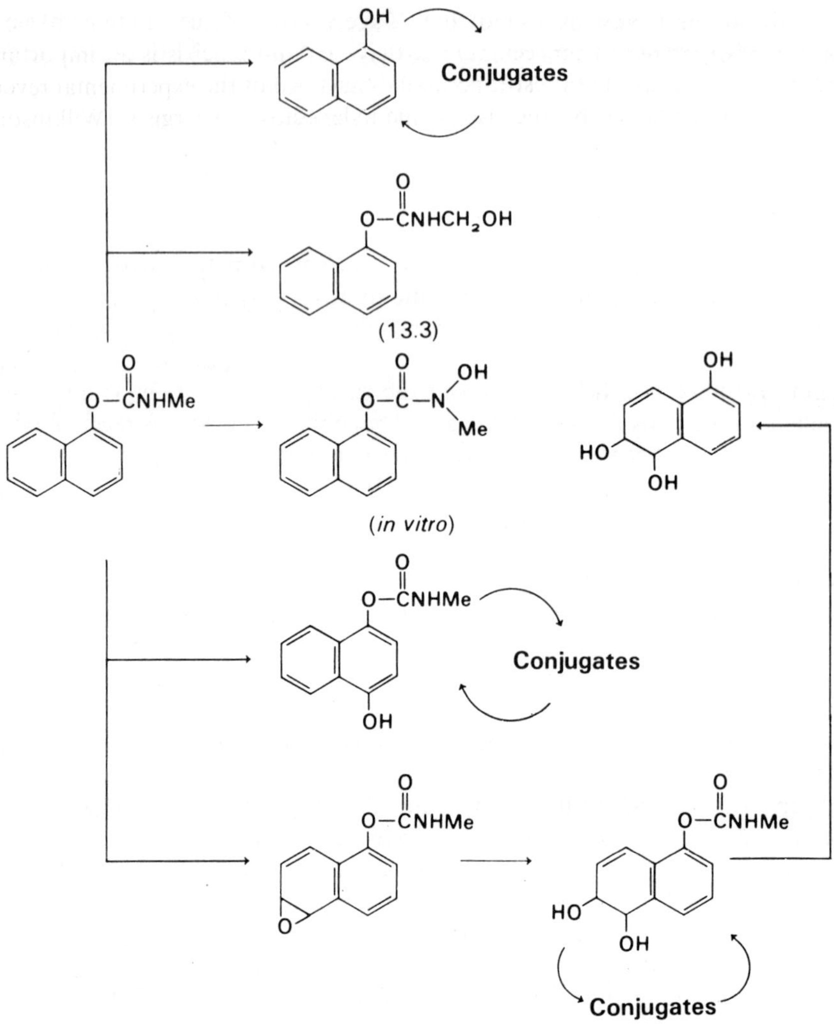

Figure 8 The metabolism of carbaryl

hydrolase' has not been identified until recently and it has been suggested (Hutson, 1972) that the reaction is oxidative rather than hydrolytic, involving oxidation of the N-methyl group to the unstable N-methylol which would decompose to the phenol or oxime. Recently, however, Mundy and Dorough (1984) have shown that phenobarbitone pretreatment (to induce cytochrome P450) has no effect on the rate of metabolism of propoxur (2-isopropoxyphenyl-N-methylcarbamate), suggesting that oxidation is not involved. The hydrolase, however, remains uncharacterized.

Hydrolysis of carbaryl is of minor importance in insects, in which most degradation appears to proceed via oxidative processes. This is an important factor in the mechanism of resistance to carbaryl and of the experimental reversal of some resistance by the use of methylenedioxy synergists (Wilkinson, 1971). The latter are known to inhibit oxidative metabolism.

The oxime carbamates such as methomyl (30) and aldicarb (31) are similarly hydrolysed in insects, mammals and plants and the oximes conjugated with sugars and sulphate ion prior to elimination or storage. Aldicarb (31) is a highly effective systemic insecticide, acaricide and nematicide. It is also highly toxic to mammals (LD_{50} in rat, 1 mg/kg) and must be formulated as granules to reduce the hazard to applicators. The sulphur atom serves as a site for oxidative bioactivation of aldicarb to its sulphoxide in plants (Bartley et al., 1970) and insects (Bull et al., 1967). This metabolite is a more potent acetylcholinesterase

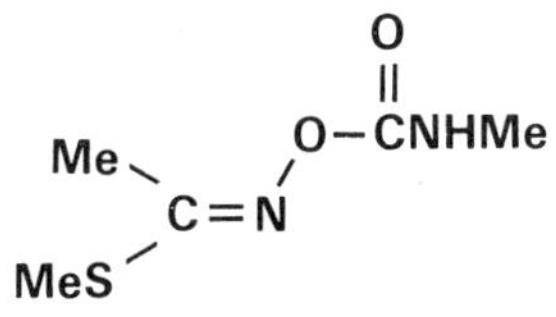

(30)

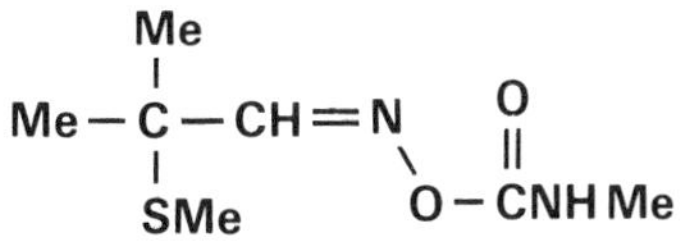

(31)

inhibitor than aldicarb and its increased water solubility aids systemicity. However, the same process occurs in mammals (Andrawes et al., 1967) and, therefore, it is not a mechanism for selectivity.

Selectivity has been achieved by N-substitution with biodegradable groups. N-Acetylation was mentioned above. N-Aryl- or N-alkyl-sulphenylation has provided an interesting series of examples synthesized by Fukuto and coworkers (Fukuto and Fahmy, 1981). Such substitution of the N–H group of, for example, carbofuran (R=H, Figure 9) markedly reduces its toxicity to mammals whilst enhancing its insecticidal activity. The latter effect is due to improved penetration followed by thiol-mediated cleavage of the N-sulphenyl bond to afford carbofuran within the insect. The reduced mammalian toxicity is due to the operation of an effective direct decarbamylation. Some values for arylsulphenyl carbofurans are shown in Figure 9. These are not isolated examples; well over 50 selective experimental insecticides have been synthesized from several N-methylcarbamates. These are examples of the more general 'propesticide' concept which is described in greater detail in Chapter 2. The metabolism of carbamate insecticides in plants and animals is reviewed in more detail in Chapter 3.

Compound	LD_{50} (housefly) mg/kg	LD_{50} (mouse) mg/kg
R = H (carbofuran)	6.7	2
R = 4 – t – butylphenyl – S –	2.7	75
R = 4 – methylphenyl – S –	9.0	100–125
R = MeS –	4.9	20

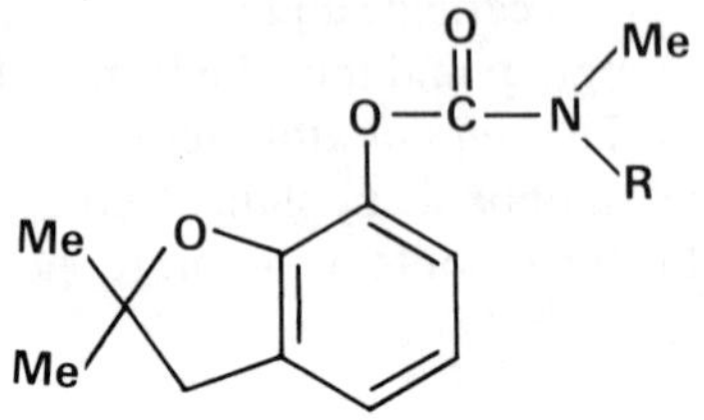

Figure 9 The effect of *N*-sulphenylation on the selective toxicity of carbofuran

Toxicology

As with the organophosphorus insecticides, the acute toxicities of the carbamate insecticides vary widely with structure. The high toxicity of some important new oxime carbamates, e.g. carbofuran and aldicarb, leave an impression that the carbamates are generally more toxic than the organophosphates; nevertheless, the economically most important carbamate, carbaryl (with an LD_{50} (rat) of *ca.* 800 mg/kg), is perfectly acceptable. Another factor in favour of the carbamates is the rapid spontaneous reactivation rate of *N*-methyl carbamylated acetylcholinesterase. The half-life is of the order of minutes whereas those for the dialkylphosphorylated enzymes range from hours to days. In addition, permanent inactivation via 'ageing' cannot occur with carbamates. Therefore, treatment of the symptoms of poisoning and the use of atropine, in conjunction with the spontaneous reactivation often allows complete recovery. Furthermore, prolonged ingestion of small amounts of carbamates does not afford a cumulative lowering of acetylcholinesterase levels. Thus there is further scope for the development of carbamate insecticides.

PYRETHRINS AND SYNTHETIC PYRETHROIDS

After many years of very limited (mostly domestic) use, the photolabile natural pyrethrins and early synthetic pyrethroids have been supplemented with highly insecticidally active photostable synthetic analogues suitable for agricultural use. The development of these pyrethroids from the discovery of activity in the flowers of a number of chrysanthemum species, structural elucidation of the active components (Staudinger and Ruzicka, 1924), optimization of this activity and commercial development in a critical society demanding high standards of safety, is already a classic example of applied science. Permethrin, cypermethrin, fenvalerate and deltamethrin form the new generation of photostable pyrethroids suitable for use in the field. They are used widely on cotton for the control of *Heliothis*. However, many other uses have been developed, e.g. oil seed crops, fruit, vegetables, animal health, public health and materials preservation.

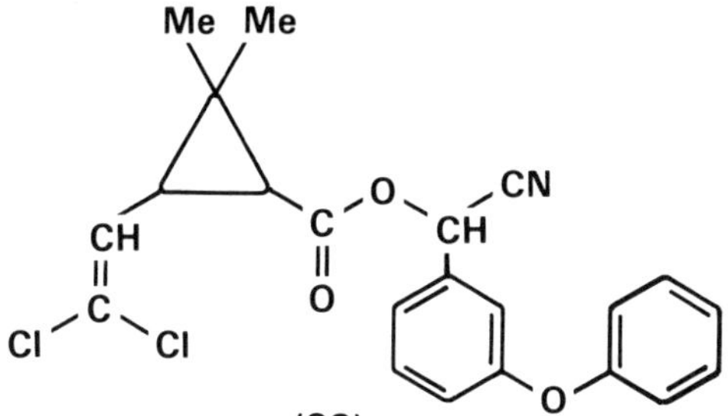

(32)

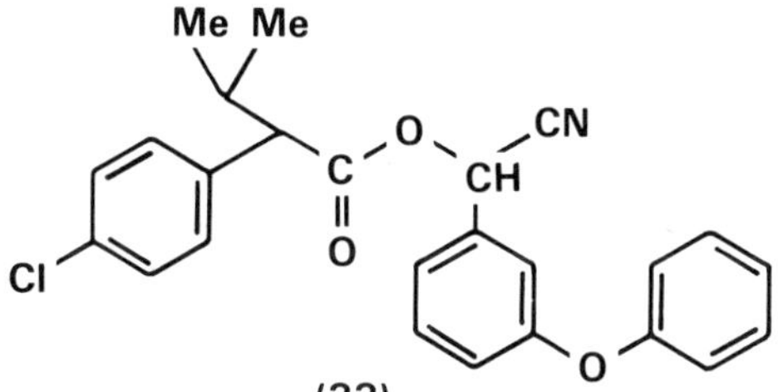

(33)

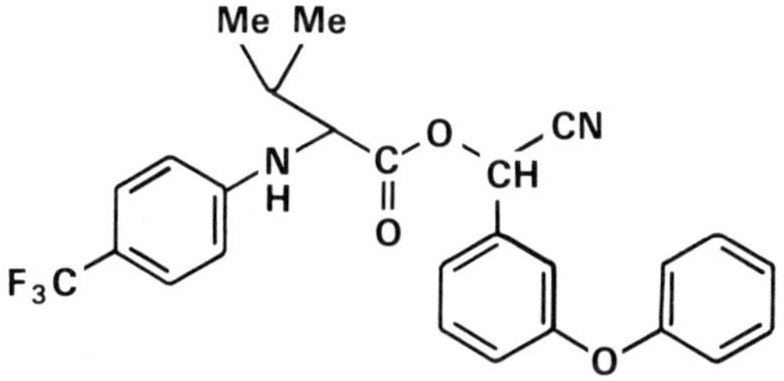

(34)

Further details of the history, properties, mode of action, metabolism and uses of the pyrethrins and pyrethroids will be found in books edited by Casida (1973), Elliott (1977) and Leahey (1985). The status of the pyrethroids as the newest class of widely used insecticides is also reflected in regular coverage in *Progress in Pesticide Biochemistry and Toxicology*: metabolism in plants and soils (Roberts, Volume 1), photochemistry (Ruzo, Volume 2), metabolism in insects (Soderlund *et al.*, Volume 3), metabolism in fish (Edwards and Millburn, this volume) and mammalian toxicology (Gray and Soderlund, this volume).

Chemistry

A brief history of the development of the photostable pyrethroids from the natural pyrethrins is given by Drs Gray and Soderlund in Chapter 5 of this volume and will be found in greater detail in the references cited above. Two examples of these pyrethroids cypermethrin (**32**) and fenvalerate (**33**) are shown above. It will be noted that fenvalerate possesses no cyclopropane ring. This surprising structure (Ohno *et al.*, 1976), based on 2-(4-chlorophenyl)-3-methylbutyric acid, has pointed the way to other interesting developments such as fluvalinate (**34**) (Henrick *et al.*, 1980).

Mode of action

The pyrethroids are potent neuroactive insecticides. In their action they are unlike phosphates and carbamates (i.e. they do not appear to react covalently with a target) but are more like DDT. Overall shape and lipophilic character are more important than chemical reactivity. They primarily affect nerve membrane sodium channels, delaying their closing. This may lead to repetitive discharges or to the blockage of nerve conduction (Narahashi, 1979; Vijverberg and van den Bercken, 1982). The α-cyano pyrethroids (e.g. cypermethrin) appear to possess some differences from the non-cyano analogues (e.g. permethrin) in their effects. As is usual with neurotoxic insecticides, it is difficult to define the ultimate cause of death of a pyrethroid-poisoned insect.

Metabolism

The most important factor of the pyrethroid insecticides in relation to their metabolism is the ester bond. Pyrethroid metabolism is dominated by ester cleavage (Hutson, 1979; Casida and Ruzo, 1980; Chambers, 1980) which is catalysed by the carboxylesterases, abundant in mammalian tissues. The reaction may also be catalysed via an oxidative mechanism (Hutson, 1979) which is more important for the *cis* isomers than for the *trans* isomers. Ester cleavage is very important in limiting the toxicity of the pyrethroids. Thus, in mammals possessing abundant capacity, toxicity is generally low, particularly for the readily hydrolysed *trans* isomers. In fish, possessing a low capacity for pyrethroid hydrolysis, relying instead on peripheral hydroxylation and conjugation (see Chapter 6), toxicity is high. Insects possess a relatively low capacity for the hydrolysis of both *cis*- and *trans*-pyrethroids and, consequently, toxicity of both groups of isomers is high.

When applied to plants, the pyrethroids are not translocated. They undergo degradation on plant surfaces, and photodegradation plays a significant part in this. Ester-cleavage products are subsequently conjugated mainly to form glycosides, some of which are complex (Mikami, *et al.*, 1985). The metabolism and residue chemistry of pyrethroids with reference to plants and soils have been reviewed (Miyamoto, 1981).

Contrary to expectation, the metabolism of the pyrethrins and some of the early pyrethroids was found to be mainly via oxidative routes (and hence the effectiveness of synergists that inhibit oxidation). Oxidation of the new synthetic pyrethroids occurs at several peripheral positions in the molecules, and it is important in ester cleavage. However, peripheral oxidation is unlikely to be significant in detoxification (except in those species, like fish, in which hydrolysis is inefficient). The role of metabolism in the mammalian toxicology of the pyrethroids is examined in further detail in Chapter 5.

Metabolism studies of several of the new pyrethroids in a variety of species have revealed a number of novel biotransformations. These are mostly conjugation reactions consequent upon the hydrolysis of the pyrethroids and , therefore, may not be directly significant in the detoxification of the parent molecule. These conjugates include those with bile acids (chicken, rat, cow), a dipeptide (duck), a diacylglycerol (rat), cholesterol (rat), xylosyl-glucose (plants) and other di- and trisaccharides (plants). These findings are the result of thorough studies using current technology. We can expect that the pyrethroids will stimulate the discovery of novel findings in other fields such as neurobiochemistry, toxicology and environmental science.

Toxicology

This subject is discussed in detail in Chapter 6.

The pyrethroids offer enormous scope for further development. Full exploitation of their stereochemistry, for example, is still to be achieved. However, intrinsic insecticidal activity is already very high, selectivity is excellent and further chemistry is complex. It remains to be seen whether existing or new incentives will be enough to encourage further commercial development of new compounds. The question of resistance, often such an incentive, is discussed in some detail in Chapter 4 of this volume.

MISCELLANEOUS

A number of individual compounds or groups of compounds possess useful insecticidal activity. Some of these are, like the pyrethrins, plant products such as nicotine and its derivatives, the isobutylamides and the rotenoids. The protein insecticides of the micro-organism *Bacillus thuringiensis* form another interesting group of natural insecticides. The defence poisons of insects and other animals have served as leads to insecticides. Nereistoxin (**35**), a product of the marine annelid worm *Lumbriconereis heteropoda*, has led to the development of Cartap (**36**) which is effective against Lepidoptera and Coleoptera. Though none of these groups has been developed to the same extent as have the carbamates or the pyrethroids, this may yet happen given that the requirement for, and the reward for, the products exists.

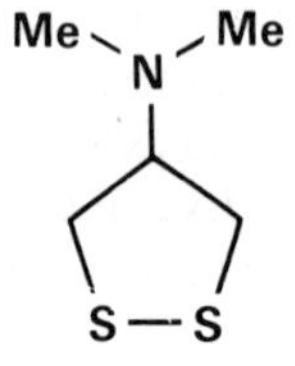

Disturbance of the insect life style or cycle may be as effective as killing one particular stage. Products of this type include hormones, hormone analogues, chemosterilants, attractants (pheromones) and repellents. Hammock and Quistad (this series, Volume 1) have reviewed the mode of action and metabolism of juvenile hormones, juvenoids and other insect growth regulators. One class of this type, inhibitors of chitin biosynthesis exemplified by diflubenzuron (Dimilin), is further reviewed by Hajjar in this volume (Chapter 7). The high specificity of insect repellents and attractants should be very effective for selected uses. They have the added appeal that they exert no overt neurotoxic action. Our expectation that this general phenomenon—the chemical mediation of insect behaviour—will be widely exploited in due course had led to its inclusion as Chapter 8 in this volume.

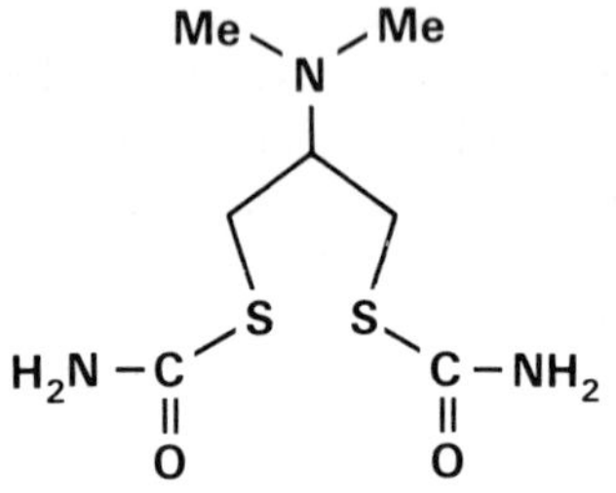

The formamidines, including chlordimeform and amitraz, exhibit pronounced acaricidal and ovicidal activity. Chlordimeform undergoes *N*-demethylation in plants, animals and soil and the resulting derivative is also insecticidally active. The mode of action of these compounds is briefly discussed in Chapter 2 of this volume.

Organotin compounds are also acaricidal. Cyhexatin and fenbutatin oxide have good residual activity against mites. As an example, cyhexatin was metabolized in rats to small amounts of dicyclohexyltin oxide, cyclohexylstannoic acid and inorganic tin (Blair, 1974). These compounds were also detected in orchard crops.

CONCLUSIONS

The foregoing introductory chapter briefly reviews the current status of the different classes of insecticides, including their relative commercial importance, their chemistry, mode of action, metabolism and toxicology. The detail with which each section has been covered has been intentionally varied. This is partly because information on some classes is well documented elsewhere but also to take account of the material presented in the remainder of this volume.

REFERENCES

Akkermans, L. M. A., Van den Bercken, J., Van der Zalm, J. M., and van Straaten, H. W. M. (1974). 'Effects of dieldrin (HEOD) and some of its metabolites on synaptic transmission in the frog motor end-plate', *Pestic. Biochem. Physiol.*, **4**, 313–324.

Andrawes, N. R., Dorough, H. W., and Lindquist, D. A. (1967). 'Degradation and elimination of Temik in rats', *J. Econ. Entomol.*, **60**, 979–987.

Anon (1983). 'ECN Special Report', *European Chemical News*, 4 July, p. 14.

Bartley, W. J., Andrawes, N. R., Chancey, E. L., Bagley, W. P., and Spurr, H. W. (1970). 'The metabolism of Temik aldicarb pesticide in the cotton plant', *J. Agric. Food Chem.*, **18**, 446–453.

Bedford, C. T., and Hutson, D. H. (1976). 'The comparative metabolism in rodents of the isomeric insecticides dieldrin and endrin', *Chem. and Ind.*, pp. 440–447.

Bedford, C. T., and Robinson, J. (1972). 'The alkylating properties of organophosphates', *Xenobiotica*, **2**, 307–337.

Beynon, K. I., Hutson, D. H., and Wright, A. N. (1973). 'The metabolism and degradation of vinyl phosphate insecticides', *Residue Rev.*, **47**, 55–142.

Blair, E. H. (1974). 'Biodegradation of tricyclohexyltin hydroxide', *Environ. Qual. Safety*, **3**, 406–409.

Bowman, J. S., and Casida, J. E. (1952). 'Metabolism of the systemic insecticide *O,O*-diethyl *S*-ethylthiomethyl phosphorodithioate (Thimet) in plants', *J. Agric. Food Chem.*, **5**, 192–197.

Brooks, G. T. (1974). *Chlorinated Insecticides, Volume I—Technology and Application*, CRC Press, Cleveland and Boca Raton, Fla.

Brooks, G. T. (1979). 'The metabolism of xenobiotics in insects', in *Progress in Drug Metabolism* (eds J. W. Bridges and L. F. Chasseaud), Vol. 3, pp. 151–214, Wiley, Chichester.

Büchel, K. H. (1983). *Chemistry of Pesticides* (ed. K. H. Büchel), Wiley, Chichester.

Bull, D. L., and Lindquist, D. A. (1964). 'Metabolism of 3-hydroxy-*N,N*-dimethylcrotonamide dimethyl phosphate by cotton plants, insects and rats', *J. Agric. Food Chem.*, **12**, 310–317.

Bull, D. L., Lindquist, D. A., and Coppedge, J. R. (1967). 'Metabolism of 2-methyl-2-(methylthio)propionaldehyde *O*-(methylcarbamoyl)oxime (Temik, UC-21149) in insects', *J. Agric. Food Chem.*, **15**, 610–616.

Casida, J. E. (1973). *Pyrethrum, the Natural Insecticide* (ed. J. E. Casida), Academic Press, New York.

Casida, J. E., and Ruzo, L. O. (1980). 'Metabolic chemistry of pyrethroid insecticides', *Pestic. Sci.*, **11**, 257–269.

Chambers, J. (1980). 'An introduction to the metabolism of pyrethroids', *Residue Rev.*, **73**, 101–124.

Corbett, J. R., Wright, K., and Baillie, A. C. (1984). *The Biochemical Mode of Action of Pesticides*, Academic Press, London.

Elliott, M. (1977). *Synthetic Pyrethroids* (ed. M. Elliott), American Chemical Society (Symposia Series No. 42), Washington DC.

Eto, M. (1974). *Organophosphorus Pesticides: Organic and Biological Chemistry*, CRC Press, Boca Raton, Florida.

Fest, C., and Schmidt, K.-J. (1982). *The Chemistry of Organophosphorus Pesticides*, Springer-Verlag, Heidelberg.

Forrest, J., Stephenson, O., and Waters, W. A. (1946). 'Chemical investigations of the insecticide "DDT" and its analogues. Part I. Reactions of "DDT" and Associated Compounds', *J. Chem. Soc.*, **1946**, 333–339.

Fraser, J., Harrison, I. R., and Wakerley, S. B. (1968). 'Synthesis and insecticidal activity of N-acyl-N-methyl carbamates', *J. Sci. Food Agric., Suppl.*, **19**, 8–12.

Fukuto, T. R., and Fahmy, M. A. H. (1981). 'Sulfur in propesticide action', in *Sulphur in Pesticide Action and Metabolism* (eds J. D. Rosen, P. S. Magee and J. E. Casida) American Chemical Society Symposia Series (No. 158), pp. 35–49.

Geissbühler, H. (1982). 'How safe are pesticides?', *Experientia*, **38**, 890–895.

Gunn, D. L. (1975). In *Foreign Compound Metabolism in Mammals*, Vol. 3, pp. 1–82, The Chemical Society, London.

Gysin, H. (1954). 'Some new insecticides', *Chimia (Switz.)*, **8**, 205–210.

Haller, H. L., Bartlett, P. D., Drake, N. L., Newman, M. S., Cristol, S. J., Eakes, C. M., Hayes, R. A., Kilmer, G. W., Magerlein, B., Mueller, G. P., Schneider, A., and Wheatly, W. (1945). 'The chemical composition of technical DDT', *J. Am. Chem. Soc.*, **67**, 1591–1595.

Hansch, C., and Deutsch, E. W. (1966). 'The use of substituent constants in the study of structure-activity relationships in cholinesterase inhibitors', *Biochim. Biophys. Acta*, **126**, 117–128.

Hayes, W. J. (1975). *The Toxicology of Pesticides*, Williams and Wilkins, Baltimore.

Henrick, C. A., Garcia, B. A., Staal, G. B., Cerf, D. C., Anderson, R. J., Gill, K., Chinn, H. R., Labovitz, J. N., Leippe, M. M., Woo, S. L., Carney, R. L., Gordon, D. C., and Kohn, G. K. (1980). '2-Anilino-3-methylbutyrates and 2-(isoindolin-2-yl)-3-methylbutyrates, the novel groups of synthetic pyrethroid esters not containing a cyclopropane ring, *Pestic. Sci.*, **11**, 224–241.

Hutson, D. H. (1972). 'Mechanisms of biotransformation' in *Foreign Compound Metabolism in Mammals*, Vol. 2, pp. 328–397, The Chemical Society, London.

Hutson, D. H. (1979). 'The metabolic fate of synthetic pyrethroid insecticides in mammals', in *Progress in Drug Metabolism* (eds J. W. Bridges and L. F. Chasseaud), Vol. 3, pp. 215–252, Wiley, Chichester.

Hutson, D. H. (1981). 'The metabolism of insecticides in man', in *Progress in Pesticide Biochemistry* (eds D. H. Hutson and T. R. Roberts), Vol. 1, pp. 287–333, Wiley, Chichester.

Johnson, M. K. (1982). 'The target for initiation of delayed neurotoxicity by organophosphorus esters: biochemical studies and toxicological applications', *Rev. Biochem. Toxicol.*, **4**, 141–212.

Jones, D. M., Bennett, D., and Elgar, K. E. (1978). 'Deaths of owls traced to insecticide-treated timber', *Nature*, **272**, 52.

Kuhr, R. J., and Dorough, H. W. (1976). *Carbamate Insecticides: Chemistry, Biochemistry and Toxicology*, CRC Press, Boca Raton, Florida.

Kurihara, N., Tanaka, K., and Nakajima, M. (1979). 'Mercapturic acid formation from lindane in rats', *Pestic. Biochem. Physiol.*, **10**, 137–150.

Lange. W., and von Krueger, G. (1932). 'Uber Ester der Monofluorphosphorsäure', *Chem. Ber.*, **65**, 1598–1601.

Leahey, J. P. (1985). *The Pyrethroid Insecticides* (ed. J. P. Leahey), Taylor and Francis Ltd, London.

Leesch, J. G., and Fukuto, T. R. (1972). 'The metabolism of Abate in mosquito larvae and houseflies', *Pestic. Biochem. Physiol.*, **2**, 223–235.

Lewis, M., and Woodburn, A. (1984). Wood, Mackenzie and Co. Agrochemical Service—Update of the Agrochemicals Products Section.

Löfroth, G. (1970). 'Alkylation of DNA by dichlorvos', *Naturwissenschaften*, **57**, 393–394.

Main, A. R. (1979). 'Mode of action of anticholinesterases', *Pharmac. Ther.*, **6**, 579–628.

Matsumura, F. (1975). *Toxicology of Insecticides*, Plenum Press, New York.

Menzer, R. E., and Casida, J. E. (1965). 'Nature of toxic metabolites formed in mammals, insects and plants from 3-(dimethoxyphosphinyloxy)-*N*,*N*-dimethyl-*cis*-crotonamide and its *N*-methyl analog', *J. Agric. Food Chem.*, **13**, 102–112.

Mikami, N., Wakabayashi, N., Yamada, H., and Miyamoto, J. (1985). 'The metabolism of fenvalerate in plants: conjugation of the acid moiety', *Pestic. Sci.*, **16**, 46–58.

Miyamoto, J. (1981). 'The chemistry, metabolism and residue analysis of synthetic pyrethroids', *Pure appl. Chem.*, **53**, 1967–2022.

Miyamoto, J., and Kearney, P. C. (eds) (1983). *Pesticide Chemistry: Human Welfare and the Environment*. IUPAC - Pergamon Press, Oxford.

Motoyama, N., and Dauterman, W. C. (1972). '*In vitro* metabolism of azinphosmethyl in susceptible and resistant houseflies', *Pestic. Biochem. Physiol.*, **2**, 113–122.

Motoyama, N., Hayashi, A., and Dauterman, W. C. (1983). 'The presence of two forms of glutathione *S*-transferases with distinct substrate specificity in OP-resistant and -susceptible housefly strains', in *Pesticide Chemistry, Human Welfare and the Environment* (eds J. Miyamoto and P. C. Kearney), Vol. 3, pp. 197–202, Pergamon, Oxford.

Müller, P. (1955). '*DDT—Das Inzektizid Dichlor-diphenyltrichloräthan und seine bedeutung*', Vol. 1, Birkhäuser, Basle-Stuttgart.

Mundy, W. R., and Dorough, H. W. (1984). 'Pathway of carbamate ester hydrolysis in rats', *The Toxicologist*, **4**, Abstr. 366.

Narahashi, T. (1979). 'Nerve membrane ionic channels as the target site of insecticides', in *Neurotoxicology of Insecticides and Pheromones* (ed. T. Narahashi), pp. 211–243, Plenum Press, New York.

Ohno, N., Fujimoto, K., Okuno, Y., Mizutani, T., Hirano, M., Itaya, N., Honda, T., and Yoshioka, H. (1976). '2-Arylalkanoates, a new group of synthetic pyrethroid esters not containing cyclopropanecarboxylates', *Pestic. Sci.*, **7**, 241–246.

Shankland, D. L., and Schroeder, M. E. (1973). 'Pharmacological evidence for a discrete neurotoxic action of dieldrin (HEOD) in the American cockroach, *Periplaneta americana* (L.)', *Pestic. Biochem. Physiol.*, **3**, 77–82.

Shires, S. W. (1983). 'Pesticides and honey bees—case studies with RIPCORD and FASTAC', *SPAN*, **26**, 118–120.

Shires, S. W., Le Blanc, J., Murray, A., Forbes, S., and Debray, P. (1984). 'A field trial to assess the effects of a new pyrethroid insecticide, WL85871, on foraging honey-bees (*Apis mellifera* L.) in oil seed rape', in press.

Slade, R. E. (1945). 'The γ-isomer of hexachlorocyclohexane (gammexane). An insecticide with outstanding properties', *Chem. Ind. (London)*, **40**, 314–319.

Soloway, S. B. (1965). 'Structure activity relationships with organochlorines', *Adv. Pest Control Res.*, **6**, 85–126.

Spindler, M. (1983). 'DDT: health aspects in relation to man and risk/benefit assessment based thereupon', *Residue Rev.*, **90**, 1–34.

Staudinger, H., and Ruzicka, L. (1924). 'Insektentotende stoffe, I-V and X', *Helv. Chim. Acta*, **7**, 177–201, 201–211, 212–235, 236–244 and 245–259.

Stephenson, R. R. (1983). 'Pesticides and freshwater animals—a case study with RIPCORD', *SPAN*, **26**, 121–122.

Stephenson, R. R., Choi, S. Y., and Olmos-Jerez, A. (1984). 'Determination of the toxicity and hazard to fish of a rice insecticide', *Crop Protection*, in press.

Stetter, J. (1983). 'Insecticidal chlorohydrocarbons', in *Chemistry of Pesticides* (ed K. H. Büchel), Wiley, Chichester.

Storck, W. J. (1984). 'Pesticides head for recovery', *Chem. Engng. News*, April 9, 35–57.

Thorpe, E., and Walker, A. I. T. (1973). 'The toxicology of dieldrin (HEOD). II. Comparative long-term oral toxicity studies in mice with dieldrin, DDT, phenobarbitone, β-BHC and γ-BHC', *Food Cosmet. Toxicol.*, **11**, 433–442.

Vijverberg, H. P. M., and van den Bercken, J. (1982). 'Action of pyrethroid insecticides on the vertebrate nervous system', *Neuropathol. appl. Neurobiol.*, **8**, 421–440.

Wilkinson, C. F. (1971). 'Effects of synergists on the metabolism and toxicity of anticholinesterases', *Bull. W.H.O.*, **44**, 171–190.

Wilkinson, C. F. (1976). *Insecticide Biochemistry and Physiology* (ed. C. F. Wilkinson), Plenum Press, New York.

Wilson, I. B., and Ginsburg, S. (1955). 'A powerful reactivator of alkylphosphate-inhibited acetylcholinesterase', *Biochim. Biophys. Acta*, **18**, 168–170.

Wooder, M. F., Wright, A. S., and King, L. J. (1977). '*In vivo* alkylation studies with dichlorvos at practical use concentrations', *Chem. – Biol. Interact.*, **19**, 25–46.

Insecticides
Edited by D. H. Hutson and T. R. Roberts
© 1985 John Wiley & Sons Ltd

CHAPTER 2

Proinsecticides

J. Drabek and R. Neumann

INTRODUCTION

The environmental modification of xenobiotics normally leads to the formation of less active compounds, but occasionally activity can be enhanced or an inactive substance can be changed into an active one. This review is the first attempt to summarize such activation processes in the field of insecticides and acaricides.

In 1958 Albert introduced the term prodrug for a substance which has to be modified to become a true drug. By analogy we refer here to substances which have to be transformed in order to show or to enhance their insecticidal or acaricidal activity. This definition also includes compounds that are slightly active on their own. It avoids the problems that arise from the experimental difficulties of measuring the low intrinsic activity of a proinsecticide. Such compounds can be activated in various ways. The process can be non-catalytic (hydrolysis, thiolysis, elimination, photolysis, cyclization, rearrangement, etc.) or by enzymatic catalysis (catalysed by oxidases, carboxylases, amidases, peptidases, etc.). Depending on the compound and on the activation reaction, the transformation of a proinsecticide to an insecticide can occur at various places in the environment (in or on the plant, in soil, water or in the insect itself). In the insect the reaction may take place almost anywhere or it may be confined to certain selected places by barriers to penetration or limited by enzyme catalytic requirements to specific places such as the fat body, the haemolymph, the corpora allata, etc.

In this review the compounds are ordered according to the chemical group to which they belong and the activation reaction is given, if it is known. A proinsecticide may have more favourable characteristics than its corresponding insecticide from the standpoints of toxicity, selectivity, stability, biodegradability, mobility, persistence, solubility, handling, etc. and some proinsecticides have been intentionally developed as commercial products.

PROCARBAMATES

Introduction

Insecticidally active carbamates possess the structure shown in Figure 1. In Figure 1 R^1 is H or CH_3 and R is a substituted phenyl, naphthyl, or a residue of a heterocycle, or an imine. A substitution at R^1 by groups other than those indi-

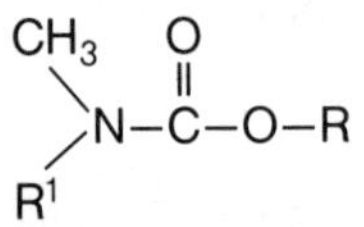

$$CH_3 \quad O \atop \diagdown \quad \parallel \atop N-C-O-R \atop \diagup \atop R^1$$

Figure 1 Structure of insecticidal carbamates

cated leads to substantially less active or inactive compounds. However, there are exceptions too. If R^1 is an acyl, phosphorothioyl, sulphenyl, sulphinyl, trialkylsilyl, iminomethyl, chlorocarbonyl, substituted oxocarbonyl, substituted aminocarbonyl, oxalyl, alkoxymethyl, or acyloxymethyl group, the insecticidal activity remains unexpectedly high. These compounds may react with a nucleophile in a way which leads to the corresponding monomethyl carbamate.

The transformation of N-acyl, N-phosphorothioyl and N-sulphenyl carbamate to the corresponding monomethyl carbamate has been proved and in other cases it is probable. In the following chapter these particular groups of procarbamates will be discussed.

N-Acyl-N-methylcarbamates

The oldest group of procarbamates is the N-acyl-N-methylcarbamates. Their structure is shown in Figure 1; R^1 is an aliphatic acyl group. Compounds of this type were first described by Fraser et al. (1965). Derivatization of N-methylcarbamate insecticides such as carbaryl, mexacarbate, Landrin[R], and propoxur with different aliphatic acyl moieties produced compounds that were much less toxic to mice than the parent carbamates. The N-acyl-N-methylcarbamates are quite effective insecticides and are often equal or superior to their parent compounds in their activity against insects. The insecticidal, toxicological, and anticholinesterase properties of some N-acyl-N-methylcarbamates are shown in Tables 1 and 2.

Because of the weaker anticholinesterase activity of the N-acyl-N-methylcarbamates it has been assumed that the parent N-methylcarbamate, generated in vivo is the actual toxicant. Support for this hypothesis was provided by Miskus

Table 1 Insecticidal activity, toxicity and cholinesterase inhibition of N-acyl-N-methylcarbamates

$$CH_3\text{-}N\text{-}C\text{-}O\text{-}R$$
$$\overset{\mid}{R^1}\ \overset{\parallel}{O}$$

R^1	R	Cydia pomonella	Phaedon cochleariae	Megoura viciae	Plutella maculip.	oral LD_{50}mice (mg/kg)	I_{50} (M) bee brain AChE
H	1-Naphthyl	0.0025	0.03	0.0125	0.02	700	5×10^{-8}
CH_3CO	1-Naphthyl	0.01	0.0175	0.0175	0.05	3000	5×10^{-6}
C_2H_5CO	1-Naphthyl	0.01	0.05	0.1	0.5	3000	9×10^{-7}
$n\text{-}C_3H_7CO$	1-Naphthyl	0.01	0.05	0.1	0.5	3000	7.5×10^{-7}
H	4-Dimethyl-amino-3,5-xylyl	0.0015	0.06	0.01	0.06	42	—
CH_3CO	4-Dimethyl-amino-3,5-xylyl	0.001	0.005	0.125	0.05	550	—
C_2H_5CO	4-Dimethyl-amino-3,5-xylyl	0.0015	0.015	0.125	0.035	400	—
$n\text{-}C_3H_7CO$	4-Dimethyl-amino-3,5-xylyl	0.008	0.02	0.05	0.045	150	—

The median lethal concentration columns (Cydia pomonella, Phaedon cochleariae, Megoura viciae, Plutella maculip.) are grouped under the heading "Median lethal concentration (% a.i.)".

See Fraser et al., 1965.

Table 2 Biological activity of carbamates of the structural formula

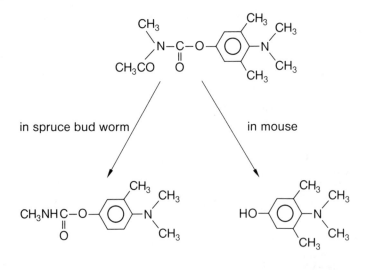

R	I_{50} (M) *Musca domestica* AChE	Topical LD_{50} *Musca domestica* (μg/g)	LC_{50} *Culex fatigans* larvae (ppm)	Oral LD_{50} mice (mg/kg)
H	4.5×10^{-7}	90	0.038	16
CH_3CO	4.0×10^{-5}	235	0.028	600–900
C_2H_5CO	1.1×10^{-4}	150	0.034	1000
i-C_3H_7CO	4.4×10^{-5}	85	0.018	250–1000
i-C_4H_9CO	—	75	0.057	250–1000

See Fahmy *et al.*, 1966.

et al. (1969), who demonstrated that *N*-acetyl-*N*-methyl-*O*-(4-dimethylamino-3,5-dimethylphenyl)carbamate was converted in spruce budworm into relatively large amounts of *N*-methyl-*O*-(4-dimethylamino-3,5-dimethylphenyl)carbamate. In contrast the *N*-acetyl-*N*-methyl-*O*-(4-dimethylamino-3,5-dimethylphenyl)carbamate was detoxified in the mouse through hydrolysis of the carbonyl–phenoxy bond to non-toxic products (Figure 2).

Figure 2 Hydrolysis of *N*-acetyl-*N*-methyl-*O*-(4-dimethylamino-3,5-dimethylphenyl)carbamate in the spruce budworm and in the mouse

Despite their good insecticidal activity none of the N-acyl-N-methylcarbamates have been developed as commercial insecticides. Recently Union Carbide Chem. Co. examined a new N-acyl derivative of carbofuran in field tests under the code No. UC-54229 (Kaplan, 1980) (Figure 3).

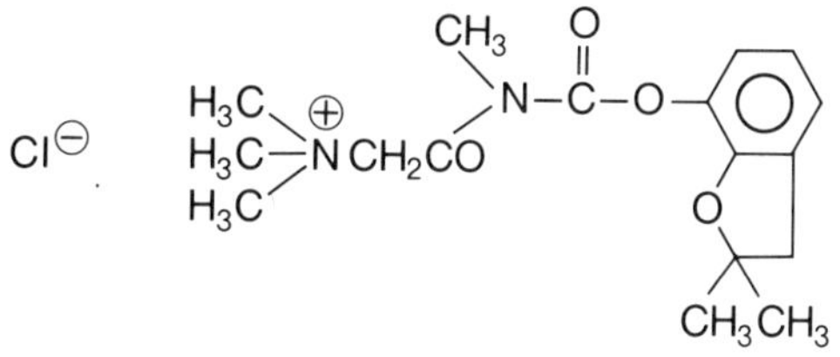

UC-54229
LD$_{50}$ rat p.os: 118 mg/kg

Figure 3 Chemical structure of UC-54229

N-Dialkoxyphosphinothioyl-N-methylcarbamates

R^1 in Figure 1 (alkyl O)$_2$P-
$$\underset{S}{\overset{\|}{}}$$

Compounds of this type were studied by Fahmy and co-workers (1970). The starting point was the selectivity of the safe organophosphorus insecticide malathion, where differences in metabolism between houseflies and mice illustrated a diminished toxicity to mammals.

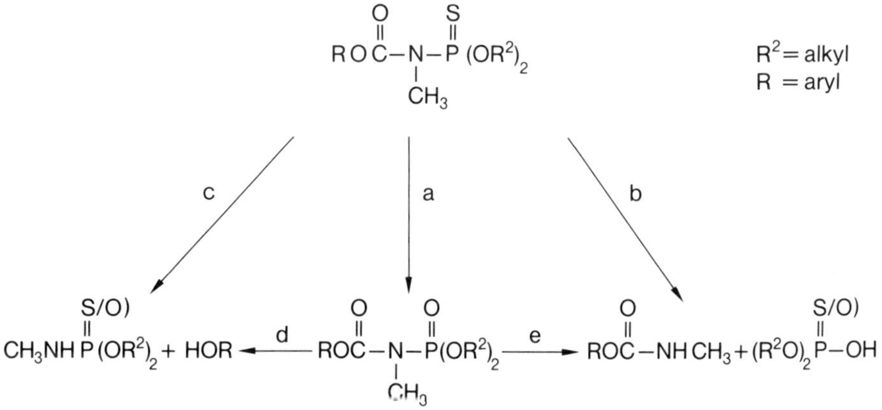

path: c or a, d mouse
path: b or a, e housefly

Figure 4 Metabolism of N-dialkylphosphinothioyl-N-methylcarbamates (From Fukuto, 1976, reproduced by permission of John Wiley & Sons Ltd)

It was expected that substitution of the hydrogen on the carbamyl nitrogen atom of carbamate esters for a dialkylphosphinothioyl moiety would produce derivatives with differences in metabolism between insects and mammals. The rationale for the approach is illustrated in Figure 4. On the basis of this rationale a number of N-dialkoxyphosphinothioyl derivatives were prepared and examined for toxicity to insects and mice (Table 3).

The results clearly showed that the N-dialkoxyphosphinothioyl-N-methylcarbamates were substantially less toxic to the white mouse whilst at the same time (with one exception) both they and the parent carbamates showed approximately equal activity against houseflies. The anticholinesterase activities of the phosphorylated N-methylcarbamates were clearly lower than those of the parent carbamates, indicating that the derivatives *per se* were not directly responsible for housefly toxicity. The differences in the toxicity to the mouse and the activity against housefly could be explained by differences in the metabolism.

Thus the last compound in Table 3 was metabolized in houseflies mainly to carbofuran and related oxidation products containing the intact N-methylcarbamate. Degradation to phenolic products was the principal route of

Table 3 Toxicological properties of N-dialkoxyphosphinothioyl-N-methylcarbamates

$$\underset{R^1}{\overset{CH_3}{>}}N-\overset{\overset{O}{\|}}{C}-O-R$$

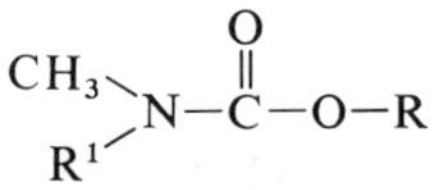

R	R^1	I_{50} (M) housefly AChE	LD$_{50}$ (mg/kg) housefly Snaidm	LD$_{50}$ (mg/kg) mice oral
3-Isopropylphenyl	H	3.7×10^{-7}	41	16
3-Isopropylphenyl	P(S)(OCH$_3$)$_2$	6.6×10^{-5}	32.5	760
3-Isopropylphenyl	P(S)(OC$_2$H$_5$)$_2$	2.9×10^{-6}	67.5	400–500
2-Isopropoxyphenyl	H	6.9×10^{-7}	22	24
2-Isopropoxyphenyl	P(S)(OCH$_3$)$_2$	8.4×10^{-6}	32	1400
2-Isopropoxyphenyl	P(S)(OC$_2$H$_5$)$_2$	1.4×10^{-5}	37	1700
CH$_3$SC(CH$_3$)$_2$CH=N–	H	8.4×10^{-5}	5.5	0.3–0.5
CH$_3$SC(CH$_3$)$_2$CH=N–	P(S)(OCH$_3$)$_2$	2.9×10^{-4}	95	170
2,2-Dimethyl-2,3-dihydro benzofuranyl (7)	H	2.5×10^{-7}	6.7	2
2,2-Dimethyl-2,3-dihydro benzofuranyl (7)	P(S)(OCH$_3$)$_2$		13	150–190

See Fukuto, 1976; reproduced by permission of John Wiley & Sons, Inc.

metabolism in rodents (Krieger *et al.*, 1976). None of the compounds of this type has been developed as a commercial insecticide.

N-Sulphenyl-*N*-methylcarbamates

From a practical point of view the most interesting procarbamates are the sulphenylated carbamates. In this case R^1 in Figure 1 is R^2–S–, where R^2 is an organic residue. The first patent application which disclosed compounds of this type was filed by Chevron Chem. Co. (Brown and Kohn, 1969). Successively patent applications by the following companies were published: ICI, Bayer, Sumitomo, Ciba-Geigy, Union Carbide, FMC, Hoffmann-La Roche, Upjohn, Du Pont and Otsuka Chem. Co. Important research work on sulphenylated carbamates which led to the discovery of highly active compounds as well as to the elucidation of their mode of action has been done by Fukuto and his co-workers at the University of California. In view of the large amount of literature (especially patents) which deals with this subject, this chapter is limited to the most important facts. Table 4 shows the principal substituents used for the sulphenylation of *N*-methylcarbamates. For the synthesis of *N*-sulphenyl-*N*-methylcarbamates see Black *et al.*, 1973a; Kuehle and Klauke, 1977; Fahmy *et al.*, 1978; Hatch III, 1978; Dutton *et al.*, 1981b.

Table 4 Principal R^2-S-substituents in *N*-sulphenyl-*N*-methylcarbamates

$\begin{array}{c} \text{CH}_3 \\ \diagdown \\ \diagup \\ R^1 \end{array} \overset{\overset{\displaystyle O}{\|}}{N}-C-O-R \quad$ (for R see Figure 1) $R^1 =$	Reference
alkyl-S-; aryl-S-	Cheng and Casida, 1973 Brown and Kohn, 1969 Black *et al.*, 1973a
alkyl-O-C-S- $\|$ O	Stetter and Homeyer, 1976
(alkyl)$_2$N-S-	Black and Fukuto, 1973 Gemrich II *et al.*, 1978 Fukuto and Fahmy, 1981
alkyl-S-S-; aryl-S-S-	Brown, 1970b D'Silva, 1976a,b Fukuto and Fahmy, 1981
R^2OC-N-(S)$_{(1,2)}$- $\| \|$ $O $ alkyl	Brown, 1970a Fahmy *et al.*, 1974 Fahmy *et al.*, 1978 Fukuto and Fahmy, 1981 Drabek and Bachmann, 1983

R^2 = alkyl, phenyl; heterocycle-, imine-residue, identical or different to R

Table 4   *(continued)*

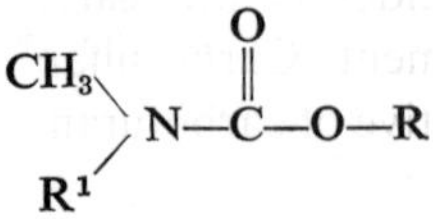

$$CH_3 \diagdown \underset{R^1}{\diagup} N - \overset{\overset{\textstyle O}{\|}}{C} - O - R$$

(for R see Figure 1)

$R^1 =$		Reference

R^2SO_2-N-S-
       |
     alkyl
	R^2 = alkyl, aryl	Kuehle and Klauke, 1977
		Fukuto and Fahmy, 1981
		Hartmann *et al.,* 1982

R^2CO-N-S-
      |
   alkyl
	R^2 = alkyl, aryl	Brown and Kohn, 1970
		Drabek and Bachmann, 1983

(alkyl)$_2$N-CO-N-S-
           |
       alkyl
		Caleb, 1980
		Drabek and Bachmann, 1983

(alkyl-O)$_2$P-N-S-; alkylen $\underset{O}{\overset{O}{\diagup}}\overset{\|}{P}$-N-S- (S)O alkyl O(S) alkyl		CIBA-GEIGY Ltd, 1976 Mitsubishi, 1980 Dutton *et al.,* 1981a
RO-C-N-S-alkylen-S-N-S- || | | O alkyl alkyl		Sauers, 1978c

aryl-N=CH-N-CO-N-S$_{1-2}$-N-S-
       | | |
    CH$_3$ CH$_3$ CH$_3$
		D'Silva, 1980

alkyl O-C-(CH$_2$)$_{1-2}$-N-S; NC(CH$_2$)$_{1-2}$-N-S-
    || | |
    O   alkyl   alkyl
      [NC(CH$_2$)$_{1-2}$]$_2$N-S-
		Otsuka Chem. Co., 1980a

R^2O-(S)$_{1-4}$-	R^2 = 1-24C hydrocarbyl	Kawata *et al.,* 1981

    In comparison with their parent compounds *N*-sulphenyl-*N*-methylcarbamates generally show lower mammalian toxicity, better residual insecticidal activity and lower phytotoxicity. On the other hand they often show less systemic insecticidal activity and diminished storage stability (they are often stabilized by epoxides).

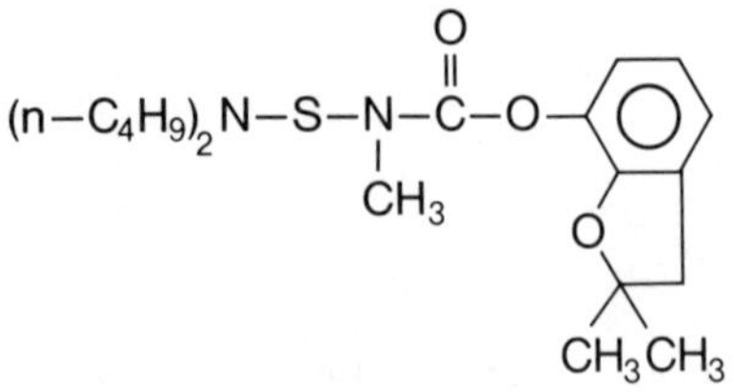

$$(n-C_4H_9)_2N-S-\underset{CH_3}{N}-\overset{\overset{\textstyle O}{\|}}{C}-O-\text{(aryl ring structure)}$$

Carbosulfan

From the group of *N*-sulphenyl-*N*-methylcarbamates carbosulfan, thiodicarb and furathiocarb have been developed to commercial insecticides; benfuracarb, and U-56295 are experimental insecticides under development. Carbosulfan (FMC-35001, Marshal[R], Advantage[R], a sulphenylated derivative of carbofuran, is a foliar, soil contact, and systemic insecticide with substantially lower mammalian toxicity than carbofuran (acute oral LD_{50} for rats: 209 mg/kg; carbofuran 8 mg/kg) (Black and Fukuto 1973; Maitlen and Sladen 1979).

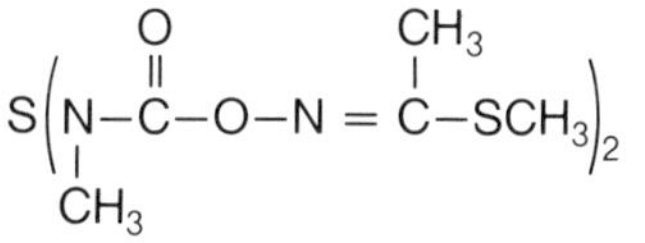

$$S\left(\underset{\underset{CH_3}{|}}{N}-\overset{\overset{O}{\|}}{C}-O-N=\overset{\overset{CH_3}{|}}{C}-SCH_3\right)_2$$

Thiodicarb

Thiodicarb (CGA 45156, UC-51762, Lepicron[R], Larvin[R]) is a contact and stomach insecticide with good activity against lepidopterous insects and low mammalian toxicity (acute oral LD_{50} for rats: 431 mg/kg; Drabek, 1974; Sousa *et al.*, 1977; Buholzer *et al.*, 1980; Yang and Thurman, 1981).

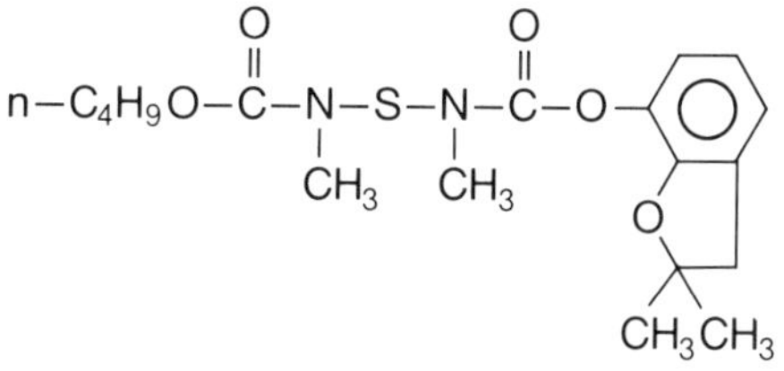

$$n-C_4H_9O-\overset{\overset{O}{\|}}{C}-\underset{\underset{CH_3}{|}}{N}-S-\underset{\underset{CH_3}{|}}{N}-\overset{\overset{O}{\|}}{C}-O-\text{[furanyl ring] } CH_3CH_3$$

Furathiocarb

Furathiocarb (CGA 73102, Deltanet[R], Promet[R]) is a contact and systemic soil insecticide with an acute oral LD_{50} for rats of 53–106 mg/kg (Drabek and Boeger, 1977; Bachmann and Drabek, 1981).

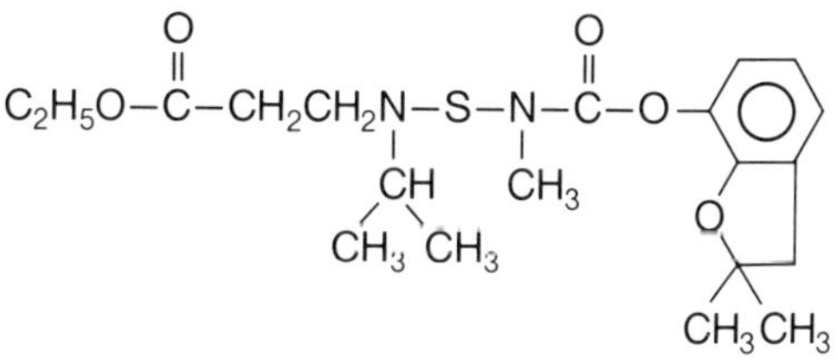

$$C_2H_5O-\overset{\overset{O}{\|}}{C}-CH_2CH_2\underset{\underset{\overset{|}{CH}}{|}}{N}-S-\underset{\underset{CH_3}{|}}{N}-\overset{\overset{O}{\|}}{C}-O-\text{[furanyl ring] } CH_3CH_3$$

Benfuracarb

Benfuracarb (OK-174, Oncol[R]) is an experimental broad spectrum insecticide with an acute oral LD_{50} for male rats of 138 mg/kg (Otsuka Chem. Co., 1980a; Goto *et al.*, 1983).

U-56295, a sulphenylated methomyl derivative, is especially active against lepidopterous insects. Its acute oral LD_{50} for rats is *ca.* 8000 mg/kg (Dutton *et al.*, 1981a).

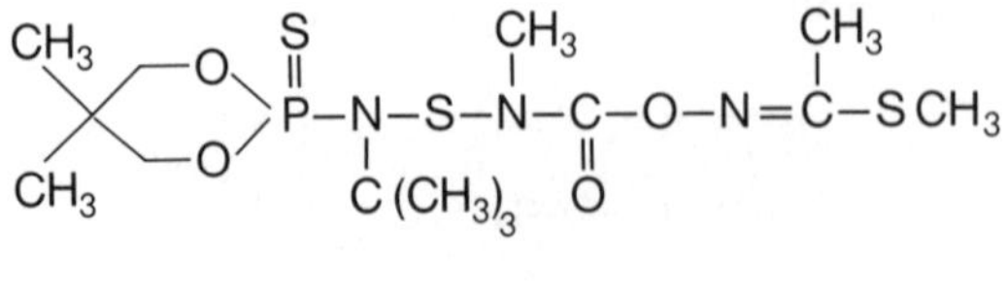

U−56295

Comparative data on the mammalian toxicity, residual insecticidal activity and phytotoxicity of some *N*-sulphenyl-*N*-methylcarbamates and of their parent compounds are shown in Tables 5, 6 and 7. On an overall basis, the sulphenylated-*N*-methylcarbamates are poorer acetylcholinesterase inhibitors than the parent compounds (Black *et al.*, 1973a; Drabek and Bachmann, 1983).

Table 5   Mammalian toxicity of sulphenyl-*N*-methylcarbamates

Sulphenyl-*N*-methylcarbamate	LD_{50} rat p. os (mg/kg)		Parent-*N*-methylcarbamate
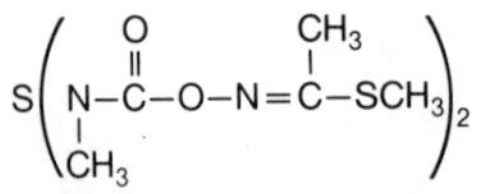 Thiodicarb	431	25	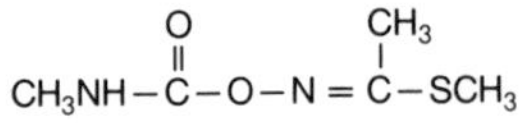 Methomyl
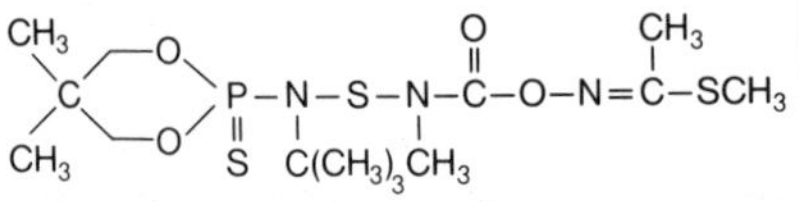 U−56295	>8000		
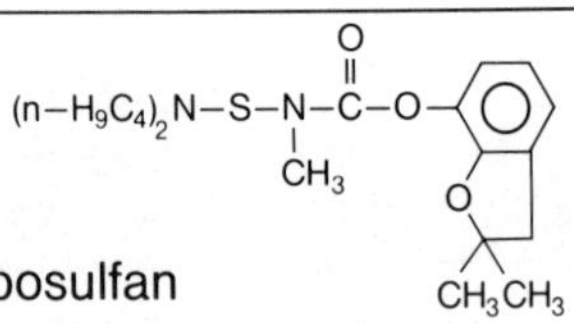 Carbosulfan	100− 250	8	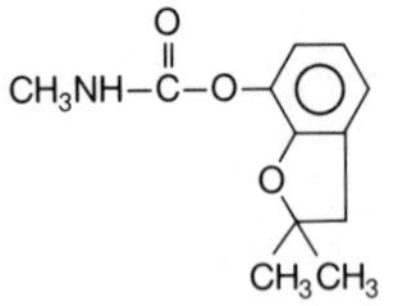 Carbofuran
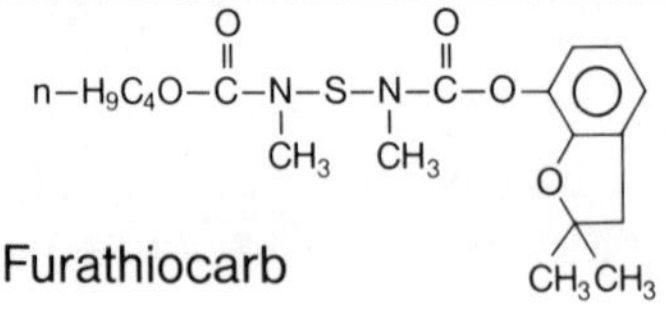 Furathiocarb	106		

Table 6 Residual effect of thiodicarb against *Spodoptera littoralis* on Egyptian cotton

Days post application		0	3	5	7
		% Mortality (*S. littoralis*)			
	g.a.i./ha				
Thiodicarb	500	100	90	94	87
Methomyl	500	100	3	0	0
Check	—	0	0	0	0

Table 7 Phytotoxicity of thiodicarb and methomyl on cotton (field trials, Brazil, 1976)

Compound	g.a.i./ha	trial no.	Phytotoxicity* (1–10)
Thiodicarb	750	1	1.75
		2	1.75
		3	3.0
Methomyl	500	1	6.0
		2	5.25
		3	7.0

*1 = no injury; 10 = plant dead.

At least a part of the discrepancy between lower mammalian toxicity and equal insecticidal activity of the sulphenylated carbamates in comparison with their parent compounds is a consequence of different metabolic pathways in insects and mammals. The sulphenyl group on the *N*-methylcarbamates allows reactions leading to less toxic products in mammals whilst the toxic parent *N*-methylcarbamate is formed in the insect (Black *et al.*, 1973a,b; Fahmy *et al.*, 1974). This suggestion has been confirmed with carbosulfan by Marsden *et al.* (1982), who found that carbosulfan (CS) was metabolized by two primary pathways in the rat; by oxidation of the sulphur to give carbosulfan-sulphone (CS-sulphone) and eventually sulphamide (SAM) and by N–S bond cleavage to give carbofuran (CF) (Figure 5). The oxidative conversion of carbosulfan to carbosulfan-sulphone and its subsequent degradation to sulphamide is undoubtedly a detoxification process since sulphonyl derivatives of carbofuran are non-toxic to insects and mammals (Kinoshita and Fukuto, 1980). A contribution to the lower mammalian toxicity of carbosulfan, relative to carbofuran could also be the high stability of carbosulfan in the rat stomach connected with its conversion to the less toxic carbosulfan-polysulphide derivatives (Umetsu and Fukuto, 1982). In contrast to the rat, N–S bond cleavage to carbofuran appeared to be the only primary metabolic pathway for carbosulfan in the housefly (Marsden *et al.*, 1982). The N–S bond cleavage of sulphenylated carbofuran *N,N'*-thiodicarbamate derivatives in the housefly appears to take place soon after application to the cuticle with the release of a sufficient quantity of carbofuran to cause toxic effects (Collins *et al.*, 1980).

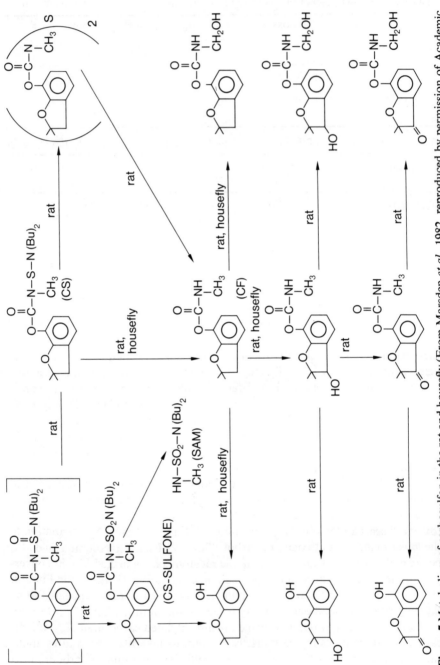

Figure 5 Metabolism of carbosulfan in the rat and housefly (From Marsden *et al.*, 1982, reproduced by permission of Academic Press Inc.)

N-Sulphinyl-N-methylcarbamates

Compounds of this type possess the following structure:

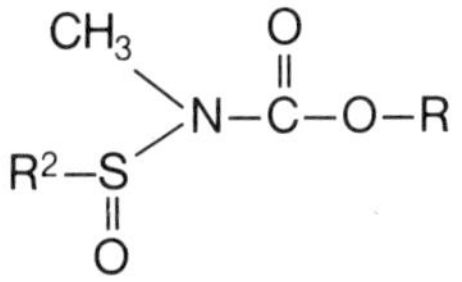

$$CH_3 \\ >N-\overset{\overset{O}{\|}}{C}-O-R \qquad \text{(for R see Fig. 1.)} \\ R^2-\underset{\underset{O}{\|}}{S}$$

In 1972 Ueda *et al.* applied for a patent for compounds where R^2 is alkyl or aryl.

Recently new types of N-sulphinyl-N-methylcarbamates were synthesized by Fahmy and Fukuto (1981). In a smooth reaction an insecticidal carbamate reacts with thionyl chloride and a base with the formation of the corresponding N-chlorosulphinyl-N-methylcarbamate. The intermediate I reacts with different

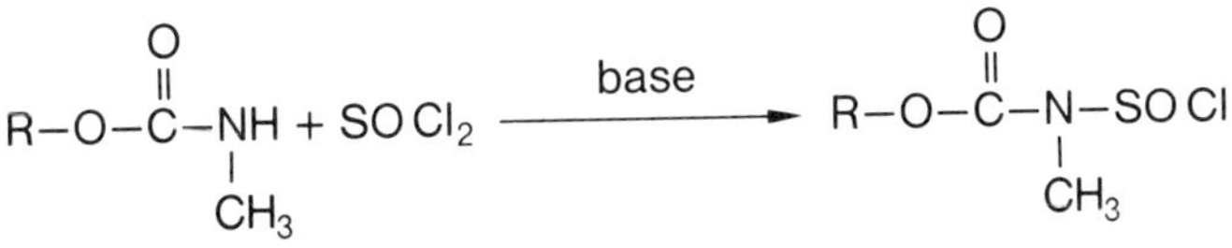

$$R-O-\overset{\overset{O}{\|}}{C}-\underset{\underset{CH_3}{|}}{N}H + SOCl_2 \xrightarrow{\text{base}} R-O-\overset{\overset{O}{\|}}{C}-\underset{\underset{CH_3}{|}}{N}-SOCl$$

$$\text{(I)}$$

nucleophiles (alcohols, phenols, thiols, N-alkylcarbamates, sulphonamides) to give a variety of N-sulphinyl-N-methylcarbamates, which show good insecticidal activity and lower mouse toxicity than the parent compounds (Table 8).

More recently N-sulphinyl-N-methylcarbamates, where R^2 are glycerol,

sugar, or $-(CH_2)_n-\langle \rangle N$ moieties were described. Derivatization with

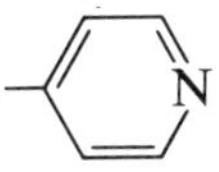

protected sugar molecules was carried out to explore the possibility of designing phloem-mobile, therefore downward-moving derivatives (Fukuto, 1983; Jojima *et al.*, 1983). A survey of patent applications containing sulphinyl-N-methylcarbamates is presented in Table 9.

Nothing is known about the mode of action and metabolism of N-sulphinyl-N-methylcarbamates. From their reduced mammalian toxicity and from the facts that:

(1) N-sulphinyl-N-methylcarbamates easily released the parent carbamates in aqueous dioxane with a trace of acid; and

(2) upon methanolysis the cleavage of the N–SO linkage occurred:

it can be assumed that N-sulphinyl-N-methylcarbamates can be metabolized to the corresponding N-methylcarbamates *in vivo* (Fukuto, 1983).

Table 8 Insecticidal activity and toxicity of sulphinyl derivatives of carbofuran and methomyl

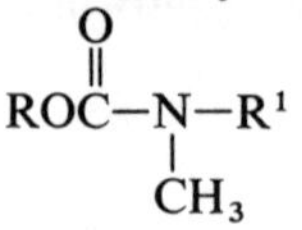

$$\underset{\underset{CH_3}{|}}{ROC-N-R^1}$$

(with $\overset{O}{\overset{\|}{}}$ on the carbonyl carbon)

R	R^1	LD$_{50}$	
		Musca domestica ($\mu g/g$)	mice, oral (mg/kg)
	$-\overset{O}{\overset{\|}{S}}-OC_6H_{13}\text{-}n$	13	280
	$-\overset{O}{\overset{\|}{S}}-O-\!\!\bigcirc$	6.4	42
	$-\overset{O}{\overset{\|}{S}}-S-\!\!\bigcirc$	9	70
	$-\overset{O}{\overset{\|}{S}}-N(CH_3)CO_2C_7H_{15}\text{-}n$	13.5	150
	$-\overset{O}{\overset{\|}{S}}-N(CH_3)SO_2N(C_4H_9\text{-}n)_2$	12	

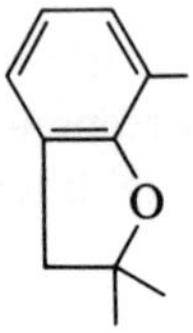

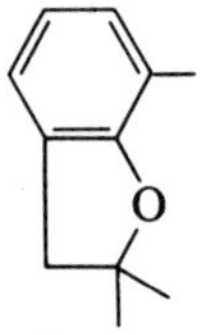

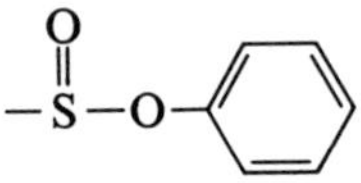

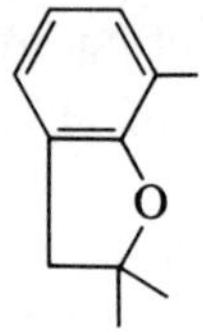

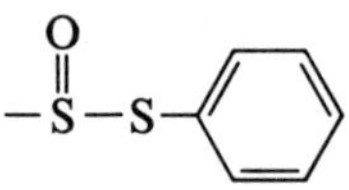

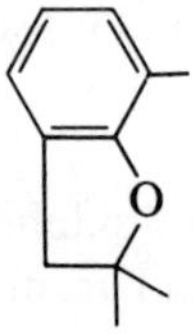

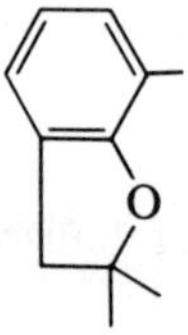

Table 8 (*continued*)

R	R^1	LD_{50}	
		Musca domestica ($\mu g/g$)	mice, oral (mg/kg)
(benzofuran structure)	$-\overset{\overset{\displaystyle O}{\|\|}}{S}-N(C_2H_5)SO_2CH_3$	12.5	75
(benzofuran structure)	$-\overset{\overset{\displaystyle O}{\|\|}}{S}-N(CH_3)SO_2-$ 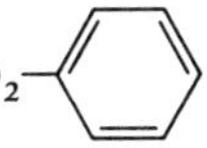 (phenyl)	9.5	35
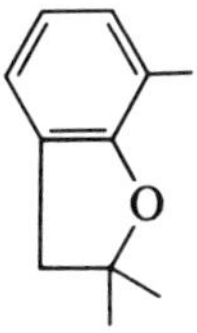 (benzofuran structure)	H(carbofuran)	6.7	11
$CH_3S(CH_3)C{=}N-$	$-\overset{\overset{\displaystyle O}{\|\|}}{S}-N(C_2H_5)\cdot CO_2C_3H_7\text{-}n$	7	
$CH_3S(CH_3)C{=}N-$	$-\overset{\overset{\displaystyle O}{\|\|}}{S}-SC_4H_9\text{-}t$	5.5	135
$CH_3S(CH_3)C{=}N-$	$-\overset{\overset{\displaystyle O}{\|\|}}{S}-N(CH_3)SO_2N(C_4H_9\text{-}n)_2$	9	
$CH_3S(CH_3)C{=}N-$	H(methomyl)	3.7	10

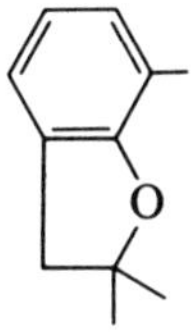

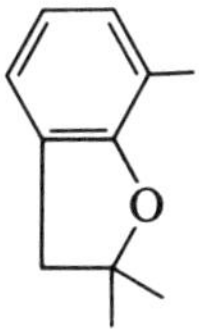

Table 9   Types of N-sulphinyl-N-methylcarbamates found in patent publications

$$\begin{matrix} & O \\ & \| \\ ROC&-N-R^1 \\ & | \\ & CH_3 \end{matrix} \qquad (R = \text{a phenyl, enol or oxime radical})$$

R^1	Reference
$\begin{matrix} O \\ \| \\ -S-N(CH_3)-COOR \end{matrix}$	Fahmy and Fukuto, 1979b,d
$\begin{matrix} O & & O \\ \| & & \| \\ -S-OR^2; & -S-SR^2 \end{matrix}$ ($R^2 = $ 1–20C hydrocarbyl)	Fahmy and Fukuto, 1979a,c, 1980a
$\begin{matrix} O & & O \\ \| & & \| \\ -S-N(R^2)-SO_2R^3; & -S-N(R^4)-SO_2N(R^3)_2 \end{matrix}$ ($R^2 = $ 1–4C alkyl; $R^3 = $ alkyl, aryl; $R^4 = $ 1–20C hydrocarbyl)	Fahmy and Fukuto, 1979c
$\begin{matrix} O \\ \| \\ (-S-X)_{(n-y)}-R^2 \end{matrix}$ ($R^2 = $ polyol or polythiol-sugar and glycerol derivatives; X = O,S; $y = $ number of unreacted hydroxyl or thiol groups; $n > y$)	Fahmy et al., 1980
$\begin{matrix} O \\ \| \\ -S-N(R^2)-COOR^3 \end{matrix}$ ($R^2 = $ 1–8C alkyl or phenyl; $R^3 = $ 1–12C hydrocarbyl or heterocycle)	Fahmy and Fukuto, 1980b
$\begin{matrix} O \\ \| \\ -S-N(R^2)-P(X)(OR^3)(OR^4) \end{matrix}$ ($R^2 = $ 1–12C alkyl; 3–6C cycloalkyl; $R^3, R^4 = $ 1–6C alkyl, or 5–6 membered ring; X = O,S)	Fukuto and Ohta, 1982

Table 9 (*continued*)

R¹	Reference

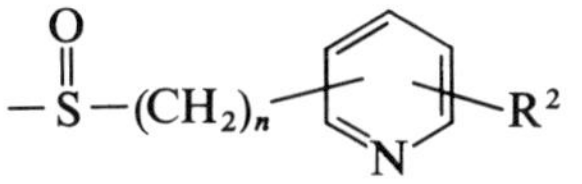

$$-\overset{\overset{\displaystyle O}{\|}}{S}-(CH_2)_n\!-\!\!\langle\text{pyridine}\rangle\!-\!R^2$$

(R^2 = H, 1–4C alkyl, halogen; n = 1–5)

Fahmy *et al.*, 1981

$$-\overset{\overset{\displaystyle O}{\|}}{S}-N(R^2)(R^3)$$

(R^2 = H, alkyl; R^3 = alkyl, alkoxy)

CIBA–GEIGY Ltd.
1979a,b

$$-\overset{\overset{\displaystyle O}{\|}}{S}-S-(CH_2)_n-O-\overset{\overset{\displaystyle O}{\|}}{S}-N(R^2)-CO_2R$$

(R^2 = alkyl) n = 2–6

Otsuka Chem. Co., 1980c

$$-\overset{\overset{\displaystyle O}{\|}}{S}-N(R^2)(R^3)$$

(R^2, R^3 = -CH$_2$CN; -CH$_2$-COOR^4;
R^4 = alkyl, cycloalkyl)

Otsuka Chem. Co., 1980b

$$-\overset{\overset{\displaystyle O}{\|}}{S}-N(R^2)P(X)(R^3)(R^4)$$

(R^2 = alkyl, cycloalkyl, phenyl, alkoxycar-
bamoylmethyl; R^3, R^4 = alkoxy; X = O,S)

Otsuka Chem. Co., 1981
Fukuto and Fahmy, 1981

N-Sulphonyl-*N*-methylcarbamates

In contrast to *N*-sulphenyl- and *N*-sulphinyl-*N*-methylcarbamates neither *N*-alkylsulphonyl nor *N*-arylsulphonyl-*N*-methylcarbamates show insecticidal activity. Obviously the SO_2–N bond is too stable to be cleaved under physiological conditions (Cheng and Casida, 1973; Kinoshita and Fukuto, 1980; Drabek *et al.*, unpublished experiments).

Other types of probable procarbamates

Other types of probable procarbamates are compounds of the structural formula:

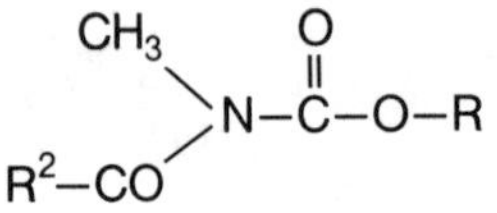

$$\begin{array}{c} CH_3 \\ \diagdown \\ N-\overset{\overset{\textstyle O}{\|}}{C}-O-R \qquad \text{(for R see Fig. 1.)} \\ \diagup \\ R^2-CO \end{array}$$

where R^2 is a Cl, alkoxy, aryloxy, aralkoxy, arylthio, amino, oximino, dialkoxy phosphoryl, imidazolyl, etc. group. Compounds of this type are described exclusively in the patent literature (Kuehle *et al.*, 1971a,b; Hartmann *et al.*, 1978; Heywang *et al.*, 1980, 1981, 1982; Drabek and Boeger, 1980a,b, 1981; Mauer *et al.*, 1981).

Nothing has yet been published about the mode of action and metabolism of these types of compounds. From the fact that they show a similar spectrum of activity to that of the parent carbamates and that they are much weaker cholinesterase inhibitors *in vitro* than the parent compounds, we assume that they can be metabolized to the parent *N*-methylcarbamates *in vivo*. Structures and properties of some carbofuran derivatives of this type are shown in Table 10.

The following types of insecticides, which are probably procarbamates have been described mainly in the patent literature.

N-*Trialkylsilyl*-N-*methylcarbamates*

(Fishbein and Zielinsky 1965; Sauers, 1978a,b; Acker *et al.*, 1980).

Typical examples of this type of compound are the trimethylsilyl derivatives of methomyl and cloethocarb.

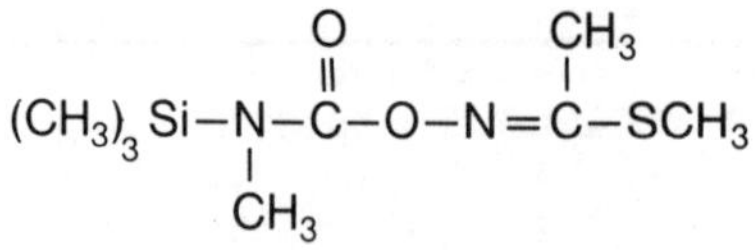

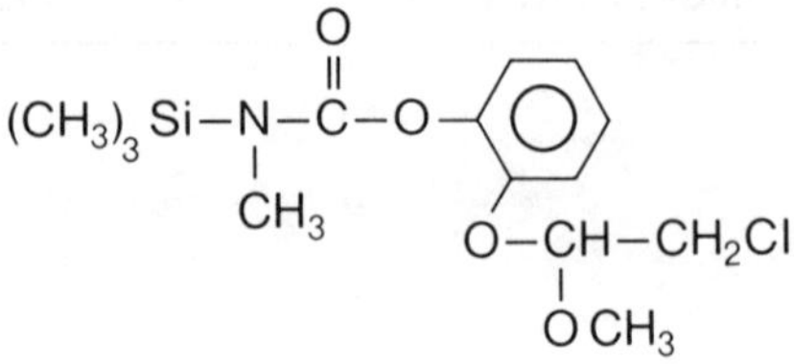

$$(CH_3)_3 Si-\underset{\underset{\textstyle CH_3}{|}}{N}-\overset{\overset{\textstyle O}{\|}}{C}-O-N=\underset{}{\overset{\overset{\textstyle CH_3}{|}}{C}}-SCH_3$$

$$(CH_3)_3 Si-\underset{\underset{\textstyle CH_3}{|}}{N}-\overset{\overset{\textstyle O}{\|}}{C}-O-\!\!\left\langle\!\!\bigcirc\!\!\right\rangle$$
$$\underset{\underset{\textstyle O\,CH_3}{|}}{O-CH-CH_2Cl}$$

Table 10 Insecticidal activity, toxicity and cholinesterase inhibition of various procarbofurans

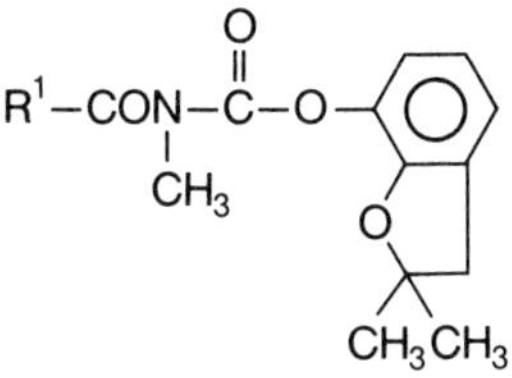

$$R^1-CON-\overset{\overset{O}{\|}}{C}-O-\text{(benzofuran ring)}$$
with N–CH_3 substituent and gem-dimethyl (CH_3 CH_3) on the furan ring

R^1	MOLAR Ac ChEI REL. I_{50}*	INSECTICIDAL ACTIVITY				LD_{50} RAT p. os mg/kg
		LC_{80-100} ppm		LC_{90-100} ppm		
		Anthonomus grandis Ad.	Laspeyresia pomonella L–1	Aphis craccivora		
				Contact	Systemic	
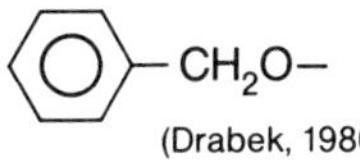 ⬡–CH_2O– (Drabek, 1980a)	0,2	400	50	3	0,75	225
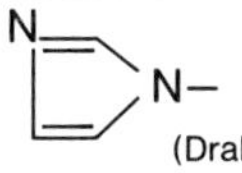 N-pyrrole N– (Drabek, 1980)	–	400	3	3	3	20
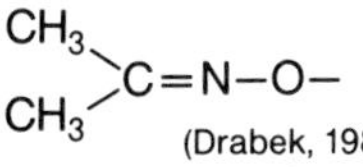 $\frac{CH_3}{CH_3}$C=N–O– (Drabek, 1982)	1	400	12,5	3	0,75	34
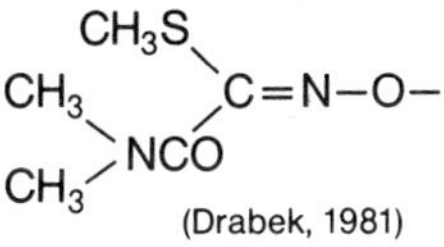 CH_3S, CH_3, CH_3 NCO C=N–O– (Drabek, 1981)	2,5	200	12,5	12,5	12,5	183
 CH_3–⬡–S– (Kuehle et al., 1971)	–	400	>400	12,5	400	–
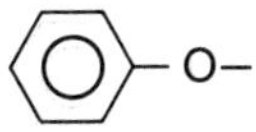 ⬡–O– (Kuehle et al., 1971)	–	>400	3	3	3	–
CARBOFURAN	100	50	3	12,5	12,5	8

$$* \text{Rel. } I_{50} = \frac{I_{50} \text{ CARBOFURAN}}{I_{50} \text{ COMPOUND}} \times 100$$

N-*Iminomethyl*-N-*methylcarbamates*

(Pissiotas and Dürr, 1971; Pissiotas, 1972). These compounds possess the following structural formula:

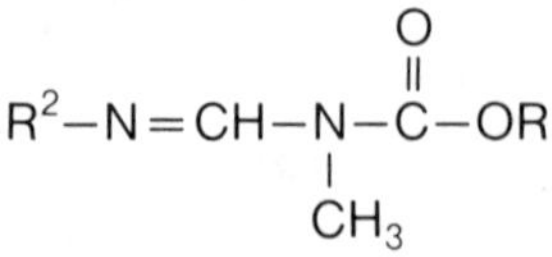

$$R^2-N=CH-N-\overset{\overset{\displaystyle O}{\|}}{C}-OR$$
$$\underset{CH_3}{|}$$

(for R see Fig. 1)

where R^2 is alkyl, alkenyl, alkinyl, aralkyl, cycloalkyl, or aryl. The insecticidal and toxicological properties of some compounds of this type are shown in Table 11.

Table 11   Insecticidal activity and toxicity of N-iminomethyl-N-methylcarbamates

Structure	Insecticidal activity (LC$_{90-100}$ ppm)		Toxicity (LD$_{50}$ rat p. os mg/kg)
	L. decemlineata L-3	*A. fabae*	
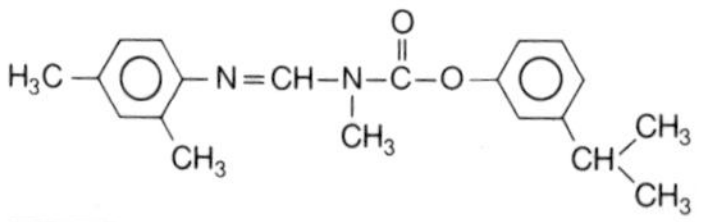 H$_3$C—⬡—N=CH—N—C(=O)—O—⬡—CH(CH$_3$)CH$_3$ (CH$_3$)	100	50	360
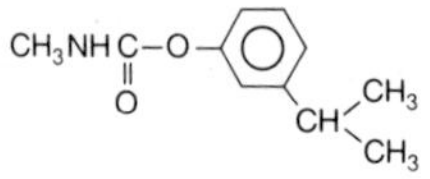 CH$_3$NH C(=O)—O—⬡—CH(CH$_3$)CH$_3$	—	—	16
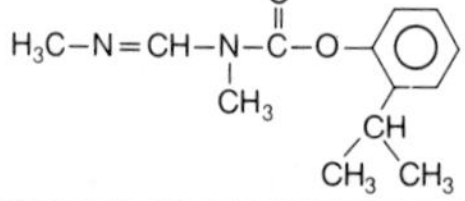 H$_3$C—N=CH—N—C(=O)—O—⬡—CH(CH$_3$)CH$_3$ (CH$_3$)	100	10	650
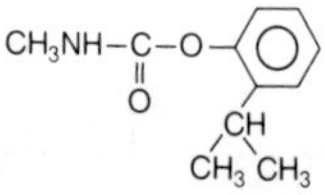 CH$_3$NH—C(=O)—O—⬡—CH(CH$_3$)CH$_3$ (Mipcin)	—	—	403—485
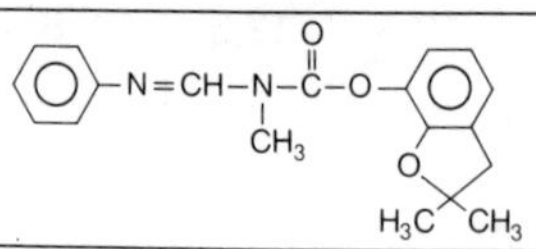 ⬡—N=CH—N—C(=O)—O—⬡ (CH$_3$) (H$_3$C CH$_3$)	100	50	230
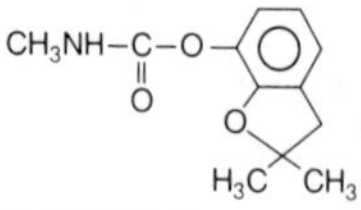 CH$_3$NH—C(=O)—O—⬡ (carbofuran) (H$_3$C CH$_3$)	50	10	8

N,N-*Dimethylcarbamates*

Hodgson and Casida (1960,1961) studied the metabolism of 45 dimethylcarbamates including the insecticides dimetilan, Isolan[R] and Pyrolan[R] by rat liver enzymes. N,N-Dimethylcarbamates formed products which could be extracted into organic solvent and which decomposed in strong acid to yield formaldehyde. This experimental approach provided strong evidence that N-dealkylation of the dimethylcarbamates proceeded through an N-methyl-N-hydroxymethyl intermediate.

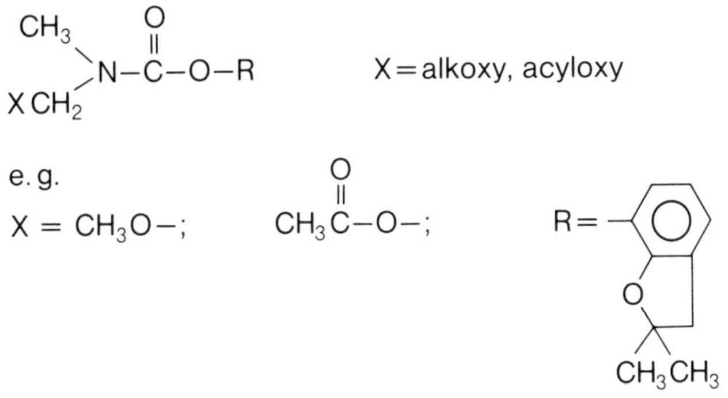

$$X = \text{alkoxy, acyloxy}$$

e.g.

$X = CH_3O-; \qquad CH_3\overset{\displaystyle O}{\overset{\|}{C}}-O-; \qquad R =$

Figure 6 N-alkoxymethyl- and N-acyloxymethyl-N-methyl-carbamates

N-*Alkoxymethyl*- and N-*acyloxymethyl*-N-*methylcarbamates*

These compounds (Figure 6) are probably demethylated by a similar mechanism as dimethylcarbamates.

Photoactivation of carbamates

A different concept of a procarbamate was presented at the ACS-Meeting in Chicago 1977 (Fu *et al.*, 1977). Insecticidal carbamates were detoxified by substitution of hydrogen on the carbamoyl nitrogen by the photosensitive *o*-nitrobenzyl group, which, under laboratory and field conditions, releases the parent N-methylcarbamate.

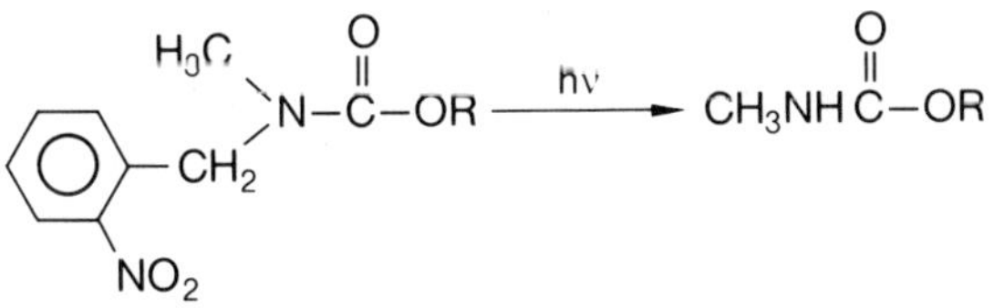

FORMAMIDINES

Using inhibitors of oxidative metabolism it was shown that chlordimeform, and chloromethiuron have to be activated to be toxic against the southern tick *Boophilus microplus* (Knowles and Roulston, 1973) (Table 12). The results in Table 12 show that C-8520, the *N*-monomethyl analogue of chlordimeform is synergized by these inhibitors. Antagonism with chlordimeform and synergism with its *N*-monomethyl derivative strongly suggest that the active compound against the southern cattle tick is the *N*-monomethyl compound. This idea is supported by the work of Knowles and Schuntner (1974), who found the demethyl analogue to be a major metabolite of chlordimeform in the cattle tick.

Table 12   Toxicity of chlordimeform and its monomethyl analogue (C-8520) to southern cattle tick larvae: alone and in the presence of piperonyl butoxide (Pb) and Tropital

		% Mortality		
	% Conc.	alone	+Pb (0.2%)	+Tropital (0.2%)
Chlordimeform	0.2	87.2	1.9	27.1
	0.02	85.5	0	26.0
	0.01	68.5	0	23.8
C-8520	0.2	75.7	76.9	91.7
	0.02	61.6	69.9	100.0
	0.01	49.3	75.6	99.0

Modified after Knowles and Roulston, 1973.

In insects, however, there is no toxicological evidence available which would indicate such an oxidative activation. Crecelius and Knowles (1976) showed chlordimeform and its demethyl analogue to be equally active against the cabbage looper *Trichoplusia ni* whether they were synergized with piperonyl butoxide or not. Chlordimeform itself was even 4–5 times more active than the demethyl analogue. On the other hand, there is extensive evidence on the biochemical level that at one of the most probable targets of the formamidines (the octopamine-sensitive adenylate cyclase), only the demethyl analogue acts as an agonist (Murdoch and Hollingworth, 1980; Evans and Gee 1980; Orchard *et al.* 1982; Hollingworth and Johnstone, 1983). The formamidine amitraz is rapidly cleaved to *N*-2,4-dimethylphenyl-*N'*-methylformamidine in ticks (Schuntner and Thompson, 1978). This cleavage may be hydrolytic in nature, and thus may give an explanation that piperonyl butoxide synergizes the activity of both compounds (Knowles and Roulston, 1973).

N-SULPHENYL-N'-ARYLFORMAMIDINES

Analogous to N-sulphenyl-N-methylcarbamates, N-sulphenyl-N'-arylform-amidines of the chlordimeform type possess insecticidal and acaricidal activity which is similar to that of the parent formamidine. Table 13 shows types of insecticidal and acaricidal N-sulphenyl-N'-arylformamidines described in the literature. Insecticidal, acaricidal and toxicological properties of N-sulphenyl-N'-arylformamidines tested in field trials are presented in Table 14.

Table 13 Types of insecticidal and acaricidal N-sulphenyl-N'-arylformamidines described in the literature

$$Y-C_6H_3(X)-N=CH-N(CH_3)(S-R)$$

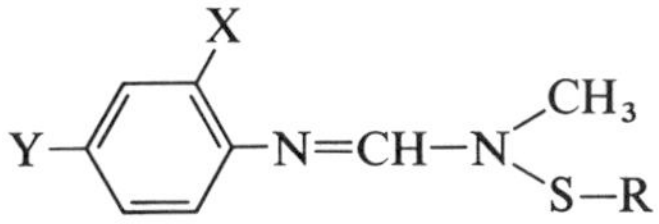

X = CH_3, Cl	Y = CH_3, Cl, Br	R =	Reference
		aryl	Gemrich II et al., 1976
		$-N(CH_3)-CH=N-C_6H_3(X)-Y$	Knowles and Gayen, 1982
		$-N(CH_3)-P(O\ alkyl)_2\ (\parallel O)$	Boeger and Drabek, 1976
		$-N(alkyl)_2$	Drabek and Boeger, 1975b
		$-N(alkyl)-CO_2alkyl$	Boeger and Drabek, 1975a,b
		$-N(alkyl)-CO-N(alkyl)_2$	Boeger and Drabek, 1975d
		$-SCCl_3$	Parham et al., 1975
		$-C(CH_3)_2CN$	Boeger and Drabek, 1975c

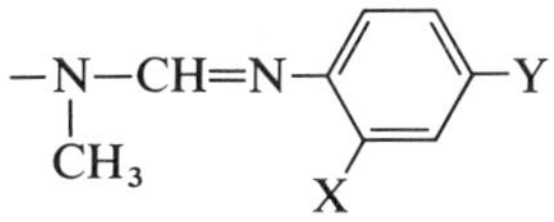

The similar activities demonstrated by analogues N'-aryl-substituted form-amidines regardless of the N-arylthio substituent, lead to one possible conclusion: that some of the observed activities may be derived from a common metabolite, perhaps the parent N-aryl-N'-methylformamidine (Gemrich II et al.,

1976). This suggested mode of action of *N*-sulphenyl-*N'*-arylformamidines has been confirmed by metabolic studies of four thiobisformamidines of the following structure:

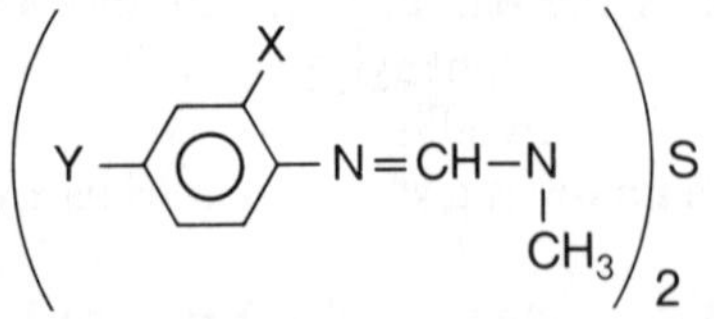

$$\left(Y-\!\!\left\langle\bigcirc\right\rangle\!\!\overset{\displaystyle X}{-}N{=}CH{-}\underset{\underset{CH_3}{|}}{N} \right)_{\!2} S$$

X = Me, Me, Cl, Cl;   Y = Me, Cl, Me, Cl

in *Diatraea grandiosella* by Knowles and Gayen (1982). The initial reaction was cleavage of the thiobisformamidine to the simple formamidine followed by cleavage to the respective formanilide with subsequent deformylation to yield the substituted anilines (Figure 7). Similar metabolism was found with the *N,N'*-thiobis-(*N*-methyl-*N'*-2,4-dimethylphenyl)formamidine in *Tetranychus urticae* (Franklin and Knowles, 1982). It is interesting to note that *N*-sulphenyl-*N'*-arylformamidines increase meat, milk, egg or wool production in livestock (Upjohn Co., 1981).

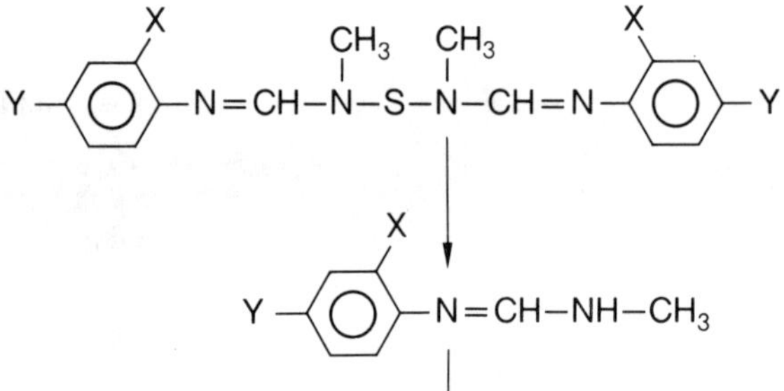

Figure 7 Metabolic paths for thiobisformamidines in *Diatraea grandiosella* (From Knowles and Gayen, 1982, reproduced by permission of Entomological Society of America)

Table 14 Acaricidal and insecticidal activities and toxicity of N-sulphenyl-N'-aryl form-amidines

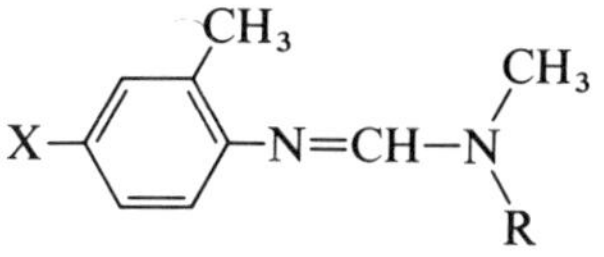

X	R	Tetranychus urticae adults (225)	eggs (30)	Heliothis zea larvae (25)	eggs (16)	Trichopulsia ni larvae (16)	eggs (64)	Oral LD$_{50}$ rats (mg/kg)
Cl	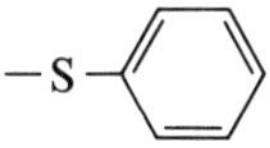 —S—⟨phenyl⟩	99	72	74	97	84	100	132
CH$_3$	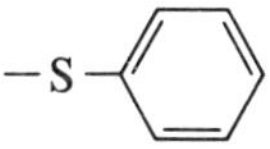 —S—⟨phenyl⟩	100	7	54	100	11	36	113
Cl	-SCCl$_3$	37	86	55	92	68	31	1000
CH$_3$	-SCCl$_3$	82	40	0	27	0	55	297
Cl	CH$_3$ (chlordimeform)	78	82	74	100	89	19	170–364
CH$_3$	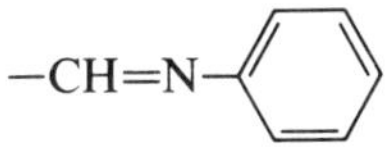 —CH=N—⟨phenyl⟩ (amitraz)	97	100	12	45	18	58	938

See Parham et al., 1975.

THIOUREAS

It has been suggested that the acaricide chloromethiuron has a common mode of action with that of chlordimeform (Schuntner and Thompson, 1979). It also has the same toxicological characteristics against *Boophilus microplus* (Table 15). It is tempting to conclude that the demethyl analogue (formed by oxidative dealkylation) is the active form of chloromethiuron since piperonyl butoxide antagonizes the parent compound and slightly synergizes the metabolite.

Table 15   Toxicity of chloromethiuron to cattle tick larvae alone and with piperonyl butoxide (Pb)

	LC_{50} (mg/1)	
	alone	+Pb (100 mg/l)
Chloromethiuron	220	7300
N-Demethyl-chloromethiuron	54	35

Modified after Schuntner and Thompson, 1979.

PROPYRETHROIDS

3-(2,2-Dihalovinyl)-2,2-dimethylcyclopropanecarboxylic acid, its chloride and esters can be easily brominated to the corresponding 1,2-dibromo-2,2-di-haloethyl derivatives. In the vinyl chain of the acid moiety brominated deriva-

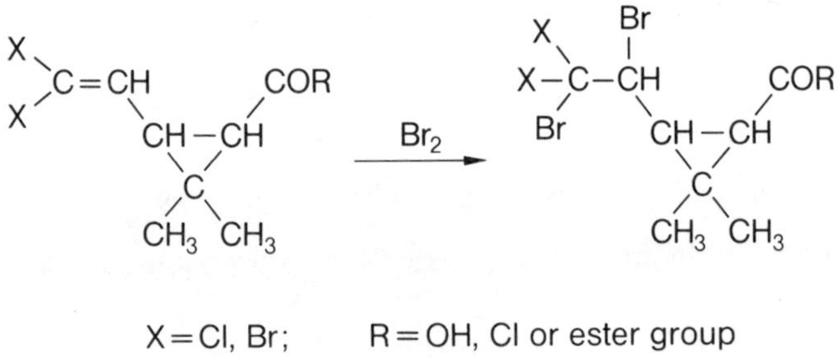

$X = Cl, Br;$ $R = OH, Cl$ or ester group

tives of permethrin, cypermethrin and deltamethrin show insecticidal activity, which on a molecular basis is comparable to that of the corresponding dihalovinyl analogues (Martel *et al.*, 1976a,b; Drabek *et al.*, 1977; Ackermann *et al.*, 1980). This activity is hardly compatible with the structure, because the presence of the vinyl group in the acid moiety determines to a great extent the biological activity of such esters.

It is known that the organophosphorus insecticide naled is transformed in the presence of metals, -SH groups or other reducing groups to dichlorvos, which is the actual toxicant. It was expected that similar debromination which leads to

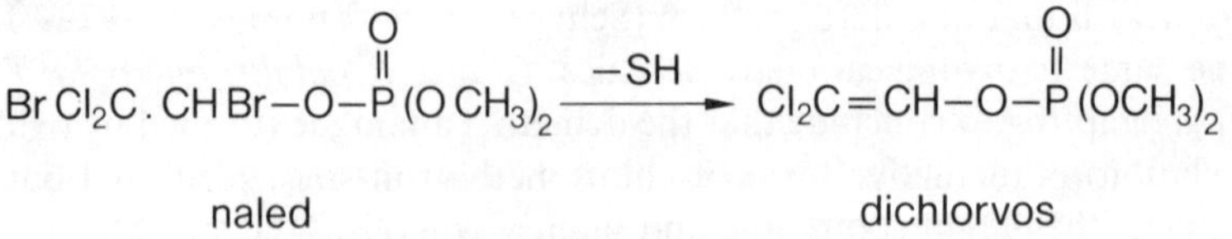

naled   dichlorvos

the corresponding vinyl esters could be responsible for the high insecticidal activity of brominated pyrethroids. *In vitro*, brominated cypermethrin (tralocythrin) can be easily debrominated to cypermethrin by cystein or H_2S (Drabek, unpublished).

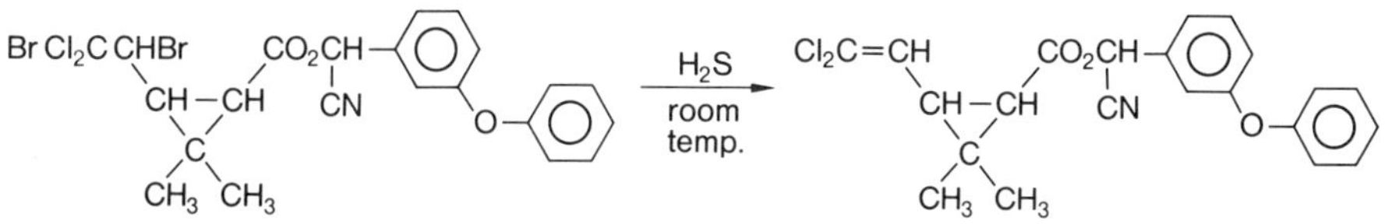

Metabolic studies with brominated deltamethrin (tralomethrin) and $(1R,\alpha S)$-*cis*-cypermethrin (tralocythrin) carried out by Cole *et al.*, (1982a,b) in rats and on cotton and bean foliage, confirmed the debromination theory. It was found that in rats both compounds undergo rapid and almost complete debromination to form deltamethrin and $(1R,\alpha S)$-*cis*-cypermethrin, respectively. Deltamethrin and cypermethrin are then hydroxylated at the 2',4' and 5 positions of the alcohol moiety and the methyl group *trans* to the carboxylate linkage. Ester cleavage reactions with deltamethrin and cypermethrin and further metabolism of the cleavage products yield series of alcohols and carboxylic acids and their glucuronide, glycine, and sulphate conjugates. The cyano fragment is retained for several days in the stomach and skin. Toxicity studies with mice provided evidence that intracerebrally administered tralomethrin and tralocythrin may be activated by debromination in the brain (Figure 8).

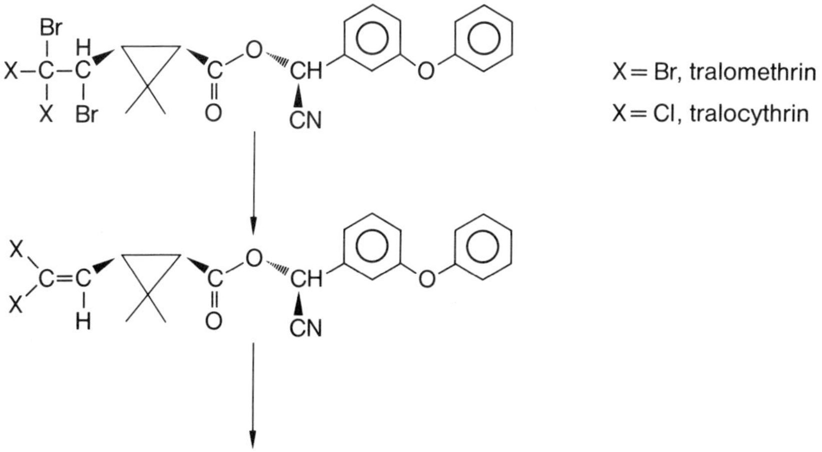

X = Br, tralomethrin

X = Cl, tralocythrin

Figure 8 Metabolism of tralomethrin and tralocythrin in rats. Further metabolism is similar to that of deltamethrin and cypermethrin

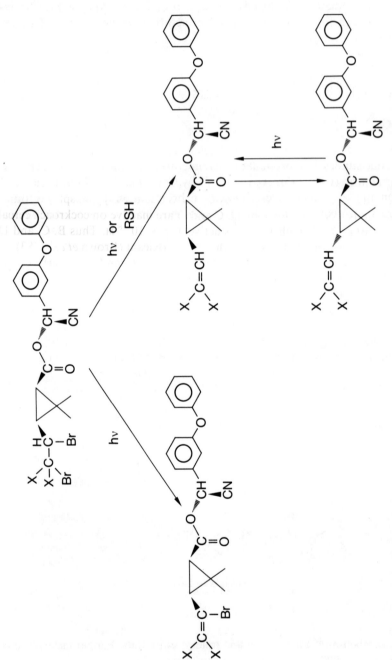

Figure 9 The major initial metabolites of tralomethrin and tralocythrin on cotton and bean foliage (From Cole *et al.*, 1982a)

On cotton and bean foliage tralomethrin and tralocythrin are degraded to deltamethrin and (1R,αS)-cis-cypermethrin by debromination and to trans-deltamethrin and trans-cypermethrin by isomerization. Dehydrobromination was less significant than debromination. With the exception of debromination- and dehydrobromination-reactions the degradation processes appear to be essentially the same as those of deltamethrin and (1R,αS)-cis-cypermethrin (Cole et al., 1982a; Figure 9; compare also Irwing and Fraser, 1984).

The potent insecticide 4-chlorophenyl-cyclopropyl-ketoxime-(3-phenoxy)-benzyl O-ether (A) can be reduced to hydroxylamine ether (B) and converted into N-methyl (C) and N-formyl (D) derivatives of B which have a moderate insecticidal activity. Houseflies and cabbage loopers oxidize and conjugate C probably to an N-hydroxymethyl glucoside and hydrolyse D to B. These insects and mouse liver microsomes also oxidize B and A. Piperonyl butoxide retards housefly and cockroach knockdown by C, and S,S,S-tributyl phosphorotrithioate delays housefly knockdown by D. C and D are inactive on cockroach cercal sensory nerve at 10^{-5} M whereas A is active at 5×10^{-8} M. Thus B, C, and D appear to be propyrethroids and A is the actual toxicant (Brown et al., 1983).

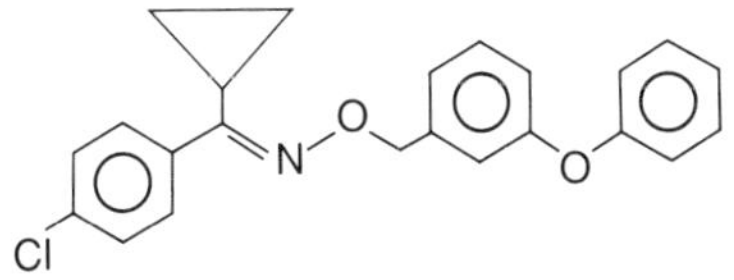

A

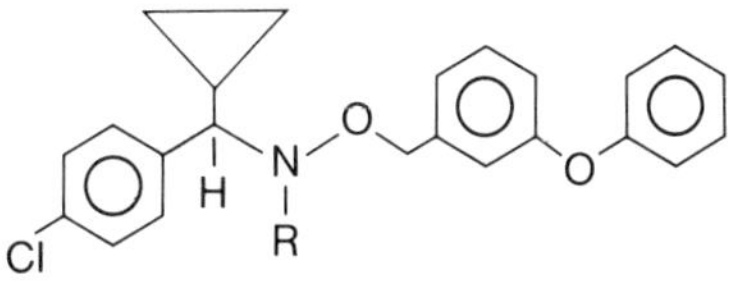

R = H (B)
R = CH_3 (C)
R = CHO (D)

Of the propyrethroids tralomethrin (Scout[R]; RU-25474, HAG-107) has been developed as a commercial insecticide.

PRONEREISTOXINS

Nereistoxin, a poison found in the common marine worm *Lumbriconereis heteropoda* shows strong insecticidal activity. Direct neurophysiological measurements indicate that nereistoxin blocks synaptic, but not axonal, transmission of cockroach nerves. Vertebrate studies (Deguchi et al., 1971; Sakai and Sato, 1972) indicate that a major site of nereistoxin action is on the post-synaptic membrane. However, the precise nature of its interaction with the acetylcholine receptor is still unknown. Using nereistoxin as a model, many compounds with insecticidal activity which can be regarded as pronereistoxins were synthesized. Table 16 shows nereistoxin derivatives which have been developed as commercial insecticides or which are experimental insecticides in a late stage of development.

Table 16 Structure and toxicity of pronereistoxins

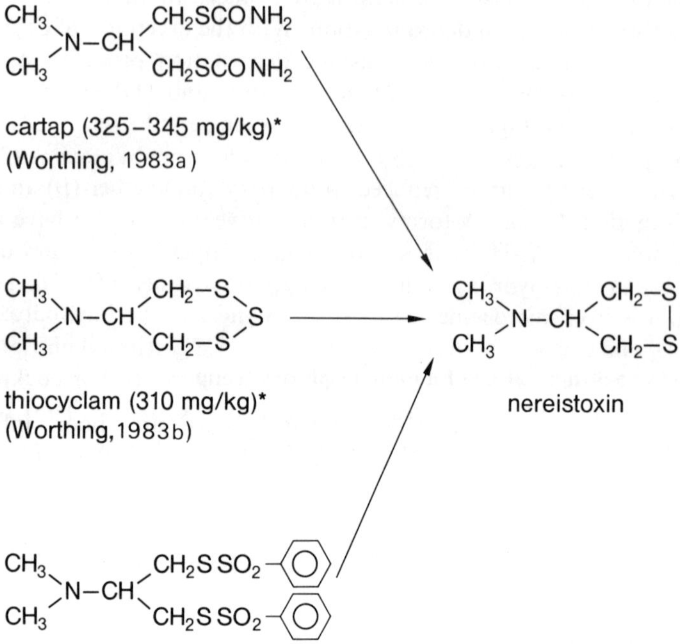

CH₃\N—CH⟨CH₂SCO NH₂ / CH₃/ CH₂SCO NH₂

cartap (325–345 mg/kg)*
(Worthing, 1983a)

CH₃\N—CH⟨CH₂—S / CH₃/ CH₂—S⟩S — — → CH₃\N—CH⟨CH₂—S / CH₃/ CH₂—S

thiocyclam (310 mg/kg)*   nereistoxin
(Worthing, 1983b)

CH₃\N—CH⟨CH₂S SO₂—⟨O⟩ / CH₃/ CH₂S SO₂—⟨O⟩

bensultap (1120 mg/kg)*
(Sakai 1982, Konishi 1970)

*Acute oral toxicity (LD$_{50}$) on rats

The results of the quantitative determination of nereistoxin, which was found in considerable amounts after the injection of cartap, bensultap, 1,3-dithio-cyano-2-(N,N-dimethylamino)propane into *Periplaneta americana*, illustrate the pronereistoxin character of these compounds (Sakai and Sato, 1972). Cartap was also proved to be converted into nereistoxin in rice paddy water.

PROINSECTICIDAL ORGANOPHOSPHATES

The majority of organophosphorus insecticides, except phosphates, inhibit acetylcholinesterase only slightly, unless they are activated. Their activity as effective anticholinesterases *in vivo* is the net result of competing biochemical processes: activation and detoxification. Many excellent reviews dealing with the activation and detoxication of organophosphorus insecticides have been published (O'Brien, 1967; Hathway *et al.*, 1970; Dauterman, 1971; Eto, 1974).

In this Chapter only reactions which lead to a (bio)transformation of an *in vitro* essentially inactive or only slightly active acetylcholinesterase inhibitor into an active insecticide are mentioned. Compounds and reactions already described in the review literature are mentioned very briefly.

Dauterman (1971) and Eto (1974) classified the activation reactions into the following types:

(1) oxidative desulphuration of a thiophosphoryl group; formation of phosphoryl compounds;
(2) thioether oxidation;
(3) activation of phosphorothiolates;
(4) oxidation of amides;
(5) miscellaneous non-oxidative activations.

Oxidative desulphuration of a thiophosphoryl group

In the pure state, phosphorothionate esters are poor inhibitors of cholinesterases, whereas the corresponding oxoanalogues are strong cholinesterase inhibitors. The insecticidal activity of the thiono compounds is

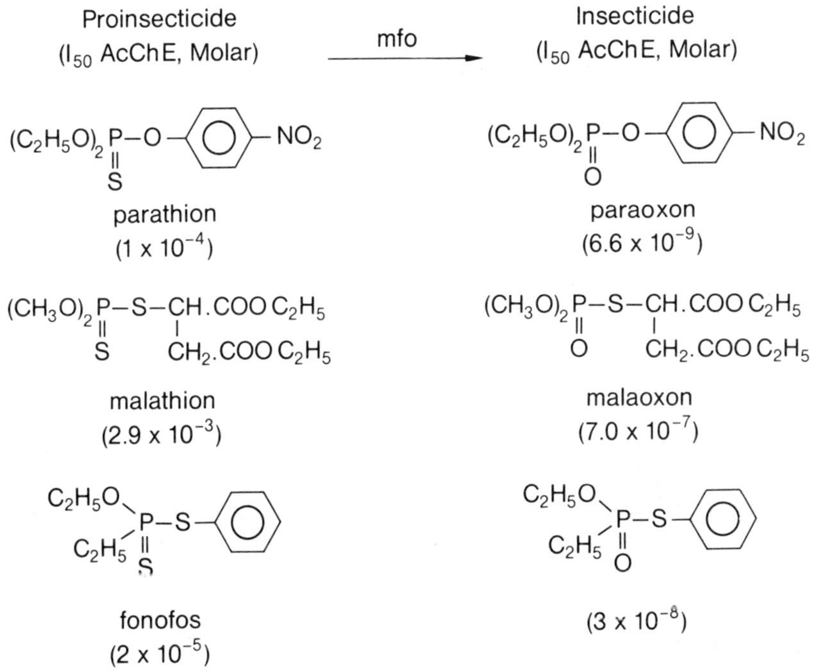

Figure 10 Increase of acetylcholinesterase inhibition by oxidative desulphuration of organophosphorus insecticides (From Eto, 1974)

the result of the oxidative desulphuration of the thionophosphoryl group *in vivo*. The conversion of thiono into oxo esters in animals is caused by microsomal mixed-function oxidase systems. In plants, peroxidases and/or light may play a role in the thiono–oxo transformation. Oxidative desulphuration of the thiophosphoryl sulphur atom has been demonstrated with a wide variety of organophosphorus insecticides such as parathion, fenitrothion, diazinon, malathion, dimethoate, azinphosmethyl, fonofos, etc. (Figure 10). It is possible that in the microsomal oxidation of phosphorothionate cyclic intermediates are formed first as was postulated for parathion and fonofos (McBain *et al.*, 1971; Ptashne and Neal, 1972; Figure 11).

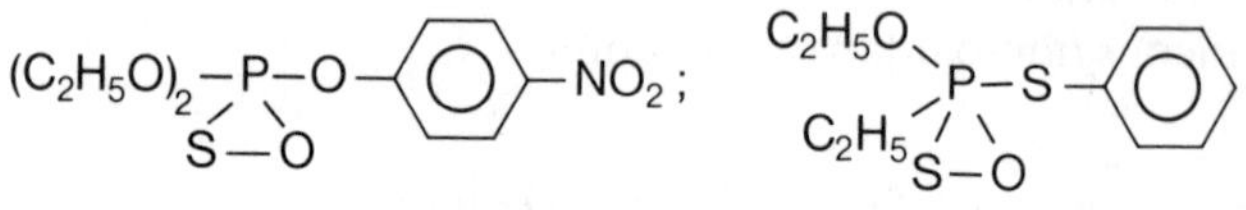

Figure 11 Cyclic intermediates in microsomal oxidation of parathion and fonofos

Thioether oxidation

The thioether oxidation of certain organophosphorus compounds has been demonstrated *in vivo* in plants, mammals, and insects. In general the initial oxidation of most thioether-containing organophosphates *in vivo* results in a rapid conversion to the sulphoxide and then to a slow conversion to the sulphone. This results in the accumulation of the sulphoxide which is probably the principal toxicant. Thioether oxidation results principally in an increase in the inhibition of acetylcholinesterase, but this activation is not as high as in the case of oxidative desulphuration (for oxidative activation of disulfoton see Figure 12).

The oxidation of a thioether group attached to a phenyl group in dialkyl phenyl phosphates causes a great increase in the electron-withdrawing properties (increase of σ value) of the phenyl group. This results in the enhancement of hydrolysability and of anti-acetylcholinesterase activity. In this case the toxicant is not the thioether, but the product resulting from this 'lethal synthesis' —the sulphoxide (Figure 13).

Thioether oxidation is found in such organophosphorus insecticides as demeton, carbophenthion, disulfoton, fenthion, phenamiphos, thiomethon, vamidothion etc.

Activation of phosphorothiolates

Regardless of the nature of the substituent, phosphorothiolates have considerably higher anticholinesterase activity than the corresponding phosphorothio-

$(C_2H_5O)_2-P-S\ CH_2CH_2SC_2H_5$ Disulfoton
$\parallel$
S

$I_{50} = >1 \times 10^{-4}M$

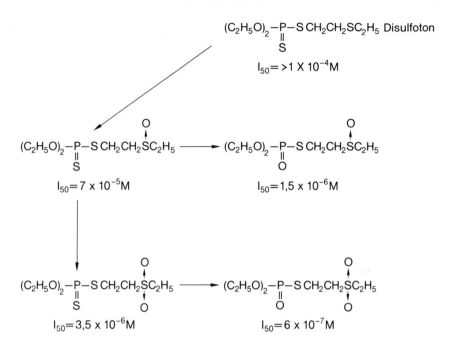

$(C_2H_5O)_2-P-S\ CH_2CH_2SC_2H_5$
$\parallel$
S

$I_{50} = 7 \times 10^{-5}M$

$(C_2H_5O)_2-\overset{O}{\underset{\parallel}{P}}-S\ CH_2CH_2\overset{O}{\underset{\downarrow}{S}}C_2H_5$

$I_{50} = 1,5 \times 10^{-6}M$

$(C_2H_5O)_2-P-S\ CH_2CH_2\overset{O}{\underset{\downarrow}{S}}C_2H_5$
$\parallel$
S

$I_{50} = 3,5 \times 10^{-6}M$

$(C_2H_5O)_2-\overset{O}{\underset{\parallel}{P}}-S\ CH_2CH_2\overset{O}{\underset{\downarrow}{S}}C_2H_5$

$I_{50} = 6 \times 10^{-7}M$

Figure 12 Activation of disulfoton by thioether oxidation (From Metcalf *et al.*, 1957, Reproduced by permission of John Wiley & Sons, Inc.)

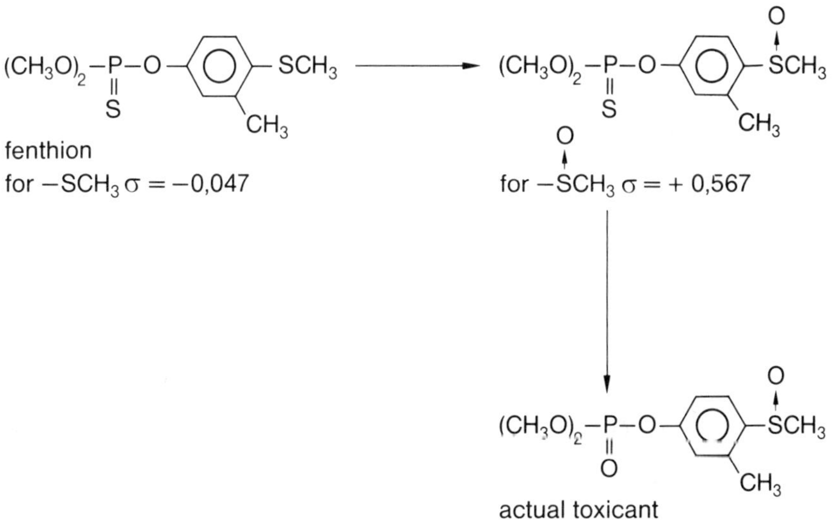

$(CH_3O)_2-P-O-\langle\bigcirc\rangle-SCH_3$
$\parallel$
S
CH_3

fenthion
for $-SCH_3\ \sigma = -0,047$

$(CH_3O)_2-P-O-\langle\bigcirc\rangle-\overset{O}{\underset{\downarrow}{S}}CH_3$
$\parallel$
S
CH_3

for $-SCH_3\ \sigma = +0,567$

$(CH_3O)_2-P-O-\langle\bigcirc\rangle-\overset{O}{\underset{\downarrow}{S}}CH_3$
$\parallel$
O
CH_3

actual toxicant

Figure 13 Activation of fenthion by thioether oxidation

nates. Thermal or photocatalytic thiono–thiolo rearrangement reactions form mainly the S-alkylthiolates during the manufacture of thionophosphate pesticides.

In some cases the thiono–thiolo rearrangement decreases the insecticidal activity as it does with parathion and methylparathion. On the other hand it increases the systemic activity of demeton. In other cases the introduction of an

$$
\begin{array}{c}
C_2H_5O \\
\diagdown \\
C_2H_5O \diagup \underset{\parallel}{P}-O\,CH_2CH_2S\,C_2H_5 \\
S
\end{array}
\quad\longrightarrow\quad
\begin{array}{c}
C_2H_5O \\
\diagdown \\
C_2H_5O \diagup \underset{\parallel}{P}-S\,CH_2CH_2S\,C_2H_5 \\
O
\end{array}
$$

S-alkyl instead of an O-alkyl group into a phosphorus ester causes a dramatic increase in the insecticidal activity. This is the case with O-ethyl-O-aryl-S-propyl or S-sec-butyl phosphorothiolates, O-ethyl-S,S-dipropyl phosphoro-dithiolate and O,S-dimethylamido-phosphorothiolate.

O-*Ethyl* O-*phenyl* S-n-*propyl-phosphorothiolates*

In the beginning of the 1970s many patent applications appeared for compounds of the structural formula I. Some of these have subsequently been

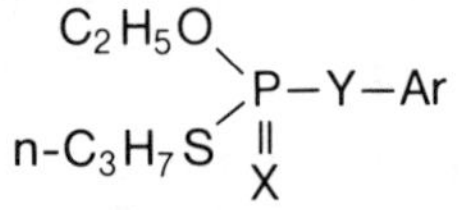

$$
\begin{array}{c}
C_2H_5O \\
\diagdown \\
n\text{-}C_3H_7S \diagup \underset{\underset{X}{\parallel}}{P}-Y-Ar
\end{array}
\qquad X, Y = O \text{ or } S
$$

I

developed as commercial insecticides e.g.:

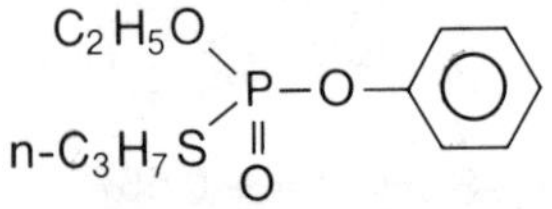

Heterophos; LD$_{50}$ rat p. os: 295 mg/kg
(Fest 1981; Hart and O'Brien, 1976)

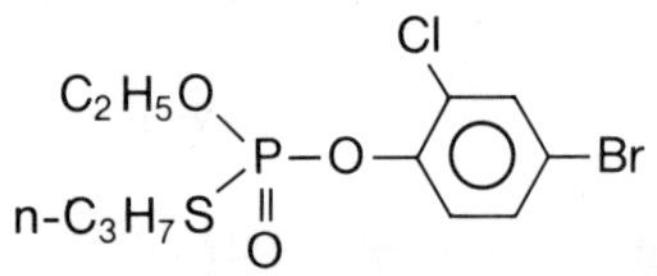

Profenofos; LD$_{50}$ rat p. os: 350 mg/kg
(Beriger and Drabek, 1971; Drabek and Flueck, 1979)

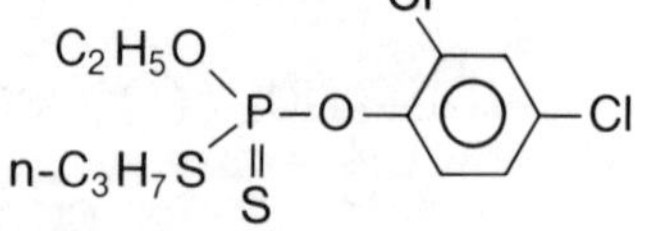

Prothiophos; LD$_{50}$ rat p. os: 1000 mg/kg
(Kishino *et al.*, 1970; Kudamatsu, 1976)

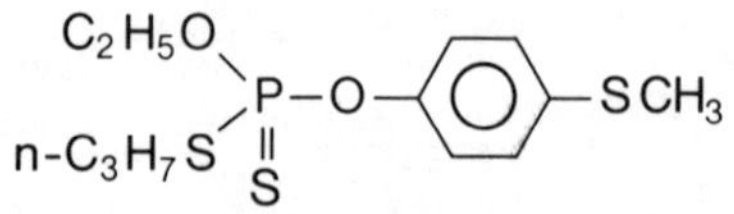

Sulprofos; LD$_{50}$ rat p. os: 100–300 mg/kg
(Kishino *et al.*, 1970)

The following experimental insecticides were also in field evaluation tests:

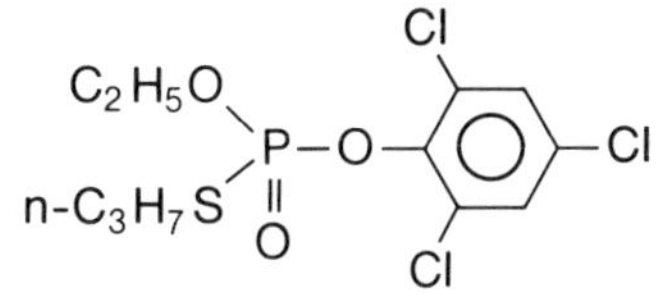

Trifenofos; LD_{50} rat p. os: 270 mg/kg

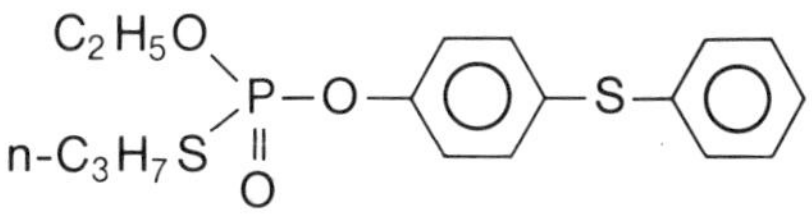

RH-0994 (Wayne, 1981)

All these compounds show improved activity against lepidopterous insects compared to the *O,O*-diethyl- and *O,O*-dimethyl-phosphorothionates.

Hart and O'Brien (1976) determined the phosphorylation, dissociation and bimolecular rate constants for six *O*-ethyl *O*-aryl-*S*-*n*-propyl-phosphorothiolates (aryl = phenyl; 2,4,6-trichlorophenyl; 4-chlorophenyl; 4-methylphenyl; 4-carbomethoxyphenyl and 4-nitrophenyl) in the presence of substrate. With the exception of the *p*-nitrophenyl derivative these compounds were found to be rather weak inhibitors of acetylcholinesterase *in vitro*. This finding, in combination with the generally poor correlation between inhibitory potency and toxicity, led to the suggestion that the activity of these compounds *in vivo* might be a function of several factors other than anticholinesterase action at the target enzyme level. Segall and Casida (1982) used peracid oxidation as a chemical model to study the activation reactions of phosphorothiolates. They found that profenofos with three equivalents of *m*-chloroperbenzoic acid (MCPBA) gave as a main product VI (Figure 14). Oxidative conversion of profenofos is most easily explained by envisaging a sequence of oxidation, rearrangement and further oxidation reactions. From the compounds shown in Figure 14 only II (not isolated) was claimed to be the phosphorylating agent. It could be expected that *in vivo* the initial cytochrome P450-catalysed oxidation forms the sulphoxide capable of phosphorylating acetylcholinesterase or other sites in the poisoning sequence. The reaction of phosphorothiolates with *m*-chloroperbenzoic acid appears to be generally applicable since a variety of phosphorothiolates give analogous products to those of profenofos. In further studies Wing *et al.* (1983) found that the microsomal mixed-function oxidase converts the (−) isomer of profenofos to a 34-fold stronger acetylcholinesterase inhibitor *in vitro*, while the (+) isomer is 2-fold deactivated. Similar activation was also found *in vivo* with *O*-[1-(4-chlorophenyl)-4-pyrazolyl]-*O*-ethyl-*S*-propyl phosphorothiolate (Kono *et al.*, 1983).

The above-mentioned studies show that activation reactions are at least partly responsible for the toxic effects caused by *O*-ethyl-*O*-aryl-*S*-*n*-propyl-phosphorothiolates. It is probable that similar activation reactions take place also with *O*-ethyl-*O*-aryl-*S*-*sec*-butyl phosphorothiolates and with *O*-ethyl-*S,S*-di-*n*-propyl-phosphorothiolates (ethoprophos).

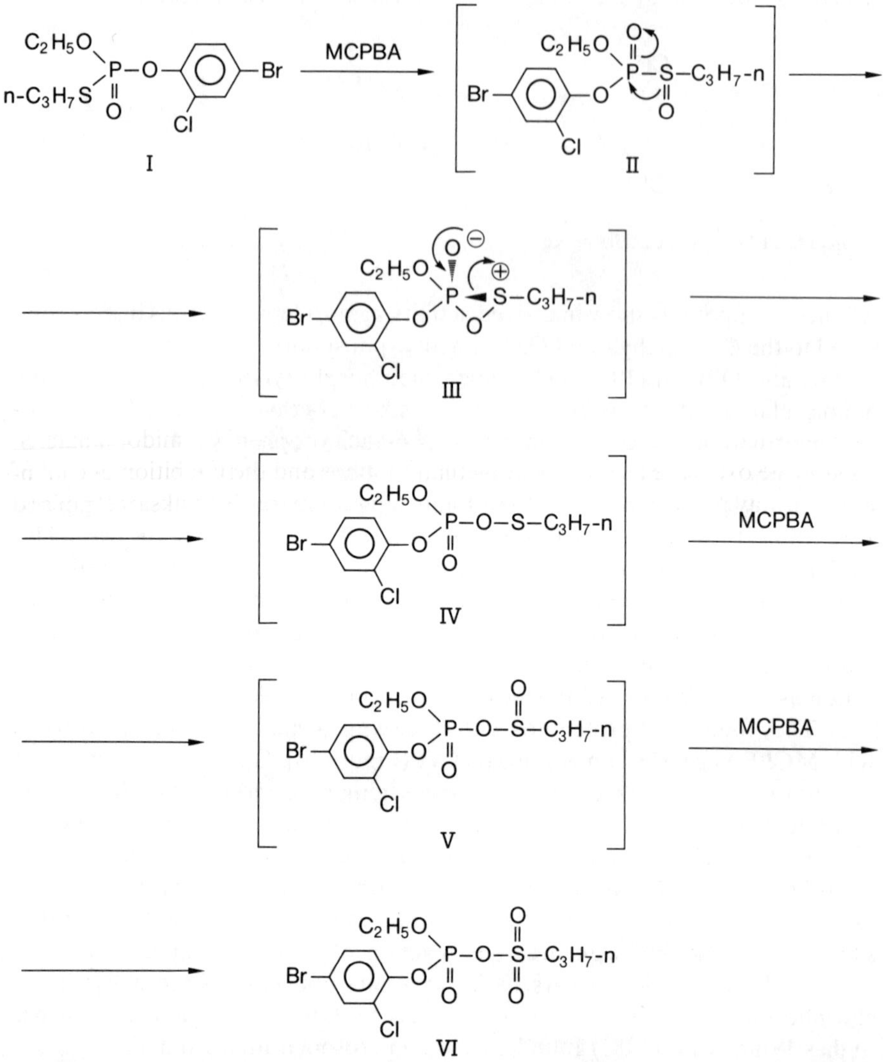

Figure 14 Oxidative activation of profenofos (From Segall and Casida, 1982)

Methamidophos and Acephate

Both are broad spectrum contact and systemic commercial insecticides. It has been proved that acephate is metabolized *in vivo* to methamidophos (Tucker, 1976; Kao and Fukuto, 1977). So acephate is a true proinsecticide.

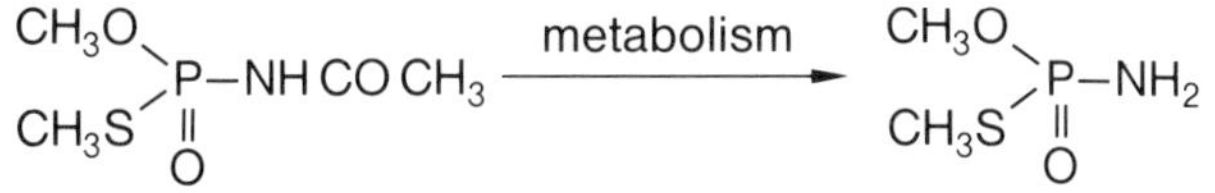

$$\begin{array}{c} CH_3O \\ \diagdown \\ CH_3S \diagup \end{array} P-NHCOCH_3 \xrightarrow{\text{metabolism}} \begin{array}{c} CH_3O \\ \diagdown \\ CH_3S \diagup \end{array} P-NH_2$$

Methamidophos is only a moderate inhibitor of acetylcholinesterase *in vitro*. Remarkably, it is activated by treatment with equimolar amounts of *m*-chloroperbenzoic acid (MCPBA). The conversion of methamidophos into an active inhibitor proceeds rapidly (methamidophos inhibits cholinesterase only 18% at 1.4×10^{-4} M before treatment with MCPBA, but almost completely in 15 min after the treatment). The *in vitro* activation of methamidophos by microsomal mixed-function oxidase is not as remarkable as the activation by peracid oxidation. It has been suggested that a similar activation occurs in insects. According to Eto and co-workers (Eto *et al.*, 1977) the most probable (not isolated) active intermediate is *O,S*-dimethyl phosphoramidothiolate *S*-oxide. The oxidative activation of methamidophos and the inhibition of cholinesterase may proceed as is shown in Figure 15 (compare also Suksayretrup and Plapp, 1977).

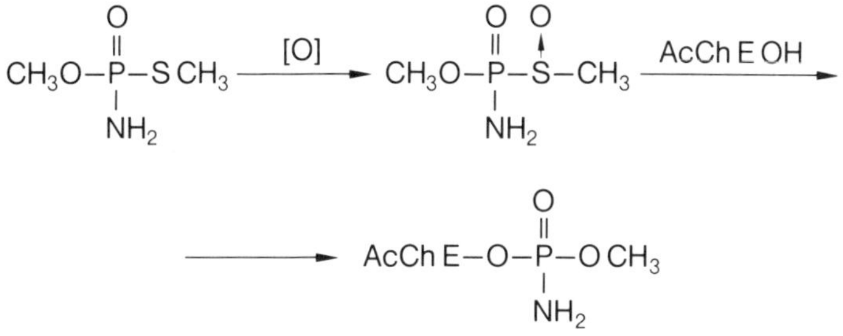

$$CH_3O-\overset{\overset{\displaystyle O}{\|}}{\underset{\underset{\displaystyle NH_2}{|}}{P}}-SCH_3 \xrightarrow{[O]} CH_3O-\overset{\overset{\displaystyle O}{\|}}{\underset{\underset{\displaystyle NH_2}{|}}{P}}-\overset{\overset{\displaystyle O}{\uparrow}}{S}-CH_3 \xrightarrow{AcChE\ OH}$$

$$\longrightarrow AcChE-O-\overset{\overset{\displaystyle O}{\|}}{\underset{\underset{\displaystyle NH_2}{|}}{P}}-OCH_3$$

Figure 15 Proposed mechanism for AChE inhibition by methamido-phos after activation

The proinsecticidal nature of the mode of action of methamidophos was not supported by the findings of Gray *et al.*, 1982, who suggested that, in spite of its only moderate inhibitory activity, methamidophos itself is toxic due to its *in vivo* stability coupled with the large quantities of the inhibitor which reach the nervous system.

Oxidation of amides

The insecticides di(tetramethylphosphoroamidic)anhydride (schradan) and tetramethylphosphorodiamidic fluoride (dimefox) are almost completely inactive as acetylcholinesterase inhibitors *in vitro*. Schradan is oxidized enzymatically or chemically into a strong anticholinesterase agent, which is 10^5 times

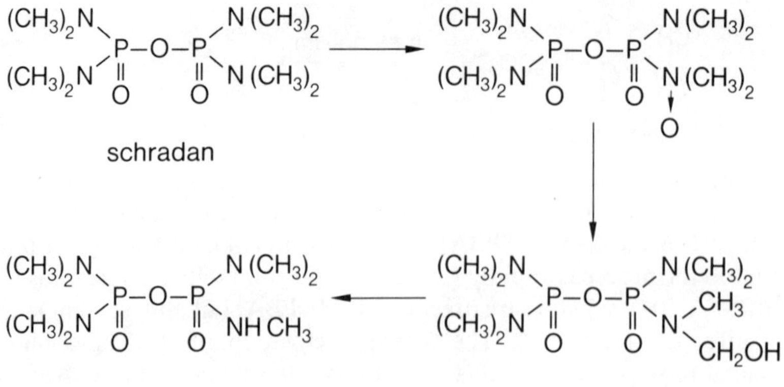

Figure 16 Proposed scheme for biological oxidation of schradan

more active than the parent compound. The bio-oxidation results in
N-demethylation. Although the intermediates are not adequately characterized,
it is suggested that schradan is metabolized according to the scheme shown in
Figure 16 (Eto, 1974, pp. 149, 165, 166).

Dicrotophos and monocrotophos are systemic and contact insecticides and
acaricides with broad spectra of activity. In its metabolic pathway the N-methyl
group of dicrotophos is oxidized initially to the N-methylol group, which then
undergoes either conjugation or elimination as formaldehyde. Thus by oxidation demethylation dicrotophos is activated to the more potent insecticide

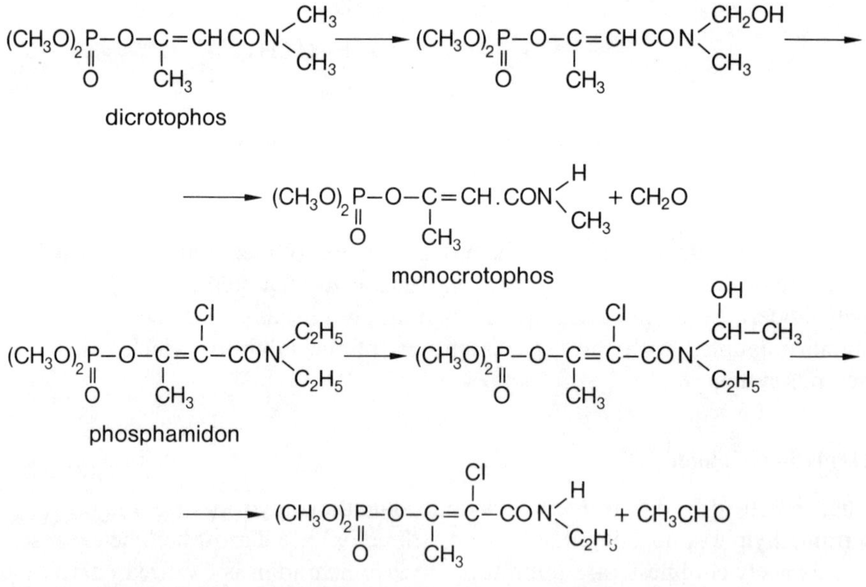

Figure 17 Metabolic activation of dicrotophos and phosphamidon

monocrotophos (see Figure 17). Monocrotophos may be further metabolized in the same way with the formation of the unsubstituted amide, which is also active as an acetylcholinesterase inhibitor and insecticide (Bull and Lindquist, 1964; Menzer and Casida, 1965; Eto, 1974; p. 167). Similar to the case of dicrotophos, the N-ethyl group of phosphamidon is also oxidatively eliminated as acetaldehyde through a hydroxylation on the α-carbon atom (Figure 17). The oxidative N-dealkylation occurs faster with *cis*-phosphamidon than with the *trans*-isomer. The N-monoethyl phosphamidon is biologically more active than the parent compound (Eto, 1974; p. 167).

Miscellaneous non-oxidative activation

Dichlorvos, a very active organophosphorus insecticide, is probably the actual toxicant of the proinsecticides trichlorphon, butonate and naled (Figure 18).

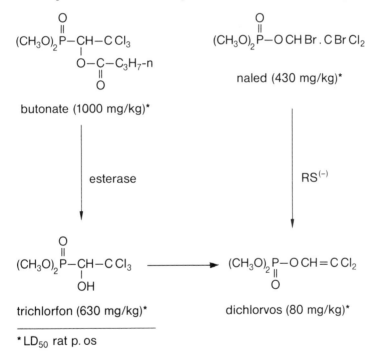

Figure 18 Dichlorvos proinsecticides. *LD_{50} rat p. os. (After Eto, 1974)

Trichlorphon is a poor inhibitor of acetylcholinesterase, but it is readily converted into the strong acetylcholinesterase inhibitor dichlorvos under physiological pH conditions. The transformation proceeds non-enzymatically and at room temperature it takes only 7.5 hours for 50% conversion at pH 7. Thus, in

the sense of its mode of action, trichlorphon may be regarded as a proinsecticide, although no evidence has been obtained for its *in vivo* conversion into
dichlorvos (Eto, 1974; p. 269).

Butonate is converted enzymatically into trichlorphon by the action of
esterases in insects and plants (Dedek, 1968). Naled, a fast-acting non-systemic
contact and stomach insecticide, reacts readily with thiol groups or other natural constituents forming dichlorvos via debromination (Casida, 1972). Cysteine
for example converts naled almost instantaneously into dichlorvos. Naled is
also debrominated in the presence of metals and sunlight.

FLUOROACETIC ACID PRECURSORS

Schrader (1947) discovered the insecticidal activity of esters, and carbamates of
2-fluoroethanol. These compounds are first hydrolysed and the 2-fluoroethanol
is then oxidized by dehydrogenases to fluoroacetic acid (Bartlett, 1952).
Fluoroacetic acid (LD_{50} rat p. os: 1–2 mg/kg) is itself still a procide, since it only
exerts its lethal action as an inhibitor of the aconitase after being transformed to
a α-fluorocitrate and causing a lethal accumulation of citrate in specific tissues.

ω-Fluoroalkyl- and ω-fluoroacyl-containing lipids were evaluated as delayed
toxicants against *Reticulitermes flavipes* (Prestwich *et al.*, 1983b). The catabolic
successive loss of two carbon units from the ω-fluoro fatty acids with an even
number of carbon atoms finally yielded fluoroacetate (Goldman, 1969).

A study of the effects of 29-fluorophytosterols on *Manduca sexta* showed
their proinsecticidal properties (Prestwich *et al.*, 1983a). The insect dealkylates
phytosterols on C-24 with the release of fluoroacetate. This dealkylation cannot
take place in mammals and therefore mammals are not intoxicated.

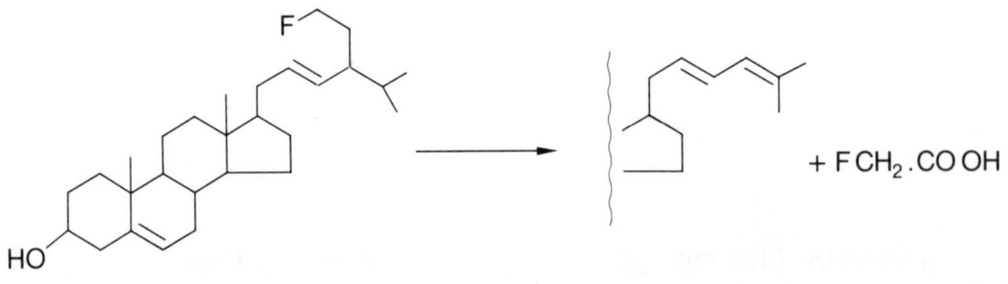

29-fluorostigmasterol

A fluorinated mevalonate (tetrahydro-4-fluoromethyl-4-hydroxy-2,4-pyran-
2-one (F Mev)) was shown to act as an inhibitor of juvenile-hormone biosynthesis (Quistad *et al.*, 1981). Although the experimental evidence is still lacking,
one may certainly hypothesize by analogy with fluoroacetate that this compound is modified to an inhibitor of the homoisoprenoid pathway. At high
doses F Mev shows a general, non-specific toxicity, which may be explained by
its conversion to fluoroacetate.

PRECOCENE

In a recent review by Pratt (1983) evidence has been presented which makes it certain that the precocenes are activated by epoxidation to short-lived intermediates which either hydrate to inactive diols or alkylate macromolecules which leads eventually to cell death (Figure 19). The half-life of the precocene-I epoxide is about 14 seconds at 25 °C. This relatively low reactivity of the epoxide might be advantageous, since it allows the epoxide to diffuse away from the oxidizing enzyme and therefore does not inhibit its producer.

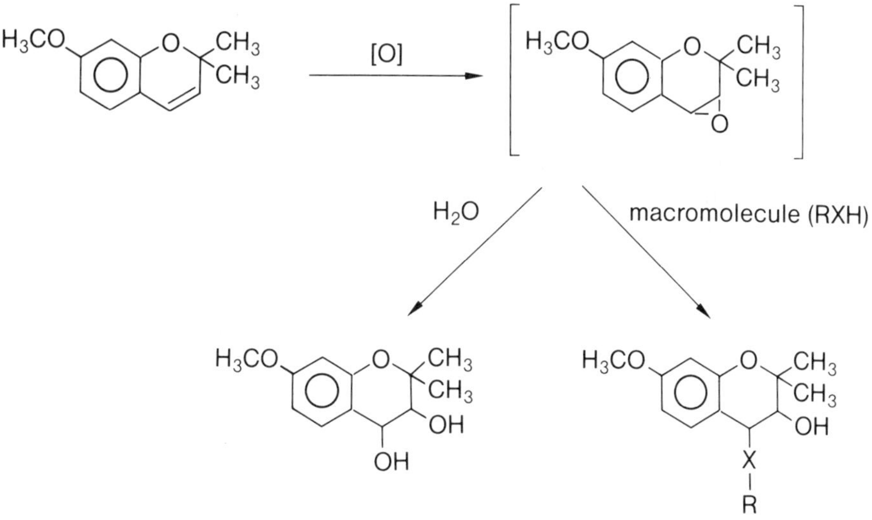

Figure 19 Precocene-1 activation

EPOXIDES OF CHLORINATED HYDROCARBONS

The chlorinated hydrocarbons aldrin, isodrin and heptachlor are metabolized through their epoxides dieldrin, endrin and heptachlor epoxide respectively (Klein and Korte, 1970). The epoxides are generally more toxic and more stable. Therefore the non-oxidized compounds may be regarded as proinsecticides. This had already been suggested earlier by Sun and Johnson (1960), who found that the unoxidized compounds were antagonized by sesamex, an epoxidation inhibitor, whereas the epoxides were synergized, probably by reduction of a further oxidative metabolism. However, this antagonism was not very strong: the effectiveness of aldrin and heptachlor was reduced by about 40%. Isodrin on the other hand was even synergized by 70%. In addition the non-oxidized compounds were found to act only after a certain latent period, during which they could have been epoxidized. Perry et al. (1958), Brooks (1960) and Joy (1976) found the same delay when measuring the convulsive properties of these insecticides in the cat central nervous system. However, though all the above-mentioned facts are evidence that epoxidation is an activation, they do not prove the inactivity of the unoxidized cyclodienes.

TRIAZINE INSECT GROWTH REGULATORS

Of the triazine insect growth regulators CGA 19255 (I), CGA 34296 (II), and CGA 72662 (III), cyromazine, the last one shows the highest activity. I is easily

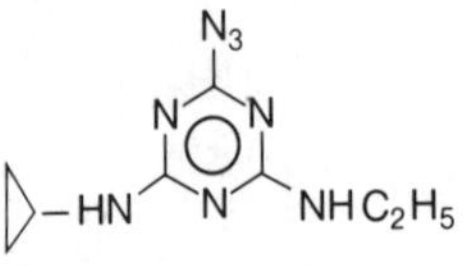

I 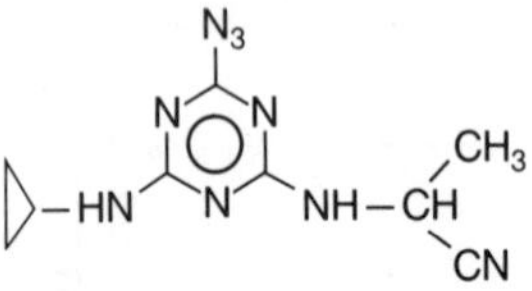   II 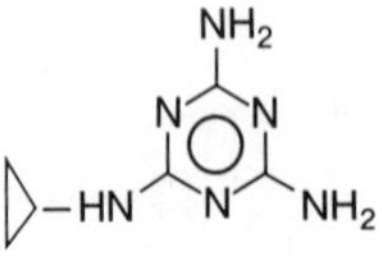   III

metabolized in two steps involving azide reduction and N-de-ethylation to cyromazine. These facts strongly support the suggestion that I and II are proinsecticides and cyromazine is the actual toxicant (Miller *et al.*, 1977, 1981; Miller and Corley 1980; Cecil *et al.*, 1981; Hart *et al.*, 1982; Ross and Brown, 1982).

ESTERS OF TOXIC ALCOHOLS, ENOLS AND ACIDS

A very common way to generate proinsecticides is by synthesizing esters of toxic alcohols or acids. The acaricides and fungicides dinocap, binapacryl, and dinobuton are based on this principle, and show a remarkably reduced mammalian toxicity and phytotoxicity in comparison to their parent dinitrophenols (Pianka and Smith, 1965; Kirby, 1966; Ilivicky and Casida, 1969).

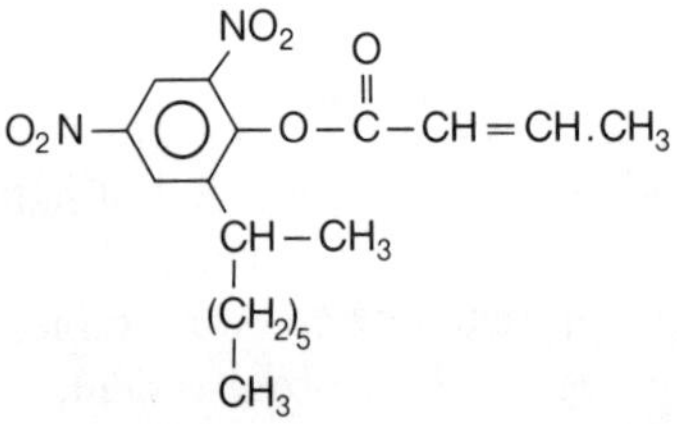

dinocap 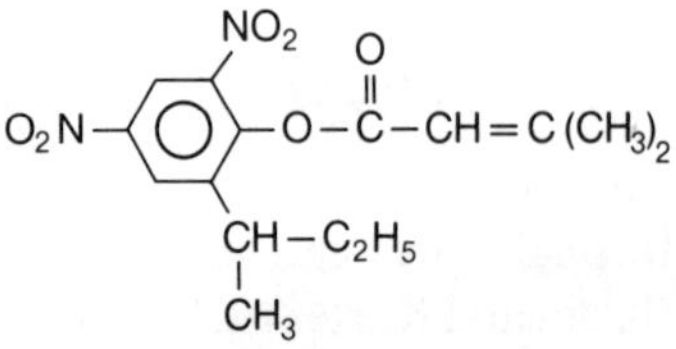   binapacryl

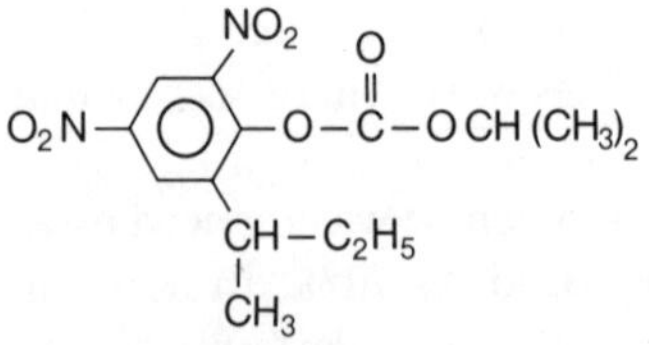

dinobuton

The acaricidal activity of certain 2-aryl-1,3-indandiones has been improved by the esterification of their enol form, e.g. UC-41305. UC-41305 shows a

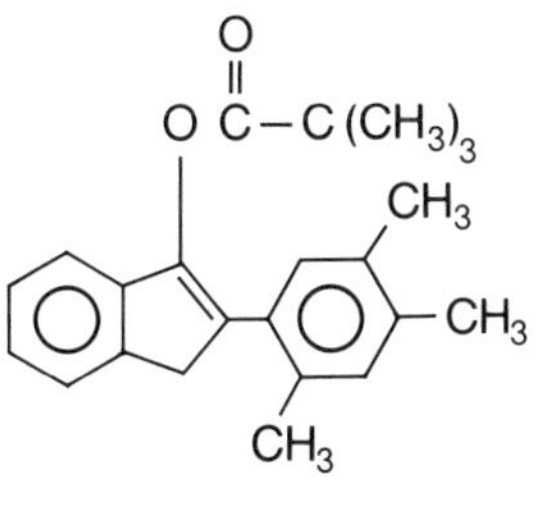

UC – 41305

steeper dose–response curve, enhanced transport properties and higher ovicidal activity than the parent dione (Durden *et al.*, 1973).

Another acaricide, the cyclopropanecarboxylate cycloprate, has been found to be hydrolysed *in vivo* to the acid which is subsequently conjugated to carnitine. This leads to a sequestration of carnitine and therefore to an inability to transport long chain fatty acids into mitochondria (Quistad *et al.*, 1979).

Another proacaricide which needs hydrolytic cleavage is fenazaflor.

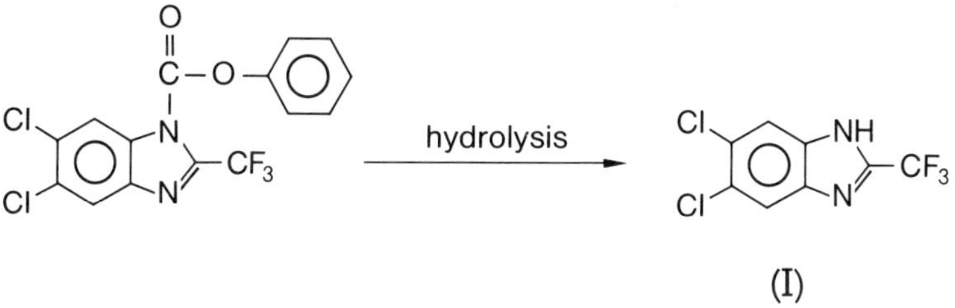

(I)

Fenazaflor is hydrolysed *in vivo* to 5,6-dichloro-2-trifluoromethyl benzimidazole (I), an uncoupler of oxidative phosphorylation (Corbett and Johnson, 1970).

PROPHEROMONES

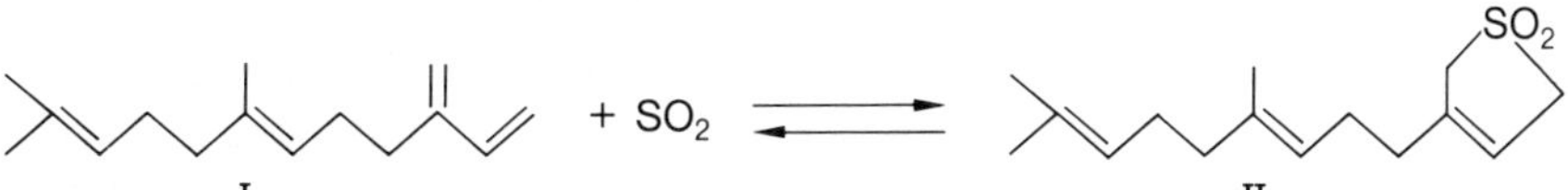

(*E*)-β-Farnesene (I) reacted with sulphur dioxide to give compound II which regenerated I at ambient temperatures in sufficient quantity to cause a weak alarm response from aphids. Adducts, such as IV and V were prepared from the pheromonal carbonyl compound III either directly by treating with alcohols or by transacetalization of the ethanol adducts. Under artificial sunlight compounds IV and V regenerated the pheromone. In field tests the attractiveness of compound IV and V to diamond-back moths persisted longer than for the parent aldehyde (III) suggesting that such adducts could be useful as prophermones (Pickett *et al.*, 1983).

(III)

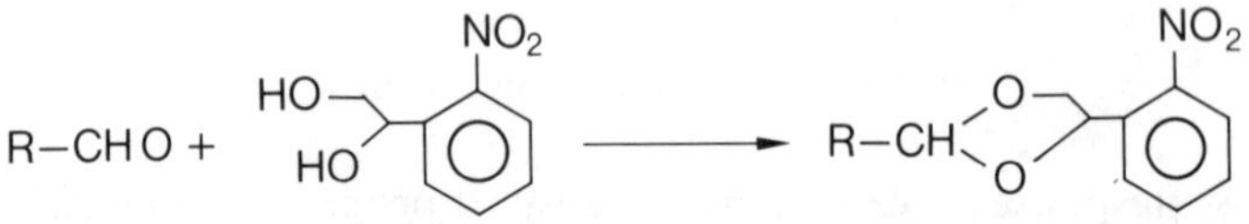

(III)   (IV)

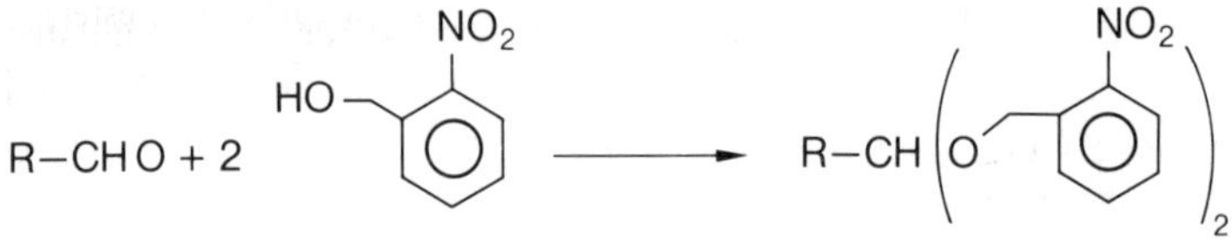

(III)   (V)

THIOCYANATES

Thiocyanates are contact insecticides with a very strong knockdown effect. Many organothiocyanates liberate hydrogen cyanide in mice and houseflies. The level of hydrogen cyanide in mouse brain and houseflies, resulting from treatment with various organothiocyanates correlates quite well with the toxicity of these hydrogen cyanide precursors. The glutathione *S*-transferases in mouse liver and houseflies liberate hydrogen cyanide from a variety of organothiocyanates, including Lethane 384[R]. These enzymes from mouse liver catalyse the attack of reduced glutathione on the thiocyanate sulphur which results in the liberation of hydrogen cyanide, oxidized glutathione, and the mercaptan moiety of the organothiocyanate. Thanite[R] reacts readily with glutathione to liberate hydrogen cyanide, even in the absence of glutathione *S*-transferases (Ohkawa *et al.*, 1972).

ACKNOWLEDGEMENTS

We have relied heavily on data from the basic research work done at certain universities and we would like especially to express our admiration for the work of the groups led by Professors Casida, Eto, Fukuto, Hollingworth and Knowles. We are grateful to the management of the R&D Department of the Agricultural Division of CIBA-GEIGY for their generous support.

REFERENCES

Acker, R., Hamprecht, G., Kiehs, K., and Adolphi, H. (BASF) (1980). Europ. pat. appl. 33,098.

Ackermann, P., Bourgeois, F., and Drabek, J. (1980). 'The optical isomers of α-cyano-3-phenoxybenzyl 3-(1,2-dibromo-2,2-dichloroethyl)-2,2-dimethylcyclopropane-carboxylate and their insecticidal activities', *Pestic. Sci.*, **11**, 169-179.

Albert, A. (1958). 'Chemical aspects of selective toxicity', *Nature*, **182**, 421–423.

Bachmann, F., and Drabek, J. (1981). 'CGA-73'102 a new soil-applied systemic carbamate insecticide', *Proc. Br. Crop Prot. Conf.—Pests & Dis.*, **11**, 51–58.

Bartlett, G. R. (1952). 'The mechanism of action of monofluoroethanol', *J. Pharmacol. exp. Ther.*, **106**, 464–467.

Beriger, E., and Drabek, J. (CIBA-GEIGY) (1971). U.S. pat. 3,992,533.

Black, A. L., Chiu, Y. C., Fahmy, M. A. H., and Fukuto, T. R. (1973a). 'Selective toxicity of N-sulfenylated derivatives of insecticidal methylcarbamate esters', *J. Agric. Food Chem.*, **21**, 747–751.

Black, A. L., Chiu, Y. C., Fukuto, T. R., and Miller, T.A. (1973b). 'Metabolism of 2,2-dimethyl-2,3-dihydrobenzofuranyl-7 N-methyl-N-(2-toluenesulfenyl) carbamate in the housefly and white mouse', *Pestic. Biochem. Physiol.*, **3**, 435–446.

Black, L. D. and Fukuto, T. R. (Univ. of California). (1973). German Off. 2,433,680.

Boeger, M., and Drabek, J. (CIBA-GEIGY) (1975a). German Off. 2,600,987.

Boeger, M., and Drabek, J. (CIBA-GEIGY) (1975b). Swiss pat. 610,181.

Boeger, M., and Drabek, J. (CIBA-GEIGY) (1975c). U.S. pat. 4,076,837.

Boeger, M., and Drabek, J. (CIBA-GEIGY) (1975d). U.S. pat. 4,098,902.

Boeger, M., and Drabek, J. (CIBA-GEIGY) (1976). U.S. pat. 4,223,027.

Brooks, G. T. (1960). 'Mechanism of resistance of the adult housefly *(Musca domestica)* to "cyclodiene" insecticides', *Nature*, **186**, 96–98.

Brown, M. S. (Chevron Res. Co.) (1970a). U.S. pat. 3,679,733.

Brown, M. S. (Chevron Res. Co.) (1970b). U.S. pat. 3,914,259.

Brown, M. A., Gammon, D. W., and Casida, J. E. (1983). 'Oxime ether pyrethroids and hydroxylamine ether propyrethroids: photochemistry, biological activity, metabolism', *J. Agric. Food Chem.*, **31**, 1091–1096.

Brown, M. S. and Kohn, G. K. (Chevron Res. Co.) (1969). U.S. pat. 3,663,594.

Brown, M. S. and Kohn, G. K. (Chevron Res. Co.) (1970). U.S. pat. 3,812,174.

Buholzer, F., Drabek, J., and Flueck, V. (1980). 'Lepicron a new cotton insecticide', *Med. Fac. Landbouww. Rijksuniv. Gent* **45**, 649–658.

Bull, D. L., and Lindquist, D. A. (1964). 'Metabolism of 3-hydroxy-N,N-dimethylcrotonamide dimethylphosphate by cotton plants, insects, and rats', *J. Agric. Food Chem.*, **12**, 310–317.

Caleb, W. H. (Du Pont Co.) (1980). U.S. pat. 4,316,911.

Casida, J. E. (1972). 'Chemistry and metabolism of terminal residues of organo-phosphorus compounds and carbamates', *Proc. 2nd Int. Congress of Pesticide Chemistry, London*, Vol. 6, pp. 295–314.

Cecil, H. C., Miller, R. W., and Corley, C. (1981). 'Feeding three insect growth regulators to white leghorn hens: residues in eggs and tissues and effects on production and reproduction', *Poultry Sci.*, **60**, 2017–2027.

Cheng, Hong-Ming, and Casida, J. E. (1973). 'Metabolites and photoproducts of 3-(2-butyl)-phenyl N-methylcarbamate and N-benzenesulfenyl-N-methylcarbamate', J. Agric. Food Chem., 21, 1037–1047.

CIBA-GEIGY Ltd. (1976). Belg. pat. 860,894.

CIBA-GEIGY Ltd. (1979a). German Off. 3,040,439.

CIBA-GEIGY Ltd. (1979b). German Off. 3,040,469.

Cole, L. M., Casida, J. E., and Ruzo, L. O. (1982a). 'Comparative degradation of the pyrethroids tralomethrin, tralocythrin, deltamethrin, and cypermethrin on cotton and bean foliage', J. Agric. Food Chem., 30, 916–920.

Cole, L. M., Ruzo, L. O., Wood, E. J., and Casida, J. E. (1982b). 'Pyrethroid metabolism: comparative fate in rats of tralomethrin, tralocythrin, deltamethrin, and (1R, aS)-cis-cypermethrin', J. Agric. Food Chem., 30, 631–636.

Collins, C., Kennedy, J. M., Fahmy, M. A. H., and Miller, T. (1980). 'Mode of action of sulfenylated carbamates: rapid conversion of N,N-thiodicarbamates to parent carbamate measured by neurophysiological bioassay', Pestic. Biochem. Physiol., 13, 158–163.

Corbett, J. R., and Johnson, E. R. (1970). 'Biochemical mode of action of the acaricide fenazaflor', Pestic. Sci., 1, 120–123.

Crecelius, C. S., and Knowles, C. O. (1976). 'Toxicity, penetration, and metabolism of chlordimeform and its N-demethyl metabolite in cabbage looper larvae', J. Agric. Food Chem., 24, 1019–1023.

Dauterman, W. C. (1971). 'Biological and nonbiological modifications of organophosphorus compounds', Bull. W.H.O., 44, 133–150.

Dedek, W. (1968). 'Abbau und Rückstande von 32 P-butonat in Früchten', Z. Naturforsch., 23b, 504–506.

Deguchi, T., Narahashi, T., and Haas, H. G. (1971). 'Mode of action of nereistoxin on the neuromuscular transmission in the frog', Pestic. Biochem. Physiol., 1, 196–204.

Drabek, J. (CIBA-GEIGY) (1974). Brit. pat. 1,489,969.

Drabek, J., Ackermann, O., Farooq, S., Gsell, L., Kristiansen, O., and Meyer, W. (CIBA-GEIGY) (1977). German Off. 2,805,226.

Drabek, J., and Bachmann, F. (1983). 'Proinsecticides: structure-activity relationships in carbamoyl sulfenyl N-methylcarbamates', IUPAC Pesticide Chemistry, Human Welfare and the Environment, Vol. 1, pp. 271–277, Pergamon Press, Oxford.

Drabek. J., and Boeger, M. (CIBA-GEIGY) (1975a). German Off. 2,621,077.

Drabek. J., and Boeger, M. (CIBA-GEIGY) (1975b). Swiss pat. 611,122.

Drabek. J., and Boeger, M. (CIBA-GEIGY) (1977). U.S. pat. 4,342,778.

Drabek. J., and Boeger, M. (CIBA-GEIGY) (1980a). German Off. 3,138,702.

Drabek. J., and Boeger, M. (CIBA-GEIGY) (1980b). U.S. pat. 4,386,101.

Drabek. J., and Boeger, M. (CIBA-GEIGY) (1981). Europ. pat. appl. 62,003.

Drabek, J., and Flueck, V. (1979). 'Synthesis and properties of O-aryl O-alkyl S-alkyl phosphorothioates', IUPAC Pesticide Chemistry, Advances in Pesticide Science, Vol. 2, pp. 130–134, Pergamon Press, Oxford.

D'Silva, T. D. J. (Union Carbide Co.) (1976a). U.S. pat. 4,081,550.

D'Silva, T. D. J. (Union Carbide Co.) (1976b). U.S. pat. 4,138,423.

D'Silva, T. D. J. (Union Carbide Co.) (1980). U.S. pat. 4,369,189.

Durden, J. A., Soussa, A. A., and Stephen, J. F. (1973). 'Acaricidal enol esters of 2-aryl-1,3-indandiones, structure–activity relationships', Abstr. Papers, Am. Chem. Soc. 165 Meet. PEST 31.

Dutton, F. E., Gemrich II, E. G., Lee, B. L., Nelson, S. J., Parham, P. H., and Seaman, W. J. (1981a). 'Insecticidal phosphoramidothio derivates of the carbamate methomyl', J. Agric. Food Chem., 29, 1111–1114.

Dutton, F. E., Gemrich II, E. G., Lee, B. L., Nelson, S. J., Parham, P. H., and Seaman, W. J. (1981b). 'Insecticidal phosphoramidothio derivates of the carbamate methomyl', *J. Agric. Food Chem.*, **29**, 1114–1118.

Eto, M. (1974). *Organophosphorus Pesticides: Organic and Biological Chemistry*, CRC-Press, Inc., Cleveland, Ohio.

Eto, M., Okabe, S., Ozoe, Y., and Maekawa, L. (1977). 'Oxidative activation of O,S-dimethyl phosphoroamidothiolate', *Pestic. Biochem. Physiol.*, **7**, 367–377.

Evans, P. D., and Gee, J. D. (1980). 'Action of formamidine pesticides on octopamine receptors', *Nature*, **287**, 60–62.

Fahmy, M. A. H., Chiu, Y. C., and Fukuto, T. R. (1974). 'Selective toxicity of *N*-substituted biscarbamoyl sulfides', *J. Agric. Food Chem.*, **22**, 59–62.

Fahmy, M. A. H., and Fukuto, T. R. (Univ. of California) (1979a). U.S. pat. 4,262,015.

Fahmy, M. A. H., and Fukuto, T. R. (Univ. of California) (1979b). U.S. pat. 4,298,617.

Fahmy, M. A. H., and Fukuto, T. R. (Univ. of California) (1979c). U.S. pat. 4,308,274.

Fahmy, M.A.H., and Fukuto, T.R. (Univ. of California)(1979d). Europ. pat. appl. 16,590.

Fahmy, M. A. H., and Fukuto, T. R. (Univ. of California) (1980a). U.S. pat. 4,298,527.

Fahmy, M. A. H., and Fukuto, T. R. (Univ. of California) (1980b). U.S. pat. 4,315,027.

Fahmy, M. A. H., and Fukuto, T. R. (1981). '*N*-Sulfinylated derivates of methylcarbamate esters', *J. Agric. Food Chem.* **29**, 567–572.

Fahmy, M. A. H., Fukuto, T. R., and Jojima, T. (Univ. of California) (1980). U.S. pat. 4,315,026.

Fahmy, M. A. H., Fukuto, T. R., and Jojima, T. (Univ. of California) (1981). U.S. pat. 4,350,669.

Fahmy, M. A. H., Fukuto, T. R., Myers, R. O., and March R. B. (1970). 'The selective toxicity of new N-phosphorothioylcarbamate esters', *J. Agric. Food Chem.*, **18**, 793–796.

Fahmy, M. A. H., Mallipudi, N. M., and Fukuto, T. R. (1978). 'Selective toxicity of N,N'-thiodicarbamates', *J. Agric. Food Chem.*, **26**, 550–557.

Fahmy, M. A. H., Metcalf, R. L., Fukuto, T. R. and Hennessy, D. J. (1966). 'Effects of deuteration, fluorination, and other structural modifications on the carbamyl moiety upon the anticholinesterase and insecticidal activities of phenyl N-methyl carbamates', *J. Agric. Food Chem.*, **14**, 79–83.

Fest, Ch. (1981). 'Insektizide Phosphorsäureester', in *Chemie der Pflanzenschutz- und Schädlingsbekämpfungsmittel* (ed. R. Wegler), Vol. 6, pp. 329–344, Springer Verlag, Berlin.

Fishbein, L., and Zielinsky Jr., E. L. (1965). 'Gas chromatography of trimethylsilyl derivatives I. Pesticidal carbamates and ureas', *J. Chromatogr.*, **20**, 9–14.

Franklin, E. J., and Knowles, C. O. (1982). 'Absorption and metabolism of xylyl formamidine acaricides by two spotted spider mites', *Abstr. Papers Am. Chem. Soc. 184 Meet.* PEST 42.

Fraser, J., Clinch, P. G., and Reay, R. C. (1965). '*N*-Acylation of *N*-methylcarbamate insecticides and its effect on biological activity', *J. Sci. Food Agric.* **16**, 615–618.

Fu, W. Y., Haag, W. G., and Rodgers, P. D. (1977). 'Photoactivated carbamates. Properties of N-(*ortho*-nitrobenzyl) substituted carbamates', *Abstr. Papers Am. Chem. Soc. 174 Meet.* PEST 7.

Fukuto, T. R. (1976). 'Carbamate Insecticides', in *The Future for Insecticides: Needs and Prospects* (eds R. L. Metcalf and J. J. McKelvey, Jr.), pp. 313–346, John Wiley, London.

Fukuto, T. R. (1983). 'Structure-activity relationships in derivates of anticholinesterase insecticides', in *IUPAC Pesticide Chemistry, Human Welfare and the Environment*, Vol. 1, pp. 203–212, Pergamon Press, Oxford.

Fukuto, T. R., and Fahmy, M. A. H. (1981). 'Sulfur in propesticide action, *ACS symposium Series*, **158**, 35–49, Am. Chem. Soc., Washington, DC.

Fukuto, T. R., and Ohta, H. (Univ. of California) (1982). U.S. pat. 4,410,518.

Gemrich II, E. G., Kangas, G., and Rizzo, V. L. (1976). 'Insecticidal and miticidal activity of arylthioformamidines', *J. Agric. Food Chem.*, **24**, 593–595.

Gemrich II, E. G., Lee, L. B., Nelson, S. J., and Rizzo, V. L. (1978). 'Insecticidal aminothioderivates of the pesticidal carbamate methomyl', *J. Agric. Food Chem.*, **26**, 391–395.

Goldman, P. (1969). 'The carbon-fluorine bond in compounds of biological interest', *Science*, **164**, 1123–1130.

Goto, T., Tanaka, A. K., Yasudomi, N., Osaki, N., Jida, S., and Umetsu, N. (1983). 'OK-174, A new broad-spectrum carbamate insecticide', *Proc. 10th Int. Congress of Plant Protection, Brighton, England*, 360–367.

Gray, A. J., Thompson, C. M., and Fukuto, T. R. (1982). 'Distribution and excretion of [$^{14}CH_3S$] methamidophos after intravenous administration of a toxic dose and the relationship with anticholinesterase activity', *Pestic. Biochem. Physiol.*, **18**, 28–37.

Hart, G. J., and O'Brien, R. D. (1976). 'Dissociation and phosphorylation constants for inhibition of acetylcholinesterase by a series of novel *O*-ethyl *O*-phenyl *S-n*-propyl phosphorothioates', *Pestic. Biochem. Physiol.*, **6**, 85–90.

Hart, R. J., Vavey, W. A., Ryan, K. J., Strong, M. B., Moore B., Thomas, P. L., Boray, J. C., and von Orelli, M. (1982). 'CGA-72′662—a new blowfly insecticide', *Australian Vet. J.*, **59**, 104–109.

Hartmann, A., Heywang, G., Kuehle, E., Hamman, I., and Homeyer, B. (1982). Europ. pat. appl. 92,721.

Hartmann, A., Kuehle, E., and Hamman, I. (Bayer Ltd.) (1978). Europ. pat. appl. 5,246.

Hatch III, C. E. (1978). 'Synthesis of *N,N*-dialkylaminosulfenylcarbamate insecticides via carbamoyl fluorides', *J. Org. Chem.*, **43**, 3953–3957.

Hathway, D. E., Brown, S. S., Chasseaud, L. F., and Hutson, D. E. (1970). *Foreign Compound Metabolism in Mammals*, Vol. 1, The Chemical Society, London.

Heywang, G., Kuehle, E., and Hammann I. (Bayer Ltd.) (1980). German Off. 3,028,231.

Heywang, G., Kuehle, E., and Hammann I., and Homeyer, B. (Bayer Ltd.) (1981). Europ. pat. appl. 43,978.

Heywang, G., Lockhoff, O., Kuehle, E., Hammann I., and Homeyer, B. (1982). Europ. Pat. appl. 90,242.

Hodgson, E., and Casida, J. E. (1960). 'Biological oxidation of *N,N*-dialkyl carbamates', *Biochim. Biophys. Acta*, **42**, 184–186.

Hodgson, E., and Casida, J. E. (1961). 'Metabolism of *N,N*-dialkyl carbamates and related compounds by rat liver', *Biochem. Pharmacol.*, **8**, 179–191.

Hollingworth, R. M., and Johnstone, E. M. (1983). 'Pharmacology and toxicology of octopamine receptors in insects', in *IUPAC Pesticide Chemistry, Human Welfare and the Environment*, Vol. 1, pp. 187–192. Pergamon Press, Oxford.

Ilivicky, J., and Casida, J. E. (1969). 'Uncoupling action of 2,4-dinitrophenols, 2-trifluoromethylbenzimidazols and certain other pesticide chemicals upon mitochondria from different sources and its relation to toxicity', *Biochem. Pharmacol.*, **18**, 1389–1401.

Irwing, S. N., and Fraser, T. E. M. (1984). 'Insecticidal activity of tralomethrin: electrophysiological assay reveals that it acts as a propesticide', *J. Agric. Food Chem.*, **32**, 111–113.

Jojima, T., Fahmy, M. A. H., and Fukuto, T. R. (1983). 'Sugar, glyceryl, and (pyridyloxy)-sulfinyl derivatives of methyl-carbamate insecticides', *J. Agric. Food Chem.* **31**, 613–620.

Joy, R. M. (1976). 'Convulsive properties of chlorinated hydrocarbon insecticides in the cat central nervous system', *Toxicol. appl. Pharmacol.*, **35**, 95–106.

Kaplan, B. W. (Union Carbide Co.) (1980). U.S. pat. 4,201,786.

Kao, T. S., and Fukuto, T. R. (1977). 'Metabolism of O,S-dimethyl propionyl and hexamoylphosphoramidothioate in the housefly and white mouse', *Pestic. Biochem. Physiol.*, **7**, 83.

Kawata, M., Umetsu, N., and Fukuto, T. R. (Univ. of California) (1981). U.S. pat. 4,394,383.

Kinoshita, Y., and Fukuto, T. R. (1980). 'Insecticidal properties of *N*-sulfonyl derivates of propoxur and carbofuran', *J. Agric. Food Chem.*, **28**, 1325–1327.

Kirby, A. H. M. (1966). 'Dinitroalkylphenols: versatile agents for control of agricultural pests and diseases', *World Rev. Pest Control*, **5**, 30–44.

Kishino, S., Kudamatsu, A., Kawasaki, K., Takase, I., Shiokawa, K., and Yamaguchi, S. (Bayer Ltd.) (1970). German Off. 2,111,414.

Klein, W., and Korte, F. (1970). 'Metabolismus von Chlorkohlenwasserstoffen', in *Chemie der Pflanzenschutz- und Schädlingsbekämpfungsmittel* (ed. R. Wegler), Vol. 1, pp. 199–218, Springer-Verlag, Berlin.

Knowles, C. O., and Gayen, A. K. (1982). 'Penetration and metabolism of thiobisformamidines by southwestern corn borer larvae', *J. Econ. Entomol.*, **75**, 232–236.

Knowles, C. O., and Roulston, W. J. (1973). 'Toxicity to *Boophilus microplus* of formamidine acaricides and related compounds, and modification of toxicity by certain insecticide synergists', *J. Econ. Entomol.*, **66**, 1245–1251.

Knowles, C. O., and Schuntner, C. A. (1974). 'Effect of piperonyl butoxide on the absorption and metabolism of chlordimeform by larvae of the cattle tick *Boophilus microplus*', *J. Austr. Entomol Soc.*, **13**, 11–16.

Konishi, K. (1970). 'Studies on organic insecticides part XIII. Synthesis of nereistoxin and related compounds VI', *Agric. Biol. Chem.*, **34**, 1549–1560.

Kono, Y., Sato, Y., and Okada, A. (1983). 'Activation of an O-ethyl S-n-propyl phosphorothiolate, TIA-230 in central nerve of *Spodoptera* larvae', *Pestic. Biochem. Physiol.*, **20**, 225–231.

Krieger, R. I., Lee, P. W., Fahmy, M. A. H., Chen, M., and Fukuto, T. R. (1976). 'Metabolism of 2,2-dimethyl-2,3-dihydrobenzofuranyl-7 *N*-dimethoxyphosphinothioyl-*N*-methyl-carbamate in the house fly, rat, and mouse', *Pestic. Biochem. Physiol.*, **6**, 1–9.

Kudamatsu, A. (1976). 'Tokuthion a new organophosphorus insecticide with low mammalian toxicity', *Japan Pesticide Information* No. **26**, 14–17.

Kuehle, E., and Klauke, E. (1977). 'Fluorierte Isocyanate und deren Derivate als Zwischenprodukte für biologisch aktive Wirkstoffe', *Angew. Chemie*, **89**, 794–804.

Kuehle, E., Siegle, P., and Behrenz, W., (Bayer Ltd.) (1971a). German Off. 2,142,496.

Kuehle, E., Siegle, P., Behrenz, W., and Hammann, I. (Bayer Ltd.) (1971b). German Off. 2,132,936.

Maitlen, E. G., and Sladen, N. A. (1979). 'Effectiveness of granular and sprayable formulations of FMC-35001', *Proc. Br. Crop. Prot. Conf. – Pests & Dis.*, **10**, 557–564.

Marsden, O. J., Kuwano, E., and Fukuto, T. R. (1982). 'Metabolism of carbosulfan [2,3-dihydro-2,2-dimethylbenzo-furan-7-yl(di-n-butylaminothio) methylcarbamate] in the rat and house fly', *Pestic. Biochem. Physiol.*, **18**, 38–48.

Martel, J., Tessier, J., and Demoute, J. P. (1976a). German Off. 2,742,546.

Martel, J., Tessier, J., and Demoute, J. P. (1976b). German Off. 2,742,547.

Mauer, F., Hammann, I., Homeyer, B., and Behrenz, W. (Bayer Ltd.) (1981). Europ. pat. appl. 43,917.

McBain, J.B., Yamamoto, I., and Casida, J. E. (1971). 'Mechanism of activation and deactivation of dyfonate by rat liver microsomes', *Life Sci.*, **10, II**, 947–954.

Menzer, R. E., and Casida, J. E. (1965). 'Nature of toxic metabolites formed in mammals, insects, and plants from 3-(dimethyloxyphosphonyloxy)-*N,N*-dimethyl-*cis*-crotonamide and its *N*-methyl analog', *J. Agric. Food Chem.*, **13**, 102–112.

Metcalf, R. L., Fukuto, T. R., and March, R. B. (1957). 'Plant metabolism of dithio--Systox and Thimet', *J. Econ. Entomol.* **50**, 338–345.

Miller, R. W., and Corley, C. (1980). 'Feed-through efficacy of CGA-19'255 and CGA-72'662 against manure-breeding flies and other arthropods and residues in feces, eggs, and tissues of laying hens', *Southwest. Entomol.*, **5**, 144–148.

Miller, R. W., Corley, C., Cohen, C. F., Robbins, W. E., and Marks, E. P. (1981). 'CGA-19'255 and CGA-72'662: efficacy against flies and possible mode of action and metabolism', *Southwest. Entomol.*, **6**, 272–278.

Miller, R. W., Corley, C., and Pickens, L. G. (1977). 'CGA-19'255: a promising new insect growth regulator yested as a poultry feed additive for control of fliies', *Southwest. Entomol.*, **2**, 197–201.

Miskus, R. P., Andrews, T. L., and Look, M. L. (1969). 'Metabolic pathways affecting toxicity of *N*-acetyl Zectran', *J. Agric. Food Chem.*, **17**, 842–844.

Mitsubishi Chem. Ind. KK, (1980). Jap. pat. appl. J 5-7046-992.

Murdoch, L. L., and Hollingworth, R. M. (1980). 'Octopamine like actions of formamidines in the firefly light organ', in *Insect Neurobiology and Insecticide Action (Neurotox 79)*, pp. 415–422, Soc. Chem. Ind., London.

O'Brien, R. D. (1967). *Insecticides Action and Metabolism*, Academic Press, New York.

Orchard, I., Singh, G. J. P., and Longhton, B. G. (1982). 'Action of formamidine pesticides on octopamine receptors in locust fat body', *Comp. Biochem. Physiol.*, **73c**, 331–334.

Ohkawa, H., Ohkawa, R., Yamamoto, I., and Casida, J. E. (1972). 'Enzymatic mechanism and toxicological significance of hydrogen cyanide liberation from various organothiocyanates and organonitriles in mice and houseflies', *Pestic. Biochem. Physiol.*, **2**, 95–112.

Otsuka Chem. Co. (1980a). Belg. pat. 890,162.

Otsuka Chem. Co. (1980b). Jap. pat. appl. J 5-7045-171.

Otsuka Chem. Co. (1980c). Jap. pat. appl. J 5-7109-764.

Otsuka Chem. Co. (1981). Jap. pat. appl. J 5-7179-193.

Parham, P. H., Gemrich II, E. G., and Weddon, T. E. (1975). 'Sulfenylated formamidines as insecticides and acaricides', *Proc. Br. Crop Prot. Conf.—Pests & Dis.*, **8** 653–658.

Perry, A. S., Mattson, A. M., and Buckner, A. J. (1958). 'The metabolism of heptachlor by resistant and susceptible houseflies', *J. Econ. Entomol.*, **51**, 346–351.

Pianka, M., and Smith, C. B. F. (1965). 'Dinobuton a new acaricide', *Chem. & Ind.* 1216.

Pickett, J. A., Dawson, G. W., Griffiths, D. C., Lui, X., Macaulay, E. D. M., Williams, I. H., and Woodcock, C. M. (1983). 'Stabilizing pheromones for field use: propheromones', in *Proc. 10th Int. Congress of Plant Protection, Brighton, England*, Vol. 1, p. 271.

Pissiotas, G. (CIBA-GEIGY) (1972). German Off. 2,350,695.

Pissiotas, G., and Dürr, D. (CIBA-GEIGY) (1971). German Off. 2,259,218.

Pratt, G. E. (1983). 'The mode of action of pro-allatocins', in *Natural Products for Innovative Pest Management*, Vol. 2 of *Current Themes in Tropical Science* (eds D. L. Whithead and W. S. Bowers), pp. 323–353, Pergamon Press, Oxford.

Prestwich, G. D., Gayen, A. K., Phirwa, W., and Kilne, T. B. (1983a). '29-Fluorophytosterols: novel pro-insecticides, which cause death by dealkylation', *Bio-Technology*, **1**, 62–65.

Prestwich, G. D., Mauldin, J. K., Engstrom, J. B., Carvalho, J. F., and Cupo, D. Y. (1983b). 'Comparative toxicity of fluorinated lipids and their evaluation as bait-block toxicants for the control of *Reticulitermes spp.* (Isoptera: Rhinotormitidae)', *J. Econ. Entomol.*, **76**, 690–695.

Ptashne, K. A., and Neal, R. A. (1972). 'Reaction of parathion and malathion with peroxytrifluoroacetic acid, a model system for the mixed function oxidases', *Biochem.*, **11**, 3224–3228.

Quistad, G. B., Cerf, D. C., Schooley, D. A., and Staal, G. B. (1981). 'Fluoromevalonate acts as an inhibitor of insect juvenile hormone biosynthesis', *Nature*, **289**, 176–177.

Quistad, G. B., Staiger, L. E., Schooley, D. A., Sparks, R. G., and Hammock, B. D. (1979). 'The possible role of carnitine in the selective toxicity of the miticide cycloprate', *Pestic. Biochem. Physiol.*, **11**, 159–165.

Ross, D. C., and Brown, T. M. (1982). 'Inhibition of larval growth in *Spodoptera frugiperda* by sublethal dietary concentrations of insecticides', *J. Agric. Food Chem.*, **30**, 193–196.

Sakai, M. (1982). 'Ti-78 (Bensultap): a new low mammalian toxic insecticide related to nereistoxin', *Vth Int. Congress of Pesticide Chemistry (IUPAC)*, Kyoto, Abstr. Vol., Posters IIa 2, and IV-29.

Sakai, M., and Sato, Y. (1972). 'Metabolic conversion of the nereistoxin-related compounds into nereistoxin as a factor of their insecticidal action', *2nd Int. Congress of Pesticide Chemistry, London*, 455–467.

Sauers, R. F. (Du Pont) (1978a). U.S. pat. 4,146,618.

Sauers, R. F. (Du Pont) (1978b). U.S. pat. 4,150,122.

Sauers, R. F. (Du Pont) (1978c). U.S. pat. 4,160,034.

Schrader, G. (1947). 'The development of new insecticides', *British Intelligence Objectives Subcommittee, Final Report 914*, London.

Schunter, C. A., and Thompson, P. G. (1978). 'Metabolism of (14C)amitraz in larvae of *Boophilus microplus*', *Austr. J. Biol. Sci.*, **31**, 141–148.

Schunter, C. A., and Thompson, P. G. (1979). 'Toxicology and metabolism of chloromethiuron in *Boophilus microplus* larvae', *Pestic. Sci.*, **10**, 519–526.

Segall, Y., and Casida, J. E. (1982). 'Oxidative conversion of phosphorothiolates to phosphinyloxysulfonates probably via phosphorothiolate S-oxides', *Tetrahedron Letters*, **23**, 139–142.

Sousa, A. A., Frazee, J. R., Weiden, M. H. J., and D'Silva, T. D. J. (1977). 'UC-51762 a new carbamate insecticide', *J. Econ. Entomol.*, **70**, 803–807.

Stetter, J., and Homeyer, B. (Bayer Ltd.) (1976). German Off. 2,635,883. For the synthesis of intermediates see: Kuehle, E. (1970). *Synthesis 576.*

Suksayretrup, P., and Plapp Jr., F. W. (1977). 'Mechanism by which methamidophos and acephate circumvent resistance to organophosphate insecticides in the housefly', *J. Agric. Food Chem.*, **25**, 481–485.

Sun, Y.-P., and Johnson, E. R. (1960). 'Synergistic and antagonistic actions of insecticide-synergist combinations and their mode of action', *J. Agric. Food Chem.*, **8**, 261–266.

Tucker, B. V. (1976). 'Conversion of acephate to Monitor by rats', *Chevron Chem. Co. Report.*

Ueda, M., Ooba, S., Hirano, M., and Takeda, H. (1972). U.S. pat. 3,950,374.

Umetsu, N., and Fukuto, T. R. (1982). 'Alternation of carbosulfan [2,3-dihydro-2,2-dimethyl-7-benzofuranyl(di-*n*-butylaminosulfenyl) methylcarbamate] in the rat stomach', *J. Agric. Food Chem.*, **30**, 555–557.

Upjohn Co. (1981). U.S. pat. 4,356,191; 4,356,192; 4,356,194.

Wayne, I. G. (1981). 'Fate of the insecticide *O*-[4-[(4-chlorophenyl)thio]phenyl]-*O*-ethyl-*S*-propyl phosphorothionate (RH-994) in water', *J. Agric. Food Chem.*, **29**, 1146–1149.

Wing, K. D., Glickman, A. H., and Casida, J. E. (1983). 'Oxidative bioactivation of *S*-alkyl phosphorothiolate pesticides: stereospecifity of profenofos insecticide activation', *Science*, **219**, 63–65.

Worthing, Ch. E. (ed.) (1983a). *The Pesticide Manual*, p. 96, The British Crop Protection Council.

Worthing, Ch. E. (ed.) (1983b). *The Pesticide Manual*, p. 527, The British Crop Protection Council.

Yang, H. S., and Thurman, D. E. (1981). 'Thiodicarb—a new insecticide for integrated pest management', *Proc. Br. Crop Prot. Conf. Pests & Dis.*, **11**, 687–697.

Insecticides
Edited by D. H. Hutson and T. R. Roberts
© 1985 John Wiley & Sons Ltd

CHAPTER 3

A survey of the metabolism of carbamate insecticides

M. Cool and C. K. Jankowski

INTRODUCTION

The carbamate insecticides form an economically very important class of anti-cholinesterases in wide use at the present time. As a class they tend to be rapid-acting and also biodegraded fairly rapidly. These are two of the most important properties required of an insecticide. Metabolism in the target organism and non-target species usually limits toxicity and may also afford the basis of selective toxicity. Metabolism in plants will limit the build-up of unwanted residues.

The metabolism of a widely used carbamate insecticide such as matacil (4-dimethylamino-3-tolyl-N-methylcarbamate of aminocarb) is one of great interest because of its use against the spruce budworm *Choristoneura fumiferana* (Clemens). Spraying with this insecticide in Eastern Canada and New England started in 1976 as a small scale experimental programme. The programme changed in 1978, when only organophosphates were used; then again it was changed totally to matacil with 100% of the forest area sprayed in 1979 and 20% in 1980. However, there was no spraying with matacil in 1981 and only some occasional use of the compound in 1982 (up to 50% of the 1980 area).

A cursory review of the expected metabolic products of carbamate insecticides would suggest oxidation as the most likely metabolic reaction. The hydroxylation of a methyl or phenyl group is quite common. The reduction of the carbamate system is rare and, except for demethylation of zectran by microorganisms and a few other cases, has not been observed.

Hydrolysis is defined as the simple breakdown of a carbamate (to a phenol, amine and CO_2). Conjugation increases the water solubility of metabolites. Conjugation with sialic acid, polysaccharides (cellulose), silicates etc. and other biopolymers leading to storage in plants is discussed in detail later. This process is one of the numerous detoxication mechanisms possible. Dilution in the environment may lower toxicity but it cannot be considered as a detoxication pathway.

Conjugation cannot be seen as a homogenous type of reaction. This process takes place when carbamates, or more usually their degradation products (phenols for instance), react with a variety of conjugands such as silicates, phosphates, glucose and other sugars, cellulose, humic acids, lignin, cholesterol, sulphates and angelic acid.

The chemistry, biochemistry and toxicology of the carbamate insecticides were reviewed in 1976 by Kuhr and Dorough (1976). We should now like to review the metabolism of this class of insecticides with special emphasis on recent findings with matacil. An evaluation of the potential toxicity of a new class of metabolite which may accumulate in the environment is also presented.

Table 1   Chemical names of carbamate insecticides

1	**Isolan,** 1-isopropyl-3-methyl-5-pyrazolyl-N,N-dimethylcarbamate
2	**Pyrolan,** 1-phenyl-3-methyl-5-pyrazolyl-N,N-dimethylcarbamate
3	**Dimetilan,** 2-dimethylcarbamoyl-3-methyl-N,N-dimethylcarbamate
4	**Pyramat,** 2-n-propyl-4-methyl-6-pyrimidyl-N,N-dimethylcarbamate
5	**Tsumacide,** MTMC, m-tolyl-N-methylcarbamate
6	**Meobal,** 3,4-dimethylphenyl-N-methylcarbamate
7	**Landrin,** 3,4,5-trimethylphenyl-N-methylcarbamate and 2,3,5-trimethylphenyl-N-methylcarbamate
8	**Etrofolan,** mipcin, MIPC, isoprocarb, 2-isopropylphenyl-N-methylcarbamate
9	**UC 10854,** IPFC, 3-isopropylphenyl-N-methylcarbamate
10	**Promecarb,** carbamult, 3-isopropyl-5-methylphenyl-N-methylcarbamate
11	**HRS 1422,** 3,5-diisopropylphenyl-N-methylcarbamate
12	**RE 5365,** m-sec-butylphenyl-N-methylcarbamate
13	**BPMC,** bassa, O-sec-butylphenyl-N-methylcarbamate
14	**Bux,** bufencarb, metalramate, 3-[1'-(methylbutyl)]phenyl-N-methylcarbamate and 3-[1'-(ethylpropyl)]phenyl-N-methylcarbamate
15	**RE 5030,** 3-$tert$-butylphenyl-N-methylcarbamate
16	**Butcarb,** 3,5-di-$tert$-butylphenyl-N-methylcarbamate
17	**Croneton,** ethiofencarb, O-(ethylthiomethyl)phenyl-N-methylcarbamate

Table 1 (*continued*)

18	**Banol,** chlorxylam, carbanolate, 2-chloro-4,5-dimethylphenyl-*N*-methylcarbamate
19	**Baygon,** arprocarb, propoxur, unden, 2-isopropoxyphenyl-*N*-methylcarbamate
20	**Matacil,** aminocarb, 3-methyl-4-(*N,N*-dimethylamino)phenyl-*N*-methylcarbamate
21	**Zectran,** mexacarbate, 4-dimethylamino-3,5-xylyl-*N*-methylcarbamate
22	**Formetanate,** dicarzol, 3-[(dimethylamino)methylene]iminophenyl-*N*-methylcarbamate
23	**UC 34096,** formparanate, 4-[(dimethylamino)methylene]imino-*m*-tolyl-*N*-methylcarbamate
24	**Mesurol,** methiocarb, mercaptodimethur, 3,5-dimethyl-4-(methylthio)phenyl-*N*-methylcarbamate
25	**Carbaryl,** Sevin, NAC, 1-naphthyl-*N*-methylcarbamate
26	**Carbofuran,** furadan, 2,3-dihydro-2,2-dimethyl-7-benzofuranyl-*N*-methylcarbamate
27	**Mobam,** 4-benzothienyl-*N*-methylcarbamate
28	**Aldicarb,** temik, 2-methyl-2-(methylthio)-propionaldehyde-*O*-(methylcarbamoyl)oxime
29	**Thiofanox,** 3,3-dimethyl-1-(methylthio)2-butanone-*O*-(methylcarbamoyl)oxime
30	**Methomyl,** lannate, 1-methylthioacetaldehyde-*O*-(methylcarbamoyl)oxime
31	**Thiocarboxime,** 2-*S*-cyanoethyl-*N*-(methylcarbamoyl)-1-oxythioacetimidate
32	**Oxamyl,** vydate, methyl-*N,N*-dimethyl-*N*-(methylcarbomyl)-1-oxythioxamimidate
33	***N*-Acetyl-zectran,** (4-dimethylamino-3,5-xylyl)-*N*-acetyl-*N*-methylcarbamate
34	**RE 11775,** *N*-benzenesulphenyl-3-(2-butyl)phenyl-*N*-methylcarbamate
35	***N*-(2-Toluenesulphenyl)carbofuran**
36	**PSC,** *N*-(dimethoxyphosphinothioyl)-carbofuran
37	**DBSC,** *N*-(dibutylaminosulphenyl)-carbofuran
38	**MSC,** *N*-(morpholinosulphenyl)-carbofuran
39	**3-Methylphenyl-*N*-propylcarbamate,** *m*-tolyl-*N*-propylcarbamate

METABOLISM OF CARBAMATE INSECTICIDES

This chapter provides a review of the metabolism of six groups of carbamates. The first group, *N,N*-dimethylcarbamates, consists of four insecticides: three pyrazolyl and one pyrimidyl (pyramat) carbamates. The next section contains carbamates with different alkyl groups on the aromatic moiety of methylcarbamates. The following group presents heteroatom phenyl methylcarbamates (with atoms such as Cl, O, S and N on the phenyl moiety, e.g. banol, baygon, methiocarb and matacil). The section on *N*-methylcarbamates with a condensed ring system contains the most popular carbamate insecticide, carbaryl, a naphthyl carbamate-type insecticide. Finally, another popular carbamate, carbofuran, constituted by a benzofuranyl moiety, and mobam with a benzothionyl carbamate of oxime residue are presented in the last section.

A new concept in a search for a new carbamate insecticide, reflecting an evolution in the search for more efficient action toward insects and lowered toxicity toward vertebrates, is the derivatization of an already-known carbamate to derivatives such as N-acetyl-zectran and N-(2-toluenesulphenyl)carbofuran. These new compounds are contained in the final section.

After presenting the general nature of the six sections, we examine the different metabolites known and define the different metabolic pathways and relationships between different kinds of chemical structures in the whole spectrum of carbamate insecticide degradation.

Although oxidation is the most common metabolic reaction of carbamate insecticides, N-demethylation is in fact a redox process in which the N-CH_3 group undergoes oxidation (to N-CHO), but the nitrogen is reduced. The carbamate insecticides have polyfunctional structures which enable the molecule to react at the same time in both reductive and oxidative modes.

N,N-Dimethylcarbamates (Figure 1)

Hydrolysis

Three out of four carbamates of this category are easily hydrolysed. The detection of the remaining hydrolysed compounds such as isolan (O'Brien, 1967, Schlagbauer and Schlagbauer, 1972; Kuhr and Dorough, 1976), dimetilan (O'Brien, 1967; Schlagbauer and Schlagbauer, 1972) and pyramat (Schlagbauer and Schlagbauer, 1972; Kuhr and Dorough, 1976), in mammals, and only of dimetilan in insects has been shown. Dimetilan is hydrolysed to the corresponding N,N-dimethylcarbamic acid and successively degraded to dimethylamine and CO_2 in mammals (O'Brien, 1967; Schlagbauer and Schlagbauer, 1972). Isolan is hydrolysed to CO_2 in mammals (O'Brien, 1967; Schlagbauer and Schlagbauer, 1972; Kuhr and Dorough, 1976), and one of the metabolites of pyramat, 5-hydroxypyramat, is also hydrolysed by mammals to 5,6-dihydroxy-4-methyl-2-propyl pyrimidine (Schlagbauer and Schlagbauer, 1972).

N-Methyl hydroxylation

The four insecticides underwent N-methyl hydroxylation which was revealed by studies in mammals in vitro (Oonnithan and Casida, 1966; O'Brien, 1967; Schlagbauer and Schlagbauer, 1972) and in vivo (Schlagbauer and Schlagbauer, 1972). Dimetilan was also hydroxylated by insects (Zubairi and Casida, 1965; Schlagbauer and Schlagbauer, 1972; Kuhr and Dorough, 1976) and the hydroxylation occurring on the carbamoyl moiety in mammals is also observed for insects (Zubairi and Casida, 1965; Schlagbauer and Schlagbauer, 1972; Kuhr and Dorough, 1976). The N-hydroxymethyl carbamyl derivative of dimetilan was found in rat liver and in some insects (Zubairi and Casida, 1965; Schlagbauer and Schlagbauer, 1972; Kuhr and Dorough, 1976).

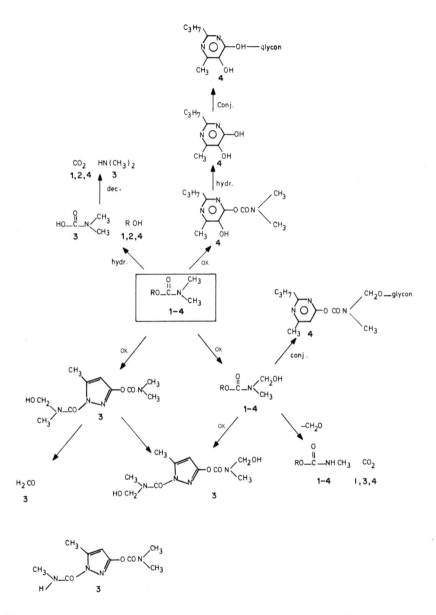

Figure 1 Metabolism of carbamates 1–4: 1, isolan; 2, pyrolan; 3, dimetilan; 4, pyramat

Oxidative N-demethylation

The oxidative degradation of isolan and pyrolan has been proved to take place readily (Oonnithan and Casida, 1966; Schlagbauer and Schlagbauer, 1972; Matsumura, 1975). Studies with rat liver and some insects shows that there is N-demethylation on the dimetilan carbamate moiety (Schlagbauer and Schlagbauer, 1972) as well as on the carbomyl moiety. This reaction has also been studied in mammals via *in vivo* experiments (Zubairi and Casida, 1965; Schlagbauer and Schlagbauer, 1972; Kuhr and Dorough, 1976).

Formaldehyde formation

Formaldehyde was produced by rat liver with all three insecticides: isolan (Casida, 1963; Schlagbauer and Schlagbauer, 1972), dimetilan (Schlagbauer and Schlagbauer, 1972) and pyramat (Hodgson and Casida, 1961; Schlagbauer and Schlagbauer, 1972). The yield, even though it is not for a major metabolic route *in vivo*, is high.

Hydroxylation

Only one insecticide, pyramat, was hydroxylated by mammals giving 5-hydroxypyramat and 5-hydroxypyramat pyrimidol (Schlagbauer and Schlagbauer, 1972; Kuhr and Dorough, 1976). This hydroxylation indicates some restriction on the oxidation of the aromatic moiety in the systems studied.

Conjugation

Pyramat affords the presence of conjugated compounds in mammals. In this case, the exocons were N-hydroxymethyl pyramat, pyramat pyrimidol and 5-hydroxypyramat pyrimidol (Schlagbauer and Schlagbauer, 1972).

Alkylphenyl-*N*-methylcarbamates (Figures 2a–e)

Hydrolysis

Only for one of the insecticides in this group, RE 5365, has the corresponding phenol derivative not been found. This case is of particular interest because of the simultaneous production of labelled CO_2 by mammals for the insecticide quoted above (Kuhr and Dorough, 1976). The others—meobal, landrin, UC 10854, HRS 1422, bux and croneton—are all easily hydrolysed by mammals (O'Brien, 1967; Knowles and Sen Gupta, 1970; Schlagbauer and Schlagbauer, 1972; Kuhr and Dorough, 1976; Nye *et al.*, 1976; Marshall and Dorough, 1977; Benson and Dorough, 1979; Dorough, 1979; Abd-Elraof *et al.*, 1981).

Similar reactions take place with insects (with UC 10854 (Kuhr and Dorough, 1976)) and microorganisms (with bux (Kuhr and Dorough, 1976)). Hydrolysis in plants occurred with etrofolan (Ogawa et al., 1977) and promecarb (Johannsen and Knowles, 1977). The phenolic compounds originating from tsumacide (Ohkawa et al., 1974; Kuhr and Dorough, 1976), RE 5030 (Kuhr and Dorough, 1976) and butacarb (Kuhr and Dorough, 1976) metabolism was found in insects, and in microorganisms metabolism with BPMC (Suzuki and Takeda, 1976a,b,c). Tsumacide (Ohkawa et al., 1974; Kuhr and Dorough, 1976), meobal (Miyamoto et al., 1969; Dorough, 1970; Miyamoto, 1970; Kuhr and Dorough, 1976), landrin (Slade and Casida, 1970; Kuhr and Dorough, 1976), UC 10854 (Schlagbauer and Schlagbauer, 1972), promecarb (Knowles and Johannsen, 1976) and HRS 1422 (Nye et al., 1976; Hathway, 1979) afford the corresponding phenol in mammals.

N-*Methyl oxidation*

The simplest case of such an oxidation, an N-hydroxymethylcarbamate derivative formation, was found in various organisms. BPMC (Suzuki and Takeda, 1976a,b,c) undergoes N-methyl oxidation in insects; tsumacide, landrin and RE 5030 (Slade and Casida, 1970; Schlagbauer and Schlagbauer, 1972; Ohkawa et al., 1974; Kuhr and Dorough, 1976) in plants; tsumacide, landrin, etrofolan, UC 10854 (Slade and Casida, 1970; Schlagbauer and Schlagbauer, 1972; Ohkawa et al., 1974; Kuhr, 1976; Kuhr and Dorough, 1976) in insects and in plants. Meobal, landrin, UC 10854, HRS 1422, bux, RE 5030, butacarb and croneton (Miyamoto et al.,1969; Dorough, 1970; Knaak, 1971; Schlagbauer and Schlagbauer, 1972; Kuhr and Dorough, 1976; Nye et al., 1976) are oxidized, producing N-hydroxymethyl metabolites, in mammals. *In vitro* studies revealed the same results with mammals for landrin, UC 10854 (Slade and Casida, 1970; Schlagbauer and Schlagbauer, 1972) and with insects for UC 10854 only (Schlagbauer and Schlagbauer, 1972).

Oxidative N-*demethylation*

N-Demethylation of etrofolan occurs in plants (Ogawa et al., 1977) and of BPMC in microorganisms (Suzuki and Takeda, 1976a,b,c).

Aryl oxidation

Plants usually favour aryl oxidation. Two insecticides, tsumacide and etrofolan (Ohkawa et al., 1974; Kuhr and Dorough, 1976; Ogawa et al., 1977), are oxidized on the aromatic moiety. Insects oxidize tsumacide and landrin (Ohkawa et al., 1974; Kuhr and Dorough, 1976; Ogawa et al., 1977). Mammals hydroxylate tsumacide (Ohkawa et al., 1974) but the hydroxylation of landrin in mammals has been demonstrated only *in vitro* (Slade and Casida, 1970).

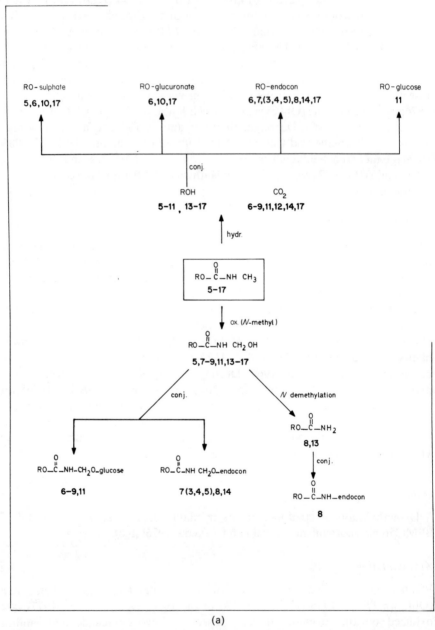

(a)

Figure 2 Metabolism of carbamates 5–17: 5, isumacide; 6, meobal; 7, landrin; 8, etrofolan; 9, UC 10854; 10, promecarb; 11, HRS 1422; 12, RE 5365; 13, bassa; 14, bux; 15, RE 5030; 16, butacarb; 17, croneton

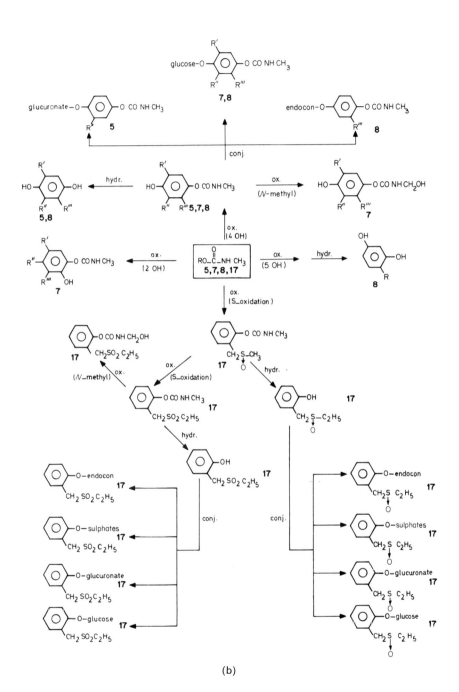

(b)

* **14** sec. pentyl   (c)

(d)

(e)

Aryl alkyl oxidation

Hydroxymethyl group formation has been found in plants with tsumacide, meobal, landrin, RE 5030, butacarb (Slade and Casida, 1970; Ohkawa *et al.*, 1974; Kuhr and Dorough, 1976), in mammals *in vivo* with tsumacide, meobal, landrin, RE 5030 and butacarb (Miyamoto *et al.*, 1969; Dorough, 1970; Miyamoto, 1970; Slade and Casida, 1970; Knaak, 1971; Ohkawa *et al.*, 1974; Kuhr and Dorough, 1976), in mammals *in vitro* with landrin (Slade and Casida, 1970) and for microorganisms with one insecticide only (BPMC (Suzuki and Takeda, 1976a,b,c)). Hydroxylation on a secondary carbon takes place only on BPMC in microorganisms (Suzuki and Takeda, 1976a,b,c). Hydroxylation on a tertiary carbon occurred with etrofolan, UC 10854, HRS 1422 (Matsumura, 1975), BPMC and bux. It also occurred in mammals *in vivo* with UC 10854 and BPMC (Knaak, 1971; Kuhr and Dorough, 1976), in plants with etrofolan and UC 10854 (Schlagbauer and Schlagbauer, 1972; Kuhr, 1976; Kuhr and Dorough, 1976; Ogawa *et al.*, 1977), in insects *in vivo* and *in vitro* with the UC 10854 (Schlagbauer and Schlagbauer, 1972; Kuhr and Dorough, 1976) and in microorganisms with the HRS 1422 and BPMC (Kuhr and Dorough, 1976; Suzuki and Takeda, 1976a,b,c). UC 10854 seems to react in a very unspecific way.

Keto group formation was observed for etrofolan in plants (Ogawa *et al.*, 1977) and for BPMC in microorganisms (Suzuki and Takeda, 1976a,b,c). Aldehyde group formation from the landrin structure has not been confirmed but this group has been suggested as part of the structure of its major degradation product in mammals *in vivo* and *in vitro*, and in plants and insects (Slade and Casida, 1970). The final step in methyl oxidation is carboxylic acid formation; this was found with meobal in mammals (Miyamoto *et al.*, 1969; Dorough, 1970; Miyamoto, 1970; Kuhr and Dorough, 1976), tsumacide and landrin in mammals and insects (Slade and Casida, 1970; Ohkawa *et al.*, 1974; Kuhr and Dorough, 1976). Croneton afforded a sulphoxidation process gradually producing sulphoxide and sulphone derivatives in mammals (Nye *et al.*, 1976; Hathway, 1979).

Conjugation

All of the expected glucuronide conjugates have been found in mammals (Slade and Casida, 1970; Knaak, 1971; Knowles and Johannsen, 1976; Kuhr and Dorough, 1976; Wilkinson, 1976; Marshall and Dorough, 1977; Dorough, 1979). The same has been observed for the sulphate conjugates (Slade and Casida, 1970; Knaak, 1971; Knowles and Johannsen, 1976; Kuhr and Dorough, 1976; Wilkinson, 1976; Marshall and Dorough, 1977; Dorough, 1979). Glucoside conjugates were found in plants in the case of meobal, landrin, etrofolan, HRS 1422, croneton (Knaak, 1971; Schlagbauer and Schlagbauer, 1972; Kuhr and Dorough, 1976; Marshall and Dorough, 1977; Ogawa *et al.*, 1977;) and in

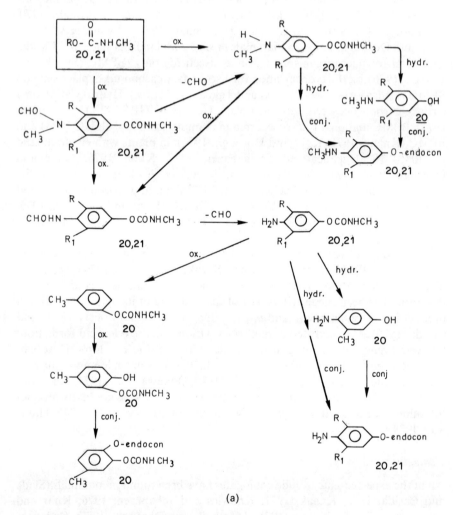

(a)

Figure 3 Metabolism of carbamates 18–24: 18, banol; 19, baygon; 20, matacil; 21, zectran; 22, formetanate; 23, UC 34096; 24, mesurol

(b)

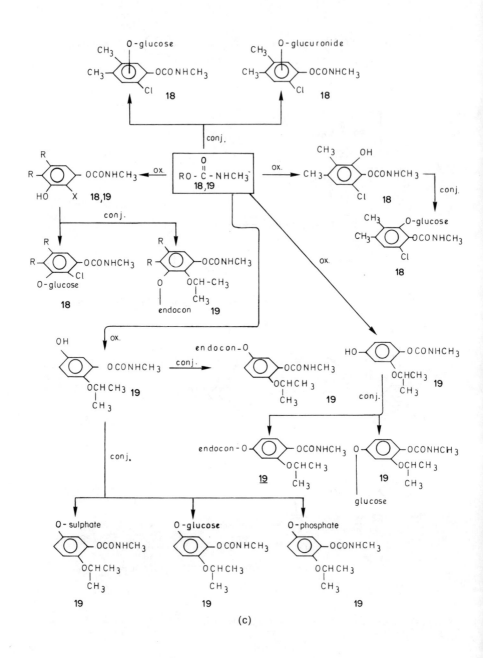

(c)

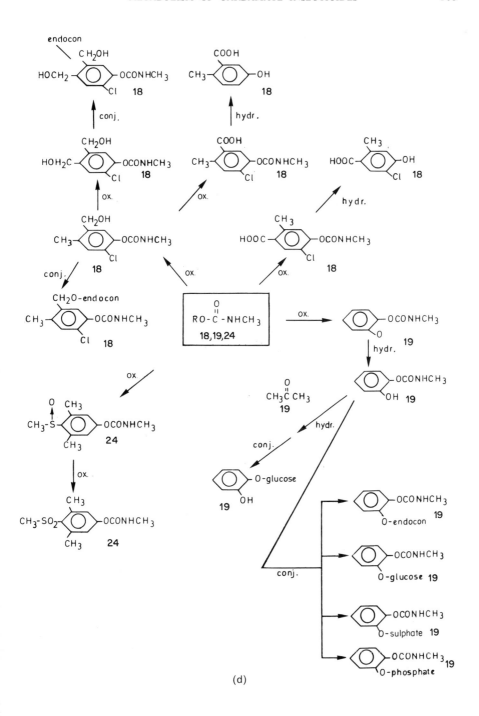

(d)

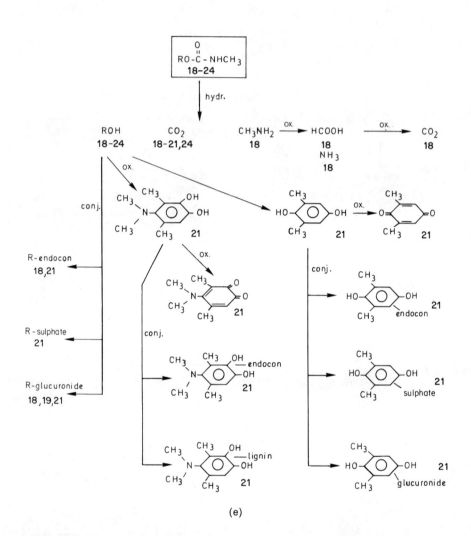

(e)

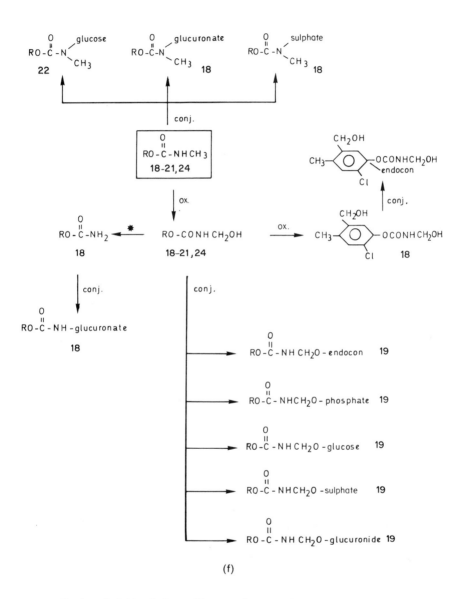

(f)

✱ N—demethylation is in reality a redox process

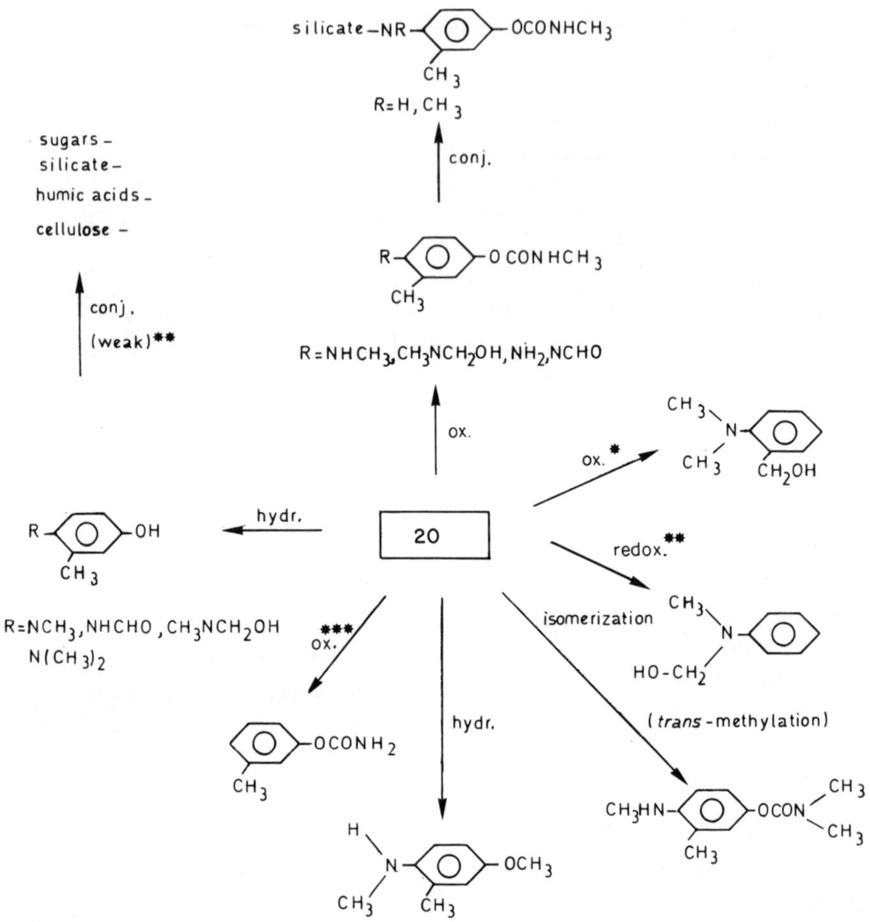

silicate—NR—⟨O⟩—OCONHCH$_3$
CH$_3$
R=H, CH$_3$

sugars —
silicate —
humic acids —
cellulose —

conj.
(weak)**

conj.

R—⟨O⟩—O CONHCH$_3$
CH$_3$

R=NHCH$_3$, CH$_3$NCH$_2$OH, NH$_2$, NCHO

ox.

ox.*

CH$_3$—N—⟨O⟩
CH$_3$ CH$_2$OH

R—⟨O⟩—OH
CH$_3$

hydr.

20

redox.**

CH$_3$—N—⟨O⟩
HO-CH$_2$

isomerization

R=NCH$_3$, NHCHO, CH$_3$NCH$_2$OH
N(CH$_3$)$_2$

ox.***

hydr.

(trans-methylation)

⟨O⟩—OCONH$_2$
CH$_3$

H—N—⟨O⟩—OCH$_3$
CH$_3$ CH$_3$

CH$_3$HN—⟨O⟩—OCON—CH$_3$
CH$_3$ CH$_3$

Matacil (20) Ferroascorbic Metabolic studies (112, 113)

* tollyl methyl oxidation
** mechanism unkown
*** deamination following demethylation (ox.)

(g)

insects with tsumacide (Kuhr and Dorough, 1976). Some unidentified conjugates have also been found in mammals with meobal (Dorough, 1970; Miyamoto, 1970), landrin (Wilkinson, 1976), bux (Knaak, 1971) and croneton (Nye *et al.,* 1976; Marshall and Dorough, 1977); in plants with landrin (Kuhr, 1976), etrofolan (Ogawa *et al.,* 1977) and UC 10854 (Kuhr, 1976); and in insects with landrin (Wilkinson, 1976). All of these appear to have quite complex polysaccharide structures.

Phenyl-substituted *N*-methylcarbamates (Figures 3a–g)

Hydrolysis

During hydrolysis, any carbamate insecticide yields an alcohol (or phenol) and carbamic acid. The carbamic acid is subsequently decomposed giving an amine and CO_2. The first reaction is a typical hydrolysis and does not have any redox character. Similarly, the formation of the carbon dioxide could be regarded as a simple hydrolysis.

Carbon dioxide is produced by mammals from practically all of these insecticides (O'Brien, 1967; Schlagbauer and Schlagbauer, 1972; Kuhr and Dorough, 1976), by insects in the case of banol, baygon, matacil and mesurol (Casida, 1963; Schlagbauer and Schlagbauer, 1972; Matsumura, 1975; Kuhr and Dorough, 1976), but in the case of plants by baygon only (Kuhr and Dorough, 1976). The corresponding phenol derivatives were found in mammals with banol, baygon, matacil, zectran and mesurol (Casida, 1963; Knaak, 1971; Schlagbauer and Schlagbauer, 1972; Matsumura, 1975; Khan *et al.,* 1976; Kuhr and Dorough, 1976; Abd-Elraof *et al.,* 1981), in mammalian tissues *in vitro* with baygon and UC 34096 (Schlagbauer and Schlagbauer, 1972; Kuhr and Dorough, 1976), in insects with banol, formetanate and mesurol (Schlagbauer and Schlagbauer, 1972; Knowles and Chang, 1979), in plants with zectran and formetanate (O'Brien, 1967; Knowles and Sen Gupta, 1970; Knaak, 1971; Schlagbauer and Schlagbauer, 1972; Meikle, 1973; Kuhr and Dorough, 1976), and in microorganisms with baygon and zectran (Bend *et al.,* 1970; Benezet and Matsumura, 1974; Gupta *et al.,* 1975; Matsumura, 1975). The more substituted phenolic products derived from banol were found in mammals (Schlagbauer and Schlagbauer, 1972); however, for the same insecticide, the dimethylamine derivatives have been found only in insects (Schlagbauer and Schlagbauer, 1972). The phenol derivative of matacil has not yet been found in mammalian tissue. The hydrolysis products of matacil and baygon were only found in mammals in *in vitro* studies (Strother, 1972; Thurlow, 1979). The formamido derivative of formetanate was found in mammals, plants, and insects (Knowles and Sen Gupta, 1970; Kuhr and Dorough, 1976; Knowles and Chang, 1979) and the phenol derivatives with the formamido or amino moieties of formetanate were found in plants and insects (Knowles and Sen Gupta, 1970; Knowles and Chang, 1979).

N-*Methyl oxidation*

The hydroxymethyl derivative resulting from *N*-methyl oxidation was found for baygon, zectran and mesurol in mammals (Knaak, 1971; Schlagbauer and Schlagbauer, 1972; Kuhr and Dorough, 1976), for banol, baygon, matacil and zectran in mammals *in vitro* (Oonnithan and Casida, 1966, 1968; Strother, 1972; Matsumura, 1975; Kuhr and Dorough, 1976; Thurlow, 1979) and for banol, baygon, matacil and zectran in insects (Fukuto and Metcalf, 1969; Shrivastava *et al.*, 1969; Schlagbauer and Schlagbauer, 1972; Meikle, 1973; Ariaratnam and Georghiou, 1975; Kuhr and Dorough, 1976). The methylformylamino derivative was found for zectran in mammals (Kuhr and Dorough, 1976), in mammals *in vitro* (Schlagbauer and Schlagbauer, 1972; Meikle, 1973), in insects (Schlagbauer and Schlagbauer, 1972; Kuhr and Dorough, 1976), in microorganisms (Benezet and Matsumura, 1974; Matsumura, 1975), and together with matacil in plants (Abdel-Wahab *et al.*, 1966; O'Brien, 1967; Schlagbauer and Schlagbauer, 1972; Kuhr and Dorough, 1976). The methylformylamino derivative of matacil has also been found in insects *in vitro* (Kuhr, 1968, 1970; Kuhr and Dorough, 1976).

The formamido derivative found only for zectran in microorganisms (Benezet and Matsumura, 1974; Matsumura, 1975) has been detected for matacil in plants (Abdel-Wahab *et al.*, 1966; O'Brien, 1967; Kuhr and Dorough, 1976), in insects (Schlagbauer and Schlagbauer, 1972; Kuhr and Dorough, 1976) as well as in insects *in vitro* (Kuhr, 1968; Meikle, 1973; Kuhr and Dorough, 1976).

The methylamino derivative was detected for zectran in mammals *in vivo* (Kuhr and Dorough, 1976; Thurlow, 1979), in insects *in vivo* (Kuhr and Dorough, 1976), in microorganisms (Casida, 1963; Benezet and Matusumura, 1974; Garg and Sethi, 1979), together with matacil in mammals *in vitro* (Oonnithan and Casida, 1966, 1968; Schlagbauer and Schlagbauer, 1972; Strother, 1972; Meikle, 1973; Kuhr and Dorough, 1976; Esaac and Matsumura, 1979; Thurlow, 1979), in insects *in vitro* (Tsukamoto and Casida, 1967; Kuhr, 1968, 1970; Meikle, 1973; Kuhr and Dorough, 1976) and with native formetanate in plants (Abdel-Wahab *et al.*, 1966; O'Brien, 1967; Knowles and Sen Gupta, 1970; Schlagbauer and Schlagbauer, 1972; Kuhr, 1976; Kuhr and Dorough, 1976). Fish readily produce the methylamino derivatives of matacil (Sundaram and Szeto, 1979).

The amino derivative of zectran was found in microorganisms (Benezet and Matsumura, 1974; Matsumura, 1975), in mammals (Knaak, 1971; Kuhr and Dorough, 1976), in insects only with the parent compounds of formetanate and baygon (O'Brien, 1967; Schlagbauer and Schlagbauer, 1972; Kuhr and Dorough, 1976; Knowles and Chang, 1979), with matacil in mammals *in vitro'* (Tsukamoto and Casida, 1967; Kuhr, 1968, 1970; Meikle, 1973; Kuhr and Dorough, 1976) and in plants (Abdel-Wahab *et al.*, 1966; O'Brien, 1967; Schlagbauer and Schlagbauer, 1972; Kuhr, 1976; Kuhr and Dorough, 1976). Fish yield the amino derivative of matacil (Sundaram and Szeto, 1979) which usually disappears after several days. The amino derivatives of formetanate resulting from the deformylation were found in both plants (Knowles and Sen Gupta, 1970; Kuhr and Dorough, 1976) and in insects (Knowles and Chang, 1979).

Aryl oxidation

Banol has been shown to undergo aryl oxidation in plants (Schlagbauer and Schlagbauer, 1972) only and an unspecified hydroxylated derivative of matacil is reported to occur in fish (Sundaram and Szeto, 1979). The pyrocatechol and its subsequent oxidized form derived from zectran were found in plants (O'Brien, 1967; Knaak, 1971; Schlagbauer and Schlagbauer, 1972; Khan *et al.*, 1976; Kuhr and Dorough, 1976). The hydroquinone derivative of zectran was identified in plants (O'Brien, 1967; Knaak, 1971; Schlagbauer and Schlagbauer, 1972; Kuhr and Dorough, 1976) and in mammals (Knaak, 1971; Meikle, 1973; Khan *et al.*, 1976). Its highly oxidized forms have been detected in plants (O'Brien, 1967; Knaak, 1971) and in mammals (Khan *et al.*, 1976).

3-Hydroxy baygon is present in plants, insects and mammals *in vitro* (Schlagbauer and Schlagbauer, 1972). However, its isomer, 4-hydroxy baygon, was found in plants (Schlagbauer and Schlagbauer, 1972; Matsumura, 1975), in insects and mammals *in vitro* (Schlagbauer and Schlagbauer, 1972) only when using specific catalysts. The other isomeric form, 5-hydroxy baygon, was identified in plants (Knaak, 1971; Kuhr, 1976), in insects (Shrivastava *et al.*, 1969; Ariaratnam and Georghiou, 1975; Matsumura, 1975; Kuhr and Dorough, 1976; Wilkinson, 1976), in mammals (Kuhr and Dorough, 1976) and in mammals *in vitro* (Matsumura, 1975). Thus the aromatic hydroxylation of baygon appears to lack specificity.

Aryl alkyl oxidation

Only insects oxidize the isopropenyl group of baygon (Fukuto and Metcalf, 1969; Shrivastava *et al.*, 1969; Schlagbauer and Schlagbauer, 1972). Mesurol was oxidized primarily to a sulphoxide derivative (Matsumura, 1975) by mammals *in vitro* (Schlagbauer and Schlagbauer, 1972; Kuhr and Dorough, 1976; Menn, 1978), by insects *in vitro* (Kuhr and Dorough, 1976; Menn, 1978) and by plants (O'Brien, 1967; Schlagbauer and Schlagbauer, 1972; Kuhr and Dorough, 1976) and subsequently to a sulphone derivative (Matsumura, 1975; Kuhr, 1976) by mammals *in vivo* (Kuhr and Dorough, 1976) and *in vitro* (Schlagbauer and Schlagbauer, 1972), by plants (Schlagbauer and Schlagbauer, 1972; Kuhr and Dorough, 1976) and by insects (Kuhr and Dorough, 1976).

In vitro studies (mammals) revealed the presence of 4-hydroxymethyl and 3,4-hydroxymethyl derivatives as well as the 3-carboxy and 4-carboxy derivatives of banol (Strother, 1972; Matsumura, 1975). The first two derivatives mentioned above were also present in plants (Schlagbauer and Schlagbauer, 1972). Insects can also produce 3,4-dihydroxymethyl-N-hydroxymethylcarbamate derivatives (Schlagbauer and Schlagbauer, 1972) even in the presence of electron-donating substituents in the 2 and 5 positions.

Reactions other than conjugation

Matacil affords a spontaneous deamination (Sundaram and Szeto, 1979) producing an aromatic derivative, occurring in fish, of uncertain structure and physiological effect. The oxidation of baygon to a 2-hydroxy derivative takes place in insects (Fukuto and Metcalf, 1969; Shrivastava *et al.,* 1969; Schlagbauer and Schlagbauer, 1972; Ariaratnam and Georghiou, 1975; Kuhr and Dorough, 1976), mammals (Schlagbauer and Schlagbauer, 1972; Kuhr and Dorough, 1976; Abd-Elraof *et al.,* 1981), mammals *in vitro* (Schlagbauer and Schlagbauer, 1972; Matsumura, 1975), and to acetone in insects *in vivo* (Fukuto and Metcalf, 1969; Shrivastava *et al.,* 1969; Schlagbauer and Schlagbauer, 1972; Ariaratnam and Georghiou, 1975; Kuhr and Dorough, 1976; Wilkinson, 1976) and *in vitro* (Schlagbauer and Schlagbauer, 1972), in mammals *in vitro* and *in vivo* (Wilkinson, 1976) and in plants (Kuhr and Dorough, 1976). In spite of the usually accepted stable character of the amide bond, the acetamide group of formetanate was oxidized in mammals (Kuhr and Dorough, 1976).

Conjugation

Glucose conjugates were formed in plants from banol (Friedman and Lemin, 1967; Schlagbauer and Schlagbauer, 1972), baygon (Casida, 1963; Knaak, 1971; Kuhr and Dorough, 1976; Wilkinson, 1976), formetanate (Kuhr, 1976) and in insects with baygon (Kuhr and Dorough, 1976; Wilkinson, 1976). Glucuronide conjugates were more abundant in mammals for banol (Schlagbauer and Schlagbauer, 1972; Kuhr and Dorough, 1976; Wilkinson, 1976), baygon (Matsumura, 1975), zectran (Khan *et al.,* 1976; Kuhr and Dorough, 1976), formetanate (Kuhr and Dorough, 1976), in mammals *in vitro* for banol (Kuhr and Dorough, 1976) and in microorganisms for zectran (Khan *et al.,* 1976). Sulphate conjugates have been detected in mammals with banol (Schlagbauer and Schlagbauer, 1972), zectran (Khan *et al.,* 1976; Kuhr and Dorough, 1976), formetanate (Kuhr and Dorough, 1976), and in insects with baygon (Wilkinson, 1976). Three phosphate conjugates of baygon were present only in insects (Wilkinson, 1976). Plants incorporated the pyrocatechol derivative of zectran in polymeric lignin molecules (Schlagbauer and Schlagbauer, 1972; Khan *et al.,* 1976). Unknown conjugates of banol were present in plants (Schlagbauer and Schlagbauer, 1972) and there was an unproven hydroxylated derivative of a matacil conjugate present in fish (Sundaram and Szeto, 1979). The N-hydroxyl methylcarbamate conjugate of baygon was detected in plants (Shrivastava *et al.*, 1969; Schlagbauer and Schlagbauer, 1972) and in insects (Schlagbauer and Schlagbauer, 1972). Conjugation of the 4- and 5-hydroxyaryl derivatives of baygon to plant polysaccharides has been confirmed in plants (Schlagbauer and Schlagbauer, 1972; Gupta *et al.*, 1975). Conjugates of the 3- and 5-hydroxyaryl derivatives of baygon were identified in insects (Shrivastava *et al.*, 1969; Schlagbauer and Schlagbauer, 1972) and the 3-hydroxyaryl derivative was also reported for mammals in *in vitro* studies (Schlagbauer and Schlagbauer, 1972). The phenol conjugates derived from baygon have been tentatively identified in insects (Fukuto and Metcalf, 1969; Matsumura, 1975) and in mammals (Matsumura, 1975). The 2-hydroxy conjugate from baygon was found in insects (Shrivastava *et al.*, 1969; Schlagbauer and Schlagbauer, 1972) and in plants (Schlagbauer and Schlagbauer, 1972). Plants produced the N-methylamino-phenol conjugate (Meikle, 1973), the phenol conjugate (Knaak, 1971; Schlagbauer and Schlagbauer, 1972; Meikle, 1973), the hydroquinone conjugate (Kuhr and Dorough, 1976) and the pyrocatechol conjugate (Schlagbauer and Schlagbauer, 1972; Khan *et al.*, 1976) derived from zectran and its formulation. The mammals also produce phenolic (O'Brien, 1967; Knaak, 1971) and hydroquinone conjugates (Knaak, 1971; Meikle, 1973), both easily detectable by the usual analytical techniques. Conjugates of the derivatives of formetanate, particularly those having a formamido group, were found in plants (Knowles and Sen Gupta, 1970); moreover, the formamido phenol in mammals (Kuhr and Dorough, 1976) and aminophenol in plants (Knowles and Sen Gupta, 1970) were also found.

Because of its important use in Canada, particularly against the spruce budworm, the metabolism of matacil has been extensively studied (Balba and Saha, 1974; Jankowski, 1978, 1979, 1980, 1981; Jankowski and Paré, 1980; Jankowski *et al.*, 1980; Cool, 1982; Cool and Jankowski, 1982). Four major derivatives of matacil, all oxidized on the $N(CH_3)_2$ group, are N-formylamino, N-methyl-formylamino, N-monodemethyl and N,N-didemethyl matacil. In environmental samples, all of these products together with the corresponding phenols have been detected and quantified by GC, GC-MS, TLC and LC methods (Jankowski, 1978, 1979, 1980, 1981; Jankowski and Paré, 1980; Jankowski *et al.*, 1980; Cool, 1982). The *in vitro* simulation of the oxidation of matacil via the vitamin $C/Fe^{2+}/EDTA$ system (ferroascorbic oxidation) has been carried out with radioactive matacil and cold vitamin C (Balba and Saha, 1974) as well as with cold matacil and radioactive vitamin C (Cool and Jankowski, 1982)—in both cases labelled with ^{14}C. As a result, the presence of new derivatives of matacil (isomatacil and 3-tolylcarbamate) as well as of the matacil phenol ((4-N-methylamino) 3-methylanisole, 2-(N-methyl N-hydroxymethylene) toluene and 2-hydroxymethyl-N,N-dimethylaniline) have been reported (Cool, 1982; Cool and Jankowski, 1982). The examination of the vitamin C fraction reveals, however, the fact that matacil also undergoes reduction, apparently the only case where the metabolism of matacil proceeds via a reductive route. Polyfunctional molecules such as matacil and particularly vitamin C can show simultaneously a reduction and an oxidation within their polyfunctional structure (Figure 3g).

The aminophenols resulting from the hydrolysis of matacil or its derivatives do not readily disappear from the environment (e.g. via conjugation systems) and they are relatively persistent (Jankowski, 1980, 1981; Jankowski et al., 1980). This information, together with the mutagenic properties of aminophenols, signals a need for more studies in this area. The aminophenols originating from matacil and its derivatives are in this respect more difficult to assimilate by the environment than phenols originating from the analogous hydrolysis of organophosphates (e.g. for fenitrothion phenol or nitrophenols which are, of course, more acidic).

N-Methylcarbamates with condensed ring system (Figures 4a–f)

Hydrolysis

The phenolic derivatives of the three insecticides carbaryl (Bend et al., 1970; Dorough, 1970; Ryan, 1971; Schlagbauer and Schlagbauer, 1972; Kuhr and Dorough, 1976; Thurlow, 1979), carbofuran (Dorough, 1970; Knaak, 1971; Schlagbauer and Schlagbauer, 1972; Kuhr and Dorough, 1976) and mobam (Schlagbauer and Schlagbauer, 1972; Matsumura, 1975; Kuhr and Dorough, 1976) have been reported for mammals as well as for plants (Fukuto and Metcalf, 1969; Knaak, 1971; Gupta et al., 1975; Miyamoto, 1975; Archer et al., 1977; Garg and Sethi, 1979; Hathway, 1979) and for insects (Casida, 1963; O'Brien, 1967; Kuhr, 1970; Schlagbauer and Schlagbauer, 1972; Guirguis and Brindley, 1975; Matsumura, 1975; Kuhr and Dorough, 1976; Chio and Metcalf, 1979; Thurlow, 1979; Ahmad et al., 1980). The same compounds from carbaryl and mobam, were found in microorganisms (Schlagbauer and Schlagbauer, 1972; Bezbariah, 1976; Khan et al., 1976; Kuhr and Dorough, 1976) and, from carbaryl and carbofuran only, in mammals (Schlagbauer and Schlagbauer, 1972; Lin et al., 1975; Matsumura, 1975; Hinderer and Menzer, 1976a; Wilkinson, 1976; Hathway, 1979; Pekas, 1979, 1980) and in fish *in vitro* (Chin et al., 1979; Gill, 1980). *In vivo* and *in vitro* studies in birds revealed only a phenolic carbaryl compound (Schlagbauer and Schlagbauer, 1972; Hinderer and Menzer, 1976b; Kuhr and Dorough, 1976).

Mammals and microorganisms produced CO_2^* from mobam (Kuhr and Dorough, 1976), carbofuran (Getzin, 1973; Kuhr and Dorough, 1976; Williams et al., 1976; Dorough, 1979) and carbaryl (Casida, 1963; O'Brien, 1967; Bend et

---

* ^{14}C-labelling of carboxylic group.

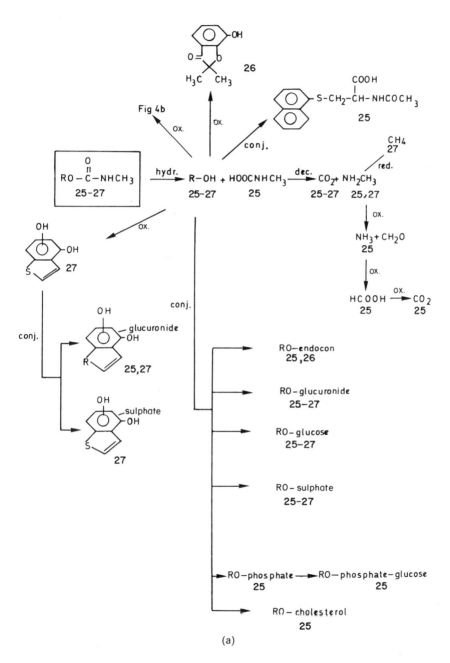

Figure 4 Metabolism of carbamates 25–27: 25, carbaryl; 26, carbofuran; 27, mobam

(b)

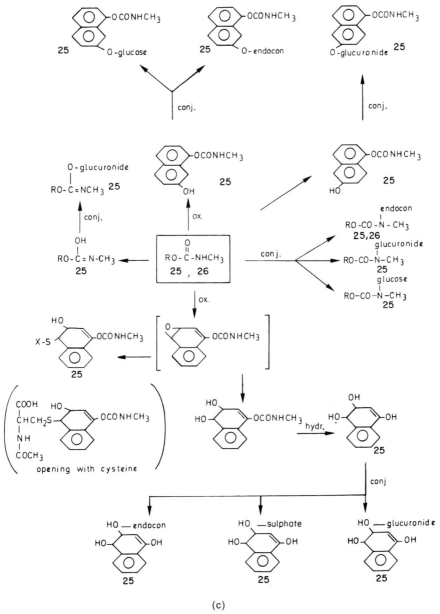

(c)

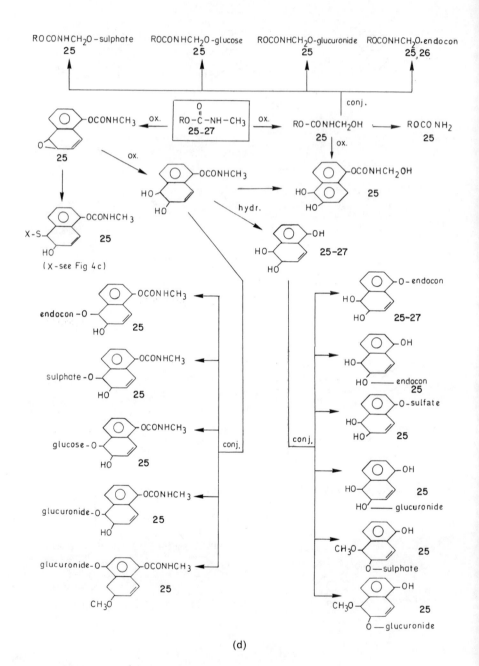

(d)

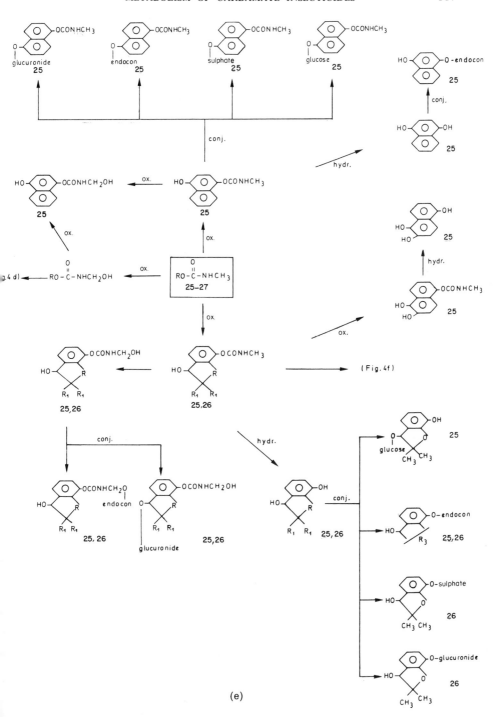

(e)

(f)

al., 1970; Kuhr and Dorough, 1976; Nye and Dorough, 1976; Rodriguez and Dorough, 1977; Benson and Dorough, 1979). Carbon dioxide was also produced by plants from carbofuran (Kuhr and Dorough, 1976) but its production has not been proved from carbaryl (Locke *et al.,* 1976). Carbaryl degradation in insects, birds and mammals *in vitro* leads directly to CO_2 (Casida, 1963; Schlagbauer and Schlagbauer, 1972; Lin *et al.,* 1975; Kuhr and Dorough, 1976; Pekas, 1980).

All hydrolysis products were found in mammals (Bend *et al.,* 1970; Schlagbauer and Schlagbauer, 1972) except for methane (reduction), which was produced exclusively by microorganisms (Schlagbauer and Schlagbauer, 1972). Methylamine was also produced by insects (Schlagbauer and Schlagbauer, 1972; Kuhr and Dorough, 1976), however, its production has not been proved in mammalian *in vitro* studies (Pekas, 1980). Formaldehyde was also reported for mammals (in the same *in vitro* studies).

The presence of the hydrolysis product of mobam (Figure 4b) was proved exclusively in mammals (Matsumura, 1975). The 1,2-naphthyldiol derivative of carbaryl was found in mammals (Casida, 1963) and microorganisms (Kazano *et al.,* 1972; Kuhr and Dorough, 1976). Mammals, plants, insects and fish *in vitro* produce the phenols derived from 3-hydroxycarbofuran. The first three organisms (and mammals *in vitro*) produce the 3-ketocarbofuran phenols (Fukuto and Metcalf, 1969; Dorough, 1970; Knaak, 1971; Schlagbauer and Schlagbauer, 1972; Kapoor and Kalra, 1975; Miyamoto, 1975; Kuhr and Dorough, 1976; Archer *et al.,* 1977; Garg and Sethi, 1979; Fuhremann and Lichtenstein, 1980; Gill, 1980). The other hydrolytic products of carbaryl were reported for insects and mammals *in vivo* and *in vitro*. Particularly, 3,4-dihydroxynaphthol (Schlagbauer and Schlagbauer, 1972) has been reported for mammals *in vivo* and *in vitro* in additon to 1,5-naphthyldiol (Bend *et al.,* 1970; Pekas, 1979, 1980) and 5,6-dihydrodihydroxynaphthol (Bend *et al.,* 1970; Dorough, 1970; Ryan, 1971; Schlagbauer and Schlagbauer, 1972; Matsumura, 1975; Kuhr and Dorough, 1976; Pekas, 1979, 1980; Chen and Dorough, 1980). In birds, the presence of 5,6-dihydrodihydroxynaphthol (Schlagbauer and Schlagbauer, 1972) has been confirmed. However, for mammals *in vitro*, only the 1,5,6-trihydroxynaphthyl derivative (Pekas, 1979, 1980) has been reported. The two derivative structures 1,5- and 1,4-naphthyldiol have not been rigorously characterized in mammalian *in vitro* studies (Lin *et al.,* 1975).

Aryl oxidation

The hydroxylation of the phenolic derivative of mobam occurs only in mammals (Schlagbauer and Schlagbauer, 1972) (Figure 4a). 3-Hydroxycarbofuran and 3-ketocarbofuran were found in mammals (Dorough, 1970; Knaak, 1971; Schlagbauer and Schlagbauer, 1972; Kuhr and Dorough, 1976; Chio and Metcalf, 1979), in fish *in vitro* (Gill, 1980) and in mammals *in vitro* (Schlagbauer and

Schlagbauer, 1972). The production of 3-hydroxycarbofuran and 3-keto-carbofuran by birds remains unproven in spite of the model study with plants and mammals (Knaak, 1971). 3-Hydroxy-N-hydroxymethylcarbofuran was produced by mammals, insects and birds (Dorough, 1970; Schlagbauer and Schlagbauer, 1972; Kuhr and Dorough, 1976). Plants and mammals produce other derivatives of carbofuran such as the 3-ketocarbofuran phenol (Dorough, 1970; Garg and Sethi, 1979). On the other hand, plants and insects afford 3-keto-2-hydroxycarbofuran (Schlagbauer and Schlagbauer, 1972; Kuhr and Dorough, 1976). Mammals *in vivo* and *in vitro*, plants, insects, microorganisms, fish *in vitro* and birds *in vivo* and *in vitro* oxidized carbaryl to 4-hydroxycarbaryl (O'Brien, 1967; Kuhr, 1968, 1970, 1976; Dorough, 1970; Ryan, 1971; Schlagbauer and Schlagbauer, 1972; Guirguis and Brindley, 1975; Lin *et al.*, 1975; Matsumura, 1975; Bezbariah, 1976; Hinderer and Menzer, 1976a,b; Khan *et al.*, 1976; Kuhr and Dorough, 1976; Wilkinson, 1976; Chin *et al.*, 1979; Hathway, 1979; Ahmad *et al.*, 1980). Mammals *in vivo* and *in vitro*, insects, plants and birds *in vivo* and *in vitro* oxidized carbaryl to 5-hydroxycarbaryl (Kuhr, 1968, 1970; Dorough, 1970; Ryan, 1971; Schlagbauer and Schlagbauer, 1972; Getzin, 1973; Guirguis and Brindley, 1975; Lin *et al.*, 1975; Matsumura, 1975; Bezbariah, 1976; Hinderer and Menzer, 1976a,b; Khan *et al.*, 1976; Kuhr, 1976; Kuhr and Dorough, 1976; Nye and Dorough, 1976; Wilkinson, 1976; Williams *et al.*, 1976; Rodriguez and Dorough, 1977; Garg and Sethi, 1979; Pekas, 1979, 1980; Ahmad *et al.*, 1980; Gill, 1980; Abd-Elraof *et al.*, 1981). Mammals *in vivo* and *in vitro*, insects, plants, microorganisms, birds and fish *in vivo* and *in vitro* also oxidized carbaryl compound to 5,6-dihydrodihydroxycarbaryl (Dorough, 1970; Kuhr, 1970; Ryan, 1971; Schlagbauer and Schlagbauer, 1972; Ariaratnam and Georghiou, 1975; Guirguis and Brindley, 1975; Lin *et al.*, 1975; Matsumura, 1975; Statham *et al.*, 1975; Hinderer and Menzer, 1976a; Kuhr and Dorough, 1976; Wilkinson, 1976; Chin *et al.*, 1979; Ahmad *et al.*, 1980; Chen and Dorough, 1980; Pekas, 1980). The hydroxylation of 1-naphthol to 1,2-naphthyldiol occurs in mammals (Casida, 1963), insects (Schlagbauer and Schlagbauer, 1972) and microorganisms (Kazano *et al.*, 1972; Bezbariah, 1976; Kuhr and Dorough, 1976). Mammals *in vivo* and *in vitro* and insects produce 3,4-dihydroperoxycarbaryl (Schlagbauer and Schlagbauer, 1972; Fukami, 1980). 3,4-Dihydroperoxycarbaryl was alleged as a possible metabolite structure in an experiment with mammals (Ryan, 1971). The same mammals have produced 6-hydroxycarbaryl (Khan *et al.*, 1976) in an impressive yield. The yield of 7-hydroxycarbaryl (Sundaram and Szeto, 1979) in plants is low; however, mammals *in vitro* synthesized 5,6-hydroxycarbaryl (Pekas, 1980), 4-hydroxymethylcarbaryl (Matsumura, 1975). A highly oxidized 5,6,7,8-tetra-hydrotetrahydroxycarbaryl derivative (Kuhr and Dorough, 1976) is reported in studies with insects. Plants *in vitro* give mostly 1,2-dihydroperoxynaphthol (Locke *et al.*, 1976). Oxidation by microorganisms yields 2-hydroxycarbaryl (Kuhr and Dorough, 1976), following all other steps shown in Figure 4b

(Kazano et al., 1972; Bezbariah, 1976; Kuhr and Dorough, 1976). 5-Hydroxy-N-hydroxymethylcarbaryl and 5,6-dihydroxy-N-hydroxymethylcarbaryl were reported, but unproven, as mammalian metabolites (Bend et al., 1970).

N-Methyl oxidation

The N-methyl oxidation of carbofuran occurs in mammals (Dorough, 1970; Kuhr and Dorough, 1976), mammals in vitro (Schlagbauer and Schlagbauer, 1972), insects (Schlagbauer and Schlagbauer, 1972; Kuhr and Dorough, 1976), birds (Kuhr and Dorough, 1976), plants (Schlagbauer and Schlagbauer, 1972), and fish in vitro (Gill, 1980). The N-methyl oxidation of carbaryl occurs in mammals (Dorough, 1970; Knaak, 1971; Ryan, 1971; Schlagbauer and Schlagbauer, 1972), mammals in vitro (O'Brien, 1967; Schlagbauer and Schlagbauer, 1972; Matsumura, 1975; Hinderer and Menzer, 1976a; Chen and Dorough, 1980), insects (O'Brien, 1967; Kuhr, 1970; Schlagbauer and Schlagbauer, 1972; Ariaratnam and Georghiou, 1975; Guirguis and Brindley, 1975; Kuhr and Dorough, 1976; Wilkinson, 1976; Ahmad et al., 1980), plants (Wilkinson, 1976), plants in vitro (Kuhr and Dorough, 1976; Locke et al., 1976), microorganisms (Matsumura, 1975; Bezbariah, 1976; Kuhr and Dorough, 1976), birds (Kuhr and Dorough, 1976), birds in vitro (Hinderer and Menzer 1976b) and fish in vitro (Gill, 1980).

Mammals oxidized carbaryl to two compounds, namely 5-hydroxy-N-hydroxymethylcarbaryl and 5,6-dihydrodihydroxy-N-hydroxymethylcarbaryl (Bend et al., 1970), but they oxidized carbofuran to a 3-hydroxy-N-hydroxymethyl carbofuran (Dorough, 1970; Schlagbauer and Schlagbauer, 1972; Kuhr and Dorough, 1976) and mobam into a sulphoxide N-hydroxymethyl derivative instead (Kuhr and Dorough, 1976). The first structure has not been rigorously proven by chemical methods. Mammalian in vitro studies show the formation of 5-hydroxy-N-hydroxymethylcarbaryl (Matsumura, 1975) and 4-hydroxy-N-hydroxymethylcarbaryl (Matsumura, 1975). Various insect species afford 3-hydroxy-N-hydroxymethylcarbofuran (Schlagbauer and Schlagbauer, 1972; Kuhr and Dorough, 1976); the same product is observed with birds (Knowles and Chang, 1979). Microorganisms produce 1-naphthylcarbamate (Matsumura, 1975; Bezbariah, 1976) as a major oxidation product.

Other oxidation products

Microorganisms produce a great number of carbaryl metabolites such as the hydroquinone (Kuhr and Dorough, 1976) and others shown in Figure 4b which are not simply produced by oxidation (Kazano et al., 1972; Bezbariah, 1976; Kuhr and Dorough, 1976) but rather by oxidation and some other parallel reactions. The isomerization of carbaryl (Figure 4c) occurs in mammals in vivo and in vitro (Schlagbauer and Schlagbauer, 1972) and in plants (Schlagbauer and

Schlagbauer, 1972). Mammals convert carbaryl into 1-(S-mercapturic acid) naphthyl (Casida, 1963) and 1-naphthyl-N-acetyl-N-methylcarbamate (O'Brien, 1967). The two glutathione conjugates and their mercapturic acid derivatives of carbaryl shown in Figures 4c and 4d were found in mammals *in vivo* and *in vitro* (Bend *et al.*, 1971; Ryan, 1971; Fukami, 1980) but the mercapturic acids were not confirmed by chemical methods (Fukami, 1980).

Conjugation

Mammals conjugate two carbamates, mobam (Schlagbauer and Schlagbauer, 1972; Kuhr and Dorough, 1976; Wilkinson, 1976) and carbofuran (Knaak, 1971; Marshall and Dorough, 1977; Dorough, 1979; Fuhremann and Lichtenstein, 1980; Pekas, 1980) with sulphate and glucuronic acid. Glucose conjugation occurs in plants with mobam (Schlagbauer and Schlagbauer, 1972), and in plants and insects with carbofuran (Knaak, 1971; Kapoor and Kalra, 1975; Kuhr, 1976; Kuhr and Dorough, 1976; Wilkinson, 1976; Marshall and Dorough, 1977; Dorough, 1979; Garg and Sethi, 1979). 3-Hydroxycarbofuran was conjugated with angelic acid in carrots (Sonobe *et al.*, 1981), a rare example of the formation of a lipophilic conjugate in plants. Conjugation to form xenobiotic lipid species (Hutson, 1982) is being noted with increasing frequency as methods for isolation and identification of metabolites become more rigorous. There are several unidentified conjugates from mammals, insects and plants derived from 3-ketocarbofuran (Kapoor and Kalra, 1975; Kuhr and Dorough, 1976), 3-hydroxycarbofuran (Hassan *et al.*, 1966; Dorough, 1970; Schlagbauer and Schlagbauer, 1972; Kapoor and Kalra, 1975; Huhtanen and Dorough, 1976; Kuhr and Dorough, 1976; Fuhreman and Lichtenstein, 1980), and 3-ketocarbofuran phenol (Fukuto and Metcalf, 1969; Dorough, 1970; Schlagbauer and Schlagbauer, 1972; Kuhr and Dorough, 1976; Marshall and Dorough, 1977). Conjugates of unknown structure, presumably of 2-hydroxycarbofuran (Schlagbauer and Schlagbauer, 1972) and of 3-hydroxycarbofuran phenols (Knaak, 1971; Marshall and Dorough, 1977; Garg and Sethi, 1979), have been found in plants and insects. Other new conjugates found in mammals and insects were formed from 3-hydroxycarbofuran phenol (Fukuto and Metcalf, 1969; Dorough, 1970; Schlagbauer and Schlagbauer, 1972; Kuhr and Dorough, 1976), N-hydroxymethylcarbofuran (Schlagbauer and Schlagbauer, 1972; Kuhr and Dorough, 1976) and 3-hydroxy-N-hydroxymethylcarbofuran (Dorough, 1970; Schlagbauer and Schlagbauer, 1972; Kuhr and Dorough, 1976). The plants, generally speaking, produce several other conjugates of unknown structure, particularly from carbofuran (Kapoor and Kalra, 1975).

Carbaryl affords 44 known conjugates: glucuronide conjugates were found in mammals (Fukuto and Metcalf, 1969; Bend *et al.*, 1970; Knaak, 1971; Schlagbauer and Schlagbauer, 1972; Matsumura, 1975; Suzuki and Takeda, 1976; Wilkinson, 1976; Chin and Sullivan, 1979; Chin *et al.*, 1979c; Chen and

Dorough, 1980), in mammals *in vitro* (Hassan *et al.,* 1966; Schlagbauer and Schlagbauer, 1972; Lin *et al.,* 1975; Hinderer and Menzer, 1976a; Chin and Sullivan, 1979; Chin *et al.,* 1979a; Pekas, 1979, 1980; Chen and Dorough, 1980), in birds (Kuhr and Dorough, 1976), in fish (Statham *et al.,* 1975) and fish *in vitro* (Chin *et al.,* 1979b). There are several mammalian glucuronide conjugates of unknown structure (O'Brien, 1967; Matsumura, 1975). Sulphate conjugates have been found in mammals (Casida, 1963; Fukuto and Metcalf, 1969; Bend *et al.,* 1970; Knaak, 1971; Schlagbauer and Schlagbauer, 1972; Matsumura, 1975; Kuhr and Dorough, 1976; Wilkinson, 1976; Chin and Sullivan, 1979; Chin *et al.,* 1979; Dorough, 1979; Chen and Dorough, 1980), in mammals *in vitro* (Hassan *et al.,* 1966; Schlagbauer and Schlagbauer, 1972; Lin *et al.,* 1975; Hinderer and Menzer, 1976a; Chin and Sullivan, 1979; Chin *et al.,* 1979a; Pekas, 1979, 1980; Chen and Dorough, 1980) in birds Schlagbauer and Schlagbauer, 1972; Kuhr and Dorough, 1976; Wilkinson, 1976), in fish *in vitro* (Chin *et al.,* 1979b) and in insects (Heenan and Smith, 1975; Matsumura, 1975). Glucose conjugates have been widely detected in plants (Knaak, 1971; Schlagbauer and Schlagbauer, 1972; Matsumura, 1975; Kuhr and Dorough, 1976; Suzuki and Takeda, 1976; Wilkinson, 1976; Rodriguez and Dorough, 1977), plants *in vitro* (Locke *et al.,* 1976) and in insects (Heenan and Smith, 1974; Ariaratnam and Georghiou, 1975; Wilkinson, 1976). The glucose conjugate of carbaryl was not confirmed in a study on plant materials (Khan *et al.,* 1976). Phosphate as well as glucophosphate conjugates have been found in insects (Heenan and Smith, 1974). Finally, a cholesterol conjugate was found in plants *in vivo* and *in vitro* (Locke *et al.,* 1976).

An unknown conjugate of 5,6-dihydrodihydroxy-1-naphthol has been reported for mammals (Bend *et al.,* 1970; Dorough, 1970), plants (Locke *et al.,* 1976) and birds (Schlagbauer and Schlagbauer, 1972). Mammals, insects, birds, and plants *in vivo* and *in vitro* seem to produce a common conjugate from 1-naphthol (Dorough, 1970; Schlagbauer and Schlagbauer, 1972; Guirguis and Brindley, 1975; Matsumura, 1975; Wilkinson, 1976). However, the structure of this conjugate also is unknown. Unknown conjugates of a hydroxycarbaryl (Dorough, 1970; Guirguis and Brindley, 1975) and naphthalene-1,5-diol (Bend *et al.,* 1970; Wilkinson, 1976) have also been detected in mammals and insects. Mammals *in vivo* and *in vitro* produce two conjugates: 4-hydroxy-carbaryl and 5,6-dihydrodihydroxycarbaryl (Dorough, 1970; Lin *et al.,* 1975). They also form complex conjugates from 5-hydroxycarbaryl (Dorough, 1970) and *in vitro*, from 3,4-dihydrodihydroxy-1-naphthol (Schlagbauer and Schlagbauer, 1972) and naphthalene-1,4-diol (Lin *et al.,* 1975) moieties. Insects afford 5-hydroxy-carbaryl (Guirguis and Brindley, 1975), 5,6-dihydrodihydroxycarbaryl (Dorough, 1970; Guirguis and Brindley, 1975), 3,4-dihydrodihydroxy-1-naphthol (Schlagbauer and Schlagbauer, 1972) and the *N*-hydroxymethyl-carbaryl (Guirguis and Brindley, 1975) conjugates. Plants *in vivo* and *in vitro* produce several other unidentified conjugates of 4-hydroxycarbaryl and

5-hydroxycarbaryl (Schlagbauer and Schlagbauer, 1972; Locke *et al.,* 1976; Rodriguez and Dorough, 1977). Not a single 5,6-dihydrodihydroxycarbaryl conjugate has been reported in plant extracts (O'Brien, 1967). Plants *in vitro* produce three other sugar conjugates (for *N*-hydroxymethylcarbaryl, 7-hydroxycarbaryl and 5,6-dihydrodihydroxy-1-naphthol (Locke *et al.,* 1976).

N-Methylcarbamates of oximes (Figures 5a–c)

Hydrolysis

The hydrolysis of all the carbamates in this class occurs in mammals (Knaak, *et al.,* 1966; Knaak, 1971; Schlagbauer and Schlagbauer, 1972; Huhtanen and Dorough, 1976; Kuhr, 1976; Kuhr and Dorough, 1976; Chang and Knowles, 1979; Chin *et al.,* 1980). Insects hydrolyse aldicarb (Schlagbauer and Schlagbauer, 1972), methomyl (Gayen and Knowles, 1981) and oxamyl (Chang and Knowles, 1979). Plants hydrolyse aldicarb (Schlagbauer and Schlagbauer, 1972) and oxamyl (Harvey *et al.,* 1978). Oxamyl is hydrolysed by microorganisms and mammals *in vitro* (Harvey and Han, 1978; Chang and Knowles, 1979); however, aldicarb hydrolysis has been reported only for mammals (Schlagbauer and Schlagbauer, 1972).

Carbon dioxide derived from oxime *N*-methylcarbamates has been reported in plants for oxamyl (Harvey *et al.,* 1978) and aldicarb (Schlagbauer and Schlagbauer, 1972), in mammals for thiofanox (Kuhr and Dorough, 1976 Chin *et al.,* 1980) and aldicarb (Kuhr and Dorough, 1976; Dorough, 1979) and in insects for aldicarb (Kuhr and Dorough, 1976; Chang and Knowles, 1978).

Other specific hydrolysis products for particular insecticides have been found for aldicarb, thiofanox, thiocarboxime and oxamyl in mammals (Dorough, 1970; Knaak, 1971; Matsumura, 1975; Kuhr and Dorough, 1976; Chang and Knowles, 1979; Chin *et al.,* 1980), and for oxamyl *in vitro* (Harvey and Han, 1978), for aldicarb (Belasco and Harvey, 1980) and oxamyl (Belasco and Harvey, 1980), in microorganisms and in plants for aldicarb (Schlagbauer and Schlagbauer, 1972; Andrawes *et al.,* 1973; Matsumura, 1975; Kuhr and Dorough, 1976; Garg and Sethi, 1979), in insects (Schlagbauer and Schlagbauer, 1972; Chang and Knowles, 1978) and in birds (Hicks *et al.,* 1972). These metabolites are, however, specific for these insecticides rather than to the general mechanism of hydrolysis.

N-*Methyl oxidation*

The *N*-methyl oxidation of aldicarb occurs in insects (Schlagbauer and Schlagbauer, 1972) and of its sulphone derivative in birds (Hicks *et al.,* 1972). The

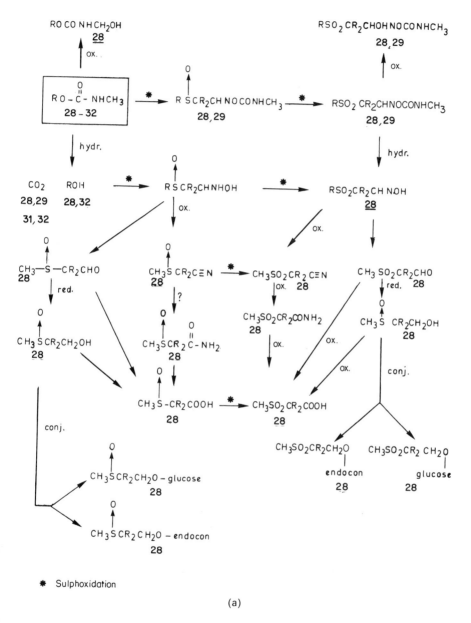

* Sulphoxidation

(a)

Figure 5 Metabolism of carbamates 28–32: 28, aldicarb; 29, thiofanox; 30, methomyl (*syn, anti*); 31, thiocarboxime; 32, oxamyl

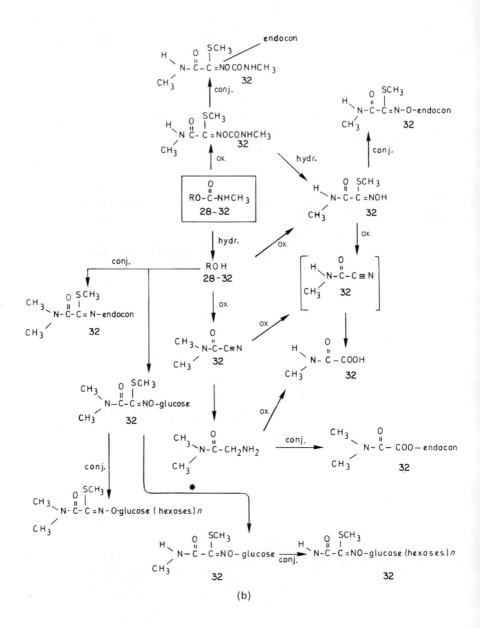

(b)

* oxidative demethylation

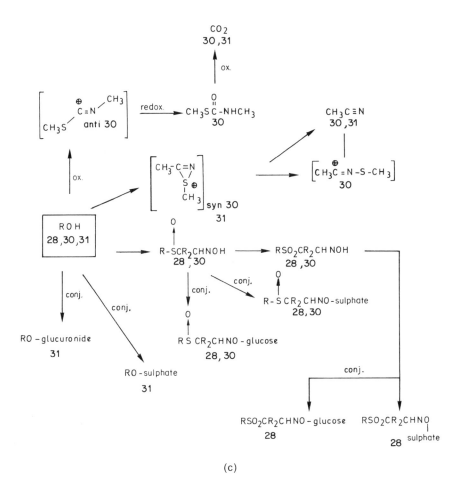

(c)

same oxidation of the sulphone derivative of thiofanox readily occurs in mammals (Kuhr and Dorough, 1976), on the carbamoyl moiety of oxamyl, in mammals *in vivo* and *in vitro* (Harvey and Han, 1978; Chang and Knowles, 1979), in insects (Chang and Knowles, 1979) and in microorganisms (Belasco and Harvey, 1980).

Sulphoxidation

Thiofanox was degraded *via* sulphoxidation in mammals, insects (Chin *et al.,* 1980) and plants (Garg and Sethi, 1979). All the various sulphoxidation products of aldicarb occur in plants (Dorough, 1970; Andrawes *et al.,* 1971, 1973; Knaak, 1971; Schlagbauer and Schlagbauer, 1972; Matsumura, 1975; Kuhr and Dorough, 1976; Garg and Sethi, 1979) and in microorganisms (Matsumura,

1975; Jones, 1976). Birds and mammals produce from aldicarb the oxime sulphoxide (Dorough, 1970; Knaak, 1971; Hicks et al., 1972; Matsumura, 1975; Kuhr and Dorough, 1976), the oxime sulphone (Dorough, 1970; Hicks et al., 1972; Matsumura, 1975; Kuhr and Dorough, 1976), the nitrile sulphone (Dorough, 1970; Hicks et al., 1973; Matsumura, 1975; Kuhr and Dorough, 1976), the aldicarb sulphone (Dorough, 1970; Knaak, 1971; Hicks et al., 1972; Schlagbauer and Schlagbauer, 1972; Matsumura, 1975; Kuhr and Dorough, 1976) and the aldicarb sulphoxide (Dorough, 1970; Knaak, 1971; Hicks et al., 1972; Schlagbauer and Schlagbauer, 1972; Matsumura, 1975; Kuhr and Dorough, 1976).

Insects, however, produce the oxime sulphoxide (Chang and Knowles, 1978), the oxime sulphone (Hicks et al., 1972), and the above sulphone (Matsumura, 1975; Chang and Knowles, 1978) and sulphoxide (Kuhr and Dorough, 1976; Chang and Knowles, 1978). Aldicarb sulphoxide was also produced by mammals in vitro and by plants in vitro (Schlagbauer and Schlagbauer, 1972; Kuhr and Dorough, 1976; Wilkinson, 1976; Krueger, 1977).

Oxidation

Plants and microorganisms oxidize aldicarb to its carboxylic acid sulphone (Schlagbauer and Schlagbauer, 1972; Jones, 1976; Kuhr and Dorough, 1976; Garg and Sethi, 1979), carboxylic acid sulphoxide (Knaak, 1971; Andrawes et al., 1973; Matsumura, 1975; Jones, 1976; Kuhr and Dorough, 1976), nitrile sulphone (Andrawes et al., 1971, 1973; Matsumura, 1975; Jones, 1976) and the nitrile sulphoxide (Andrawes et al., 1971, 1973; Knaak, 1971; Schlagbauer and Schlagbauer, 1972; Matsumura, 1975; Jones, 1976). Mammals only oxidize aldicarb to the latter three products (Dorough, 1970; Schlagbauer and Schlagbauer, 1972; Matsumura, 1975; Kuhr and Dorough, 1976). Birds afford the two nitrile derivatives (Hicks et al., 1972). The production of CO_2 from thiocarboxime (Kuhr and Dorough, 1976; Dorough, 1979) and methomyl (Huhtanen and Dorough, 1976; Hathway, 1979) in mammals, as well as the oxidative reaction during the formation of acetonitrile (Huhtanen and Dorough, 1976), are considered as proof of the degradative nature of oxidation and its role in detoxication.

Reduction

The reductive metabolism of methomyl apparently occurs only in mammals (Huhtanen and Dorough, 1976; Suzuki and Takeda, 1976); however, the two alcohol derivatives of aldicarb were found in plants (Andrawes et al., 1971, 1973; Knaak, 1971; Schlagbauer and Schlagbauer, 1972; Matsumura, 1975), in microorganisms (Jones, 1976) and in birds (Hicks et al., 1972).

Other reactions

The amide and carboxylic acid derivatives of aldicarb, with the exception of the amide sulphone, were found in microorganisms (Jones, 1976) and in plants (Schlagbauer and Schlagbauer, 1972; Andrawes *et al.*, 1973; Kuhr and Dorough, 1976; Garg and Sethi, 1979). Plants also produce the aldehyde derivatives (Matsumura, 1975), but the sulphoxide structure is tentative (Knaak, 1971; Schlagbauer and Schlagbauer, 1972). Another product of uncertain structure, from methomyl, was found in mammals and tentatively identified as acetonitrile (Huhtanen and Dorough, 1976; Kuhr and Dorough, 1976; Hathway, 1979); this is also produced by plants (Kuhr and Dorough, 1976) and insects (Black *et al.*, 1973) in significant quantities. Acetonitrile formation from thiocarboxime has been reported only in mammals (Kuhr and Dorough, 1976). Microorganisms degrade oxamyl to the amide, carboxylic acid and nitrile derivatives (Belasco and Harvey, 1980); insects, and mammals *in vivo* and *in vitro* degrade this insecticide to the carboxylic acid and nitrile derivatives (Harvey and Han, 1978; Chang and Knowles, 1979); in plants, only the nitrile and carboxylic derivatives of the dimethylamino group were found (Harvey *et al.*, 1978).

Conjugation

Plants produce a variety of conjugation products of aldicarb (Knaak, 1971; Schlagbauer and Schlagbauer, 1972; Andrawes *et al.*, 1973; Matsumura, 1975; Kuhr and Dorough, 1976; Dorough, 1979; Garg and Sethi, 1979); mammals and insects apparently produce only sulphate conjugates (Schlagbauer and Schlagbauer, 1972). Sulphate and glucuronide conjugates derived from thiocarboxime (Meikle, 1973) are synthesized by mammals; however, several sugar conjugates (Dorough, 1976; Kuhr and Dorough, 1976; Wilkinson, 1976) of oxamyl, of unidentified structures, were also found in mammals (Harvey and Han, 1978) and some glucose conjugates, in plants (Harvey *et al.*, 1978). Only microorganisms form glucose dimethylamino conjugates (Belasco and Harvey, 1980) for oxamyl.

Other carbamate insecticides (Figures 6a–c)

Hydrolysis

The phenol derivatives of *N*-acetyl-zectran, *N*-(toluenesulphenyl)carbofuran and PSC (Black *et al.*, 1973; Krieger *et al.*, 1976; Kuhr and Dorough, 1976; Yamamoto *et al.*, 1978; Hathway, 1979) were found in mammals; in mammals *in vitro*, in the case of 3-methylphenyl-*N*-propylcarbamate (Umetsu *et al.*, 1979); in insects in the case of *N*-acetyl-zectran (Kuhr and Dorough, 1976), and in plants, with DBSC and MSC (Chiu *et al.*, 1975). CO_2 was excreted by mammals from *N*-acetyl-zectran, RE 11775 and *N*-(toluenesulphenyl)carbofuran (Kuhr and Dorough, 1976) and by insects from PSC (Kuhr and Dorough, 1976).

Mammals were able to hydrolyse *N*-acetyl-zectran (Schlagbauer and Schlag-bauer, 1972; Kuhr and Dorough, 1976), RE 11775 (Kuhr and Dorough, 1976; Nishioka *et al.,* 1981), *N*-(toluenesulphenyl)-carbofuran (Black *et al.,* 1973; Fukuto, 1976; Kuhr and Dorough, 1976) and PSC (Fukuto, 1976; Krieger *et al.,* 1976; Suzuki and Takeda, 1976; Yamamoto *et al.,* 1978). Insects also hydrolysed *N*-acetyl-zectran (Schlagbauer and Schlagbauer, 1972; Fukuto, 1976; Kuhr and Dorough, 1976). *N*-(Toluenesulphenyl)carbofuran and PSC, however, are transformed into carbofuran (Black *et al.,* 1973; Fukuto, 1976; Krieger *et al.,* 1976; Kuhr and Dorough, 1976; Hathway, 1979). This transform-ation in insects is a bioactivation reaction. DBSC and MSC are simply hydrolysed by plants (Chiu *et al.,* 1975).

Oxidation

3-Methylphenyl-*N*-propylcarbamate has been shown to be oxidized by mam-mals *in vitro* (Umetsu *et al.,* 1979), and DBSC and MSC exclusively by plants (Chiu *et al.,* 1975). *N*-Acetyl-zectran is oxidized by insects (Kuhr and Dorough, 1976) and RE 11775 by mammals (Kuhr and Dorough, 1976), but *N*-(toluenesulphenyl)carbofuran produces a sulphoxide derivative (Black *et al.,* 1973; Fukuto, 1976; Kuhr and Dorough, 1976) in mammals together with a 3-hydroxy compound (Black *et al.,* 1973; Kuhr and Dorough, 1976) of uncon-firmed structure (Fukuto, 1976) and a 3-keto derivative (Black *et al.,* 1973; Fukuto, 1976; Kuhr and Dorough, 1976). This insecticide is metabolized by insects to 3-hydroxycarbofuran (Black *et al.,* 1973; Fukuto, 1976; Kuhr and Dorough, 1976), both compounds being further oxidized to 3-ketocarbofuran (Black *et al.,* 1973; Fukuto, 1976) and 3-hydroxy-*N*-hydroxymethylcarbofuran (Black *et al.,* 1973; Fukuto, 1976; Kuhr and Dorough, 1976). PSC was oxidized by mammals into a phosphinyl derivative (Fukuto, 1976; Yamamoto *et al.,* 1978), dimethoxy-*N*-methyl phosphoramidate (Hathway, 1979; Yamamoto *et al.,* 1978), a 3,6-dihydroxy compound (Yamamoto *et al.,* 1978; Hathway, 1979) and a 3-keto-6-hydroxy derivative (Yamamoto *et al.,* 1978). It is metabolized by insects to *N*-hydroxymethylcarbofuran (Krieger *et al.,* 1976) and finally into 3-ketocarbofuran (Krieger *et al.,* 1976) and 3-hydroxycarbofuran (Fukuto, 1976; Krieger *et al.,* 1976; Hathway, 1979).

Other reactions

The dimethoxy-*N*-methylphosphorothioyl derivative produced from PSC was found in mammals (Yamamoto *et al.,* 1978) and insects (Krieger *et al.,* 1976). The dimethoxy-*N*-methyl phosphoramidate was also detected in mammals (Fukuto, 1976; Yamamoto *et al.,* 1978). The NH_2-free and dealkylated car-bamate derivatives of 3-methylphenyl-*N*-propylcarbamate has been identified in mammals *in vitro* (Umetsu *et al.,* 1979). Finally, the monobutylamine derivative of DBSC was found in plants in significant yield (Kamoshita *et al.,* 1979).

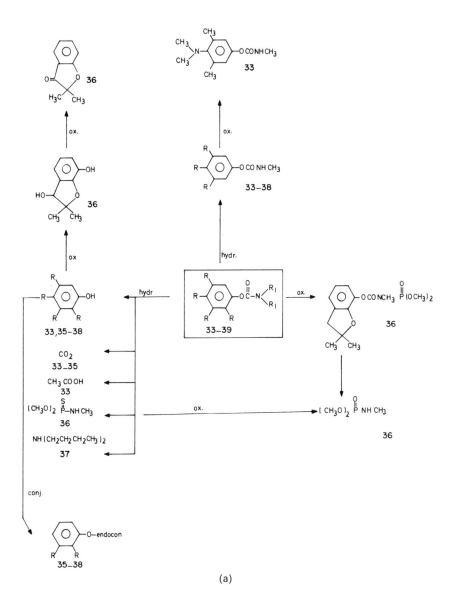

(a)

Figure 6 Metabolism of carbamates 33–39: 33, *N*-acetyl-zectran; 34, RE 11775; 35, *N*-(toluenesulphenyl)-carbofuran; 36, PSC; 37, DBSC; 38, MSC; 39, 3-methylphenyl-*N*-propylcarbamate

(b)

(c)

Conjugation

Glucuronide conjugation of this class of carbamate insecticides has been found in mammals (Wilkinson, 1976) and the glucose conjugates have been often found in plants (Chiu *et al.*, 1975). DBSC and MSC sugar conjugates were only found in metabolic studies with plants (Chiu *et al.*, 1975). Mammals conjugate a phenol derivative, a 3,6-dihydroxy derivative and a 3-keto-6-hydroxy derivative of *N*-(toluenesulphenyl)carbofuran derived from PSC (Krieger *et al.*, 1976); however, insects can transform it into an *N*-hydroxymethylcarbofuran derivative of PSC (Krieger *et al.*, 1976) only. Mammals and insects can conjugate the 3-hydroxy derivatives of *N*-(toluenesulphenyl)carbofuran and PSC (Black *et al.*, 1973; Fukuto, 1976; Krieger, *et al.*, 1976; Kuhr and Dorough, 1976; Nishioka *et al.*, 1981) as well as the 3-hydroxy-*N*-hydroxymethylcarbofuran derivatives of *N*-(toluenesulphenyl)carbofuran (Black *et al.*, 1973; Fukuto, 1976; Kuhr and Dorough, 1976).

TOXICOLOGICAL CONSIDERATIONS

The insecticides are discussed in the same order as that used above. Dimetilan has an MDL* (minimum detectable level) for anticholinesterase activity of $0.2\,\mu g$. Its oxidation products, e.g. methylcarbamate and the 2-methyl-carbamoyl derivatives, have MDL values of 2.0 and $0.2\,\mu g$ respectively compared to the hydrolytic product of the pyrazolone derivative (which has a MDL of $100\,\mu g$). The value for isolan itself is $0.01\,\mu g$ and of its hydrolysis product, $80\,\mu g$ (Oonnithan and Casida, 1968). Oxidation products are generally as toxic as the original compounds but are generally ineffective against the acetyl-cholinesterase after hydrolysis. The LD_{50} to rats (oral administration) is 90 and 25–$64\,mg/kg$ respectively for pyrolan and dimetilan (Matsumura, 1975); generally the carbamate insecticides are quite toxic to mammals.

   The MDL for landrin was $0.06\,\mu g$ and for UC 10854 and HRS 1422 were, respectively, 0.04 and $0.03\mu g$. Their oxidation products had MDL values lower than those of the original compounds (as low as 0.004 and $0.001\,\mu g$) but their phenol derivatives are 1000 times less toxic (5.0 and $6.0\,\mu g$) (Oonnithan and Casida, 1968). This is again proof of the lack of action of hydrolysis products against cholinesterase. The LD_{50} values given in M (moles/head) for tsumacide, etrofolan, UC 10854 and RE 5030 for houseflies are 2.09×10^{-7}, 1.99×10^{-7}, 4.32×10^{-8} and 6.87×10^{-8}, respectively (Kamoshita *et al.*, 1979). Tsumacide and UC 10854 show respectively LD_{50} values of 9.27×10^{-11} and 2.6×10^{-11} mole/head when ingested by houseflies. The highest LD_{50} for rats (oral) was reported for butacarb at $4000\,mg/kg$ and the lowest at $50\,mg/kg$ for

---

*The minimum detectable level (MDL) is only defined in µg of the original compound having an equivalent radiocarbon content after incubation with acetylcholinesterase from human blood. Dimetilan has an MDL for anticholinesterase activity of 0.2 µg.

UC 10854. Promecarb, bux, landrin, tsumacide, meobal and etrofolan possess LD_{50} values (rat) of 75, 87, 208, 268, 380 and 485 respectively (Kuhr and Dorough, 1976).

The MDL for banol has been reported to be 0.03 µg compared to 8.0 µg for its hydrolysis product. The phenolic compound usually used as a reference has itself an MDL of 50 µg. Matacil and zectran have MDL values of 1.0 and 0.02 µg, respectively, and their oxidation products have, more or less, the same values. The phenols derived from zectran and matacil have MDL values of 5.0 µg and 20 µg, respectively. The loss of effectiveness of the insecticide after hydrolysis has, therefore, again been observed. Mesurol (MDL 0.1 µg) its sulphoxide derivative (MDL 0.01 µg) and the sulphone derivative (MDL 0.1 µg) are somewhat bioactivated by oxidation. The phenol derivative of mesurol has an MDL value of 20 µg (Oonnithan and Casida, 1968). The ED_{50} value* for houseflies, given in moles/head, is for baygon 2.27×10^{-7} (Kamoshita et al., 1979) and its LD_{50} against houseflies is 25 mg/kg. Baygon has an LD_{50} in rats (oral administration) of 100 (Kuhr and Dorough, 1976) or 95–104 mg/kg (Matsumura, 1975). The LD_{50} values of matacil and zectran in rats were similar with 30 and 30 mg/kg (Kuhr and Dorough, 1976) or 15 and 63 mg/kg (Matsumura, 1975) respectively. Formetanate possesses an LD_{50} of 20 (Kuhr and Dorough, 1976) and mesurol, of 130 mg/kg (Matsumura, 1975; Kuhr and Dorough, 1976).

Carbofuran and mobam have quoted MDL values of 0.02 and 0.08 µg, respectively. The value for carbaryl is only 0.2 µg, but for the oxidation products of this insecticide MDL values are from 0.01 to 0.4 µg (compared to 1-naphthol at 20 µg) (Oonnithan and Casida, 1968). Another set of values proving the relatively high toxicity of some oxidation products are those for LD_{50} to rats (oral administration) of carbofuran (11 mg/kg) compared to that of its 3-hydroxy derivative at 17.9 and of its 3-keto derivative at 69 (Arunachalam et al., 1980; Fuhremann and Lichtenstein, 1980).

A MDL value of 0.22 µg for the insecticide aldicarb has been reported. The sulphoxide and the sulphone derivatives have MDL values of 0.001 and 0.005 µg, respectively. The hydrolysis products, the oxime, the oxime sulphoxide and the oxime sulphone have values around 50 µg (Oonnithan and Casida, 1968). For aldicarb, an LD_{50} value to houseflies of 6 mg/kg has been found Kuhr et al., 1980) and the LD_{50} to rats (oral) for aldicarb is reported at 1 (Kuhr and Dorough, 1976), 1–30 (Matsumura, 1975) or 0.6–0.8 mg/kg (Kuhr et al., 1980) (and is 0.5–3 for mice; Fukuto, 1976). The LD_{50} value for methomyl to rats (oral) is 21 mg/kg and for oxamyl 5 mg/kg (Kuhr and Dorough, 1976).

DBSC and MSC are derivatives of the carbofuran family. Their LD_{50} values in rats (oral) are much greater, at 1200 and 500 mg/kg respectively (Umetsu et al., 1979). N-(2-Toluenesulphenyl)carbofuran has an LD_{50} to houseflies of

---

*ED_{50} dose evoking tremor in 50% of the treated animals.

100–125 mg/kg; its sulphone derivative shows the LD_{50} at 150 mg/kg to mice and even more (~ 500) to houseflies. All oxidation products of these insecticides are toxic; N-(2-toluenesulphenyl)carbofuran is itself much more toxic to insects than to mammals.

CONCLUSIONS

The carbamate insecticides are potentially toxic to all living organisms that possess acetylcholinesterase in their nervous system. The LD_{50} is a valuable reference to this toxicity, although the values for different carbamate insecticides vary widely according to their chemical structures. Using the groupings as above, the most toxic insecticides to mammals are the N-methylcarbamates of oximes, the heteroaromatic phenyl N-methylcarbamates with condensed ring systems and the alkylphenyl-N-methylcarbamates.

Insecticides such as matacil and carbofuran are a potential danger to mammals, but compounds like DBSC, MSC and N-(2-toluenesulphenyl)carbofuran represent an improvement since they are less toxic, for example to accidentally sprayed animals. These relatively non-toxic yet effective propesticides developed by Fukuto and co-workers (Fukuto and Fahmy, 1981) represent a very useful advance in the carbamate insecticides.

Efficacy and mammalian toxicity vary widely in this class of compounds, even with the relatively simple aryl-N-methylcarbamates. It is likely that carbamate insecticides will continue to be useful for many years, though whether any will rival carbaryl, one of the most successful insecticides, remains to be seen.

ACKNOWLEDGEMENTS

This work has been supported by a research grant from the New Brunswick Health Department. One of us (K.J.) thanks the GREACE group for financial help as well as Mr R. Garvie, R. Luce, E. Luce and G. Poulin for technical assistance. Finally, helpful discussions with Professor V. Mallet are acknowledged.

We would like to thank Miss N. Boudreau and Mr P. Teyton for excellent technical help in the preparation of this manuscript.

REFERENCES

Abd-Elraof, T. K., Dauterman, W. C., and Mailman, R. B. (1981). '*In vivo* metabolism and excretion of propoxur and malathion in the rat: effect of lead treatment', *Toxicol appl. Pharmacol.*, **59**, 324–330.
Abdel-Wahab, A. M., Kuhr, R. J., and Casida, J. E. (1966). 'Fate of ^{14}C-Carbonyl-labeled aryl methylcarbamate insecticide chemicals in and on bean plants', *J. Agric. Food Chem.*, **14**, 290–298.

Ahmad, S., Forgash, A. J., and Das, Y. T. (1980). 'Penetration and metabolism of [^{14}C] carbaryl in larvae of the gypsy moth, Lymantria dispar (L.)', Pestic. Biochem. Physiol., 14, 236–248.

Andrawes, N. R., Bagley, W. P., and Herrett, R. A. (1971). 'Metabolism of 2-methyl-2-(methylthio)propionaldehyde O-(methylcarbamoyl) oxime (Temir aldicarb pesticide) in potato plants', J. Agric. Food Chem., 19, 731–737.

Andrawes, N. R., Romine, R. R., and Bagley, W. P. (1973). 'Metabolism and residues of Temir aldicarb pesticide in cotton foliage and seed under field conditions', J. Agric. Food Chem., 21, 379–386.

Archer, T. E., Stokes, J. D., and Bringhurst, R. S. (1977). 'Fate of carbofuran and its metabolites on strawberries in the environment', J. Agric. Food Chem., 25, 536–541.

Ariaratnam, V., and Georghiou, G. P. (1975). 'Carbamate resistance in Anopheles albimanus. Penetration and metabolism of carbaryl in propoxur-selected larvae', Bull. W.H.O., 52, 91–96.

Arunachalam, S., Jeyalakshmi, K., and Aboobucker, S. (1980). 'Toxic and sublethal effects of carbaryl on a freshwater catfish, Mystus vittatus (Bloch)', Archs Environ. Contam. Toxicol., 9, 307–316.

Balba, M. H., and Saha, J. G. (1974). 'Degradation of matacil by the ascorbic acid oxidation system', Bull. Environ. Contam. Toxicol., 11, 193–200.

Belasco, I. J., and Harvey, J. Jr. (1980). 'In vitro rumen metabolism of carbon-14-labeled oxamyl and selected metabolites of oxamyl', J. Agric. Food Chem., 28, 689–692.

Bend, J. R., Holder, G. M., Protos, E., and Ryan, A. J. (1970). 'Metabolism of carbaryl in the cattle tick Boophilus microplus', Aust. J. Biol. Sci., 23, 361–367.

Bend, J. R., Holder, G. M., Protos, E., and Ryan, A. J. (1971). 'Water-soluble metabolites of carbaryl (1-naphthyl N-methylcarbamate) in mouse liver preparations and in the rat', Aust, J. Biol. Sci., 24, 535–546.

Benezet, H. J., and Matsumura, F. (1974). 'Factors influencing the metabolism of mexacarbate by microorganisms', J. Agric. Food Chem., 22, 427–430.

Benson, W. H., and Dorough, H. W. (1979). 'Comparative carbamate ester hydrolysis in 4 mammalian-species'. Toxicol. appl. Pharmacol., 48, A139.

Bezbariah, B. (1976). 'A discussion of the possibilities of biodegradation of certain chemicals used or obtained during the manufacture of some pesticides', Chem. Age India, 27, 1065–1073.

Black, A. L., Chu, Y. C., Fukuto, T. R., and Miller, T. A. (1973). 'Metabolism of 2,2-dimethyl-2,3-dihydro-7-benzofuranyl N-methyl-N-(2-toluenesulfenyl)carbamate in the housefly and white mouse', Pestic. Biochem. Physiol., 3, 435–446.

Casida, J. E. (1963). 'Mode of action of carbamates', Ann. Rev. Entomol., 8, 39–58.

Chang, K.-M., and Knowles, C. O. (1978). 'Aldicarb metabolism in twospotted spider mites', J. Econ. Entomol., 71, 158–160.

Chang, K.-M., and Knowles, C. O. (1979). 'Metabolism of oxamyl in mice and twospotted spider mites', Archs Environ. Contam. Toxicol., 8, 499–508.

Chen, K.-C., and Dorough, H. W. (1980). 'Glutathione and mercapturic acid conjugations in the metabolism of naphthalene and 1-naphthyl N-methylcarbamate (carbaryl)', Drug Chem. Toxicol., 2, 331–354.

Chin, B. H., Eldridge, J. M., Anderson, J. H. Jr., and Sullivan, L. J. (1979a). 'Carbaryl metabolism in the rat. A comparison of in vivo, in vitro (tissue explant), and liver perfusion techniques', J. Agric. Food Chem., 27, 716–720.

Chin, B. H., and Sullivan, L. J. (1979). 'Carbaryl metabolism by the selected tissues of the dog via the in vitro explant-maintenance technique', J. Agric. Food Chem., 27, 419–420.

Chin, B. H., Sullivan, L. J., and Eldridge, J. E (1979b). '*In vitro* metabolism of carbaryl by liver explants of bluegill, catfish, perch, goldfish, and kissing gourami', *J. Agric. Food Chem.*, **27**, 1395–1398.

Chin, B. H., Sullivan, L. J., Eldridge, J. M., and Tallant, M. J. (1979c). 'Metabolism of carbaryl by kidney, liver, and lung from human postembryonic fetal autopsy tissue', *Clin. Toxicol.*, **14**, 489–498.

Chin, B. H., Tallant, M. J., Duane, W. C., and Sullivan, L. J. (1980). 'Metabolism of carbamate insecticide thiofanox in rats', *J. Agric. Food Chem.*, **28**, 1085–1089.

Chio, H., and Metcalf, R. L. (1979). 'Detoxication mechanisms for aldrin, carbofuran, fonofos, phorate and terbufos in four species of diabroticides', *J. Econ. Entomol.*, **72**, 732–738.

Chiu, Y. C., Black, A. L., and Fukuto, T. R. (1975). 'Thiolysis as an activation process in *N*-sulfenylated derivatives of methylcarbamate esters', *Pestic. Biochem. Physiol.*, **5**, 359–366.

Cool, M. (1982). 'Etude de la dégradation du matacil par le système d'oxydation ferro-ascorbique', Thesis, Université de Moncton, New Brunswick, Canada.

Cool, M., and Jankowski, K. (1982). 'Study of the interaction of matacil and L-ascorbic acid', *Eur. J. Mass Spectrom. Biochem., Med. Environ. Res.*, **2**, 83–88.

Dorough, H. H. (1970). 'Metabolism of insecticidal methylcarbamates in animals', *J. Agric. Food. Chem.*, **18**, 1015–1022.

Dorough, H. W. (1979). 'Metabolism of insecticides by conjugation mechanisms', *Pharmacol. Ther.*, **4**, 433–471.

Esaac, E. G., and Matsumura, F. (1979). 'Roles of flavoproteins in degradation of mexacarbate in rats', *Pestic. Biochem. Physiol.*, **10**, 67–78.

Friedman, A. R., and Lemin, A. J. (1967). 'Metabolism of 2-chloro-4,5-dimethylphenyl *N*-methylcarbamate in bean plants', *J. Agric. Food Chem.*, **15**, 642–647.

Fuhremann, T. W., and Lichtenstein, E. P. (1980). 'A comparative study of the persistence, movements, and metabolism of six carbon-14 insecticides in soils and plants', *J. Agric. Food Chem.*, **28**, 446–452.

Fukami, J. (1980). 'Metabolism of several insecticides by glutathion *S*-transferase', *Pharmacol. Ther.*, **10**, 473–514.

Fukuto, T. R. (1976). 'Carbamate insecticides', *Adv. Environ. Sci. Technol.*, **6**, 313–346.

Fukuto, T. R., and Fahmy, M. A. H. (1981). 'Sulphur in pesticide action', in *Sulphur in Pesticide Action and Metabolism* (eds J. D. Rosen, P. N. Magee and J. E. Casida), pp. 35–49, ACS Symp. Series, No. 158.

Fukuto, T. R., and Metcalf, R. L. (1969). 'Metabolism of insecticides in plants and animals', *Ann. N.Y. Acad. Sci.*, **160**, 97–111.

Garg, A. K., and Sethi, G. R. (1979). 'Efficient use of insecticides: metabolism of systemic insecticides', *Pesticides*, **13**, 17–22.

Gayen, A. K., and Knowles, C. O. (1981). 'Penetration and fate of methomyl and its oxime metabolite in insects and twospotted spider mites', *Archs Environ. Contam. Toxicol.*, **10**, 55–67.

Getzin, L. W. (1973). 'Persistence and degradation of carbofuran in soil', *Environ. Entomol.*, **2**, 461–467.

Gill, S. S. (1980). '*In vitro* metabolism of carbofuran by liver microsomes of the padifield fish *Trichogaster pectoralis*', *Bull. Environ. Contam. Toxicol.*, **25**, 697–701.

Guirguis, G. N., and Brindley, W. A. (1975). 'Carbaryl penetration into and metabolism by alfalfa leaf cutting bees, Megachile', *J. Agric. Food Chem.*, **23**, 274–279.

Gupta, K. G., Sud, R. K., Aggarwal, P. K., and Aggarwal, J. C. (1975). 'Effect of baygon (2-isopropoxyphenyl-*N*-methylcarbamate) on some soil biological processes and its degradation by a *Pseudomonas* species', *Plant Soil*, **42**, 317–325.

Harvey, J. Jr., and Han, J. C. Y. (1978). 'Metabolism of oxamyl and selected metabolites in the rat', *J. Agric. Food Chem.*, **26**, 902–910.

Harvey, J. Jr., Han, J. C. Y., and Reiser, R. W. (1978). 'Metabolism of oxamyl in plants', *J. Agric. Food Chem.*, **26**, 529–536.

Hassan, A., Zayed, S. M. A. D., and Abdel-Hamid, F. M. (1966). 'Metabolism of carbamate drugs. I. Metabolism of 1-naphthyl-*N*-methylcarbamate (Sevin) in the rat', *Biochem. Pharmacol.*, **15**, 2045–2055.

Hathway, D. E. (1979). *Foreign Compound Metabolism in Mammals*. A specialist periodical report. Volume 5, The Chemical Society, Burlington House, London.

Heenan, M. P., and Smith, J. N. (1974). 'Water-soluble metabolites of *p*-nitrophenol and 1-naphthyl *N*-methylcarbamate in flies and grass grubs. Formation of glucose phosphate and phosphate conjugates', *Biochem. J.*, **144**, 303–310.

Hicks, B. W., Dorough, H. W., and Mehendale, H. M. (1972). 'Metabolism of aldicarb pesticide in laying hens', *J. Agric. Food Chem.*, **20**, 151–156.

Hinderer, R. K., and Menzer, R. E. (1976a). 'Comparative enzyme activities and cytochrome P-450 levels of some rat tissues with respect to their metabolism of several pesticides', *Pestic. Biochem. Physiol.*, **6**, 148–160.

Hinderer, R. K., and Menzer, R. E. (1976b). 'Enzyme activities and cytochrome P-450 levels of some Japanese quail tissues with respect to their metabolism of several pesticides', *Pestic. Biochem. Physiol.*, **6**, 161–169.

Hodgson, E., and Casida, J. E. (1961). 'Metabolism of *N,N*-dialkylcarbamates and related compounds by rat liver', *Biochem. Pharmacol.*, **8**, 179–191.

Huhtanen, K., and Dorough, H. W. (1976). 'Isomerization and Beckmann rearrangement reactions in the metabolism of methomyl in rats', *Pestic. Biochem. Physiol.*, **6**, 571–583.

Hutson, D. H. (1982). 'Formation of lipophilic conjugates of pesticides and other xenobiotic compounds', in *Progress in Pesticide Biochemistry* (eds D. H. Hutson and T. R. Roberts), Vol. 2, pp. 171–184, John Wiley, Chichester.

Hutson, D. H., Hoadley, E. C., and Pickering, B. A. (1971). 'The metabolism of *S*-2-cyanoethyl-*N*-[(methylcarbamoyl)oxy]thioacetimidate, an insecticidal carbamate, in the rat', *Xenobiotica*, **1**, 179–191.

Jankowski, K. (1978). 'GC–MS analysis of aminocarb', *FPL report.*

Jankowski, K. (1979). 'GC–MS analysis of aminocarb', *FPL report.*

Jankowski, K. (1980). *N.B. Health Department Report*, pp. 11.

Jankowski, K. (1981). *N.B. Health Department Report*, pp. 27.

Jankowski, K., Garvie, R., Luce, E., and Poulin, G. (1980). 'Vitamin C extraction', *STEPEX report*, pp. 77, NRC of Canada.

Jankowski, K., and Paré, J. R. J. (1980). 'Mass spectrometric study of carbamates', *Eur. J. Mass Spectrom. Biochem., Med. Environ. Res.*, **1**, 141–147.

Johannsen, F. R., and Knowles, C. O. (1977). 'Distribution and metabolism of promecarb in corn following soil application', *Nippon Noyaku Gakkaishi*, **2**, 159–161.

Jones, A. S. (1976). 'Metabolism of aldicarb by five soil fungi', *J. Agric. Food Chem.*, **24**, 115–117.

Kamoshita, K., Ohno, I., Kasamatsu, K., Fujita, T., and Nakajima, M. (1979). 'Quantitative structure-activity relationships of phenyl *N*-methylcarbamates against the smaller brown planthooper and its acetylcholinesterase', *Pestic. Biochem. Physiol.*, **11**, 104–116.

Kapoor, S. K., and Kalra, R. L. (1975). 'Uptake and metabolism of carbofuran in maize plants', *J. Food Sci. Technol.*, **12**, 227–230.

Kazano, H., Kearny, P. C., and Kaufman, D. D. (1972). 'Metabolism of methylcarbamate insecticides in soils', *J. Agric. Food Chem.*, **20**, 975–979.

Khan, M., Gassman, M., and Haque, R. (1976). 'Biodegradation of pesticides', *CHEM-TECH*, **6**, 62–69.

Knaak, J. B. (1971). 'Biological and nonbiological modifications of carbamates', *Bull. W.H.O.*, **44**, 121–131.

Knaak, J. B., Tallant, M. J., and Sullivan, L. J. (1966). 'The metabolism of 2-methyl-2-(methylthio)propionaldehyde O-(methylcarbamoyl)oxime in the rat', *J. Agric. Food Chem.*, **14**, 573–578.

Knowles, C. O., and Chang, K.-M. (1979). 'Degradation of formetanate acaricide by twospotted spider mites', *Chemosphere*, **8**, 3–6.

Knowles, C. O., and Sen Gupta, A. K. (1970). 'Metabolism of formetanate acaricide in orange seedlings', *J. Econ. Entomol.*, **63**, 615–620.

Knowles, C. O., and Johannsen, F. R. (1976). 'Metabolism of promecarb in the rat', *J. Econ. Entomol.*, **69**, 595–596.

Krieger, R. I., Lee, P. W., Fahmy, M. A. H., Chen, M., and Fukuto, T. R. (1976). 'Metabolism of 2,2-dimethyl-2,3-dihydrobenzofuranyl-7 N-dimethoxyphosphinothioyl-N-methylcarbamate in the house fly, rat, and mouse', *Pestic. Biochem. Physiol.*, **6**, 1–9.

Krueger, A. R. (1977). 'Aldicarb sulfoxidation by plant root extracts', *Pestic. Biochem. Physiol.*, **7**, 154-160.

Kuhr, R. J. (1968). 'Metabolism of methylcarbamate insecticides by insects *in vivo* and *in vitro*', *Med. Ed. Rijksfac. Landbouwwetensch., Gent.*, **33**, 647–657.

Kuhr, R. J. (1970). 'Metabolism of carbamate insecticide chemicals in plants and insects', *J. Agric. Food Chem.*, **18**, 1023–1030.

Kuhr, R. J. (1976). 'Insecticide metabolites in and on plants', *CHEMTECH*, **6**, 316–321.

Kuhr, R. J., and Dorough, H. W. (1976). *Carbamate Insecticides: Chemistry, Biochemistry and Toxicology*, CRC Press Inc., Cleveland, Ohio.

Kuhr, R. J., Hassan, A., and Winteringham, F. P. W. (1980). 'Carbamate insecticides (Issue No. 1), Comparative summary', *Chemosphere*, **9**, 251–253.

Kuhr, R. J., and Hessney, C. W. (1977). 'Toxicity and metabolism of methomyl in the European corn borer', *Pestic. Biochem. Physiol.*, **7**, 301–308.

Lin, T. H., North, H. H., and Menzer, R. E. (1975). 'Metabolism of carbaryl (1-naphthyl N-methylcarbamate) in human embryonic lung cell cultures', *J. Agric. Food Chem.*, **23**, 253–256.

Locke, K. K., Chen, J.-Y. T., Damico, J. N., Dusold, L. R., and Sphon, J. A. (1976). 'Identification by physical means of organic moieties of conjugates produced from carbaryl by tobacco cells in suspension culture', *Archs Environ. Contam. Toxicol.*, **4**, 60–100.

Marshall, T. C., and Dorough, H. W. (1977). 'Bioavailability in rats of bound and conjugated plant carbamate insecticide residues', *J. Agric. Food Chem.*, **25**, 1003–1009.

Matsumura, F. (1975). *Toxicology of Insecticides*, Plenum Press, New York and London.

Meikle, R. W. (1973). 'Metabolism of 4-dimethylamino-3,5-xylylmethylcarbamate (mexacarbate, active ingredient of Zectran insecticide)', *Bull. Environ. Contam. Toxicol.*, **10**, 29–36.

Menn, J. J. (1978). 'Comparative aspects of pesticide metabolism in plants and animals', *EHP, Environ. Health Perspect.*, **27**, 113-124.

Miyamoto, J. (1970). 'Feature of detoxication of carbamate insecticide in mammals', in *Biochem. Toxicol. Insectic. Proc. U.S.-Jap. Coop. Sci. Program, 5th, 1969* (ed. R. D. O'Brien), pp. 115–130, Plenum Press, New York.

Miyamoto, J. (1975). 'Transformation of carbamate insecticides in the environment', *Environ. Qual. Saf.*, **4**, 134–135.

Miyamoto, J., Yamamoto, K., and Matsumoto, T. (1969). 'Metabolism of 3,4-dimethyl-phenyl N-methylcarbamate in white rats', *Agr. Biol. Chem.*, **33**, 1060–1073.

Nishioka, T., Umetsu, N., and Fukuto, T. R. (1981). 'Metabolism of dibutyl-carbon-14-labeled dibutylaminosulfenyl derivative of carbofuran in the cotton plant', *Pestic. Biochem. Physiol.*, **16**, 141-148.

Nye, D. E., and Dorough, H. W. (1976). 'Fate of insecticides administered endo-tracheally to rats', *Bull. Environ. Contam. Toxicol.*, **15**, 291–296.

Nye, D. E., Hurst, H . E., and Dorough, H. W. (1976). 'Fate of croneton (2-ethyl-thiomethylphenyl N-methylcarbamate) in rats', *J. Agric. Food Chem.*, **24**, 371–377.

O'Brien, R. D. (1967). *Insecticides Action and Metabolism*, Academic Press, New York and London.

Ogawa, K., Tsuda, M., Yamauchi, F., Yamaguchi, I., and Misato, T. (1977). 'Metabol-ism of 2-isopropylphenyl N-methylcarbamate (Mipcin, MIPC) in rice plants and its degradation in soils', *Nippon Noyaku Gakkaishi*, **2**, 51–57.

Ohkawa, H., Yoshihara, R., Kohara, T., and Miyamoto, J. (1974). 'Metabolism of m-tolyl N-methylcarbamate (Tsumacide) in rats, houseflies, and bean plants', *Agr. Biol. Chem.*, **38**, 1035–1044.

Oonnithan, E. S., and Casida, J. E. (1966). 'Metabolites of methyl- and dimethyl-carbamate insecticide chemicals as formed by rat liver microsomes', *Bull. Environ. Contam. Toxicol.*, **1**, 59-69.

Oonnithan, E. S., and Casida, J. E. (1968). 'Oxidation of methyl- and dimethyl-carbamate insecticide chemicals by microsomal enzymes and anticholinesterase activity of the metabolites', *J. Agric. Food Chem.*, **16**, 28–44.

Pekas, J. C. (1979). 'Further metabolism of naphthyl N-methylcarbamate (carbaryl) by the intestine', *Pestic. Biochem. Physiol.*, **11**, 166–175.

Pekas, J. C. (1980). Gastrointestinal metabolism and transport of pesticidal carbamates, *CRC Crit. Rev. Toxicol.*, **7**, 37–101.

Rodriguez, L. D., and Dorough, H. W. (1977). 'Degradation of carbaryl by soil microorganisms', *Archs Environ. Contam. Toxicol.*, **6**, 47–56.

Ryan, A. J. (1971). 'The metabolism of pesticidal carbamates', *CRC Crit. Rev. Toxicol.*, **1**, 33–54.

Schlagbauer, B. G. L., and Schlagbauer, A. W. J. (1972). 'Metabolism of carbamate pesticides, Literature analysis I', *Residue Rev.*, **42**, 1–84.

Shrivastava, S. P., Tsukamoto, M., and Casida, J. E. (1969). 'Oxidative metabolism of C-labeled Baygon by living houseflies and by housefly enzyme preparations', *J. Econ. Entomol.*, **62**, 483–498.

Slade, M., and Casida, J. E. (1970). 'Metabolite fate of 3,4,5- and 2,3,5-trimethylphenyl-methylcarbamates, the major constituents in Landrin insecticide', *J. Agric. Food Chem.*, **18**, 469–474.

Sonobe, H., Kamps, L. V. R., Mazzola, E. P., and Roach, J. A. G. (1981). 'Isolation and identification of a new conjugated carbofuran metabolite in carrots: angelic acid ester of 3-hydroxycarbofuran', *J. Agric. Food Chem.*, **29**, 1125–1129.

Statham, C. N., Pepple, S. K., and Lech, J. J. (1975). 'Biliary excretion products of 1-naphthyl-N-methylcarbamate-1-C (carbaryl) in rainbow trout (*Salmo Gairdneri*)', *Drug Metab. Dispos.*, **3**, 400–406.

Strother, A. (1972). '*In vivo* metabolism of methylcarbamate insecticides by human and rat liver fraction', *Toxicol. appl. Pharmacol.*, **21**, 112–129.

Sundaram, K. M. S., and Szeto, S. Y. (1979). 'A study on the lethal toxicity of aminocarb to freshwater crayfish and its *in vivo* metabolism', *J. Environ. Sci. Health, Part B*, **B14**, 589–602.

Suzuki, T., and Takeda, M. (1976a). 'Microbial metabolism of N-methylcarbamate I. Metabolism of o-sec-butylphenyl N-methylcarbamate by *Aspergillus niger* van Tieghem', *Chem. Pharm. Bull.*, **24**, 1967–1975.

Suzuki, T., and Takeda, M. (1976b). 'Microbial metabolism of N-methylcarbamate insecticide III. Time course in the metabolism of o-sec-butylphenyl N-methyl-carbamate by *Aspergillus niger* and species differences among soil fungi', *Chem. Pharm. Bull.*, **24**, 1983–1987.

Suzuki, T., and Takeda, M. (1976c). 'Microbial metabolism of N-methylcarbamate insecticide IV. New metabolites of o-sec-butylphenyl N-methylcarbamate *Cladosporium cladosporioides*', *Chem. Pharm. Bull.*, **24**, 1988–1991.

Thurlow, W. (1979). *Matacil Spray Report*, Gander Environmental Group, Gander, Nfld.

Tsukamoto, M., and Casida, J. E. (1967). 'Metabolism of methylcarbamate insecticides by the NADPH-requiring enzyme system from houseflies', *Nature*, **213**, 49–51.

Umetsu, N., Fahmy, M. A. H., and Fukuto, T. R. (1979). 'Metabolism of 2,3-dihydro-2,2-dimethyl-7-benzofuranyl(di-n-butylaminosulfenyl)(methyl) carbamate and 2,3-di-hydro-2,2-dimethyl-7-benzofuranyl (morpholinosulfenyl)(methyl) carbamate in cotton and corn plants', *Pestic. Biochem. Physiol.*, **10**, 104–119.

Wilkinson, C. F. (1976). *Insecticide Biochemistry and Physiology*, Plenum Press, New York and London.

Williams, I. H., Pepin, H. S., and Brown, M. J. (1976), 'Degradation of carbofuran by soil microorganisms', *Bull. Environ. Contam. Toxicol.*, **15**, 344-349.

Yamamoto, I., Maki, S., and Sato, K. (1978). 'Metabolism of an aryl N-propyl-carbamate *in vitro* by a rat liver enzyme system', *Nippon Noyaku Gakkaishi*, **3**, 53-54.

Zubairi, M. Y., and Casida, J. E. (1965). 'Detoxication of dimetilan in cockroaches and houseflies', *J. Econ. Entomol.*, **58**, 403–409.

Insecticides
Edited by D. H. Hutson and T. R. Roberts
© 1985 John Wiley & Sons Ltd

CHAPTER 4

Resistance to pyrethroid insecticides in arthropods

R. M. Sawicki

INTRODUCTION

The discovery of photostable pyrethroids (Elliott et al., 1973, 1974) made it possible to re-establish control of key pests of many major crops which were by then becoming increasingly difficult to control with organophosphorus and carbamate insecticides (Georghiou and Mellon, 1983). Although concern was voiced at the time (Sawicki, 1975; Lhoste, 1977; Elliott et al., 1978) about the likelihood of pyrethroid resistance if used excessively, warnings went largely unheeded, and many arthropods of major economic importance now resist pyrethroids.

Pyrethroid resistance has stimulated much interest, and a great deal of work on it has been done, especially over the last few years; some aspects of the biochemistry were recently reviewed by Soderlund et al. (1983a). The present paper brings the subject up to date, and discusses a number of topics of pyrethroid resistance in arthropods.

Definition of pyrethroid resistance

The term 'pyrethroid resistance' as used in the scientific literature is often totally divorced from practical connotations associated with control failure and includes any tolerance to a pyrethroid(s) that is greater than that of the most sensitive 'susceptible' standard or reference strain. It is normally considered to occur when either:

(1) the pyrethroid used no longer gives adequate control,
(2) the tolerance of field population(s) though effectively controlled in practice exceed(s) that of the susceptible standard in laboratory tests,
(3) laboratory selection results in increased tolerance, or
(4) cross-resistance studies demonstrate greater tolerance to pyrethroids than the susceptible standard strain.

Whenever possible the degree of resistance and its practical significance have been specified in the text.

Resistance is of practical importance only when pyrethroid(s) no longer give effective control at doses recommended for this purpose. This clearly excludes pests against which present-day pyrethroids are unlikely to be used, e.g. some species of rice leaf-hoppers. It may also include species for which resistance has only been shown in the laboratory, because the differences in tolerance obtained in such assays are not reliable indicators of the loss of effectiveness in the field. Using this criterion, only about half of the species in which pyrethroid resistance has been reported have strains no longer adequately controlled in the field. It is important to remember that with very rare exceptions, even in those arthropods, resistance does not occur throughout the geographical range of the species.

The practical significance of resistance in the remainder of pyrethroid-resistant species is difficult to evaluate since the differences in tolerance observed in the laboratory may be too small ever to become relevant in practice. Besides, several species, especially in the areas of personal or household hygiene, such as *Pediculus humanus humanus* L. or *Blatella germanica* L. are apparently no longer resistant to pyrethroids, although they frequently were, where and when synergized natural pyrethroids were used in their control. In those species resistance is therefore, at least for the time being, mainly of historical interest; if pyrethroids were to be used excessively however, they could once more regain resistance. It is to be expected that with the ever increasing use of pyrethroids, the number of resistant species listed here and their distribution will expand, and steps to control this resistance are vital.

Origins and causes of pyrethroid resistance

The origins and causes of resistance to pyrethroids are seldom known or satisfactorily documented, the exception being laboratory selection experiments specifically designed to select for pyrethroid resistance. Normally pyrethroid resistance is disclosed either through control failure or by cross-resistance studies. Since critical cross-resistance studies are seldom done before the introduction and widespread or continuous use of the insecticide to control a pest species or pest complex, the evidence about the causes of resistance obtained *a posteriori* is at best tentative. Such studies often implicate insecticides used previously to control the pest; DDT is the insecticide most often implicated, since there is strong indirect evidence that it has probably selected one or more mechanisms collectively referred to as *kdr* which confer cross-resistance between DDT and the pyrethroids in many arthropods.

With some pests, however, e.g. *Spodoptera exigua*, even tenuous speculative evidence is lacking since the history of insecticidal use, cross-resistance studies and the resistance mechanisms identified offer no clues about the origins of pyrethroid resistance in the species.

Table 1 Resistance to pyrethroids in insect vectors of diseases and household pests

Species	Level of resistance	Selection	Origin/detection	Authors
Pediculus humanus	Moderate, pyrethrins	—	Algeria, field	Nicoli and Sautet (1955)
	× 9, pyrethrins	Pyrethrins + sulphoxide	USA, lab	Cole and Clarke (1961)
Cimex lectularius	× 4, pyrethrins	—	Israel	Busvine (1958)
Cimex hemipterus	× 10, pyrethrins	—	Tanzania	Busvine (1958)
Blatella germanica	× 20, pyrethrins	Various insecticides	USA, field	Keller *et al.* (1956)
	× 34, pyrethrins	Diazinon	USA, lab	Collins (1975)
	× 75, pyrethrins	Pyrethrins	USA field/lab	Cochran (1973)
	× 20, pyrethrins	DDT	USA	Scott and Matsumura (1983)
	× 20, permethrin			
	× 10, fenvalerate			
	× 2, cypermethrin			
	× 1.5, deltamethrin			
Musca domestica	Variable—up to very strong to all pyrethroids	DDT, pyrethroids ± synergists	Europe, field/lab	See text
	× 42, permethrin	Permethrin	Canada, field	Harris *et al.* (1982)
	× variable—up to very strong	Bioresmethrin Permethrin	USA lab,	De Vries and Georghiou (1980)
Culex tarsalis	× 10 pyrethrins	DDT/pyrethrins	USA, lab	Plapp and Hoyer (1968a)
Culex quinquefasciatus				
larvae	× 4100 (1R)trans-permethrin / × 1000 (1R)cis-permethrin	permethrin (1R)trans-permethrin	USA, lab	Priester and Georghiou (1978)
larvae	× 2900 (1R)trans-permethrin / × 450 (1R)cis-permethrin	permethrin (1R)cis-permethrin	USA, lab	Priester and Georghiou (1978)
adults	× 28 (1R)trans-permethrin / × 11 (1R)cis-permethrin	permethrin (1R)trans-permethrin	USA, lab	Priester and Georghiou (1978)
adults	× 23 (1R)trans-permethrin / × 18 (1R)cis-permethrin	permethrin (1R)cis-permethrin	USA, lab	Priester and Georghiou (1978)

Anopheles arabiensis				
adults	Moderate, permethrin	DDT	Sudan, lab	Hemmingway (1981)
Anopheles gambiae				
larvae	× 4, bioallethrin	DDT	Central Africa, lab	Hemmingway (1981)
adults	Moderate, permethrin	Permethrin	Nigeria, lab	Rongsriyam and Busvine (1976)
Anopheles stephensi				
adults and larvae	× 12–18, permethrin	DDT + DMC + pip. but.	Pakistan, lab	Omer et al. (1980)
Aedes aegypti				
adults	Moderate, bioresmethrin	Various insecticides especially DDT	Thailand, field	Chadwick et al. (1977)
adults	× 18–21, permethrin	DDT	Thailand, lab	Malcolm (1981)
larvae	× 4, bioallethrin	Various insecticides	Central America	Prasittisuk and Busvine (1977)

REPORTS OF PYRETHROID RESISTANCE

Insects of medical and domestic importance (Table 1)

In the Culicidae, field resistance was first found in 1974 in Thailand during trials to control *Ae. aegypti* L. with bioresmethrin (Chadwick *et al.*, 1977). This caused surprise because pyrethroids had only been used very sparingly in Thailand. Subsequent studies (Malcolm, 1981; Malcolm and Wood, 1982) showed that it was due to gene R DDT2-py which confers resistance to the pyrethroids and DDT. It is thought that this gene had first been selected in the area by DDT during the malaria eradication campaign. Other strains of this species from several sites of South East Asia and Central America have been shown in laboratory tests to resist pyrethroids slightly (Prasittisuk and Busvine, 1977). Low levels of resistance have also been found in *An. arabiensis* Patt. and *An. gambiae* Giles after selection with permethrin (Hemmingway, 1981) and in *An. stephensi* Liston after selection with DDT (Omer *et al.*, 1980). The situation regarding *An. sacharovi* (Favr) has not yet been resolved (Davidson *et al.*, 1980; Davidson, 1984, personal communication).

In 1968 Plapp and Hoyer (1968a) observed cross-resistance between DDT and natural pyrethrins in *Culex tarsalis* Coquillet. By selecting an organophosphorus-resistant strain of *C. pipiens quinquefasciatus* Say with (1*R*)-*cis*- and (1*R*)-*trans*-permethrin, Priester and Georghiou (1978, 1979, 1980) obtained pyrethroid-resistant substrains which differed in resistance to these two isomers.

Resistance to pyrethroids in the housefly has been amply documented since it was first detected in the laboratory and in the field (Busvine, 1951; Davies *et al.*, 1958; Fine, 1961; Keiding, 1976, 1977). There are now many reports of resistance, particularly on animal farms of north-western Europe. (Künast, 1979; Bills, 1980; Chapman and Lloyd, 1981; Künast *et al.*, 1981; Sawicki *et al.*, 1981). Laboratory selection of resistant strains in central Europe (Rupes and Pinterova, 1976; Malinowski, 1980) and in the USSR (Roslatseva *et al.*, 1979) demonstrated potential for field resistance there also, and Rupes *et al.* (1983) have now confirmed its presence in Czechoslovakia. Strong resistance to permethrin was also found in Canada (Harris *et al.*, 1982; McDonald *et al.*, 1983a) and Georgia (Gullickson, 1982, personal communication), but in the rest of the United States pyrethroid resistance has been rare (Georghiou, 1983, personal communication). The relative rarity of pyrethroid resistance in houseflies in the USA is surprising since *kdr*, giving cross-resistance to pyrethroids and DDT (Milani and Franco, 1959; Milani, 1960) was demonstrated in a strain from Florida as early as 1960. In California a strongly resistant strain was obtained through prolonged selection in the laboratory (De Vries and Georghiou, 1980).

Pyrethroid resistance has also been detected in Japan and South Africa

(Taylor *et al.*, 1981), but not in the Middle East or North Africa (Taylor, 1982) although pyrethroids have been used there on a large scale for several years.

Although resistance to the synergized natural pyrethroids was relatively common in *Blatella germanica* L. in the USA (Keller *et al.*, 1956) when these insecticides were widely used for cockroach control, resistance no longer seems to be a problem (Nelson and Wood, 1982), and the synthetic pyrethrins are widely used in cockroach control. Selection with organophosphates and carbamates resulted in some cases in pyrethroid resistance (Collins, 1975, 1976); so has selection with DDT. Scott and Matsumura (1983) obtained strong resistance to some pyrethroids (Type I) but not to others (Type II) (see below) by selecting a DDT-resistant strain with DDT.

In lice (Nicoli and Sautet, 1955; Wright and Brown, 1957; Sautet *et al.*, 1959; Cole and Clarke, 1961; Wright and Pal, 1965) and in bed-bugs (Busvine, 1958) cross-resistance between the natural pyrethrins and DDT was found in the 1950s and 1960s in many parts of the world but now seems to have disappeared, and Nassif *et al.* (1980) obtained very good control of body lice with permethrin in Egypt.

It is important to remember that populations of insects, which are again easily controlled with pyrethroids, have most probably retained the potential for pyrethroid resistance originally disclosed by the natural products, and are likely to become resistant again when pyrethroids are used continuously or excessively.

Arthropods of veterinary importance (Table 2)

In South Africa cross-resistance between DDT and pyrethrins was found in *Boophilus decoloratus* Koch in 1959 (Whitehead, 1959), and nearly 20 years later the Australians (Nolan *et al.*, 1977, 1979) showed the same in *B. microplus* (Can.) and *Haemotobia irritans exigua* (De Meijera) (Schnitzerling *et al.*, 1982) which is no longer controlled in the field with fenvalerate.

Stored products pests (Table 3)

Although pyrethroid resistance has been reported in seven species of stored product pests (Table 3) there has been little control failure in the field, *viz. Sitophilus granarius* L. with synergized natural pyrethrins in England (Holborn, 1957), and *Plodia interpunctella* (Hubner) with deltamethrin in Morocco (Carle, 1983, personal communication). Long-term selection in the laboratory resulted in strong resistance in *S. granarius* (Parkin and Lloyd, 1960; Lloyd and Parkin, 1963; Lloyd and Ruczkowski, 1980; Lloyd, 1969), and cross-resistance studies demonstrated very strong resistance to some pyrethroids in the malathion-selected CTC 12 strain of *Tribolium castaneum* (Herbst.) (Champ and Dyte, 1976).

Table 2 Resistance to pyrethroids in arthropods of veterinary importance

Species	Level of resistance	Selection	Origin/detection	Author(s)
Boophilus decoloratus	× 18, pyrethrins	DDT	South Africa	Whitehead (1959)
Boophilus microplus	× 5–7, permethrin (kd)	Permethrin	Australia, lab/field	Schnitzerling *et al.* (1983)
	× 2, cypermethrin (kd)			
	× 4, permethrin (kill)			
Haemotobia irritans exigua	× 45, fenvalerate	Fenvalerate	Australia, field	Schnitzerling *et al.* (1982)
	× 29, cypermethrin			

Table 3 Resistance to pyrethroids in stored products pests

Species	Level of resistance	Selection	Origin/detection	Authors
Sitophilus granarius	× 3.5, pyrethrins	Pyrethrins + pip. but.	UK, lab	Blackith (1953)
	× 7, pyrethrins	Pyrethrins + pip. but.	UK, field	Holborn (1957)
		Pyrethrins	Lab	Lloyd (1969)
	× 148, pyrethrins + pip. but.			Lloyd and Parkin (1963)
Sitophilus oryzae	× 6, pyrethrins	DDT	Poland, lab	Cichy (1971)
Tribolium castaneum	× 3.5, pyrethrins		UK, lab	Holborn (1957)
	× 13, pyrethrins + pip. but.	Malathion	USA, field	Spiers and Zettler (1969)
	× 2– × 120, pyrethroids	Various insecticides	Australia, lab	Champ and Dyte (1976)
Tribolium confusum	× 3, bioresmethrin	Malathion	Australia, lab	Champ and Dyte (1976)
Oryzaephylus surinamensis	× 8, bioresmethrin	Malathion	Australia, lab	Champ and Dyte (1976)
Plodia interpunctella	× 2.5, pyrethrins	Malathion	USA, lab	Zettler *et al.* (1973)
	Very strong	Deltamethrin	Morocco, field	Carle (1983 pers. comm.)
Ephestia cautella	× 3.5, pyrethrins	Malathion	USA, lab	Zettler *et al.* (1973)

Pests of agricultural importance (Table 4)

Although the number of agricultural pests which resist pyrethroids is still relatively small, this number includes several key pests of major crops which have become resistant in many areas of the tropics. The exceptional effectiveness of photostable pyrethroids against Lepidoptera (Ruscoe, 1977) did not prevent at least six of this order developing resistance of significant practical importance. *Scrobipalpula obsoluta* (Meyrick) and *Keiferia lycopersicolla* (Wlsm.) (*Gnorimschema operculella*) (J. J. Hervé and D. D. Evans, 1982, personal communication) are resistant on tomatoes in Peru. Throughout much of Eastern and South Eastern Asia *Plutella xylostella* L. also resists, sometimes very strongly, the original four photostable pyrethroids, permethrin, cypermethrin, deltamethrin and fenvalerate (Cheng, 1981; Liu *et al.*, 1981; Ho *et al.*, 1983), and in Taiwan, Cheng (1981) found good correlation between the level of resistance and intensity of previous insecticidal treatment. *Spodoptera exigua* has strongly resisted photostable pyrethroids since their introduction (Holden in De Vries and Georghiou, 1980; Evans, 1982, personal communication) especially in Central America where *Trichoplusia ni* (Hübner) has also become resistant (Dittrich, 1984, personal communication). In New Zealand *Wiseana cervinata* (Walker) became refractory to control with pyrethroids soon after their introduction (Du Toit *et al.*, 1978).

Control of *Heliothis armigera* (Hübner) with permethrin deteriorated in Thailand within 2 years of the introduction of photostable pyrethroids (Wang Boonkong, 1981) and in Australia pyrethroids, introduced in 1977, failed without prior warning in 1983 against *H. armigera* on cotton and soya bean at Emerald, Central Queensland (Gunning *et al.*, 1984). About half the larvae collected from this area were at least × 10 or more resistant to fenvalerate than the most susceptible field strain, and a single selection with twice the discriminating dose increased resistance to × 50 at LD_{50}. A survey showed that pyrethroid resistance was widespread, but at low frequencies throughout much of Eastern Australia.

The widespread resistance resulting often rapidly in control failure in larvae of these moths has not been paralleled for other Lepidoptera of agricultural importance although, surprisingly, resistance in some species, e.g. *Heliothis virescens* and *H. zea* (Davies *et al.*, 1975, 1977; Harding *et al.*, 1977; Wolfenbarger *et al.*, 1977; Crowder *et al.*, 1979; Twine and Reynolds, 1980; Martinez-Carvillo and Reynolds, 1983), and *S. littoralis* (El-Guindy *et al.*, 1982b; Riskallah *et al.*, 1983) was sometimes much higher in terms of resistance ratios, i.e. LD_{50}-susceptible strain ÷ LD_{50}-resistant strain, as well as LD_{50}s in μg pyrethroid per gram larvae, than for *H. armigera*.

A 3-year survey, between 1979 and 1981, did not show significant field resistance to permethrin in *H. virescens* on cotton in the USA. Great differences in susceptibility of this insect were found from one year to another in different

Table 4   Resistance to pyrethroids in agricultural pests

Species	Level of resistance	Selection	Origin/detection	Authors
Phorodon humuli	Weak, deltamethrin	Ethyl methane sulphone	Switzerland, lab	Buchi (1981)
Myzus persicae	× 68, permethrin	Demeton-S-methyl	UK, lab	Sawicki and Rice (1978)
	× 87, cypermethrin	Demeton-S-methyl		
	× 1280, deltamethrin	Demeton-S-methyl		
	× 16, permethrin	OPs and carbamates	Australia, lab/field	Attia and Hamilton (1978)
	Strong, pyrethroids	Not known	USA, field	McClanhan and Founk (1983)
Nephotettix cincticeps	× 5, permethrin	OPs and carbamates	Taiwan, lab	Kao *et al.* (1981)
Nilaparvata lugens	× 171, permethrin	OPs and carbamates	Taiwan, lab	Chung *et al.* (1981)
	× 5, fenvalerate	OPs and carbamates		
	× ? fenvalerate	Fenvalerate	Japan, lab	Miyata (1980)
Laeodelphax striatellus	× ? fenvalerate	Fenvalerate	Japan, lab	Miyata *et al.* (1982)
Psylla pyricola	Strong	—	USA, field	Croft (1981, pers. comm.)
Leptinotarsa decemlineata	× 30, fenvalerate	Carbofuran	USA, field	Soderlund *et al.* (1983b)
Megaselia halterata	Strong, pyrethroids	Diazinon and other insecticides	UK, field	Hussey (1964)
Trialeurodes vaporariorum	× 17, resmethrin	Various insecticides	UK, glass house/lab	Wardlow *et al.* (1976)
	× 1, permethrin	Various insecticides	UK, glass house/lab	
	Very strong to all pyrethroids, resmethrin	Not known	UK and Holland, glass house/lab	Wardlow (1983, pers. comm.)
Liriomyza trifolii	Very strong, permethrin	Not known	USA glass house	Elhag and Horn (1983)
Scrobipalpula obsoluta	Strong ?	Permethrin	USA, field	Pavella (1983)
		Many insecticides	Peru, field	Hervé (1982, pers. comm.)
			Peru, field	Evans (1982, pers. comm.)
Keiferia lycopersicolla	Strong ?	Many insecticides	Peru, field	Evans (1982, pers. comm.)
Spodoptera exigua	× 400, permethrin	—	UK, lab	Fullbrook and Holden (1980)
	Strong, all pyrethroids	—	Central America, field	Evans (1982, pers. comm.)

Species	Resistance (pyrethroids)	Other resistance	Location	Reference
Spodoptera frugiperda	—			Georghiou and Mellon (1983)
Spodoptera littoralis	× 29, cypermethrin × 24, deltamethrin × 13, fenvalerate	Many insecticides and pyrethroids	Egypt, lab	El-Guindy et al. (1982b)
Plutella xylostella	× 2800, fenvalerate × 2200, decamethrin × 900, cypermethrin × 110, permethrin Strong, all pyrethroids	Many insecticides Many insecticides Many insecticides Many insecticides	Taiwan, field, lab Thailand, Phillipines, Indonesia	Sun (1982), Liu et al. (1981, pers. comm.) Hervé (1982, pers. comm.)
Heliothis virescens	× 54, permethrin × 21, fenvalerate × 5, deltamethrin	Many insecticides Methyl parathion	USA, lab	Sparks (1981)
Heliothis zea	× 70, permethrin × 18, fenvalerate	Many insecticides Methyl parathion	Mexico, lab	Wolfenbarger et al. (1981)
Wiseana cervinata	Strong	—	New Zealand, field	Du Toit et al. (1978)

parts of the 'Cotton Belt' but increased tolerance acquired during the control season disappeared quickly when the insects were no longer selected with insecticide (Nye, 1982, personal communication). However, in the Imperial Valley of California tolerance to pyrethroid increased not only during the spray season but also steadily between 1979 and 1981 (Martinez-Carvillo and Reynolds, 1983). In contrast, in South Carolina resistance to permethrin decreased between 1978 and 1980 from × 14 to × 3.5 (Brown et al., 1982). Widespread variations (15-fold) in tolerance to permethrin throughout the USA were also recorded in widely separated populations of T. ni (Shelton and Soderlund, 1983) although control in all cases was satisfactory.

Adult Colorado beetles, *Leptinotarsa decemlineata* Say from Long Island, New York, were significantly (× 39) more resistant to fenvalerate than a susceptible, field-derived population, and were also cross-resistant to permethrin, deltamethrin and fenpropathrin (Soderlund, 1984, personal communication).

Cross-resistance studies have shown pyrethroid resistance in several rice pests (Table 4), but since present day pyrethroids are unlikely to be used in rice fields these interesting findings are of no immediate practical importance.

Variants of *Myzus persicae* Sulzer with elevated levels of the esterase E4 show not only strong resistance to the pyrethroids, but retain the ability to transmit plant virus disease (Gibson et al., 1982; Rice et al., 1983).

Several pests of protected crops resist pyrethroids. By 1964, pyrethroids were no longer effective in the UK against *Megaselia halterata* (Wood) (Hussey, 1964). Recently, Wardlow (1984, personal communication) found strains of *Trialeurodes vaporariorum* (West) virtually immune to the photostable pyrethroids. Until then, pyrethroid resistance in this species under glass in the UK had been confined to non-persistent pyrethroids such as resmethrin (Wardlow et al., 1976). In the USA weekly, all-year-round treatments of chrysanthemums and celery with pyrethroids have resulted in control failure of *Liriomyza trifolii* (Burgess) (Parella, 1983).

Beneficial arthropods (Table 5)

It is encouraging to note that resistance to pyrethroids is not confined to pests. Two predatory mites, *Amblyseius fallacis* German (Croft and Wagner, 1981; Strickler and Croft, 1982; Croft et al., 1982; Croft, 1983) and *Mataseiulus occidentalis* (Nesbitt) (Hoy et al., 1980; Hoy and Knop, 1981) used in biological control in orchards are now sufficiently resistant to pyrethroids to survive during and after treatment of orchards with pyrethroids (Croft, 1983; Hoy et al., 1983). *Chrysopa carnea* (Stephens), which is exceptionally resistant to pyrethroids, also survives treatment of cotton fields in Texas (Plapp and Bull, 1978) (Table 5).

Table 5 Resistance to pyrethroids in arthropods used in integrated pest management

Species	Level of resistance	Selection	Origin/detection	Authors
Amblyseius fallacis	× 15, permethrin	Many insecticides and pyrethroids	USA, field/lab	Strickler and Croft (1981)
Metaseiulus occidentalis	× 10, permethrin	Permethrin	USA, lab	Hoy and Knop (1981)
Chrysopa caraea larvae	Very strong Pyrethroids	Natural tolerance?	USA, field/lab	Plapp and Bull (1978) Ishaaya and Casida (1981)

Table 6 Cross resistance to several pyrethroids by houseflies of different origins

Insecticide	Resistance factor of strain (susceptible = 1.0)								
	NPR[1]	[2]	I[3]	W[4]	Kdr	Super-kdr	EO.39[5]	S-kdr + EO.39	SBC[6]
Pyrethrins	178	290[2]	6	16	4	71	1	26	—
Bioresmethrin	160	71	10	200	11	58	4	1200	86
Permethrin	142	1200	9	280	10	43	2	420	35–40
Deltamethrin	200	20 µg = 10% kill	5	3000	18	600	2	20 µg = 10% kill	35

1, Danish strain selected with natural pyrethroids (Farnham, 1973).
2, Danish strain selected with resmethrin (Farnham, 1973).
3, British strain uncontrolled by natural pyrethroids/piperonyl butoxide sprays (Nicholson and Sawicki, 1982).
4, British strain resistant to pyrethroids.
5, Strains with EO.39 derived from I.
6, USA selected with bioresmethrin (De Vries and Georghiou, 1980).

Compiled from published and unpublished data by Farnham, 1973; De Vries and Georghiou, 1980; Nicholson and Sawicki, 1982; Sawicki et al., 1984.

DIFFERENTIAL RESPONSE IN PYRETHROID-SUSCEPTIBLE
AND RESISTANT STRAINS

Negative cross-resistance

Resistance to one group of insecticides sometimes increases susceptibility to compounds of other group(s). This phenomenon, termed 'negative cross-resistance', occurred in some organophosphorus (OP)-resistant strains of *Tetranychus urticae* (Koch) which were more susceptible to fenvalerate than fenvalerate-susceptible strains (Chapman and Penman, 1979, 1980), and in OP-resistant strains of *Nephotettix cincticeps* (Uhler), *Nilaparvata lugens* (Stal.) and *Laodelphax striatellus* (Fall.) (Miyata, 1980; Ozaki, 1980; Miyata *et al.*, 1982). Negative cross-resistance may be due to the presence of polymorphic enzyme(s) within these populations, one form of the enzyme degrading compounds of one chemical group much faster while conferring sensitivity to compounds of the other, and vice versa. With this type of polymorphism, selection with and resistance to compounds of one group would automatically result in increased susceptibility to the other.

Synergism

Mixtures of pyrethroids/non-pyrethroids or pyrethroids/synergists are sometimes more effective against pyrethroid-resistant than against pyrethroid-susceptible strains of an arthropod species. Thus mixtures of organophosphorus insecticides and pyrethroids even at ratios of 100:1 were mutually synergistic (Nolan and Bird, 1977), producing very high coefficients of cotoxicity (Sun and Johnson, 1960) against strains of *B. microplus* resistant to OPs (Schnitzerling *et al.*, 1982), and mixtures of chemicals of these two groups controlled the spread of sugar beet yellow virus by *M. persicae* and *Aphis fabae* (Scopoli) (Procida, 1982).

Although piperonyl butoxide sometimes increases the toxicity of the natural and synthetic pyrethroids more against pyrethroid-resistant than susceptible strains of *S. granarius* (Lloyd, 1973), *Sitophilus oryzae* L. (Carter *et al.*, 1975) or *M. domestica* (Farnham, 1973), this increased synergism has so far not warranted commercial exploitation in controlling pyrethroid-resistant *P. xylostella* (Sun and Liu, 1982, in Glynn-Jones, 1983) or *H. armigera* (Edge, 1984, personal communication). However, suppression of fenvalerate resistance in *Leptinotarsa decemlineata* Say with standard tank mixes of fenvalerate and piperonyl butoxide at the 1:4 ratio (Giudice and Silcox, 1981, in Glynn-Jones, 1983) has prompted trials of piperonyl butoxide/pyrethroid mixtures on a commercial scale.

CROSS-RESISTANCE

Pyrethroid resistant species or strains often show considerable variations in their cross-resistance spectra, not only to non-pyrethroid insecticides, but also to compounds within the pyrethroid group itself.

Within the pyrethroid group

There has been as yet no systematic study of the cross-resistance within the pyrethroids. Most of the recently published data on cross-resistance refer to a few or several of the commonly used photostable pyrethroids against strains whose resistance is almost always undefined, unstable or uncharacterized. It is, therefore, difficult to generalize about cross-resistance characteristics within pyrethroids, especially since, as pointed out by Casida *et al.* (1983) and Briggs *et al.* (1983), seemingly small structural changes sometimes have a large influence on the relative potency and species selectivity of pyrethroids. Casida *et al.* (1983) gave as example the high sensitivity of deltamethrin-resistant *P. xylostella* to the analogue of bioresmethrin, whose terminal methyl groups are replaced with bromine (houseflies with *kdr* are resistant to this analogue). The difference in relative potency is even greater between deltamethrin and its dichloro analogue, WL 48281, in a deltamethrin-selected strain of houseflies (Sawicki, in preparation). In general, arthropods resistant to a pyrethroid are resistant to other members of the group, but there are some exceptions: a strain of *N. lugens* strongly resistant to permethrin ($\times$ 171) was almost susceptible ($\times$ 5) to fenvalerate (Chung *et al.*, 1981), *B. microplus* resistant to permethrin was not resistant to cypermethrin (Schnitzerling *et al.*, 1983), and resmethrin-resistant strains of *T. vaporariorum* (Wardlow *et al.*, 1976) did not resist permethrin.

The levels of resistance to different compounds of the group, as well as the speeds of selection, vary not only between different arthropods but also between the same strains selected with different pyrethroids (Farnham, 1971; Farnham and Sawicki, 1976) (Table 6) or with different enantiomers (Priester and Georghiou, 1978). It is therefore not possible to generalize or draw meaningful conclusions about the mechanisms responsible or reasons for the existence of strong or slight resistance. On the whole, the strength of resistance is greatest, or anyway large, towards the selecting agent but the nature of the isomers used in the selection is also important. For example, the larvae of the same strain of *C. pipiens quinquefasciatus* selected with (1R)-*trans*-permethrin were more resistant to both isomers than larvae selected with the (1R)-*cis* isomer (Priester and Georghiou, 1978). According to these authors, the site of action of these compounds appears to be strongly specific. This is unlikely because the pharmacokinetics of the different isomers seem to be the same, at least in *Periplaneta americana* (Soderlund, 1979). It is probably also not so if the site of action is *kdr* as suggested by these authors, because two distinct forms of *kdr* are not likely to coexist in the same species. More probably this strong specificity was due to the selection of esterases with different selectivities towards the two isomers of permethrin.

The use of synthetic pyrethroids such as deltamethrin, cypermethrin, permethrin or fenvalerate may elicit higher levels of resistance than natural pyrethrins, allethrin or resmethrin at least in houseflies, but here also there are exceptions. A strain of *M. domestica* selected with natural pyrethroids became very resistant to these compounds (LD_{50} greater than 50 µg per insect, i.e. virtual immunity) but less so to resmethrin ($\times$ 140) (Farnham, 1971). However, when the same strain was selected with resmethrin, resistance to it reached $\times$ 52, but that to natural pyrethrins was only $\times$ 4 (Farnham, 1973). Yet, when *S. granarius* was selected with natural pyrethroids, the resistance to resmethrin ($\times$ 207) was higher than to natural pyrethroids ($\times$ 148) (Lloyd, 1973).

Such differences are likely to be caused by the preferential selectivity of one or more of the several pyrethroid resistance mechanisms present in at least some of the arthropods. Although the selector probably selects one or more factor(s) common to most compounds, it is likely to select most effectively factors specific to itself and closely related analogues. As a result, resistance is generally stronger to the selector than to other members of the group.

Insecticides from other chemical groups

Pyrethroid resistance has at times developed rapidly in pest populations, and within wide areas, e.g. in *P. xylostella* of the Far East, *M. domestica* in Western Europe, or *H. armigera* in Australia. In those areas, pyrethroids had not been used in agriculture nor extensively for fly control on animal farms. This rapid build-up of resistance suggests that pyrethroid resistance factors were present at fairly high frequencies in populations of these pests before the introduction of pyrethroids for pest control. The most plausible explanation is the assumption that such factors were already common because they conferred not only resistance to and were selected by the non-pyrethroids used in the past to control the pests, but also happened to confer cross-resistance to pyrethroids. Thus widespread use of these pesticides ensured selection of factors conferring resistance to pyrethroids.

So far, indications of cross-resistance between pyrethroids and cyclodienes have only been found in *H. armigera*, where a '*kdr*'-like resistance may involve pyrethroids and endosulfan (Gunning, 1984, personal communication).

Cross-resistance between organophosphorus insecticides and pyrethroids seems to be rare since there are few recorded instances, e.g. *M. domestica* (Sawicki *et al.*, 1984), *N. lugens* (Chung *et al.*, 1981) and the CTC 12 strain of *T. castaneum* (Lloyd *et al.*, 1976). This sparsity is surprising since esterases hydrolysing organophosphorus insecticides are also of major importance in the metabolism of pyrethroids even in susceptible strains (Jao and Casida, 1974;

Shono *et al.*, 1978; Abdel-Aal and Soderlund, 1980; Casida and Ruzo, 1980; Ishaaya and Casida, 1980).

Most pyrethroid cross-resistance has been with DDT. It is assumed to relate to a common step in their mode or site of action; pyrethroids and DDT modify the Na$^+$ channel in the nerve membrane (Narahashi, 1980, 1983; Miller and Adams, 1982; Vijverberg *et al.*, 1982) in a very similar way and reduce the speed of the closing of the gate(s), but at very different concentrations.

kdr (KNOCK-DOWN RESISTANCE)

One, perhaps the only, mechanism responsible for cross-resistance between DDT and the pyrethroids is called *kdr*. In the housefly it delays the onset of knock-down by DDT and the pyrethroids. Strictly speaking, *kdr* is a mechanism conferring resistance to DDT, its analogues and the pyrethroids only in the adult housefly, but the term is now widely used in the literature to describe pyrethroid resistance in arthropods that also happen to resist DDT. This complicates further the interpretation of studies on the resistance to, and mode of action of, DDT and pyrethroids, and of *kdr*.

In the housefly, this mechanism was first observed by Busvine (1951) who described cross-resistance between DDT and the pyrethroids, while Milani (1954), who isolated the *kdr* factor genetically from other DDT resistance mechanisms, demonstrated that it is controlled by a recessive gene on autosome 3, about 50 cross-over units from *bwb*, the gene for the brown body colour mutant. Several alleles have since been described, which according to Farnham (1977) are identical to the allele originally described by Milani. Sawicki (1978) isolated from a Danish strain the allele *super-kdr* (Table 6) which confers much stronger resistance than *kdr* to DDT, its analogues and the pyrethroids.

In houseflies, *kdr* delays the symptoms of knock-down and decreases kill by DDT and the pyrethroids. It is a non-metabolic resistance mechanism (Nicholson *et al.*, 1980b; Nicholson and Sawicki, 1982) and is unaffected by the usual DDT and pyrethroid synergists (Sawicki and Farnham, 1967). Lack of synergism of DDT with piperonyl butoxide and FDMC (2,2-*bis*-(4-chlorophenyl)-1,1,1-trifluoroethanol) has been used for detecting *kdr* in houseflies and other insects for want of better or more positive identification test.

Tsukamoto *et al.* (1965) demonstrated lowered nerve sensitivity to DDT in adult houseflies where resistance to DDT was only conferred by a recessive gene on autosome 3 (Tsukamoto and Suzuki, 1964), and subsequent studies on the central nervous system (Miller *et al.*, 1979, 1983) showed insensitivity to pyrethroids. Work on housefly larvae (Osborne and Hart, 1979) has also shown differences in the sensitivity of the nervous system of larvae of the susceptible and *kdr* strains. Since nerve insensitivity in adults and larvae is approximately the same (Osborne 1984, personal communication), other factors probably account for the large difference in response to pyrethroids between adults and larvae (see

p. 169). Housefly larvae like those of *Lucilia sericata* (Mg.) (Nicholson *et al.*, 1983) are extremely tolerant to topically applied pyrethroids (Potter, 1981, personal communication) and the very large differences in nerve sensitivity between larvae of susceptible and *kdr* flies recorded by Osborne and Hart (1979) almost certainly do not reflect pyrethroid resistance caused by *kdr* in housefly larvae.

Nature of *kdr*

Although it is accepted that *kdr* decreases the sensitivity of the nervous system to DDT and the pyrethroids (Beeman, 1982), its exact nature or mode of action has not yet been fully worked out.

Biochemical differences in the nerve membrane

Chiang and Devonshire (1982) demonstrated a modification in the phospholipid composition of the nerve membrane in houseflies with *kdr*. Arrhenius plots showed that the transition temperature in activity of the membrane-bound acetylcholinesterase of the housefly nervous system on susceptible flies is 14°, 19° in flies with the *kdr* gene and 21° in the strain with *super-kdr*. These changes are associated with a lowering in the activation energy of the acetylcholinesterase above and below these temperatures and probably reflect changes in the conformation of the nerve membrane of the three strains at different temperatures through differences in the phospholipid composition of the membrane. After digestion with the phospholipase A2 the transition temperature changed, but only in the susceptible strain where it increased from 14° to 19°, i.e. to the transition temperature of the *kdr* strain. It seems therefore that in the housefly, the resistance caused by *kdr* may be due to a modification of the phospholipids of the nerve membrane. The genetics of the results of Arrhenius plots agree with those for *kdr*; in the hybrid: susceptible × *super-kdr*, the plot is identical with that of the susceptible strain. Thus, the expression of the gene like that of *kdr* is recessive.

Differences in receptor binding

According to Chang and Plapp, *kdr* reduces binding of DDT and *cis*-permethrin to specific membrane receptors in the head of the housefly (1983a,b), *H. virescens* and *C. carnea* (1983c), but not at precisely the same site for these two chemicals. Receptor binding in these insects accounts for from 30 to over 80% of the total binding. This strongly contrasts with the receptor-like stereospecific binding of pyrethroids in mice brain tissue which accounts for a mere 2.8% of total binding (Soderlund *et al.*, 1983a). *In vivo* heads of susceptible flies bound two to three times more DDT than *kdr* flies derived from the Orlando DDT strain (Chang and Plapp, 1983a), which carries the gene for reduced penetration of insecticides (Plapp and Hoyer, 1968b) on the same autosome as *kdr*. Reduced binding of DDT and *cis*-permethrin was also found *in vitro* in *kdr* houseflies, but the chemical nature of the bound radioactivity was not established. The binding of both insecticides, apparently stable in *H. virescens*, was readily reversible in *C. carnea*, and this difference could be correlated with the selective toxicity of these insecticides to *H. virescens* and *C. carnea*. The latter has esterase(s) with exceptionally strong pyrethroid-hydrolysing activity, especially towards the *cis* compounds (Ishaaya and Casida, 1981; Bashir and Crowder, 1983).

Neurophysiological studies

From the examination of a number of neurotoxins which specifically attack clearly defined regions of the Na^+ channel in the nerve membrane, Osborne and Smallcombe (1983) concluded that pyrethroids, DDT and veratrine act on the Na^+ channel within the lipophilic environment of the membrane adjacent to, or near, the Na^+ gate proteins since polar toxins such as tetrodotoxin which binds to the outer orifice of the Na^+ channel, and scorpion venom which perturbs the voltage-dependent gates are effective against *kdr*. The *kdr* factor may constitute a change in the lipid moiety of the nerve membrane, as found by Chiang and Devonshire (1982), which decreases the solubility or binding of lipophilic drugs in the region of the voltage-sensitive channels. However, Osborne and Small-combe (1983) did not rule out a change in the composition of the Na^+ channel or membrane because proteins can effect the dissolution of lipophilic compounds in the faulty components of the membrane.

Neurophysiological studies with vertebrates and insects upon poison symptoms (Clements and May, 1977; Narahashi, 1980; Gammon *et al.*, 1982; Glickman and Casida, 1982; Casida *et al.*, 1983; Scott and Matsumura, 1983; Saldago *et al.*, 1983a,b) indicate that pyrethroids fall into two groups (Narahashi, 1980; Gammon *et al.*, 1981; Casida *et al.*, 1983). Compounds of Type I, which lack the α-cyano group, produce restlessness in the cockroach, followed by incoordination, prostration and paralysis, and repetitive discharges in the cercal sensory nerves (Gammon *et al.*, 1982) and the nerve cord (Scott and Matsumura, 1983). Type II compounds, which have the α-cyano group, produce ataxia and seizure accompanied by intense hyperactivity. Type II pyrethroids act stereospecifically to inhibit binding of *t*-butylcyclophosphorothionate, a ligand for picrotoxin binding sites to rat synaptic brain membrane. Type I pyrethroids are either much less potent or inactive. This inhibition possibly involves a closely associated site in the γ-aminobutyric acid (GABA) receptor–ionophore complex (Lawrence and Casida, 1983).

In insects, type II compounds increase the miniature excitatory post-synaptic potential rate, indicating depolarization of nerves by modification of Na^+ channels at much lower concentrations than type I compounds which are more potent at inducing repetitive firing. The results of Saldago et al. (1983a,b) indicate that larvae of kdr-resistant houseflies resist the depolarizing action of deltamethrin and aconitine, but not the K^+ channel blocker, the tetraethylammonium ion, hence kdr decreases the sensitivity of the Na^+ channel to pyrethroids and aconitine. However according to Osborne and Smallcombe (1983) kdr larvae of M. domestica are not sensitive to aconitine. There is therefore disagreement on the action of an important Na^+ channel probe on kdr. According to Lund and Narahashi (1983), the differences observed at the membrane potential level between DDT and pyrethroids of both types are explicable in terms of a continuous variation in differences of the kinetics with which the insecticides interact with the nerve membrane Na^+ channels. This might explain why some α-cyano compounds produce type I and non-α-cyano pyrethroids produce type II symptoms (Scott and Matsumura, 1983) and the lack of a clear-cut difference in cross-resistance to pyrethroids of the two groups in houseflies with and without kdr or super-kdr (Sawicki et al., in preparation).

It is obvious from the foregoing that much needs still to be done to work out the mode of action of the pyrethroids, which no longer appear to be physiologically and toxicologically homologous. Even more work will be needed to establish the homology of kdr and its action in various arthropods.

Homology of kdr in different arthropods

Kdr has been invoked sometimes without proof to explain the following.

Cross-resistance between DDT/pyrethroids

Cross-resistance has been definitely established in houseflies (Plapp and Hoyer, 1968a), H. irritans exigua (Schnitzerling et al., 1982), anophelenes (Omer et al., 1980), C. tarsalis (Plapp and Hoyer, 1968a), lice (Cole and Clarke, 1961), the bed-bug (Busvine, 1958), cattle ticks (Whitehead, 1959; Nolan et al., 1977) and has been suggested as the cause of resistance in M. persicae in Australia (Attia and Hamilton, 1978; Frazman et al., 1980) and A. fallacis (Scott et al., 1983). In the majority of resistant strains, selection by either DDT or pyrethroids increases the level of resistance to both groups, but the level of resistance varies not only between species but also between strains because of additional factors. The dominance of the gene, moreover, seems to vary between species; in the housefly it is recessive, but its analogue in Ae. aegypti DDT2-py is semi-dominant (Malcolm, 1983a,b).

In B. germanica, the DDT/pyrethroid resistance referred to as kdr by Scott and Matsumura (1981, 1983) covers allethrin, natural pyrethroids, permethrin and fenvalerate, but not deltamethrin or cypermethrin. Scott and Matsumura have tried to correlate this unusual cross-resistance to the two types of action produced by pyrethroids. This '*kdr*' factor is correlated with decreased nerve sensitivity to DDT and permethrin, resistance to DDT and pyrethroids of type I at kill, and even stronger resistance at knock-down. However, the strain strongly resists fenvalerate although its activity at the nerve level is of type II. There are thus important differences between the *kdr* of houseflies and the *kdr* of this strain of *B. germanica* which need resolving.

Absence of metabolic differences between susceptible and resistant strains

Lack of metabolic differences for pyrethroids or DDT between susceptible and resistant strains of the housefly (Nicholson *et al.*, 1980b; De Vries and Georghiou, 1981b), *S. littoralis* and *exigua* (Fullbrook and Holden, 1980) and, according to Nicholson *et al.* (1980a), *B. microplus* (Malchi strain), has been attributed to *kdr*. This agrees with the hypothesis that such resistance is due to decreased sensitivity at the site of action and not to metabolic difference(s).

Reduction of the sensitivity of the nervous system to DDT and pyrethroids

This phenomenon first observed in larvae and adults of resistant and susceptible strains of houseflies (Tsukamoto and Suzuki, 1964; Miller *et al.*, 1979; De Vries and Georghiou, 1981a), has also been reported in *An. stephensi* (Omer *et al.*, 1980), *S. littoralis* (Gammon and Holden, 1980), *B. microplus* (Nicholson *et al.*, 1980a) and *H. virescens* (Nicholson and Miller, 1985). The lowering of susceptibility varies very much with the test species, the techniques used and the part of the nervous system studied (Miller *et al.*, 1979; Osborne and Hart, 1979; Gammon, 1980; Gammon and Holden, 1980; Omer *et al.*, 1980; Beeman, 1982). It is therefore impossible to assume that decreased sensitivity of the nervous system of arthropods or cross-resistance between DDT and the pyrethroids is in all cases due to the same mechanism as *kdr* in the housefly. Thus the term '*kdr*-like' nerve insensitivity should be used instead of *kdr* when homology with the housefly *kdr* has not been fully established.

Contribution of *kdr* to pyrethroid and DDT resistance

In many arthropods, most of the resistance to DDT and/or the pyrethroids has been attributed directly or implicitly to the presence of a *kdr*-like mechanism. Yet the exact contribution of this resistance factor has only been ascertained for *kdr* and *super-kdr* in *M. domestica* (Plapp and Hoyer, 1968a; Farnham, 1973, 1977; Rupes and Pinterova, 1976; Sawicki, 1978), and for DDT2-py in *Ae. aegypti* (Malcolm, 1983a,b).

   In the housefly *kdr* and *super-kdr*, both recessive, confer stronger resistance to DDT than to most pyrethroids, but the analogous non-recessive gene DDT2-py in *Ae. aegypti* is more effective against permethrin than DDT. Both *kdr* alleles are least effective against the natural pyrethroids, and most against deltamethrin, *super-kdr* particularly so. For unknown reasons, female houseflies with *kdr* are much more resistant than the males in topical application but not in residue tests (Sawicki *et al.*, unpublished data).

In the F_1 hybrid, *kdr* × *super-kdr* resistance to pyrethroids and DDT is intermediate (Sawicki, unpublished data). *Pen*, the penetration delaying factor (Plapp and Hoyer, 1968b; Sawicki and Farnham, 1968) delays the onset of poisoning but does not increase resistance of *kdr* (Farnham, 1973), and the esterase E0.39, which confers strong resistance to some organophosphorus insecticides, increases pyrethroid resistance by a factor equivalent to the weak (*ca.* 2–4-fold) resistance conferred by E0.39 alone, when introduced into a strain homozygous for *super-kdr* (Sawicki *et al.*, 1984).

In contrast, *kdr*, though recessive, strongly increases the expression of DDT-ase even in the hybrid (Grigolo and Oppenoorth, 1966), and current work (Sawicki *et al.*, in preparation) has demonstrated exceptional intensification between *super-kdr* and the barely detectable 161 factor of resistance on autosome 2. The selection of the hybrid (*super-kdr* × autosome 2 R) with deltamethrin ultimately yielded a strain virtually immune to deltamethrin ($LD_{50} > 50$ µg per female), but the increase in resistance to other pyrethroids was far smaller. Most surprising was the comparatively small increase in LD_{50} to WL 48281, the dichloro analogue of deltamethrin; it increased only from 0.22 µg in the *super-kdr* strain to 1.4 µg per female in the hybrid. This intensification of resistance was not matched by increased metabolism of deltamethrin (Devonshire, 1982, personal communication), and the mode of action of the intensifier(s) remains unknown. Similar intensifiers may well account for the very strong resistance to pyrethroids (De Vries and Georghiou, 1981b; Lee *et al.*, 1982) in which no significant increase in metabolism has been found.

PYRETHROID METABOLISM

Soderlund *et al.* (1983c), who recently reviewed the metabolism of pyrethroids in insects, have emphasized that it is far less known and documented than in mammals, where metabolism has been amply reviewed (Casida *et al.*, 1976, 1979; Miyamoto, 1976; Hutson, 1979; Casida and Ruzo, 1980; Chambers, 1980) because insect metabolic studies are not a statutory requirement for registration.

Our knowledge and understanding of the nature and role of metabolism in resistance to pyrethroids is even more limited; most of the available information comes from a relatively small number of studies done sometimes with unresolved pyrethroids on a limited number of species often with poorly defined or characterized resistance. In most cases, only slight metabolic differences were found and the number and/or contribution of individual resistance mechanisms was not known. Unidentified metabolites, and reliance on synergists, whose mode of action is often insufficiently understood or too complex to be diagnostic, have prevented the unequivocal characterization of most of these resistance mechanisms. Also, there has been little work to quantify the role of enzymes involved *in vitro*.

In spite of these limitations, research into metabolic causes of resistance has shown a great diversity in the mechanisms involved, which range from increased hydrolytic activity of either the *cis* or *trans* isomers, through increased oxidative metabolism to possible sequestration by large amounts of a catalytically inefficient enzyme.

Musca domestica

The study of the metabolic mechanisms of resistance to pyrethroids in this species illustrates the difficulties referred to above; none of a number of metabolic systems identified has been unequivocally linked with resistance. De Vries and Georghiou (1981b) found no difference in the overall metabolism of (*1R*)-*trans*-permethrin by a strain 100–140-fold resistant to this insecticide. Mixed function oxidases (as judged from measurements of cytochrome P450, cytochrome b_5, and activity of NADPH cytochrome c reductase) were not involved; DEF (*S,S,S*-tributylphosphorotrithioate), considered to be an esterase inhibitor, synergized permethrin little. Although the resistant strain retained less and excreted more metabolites and permethrin, overall differences in metabolism were slight. De Vries and Georghiou concluded that decreased penetration coupled with *kdr*, hindered the toxicant reaching the threshold level at which poisoning occurred, so delaying the onset of uncoupled convulsion (De Vries and Georghiou, 1981a).

Permethrin resistance was also studied in the weakly resistant (2–6-fold) R2 strain, lacking *kdr*, derived from a field population treated with synergized natural pyrethroids (Nicholson and Sawicki, 1982), and fenvalerate resistance in a permethrin-selected strain with 1000-fold resistance to fenvalerate (Lee *et al.*, 1982). In neither were the metabolites or their amounts greatly different from those in susceptible insects.

The R2 strain degraded injected *trans*- and *cis*-permethrin slightly faster than the susceptible (S) strain, and both strains metabolized more of the *trans* than *cis* isomer(s). The amounts of the *cis* and *trans* acids were negligible (4% *trans* in R2 and S; 2% *cis* in R2 and S) and most metabolites (unidentified) were in the aqueous methanol phase. NIA 16388 (propyl prop-2-ynyl phenylphosphonate) was more effective than piperonyl butoxide in reducing the degradation of the *trans* isomer. Since there was little further reduction when the two synergists were applied together, NIA 16388 probably also inhibited metabolism sensitive to piperonyl butoxide, in line with its postulated dual mode of action, as an inhibitor of both esterases and mixed function oxidases (Ueda *et al.*, 1975). Both synergists also inhibited metabolism of *trans*-permethrin but less in R2 than S insects, and this is presumably why NIA 16388 conferred stronger synergism in susceptible insects.

Lee *et al.* (1982) found very little metabolism of fenvalerate in the 1000-fold resistant strain. It penetrated slightly slower into R insects, and of the 6% internal radioactivity in the S strain, 4.5% were metabolites of which the only one identified was the 2-(4-chlorophenyl)-3-methyl-butyric acid. In resistant insects, 9.5% of the 17% (almost three times more than in susceptible insects) internal radioactivity consisted of a conjugate of 2-(4-chlorophenyl)-isovaleric acid resulting from ester cleavage. Other unknown conjugates accounted for 4% of the total radioactivity. Clearly the large differences in tolerance cannot be explained solely by the moderate differences in metabolism, and it must be assumed that most of the fenvalerate resistance was caused by an unknown mechanism, possibly *kdr*. The latter confers strong resistance to fenvalerate (Sawicki *et al*, 1983, unpublished results).

Boophilus microplus

Pyrethroid resistance in the Malchi strain was examined by Nicholson *et al.* (1980a) and Schnitzerling *et al.* (1982), and their conclusions differ. According to Nicholson *et al.*, who injected the ticks, Malchi differed from other strains by decreased sensitivity of its nervous system to permethrin. Racemic *trans*-permethrin was more readily hydrolysed than racemic *cis*-permethrin in the three strains, and paraoxon increased the internal persistence of permethrin presumably by inhibiting degrading esterase(s).

According to Schnitzerling *et al.* (1982), Malchi (permethrin-selected) and DDT-R (DDT-selected) larvae resisted knock-down by permethrin, but were susceptible to knock-down by cypermethrin or deltamethrin, although the *cis* and *trans* forms of cypermethrin penetrated slower into Malchi than into the DDT-R or Y (susceptible) larvae. Both strains resisted the three pyrethroids at kill, Malchi about two to three times more than DDT-R. The *trans* form was metabolized faster than the *cis* analogue, but unlike most other arthropods, the three strains metabolized cypermethrin faster than permethrin. Malchi metabolized both permethrin and cypermethrin marginally faster than did the other strains, the major metabolites being the constituent alcohols. The addition of coumaphos reduced metabolism of cypermethrin more than of permethrin and this was reflected in the stronger synergism of cypermethrin. Schnitzerling *et al.* concluded that resistance to knock-down is caused in Malchi by insensitivity of the target, and resistance to kill, by detoxication.

It seems more likely that Malchi has two resistance mechanisms for kill: nerve insensitivity (Nicholson *et al.*, 1980a) and weak metabolic resistance, presumably hydrolytic, inhibited by coumaphos. Nerve insensitivity and hydrolytic activity were presumably selected in Malchi by permethrin, whereas in DDT-R,

DDT selected only nerve insensitivity. This may by why Malchi was slightly more resistant to pyrethroids than DDT-R, but did not differ in resistance to DDT.

Wiseana cervinata

Wiseana cervinata larvae have strong insensitivity to cypermethrin (LD_{50} *ca.* 6 µg per larva) and even stronger to permethrin (LD_{50} greater than 20 µg per larva) (Du Toit *et al.*, 1978). This is probably caused either by the ability of larvae to store large amounts of the unchanged pyrethroid in their body or by the insensitivity of their nervous system to these chemicals. Up to 0.5 µg of unchanged *cis*-cypermethrin were recovered from unaffected larvae, 3 and 24 h after application of 2.5 µg per larva (Chang and Jordan, 1982).

Permethrin and cypermethrin were metabolized by pathways similar to those in other insects (Soderlund *et al.*, 1983a), i.e. by hydrolysis and, to a lesser extent, by oxidation, inhibited by carbaryl or pirimiphos-methyl and piperonyl butoxide respectively. The primary metabolites were rapidly conjugated to water-soluble compounds. Esterase activity was mainly located in the gut, which was *ca.* eight times more active on a per gram of tissue basis than either the fat body or the carcass (Chang and Jordan, 1983). *cis*-Permethrin was less readily metabolized than *trans*-, or the two isomers of cypermethrin, and this difference in metabolic rate may explain the inherently higher toxicity of cypermethrin to larvae of this moth. The *trans* isomers were absorbed faster at low doses (0.5 µg) than the corresponding *cis* isomers, but this difference in penetration was not apparent at higher doses (5 µg per larva). Excretion was negligible even after 24 h, and Chang and Jordan (1982) believe that the body levels of unchanged pyrethroids depended on interaction between penetration and metabolism. Of several enzyme inhibitors examined, piperonyl butoxide was mainly responsible for inhibition of oxidative metabolism and had little effect on the production of either *cis*- or *trans*-3-(2,2-dichlorovinyl)-2,2-dimethylcyclopropanecarboxylic acids. This was significantly inhibited *in vitro* by 1 µM of either carbaryl or pirimiphos-methyl. Dodecylimidazole was only effective in inhibiting hydrolysis *in vivo* (Chang and Jordan, 1983). This compound also inhibited hydrolysis of permethrin in *Spodoptera littoralis*, but not in *Spodoptera exigua* (Fullbrook and Holden, 1980).

Heliothis virescens

Increased detoxication and decreased nerve sensitivity were involved in the 15–20-fold resistance of third instar field-collected larvae of *H. virescens* to topically applied [^{3}H]-*trans*-permethrin (Nicholson and Miller, 1984). Adults derived from larvae of these strains were much less resistant to pyrethroids (Miller, 1983, personal communication).

Increased metabolism was assumed from differences in internal amounts of unmetabolized permethrin, 3–6-fold greater in susceptible insects, a third less 3-phenoxybenzoic acid and unidentified polar metabolites within resistant larvae, and the greater production of unidentified ^{3}H-labelled volatile compounds by resistant insects. Neurobioassays demonstrated that threshold concentration of fenvalerate and permethrin producing an increasee in miniature post-synaptic potentials were 10–50 times greater in the resistant strains than in the susceptible strain. Resistance to cyfluthrin, never used against these strains subjected in the field to permethrin and fenvalerate, was 8–10-fold.

Leptinotarsa decemlineata

Piperonyl butoxide reduced the 39-fold resistance to fenvalerate to 1.5-fold, suggesting oxidative metabolism as the predominant resistance mechanism (Soderlund, 1984, personal communication). The insecticide was extensively metabolized by three major metabolic pathways: hydroxylation of a methyl in the acid moiety, parent ester cleavage, and diphenyl ether cleavage in the alcohol moiety. There was apparently no hydroxylation of the 4′ position of the alcohol moiety. Additional metabolites, including conjugates of both the acid and alcohol moieties, have not yet been fully identified (Soderlund *et al.*, 1983b).

Chrysopa carnea

This is by far the most tolerant insect to pyrethroids; a 7–9 mg third instar larva is unaffected by 112 μg of deltamethrin, and the LD_{50} of *trans*-permethrin is *ca.* 136 μg per larva (calculated from results by Ishaaya and Casida, 1981) (the LD_{50} for deltamethrin to a susceptible 20 mg female housefly is 0.0003 μg). Lacewing larvae metabolized by hydrolysis *ca.* 80% of *cis-* and 71% of *trans*-permethrin topically applied 2 h beforehand (Bashir and Crowder, 1983), confirming the observations of Ishaaya and Casida that the relative rate of hydrolysis of deltamethrin, the *cis* and *trans* isomers of permethrin and cypermethrin coincide with the tolerance of larvae to those pyrethroids. In the lacewing, unlike most other insects, the esterases hydrolyse *cis* isomers faster than the corresponding *trans* compounds. Phenyl saligenin cyclic phosphonate, a potent esterase inhibitor and a malathion synergist (Eto *et al.*, 1966), reduced the tolerance to *trans*-permethrin 68-fold, i.e. down to 2 μg per larva, but this relatively high residual tolerance may, according to Ishaaya and Casida, be due to nerve insensitivity or other mechanisms.

Spodoptera littoralis

Earlier work by Holden (1979) on a strain with 4-fold resistance to permethrin only, failed to show any metabolic resistance mechanism. According to Riskallah (1983) the hydrolytic activity of α and β, naphthyl acetate- and p-nitrophenyl acetate-hydrolysing esterases in a multiresistant strain, 18–49-fold resistant to cypermethrin, deltamethrin and fenvalerate, was three to six times greater than in a susceptible strain. Massive doses of DEF (S,S,S-tributylphosphorotrithioate) inhibited hydrolysis of the naphthyl esters, and in particular hydrolysis of the p-nitrophenyl acetate, and reduced, but did not eliminate resistance to the pyrethroids. Riskallah believes that increased hydrolysis, resistance and synergism by DEF are all caused by increased esterase activity in the resistant larvae. However, increased hydrolytic activity of naphthyl acetate-hydrolysing esterases need not signify increased hydrolysis of pyrethroids. Riddles *et al.* (1983a) noted two distinct forms of p-nitrophenylbutyrate-hydrolysing activity in *Boophilus microplus*, of which only one hydrolysed *trans*-permethrin. Even hydrolysis of the pyrethroid-like esters of p-nitrophenol with racemic *trans*- and *cis*-3-(2,2-dichlorovinyl)-2,2-dimethylcyclopropanecarboxylate, used to assay esterases which detoxify synthetic pyrethroids in *B. microplus* (Riddles *et al.*, 1983b), is not necessarily very specific. Similar compounds, (1R)-*trans*-chrysanthemate esters with 1- or 2-naphthol, were not hydrolysed by the pyrethroid-hydrolysing esterase E4 of *Myzus persicae* (Devonshire, 1984, personal communication).

Besides the esterases, there appears to be in *S. littoralis* residual unsynergizable resistance, which might correspond to the weak nerve insensitivity reported in this species by Gammon and Holden (1980).

Myzus persicae

Purified E4, the esterase which in *M. persicae* is linked with resistance to OPs, carbamates and pyrethroids (Devonshire, 1977; Sawicki and Rice, 1978; Devonshire and Moores, 1982) hydrolysed the (1S)-*trans* isomer of permethrin but none of the other isomers, even when assayed with much more E4 or larger amounts of these chemicals. E4 thus hydrolyses only one of the four isomers of permethrin. From the known catalytic centre activity (k_3) of E4 for α-naphthyl acetate and several insecticides, Devonshire and Moores (1982) calculated that E4 hydrolyses (1S)-*trans*-permethrin 250, 2500 and 8000 times faster than dimethyl and diethyl phosphates and methylcarbamates, respectively.

Since (1S)-*trans*-permethrin is the least toxic isomer of permethrin, Devonshire and Moores (1982) suggested that the resistance of variants of *M. persicae* with large amounts of E4 might be caused by the sequestration of pyrethroids by the large amounts of the E4 protein in resistant aphids (up to 3% of the total protein in the most resistant variants). Subsequent work (Devonshire, unpublished) has confirmed that several pyrethroids interact reversibly with the catalytic centre of E4. Sequestration by E4, as well as its insecticide-hydrolysing activity, plays a major role in resistance to organophosphorus and carbamate insecticides. Sequestration of permethrin by the larval fat body of *Lucilia sericata* has also been postulated as a major cause of the considerable tolerance of the larval instar to this pyrethroid (Nicholson *et al.*, 1983).

Studies of pyrethroid metabolism and pharmacokinetics have generally used a racemate, i.e. a simple pair of enantiomers, either (1*RS*)-*cis*- or (1*RS*)-*trans*-, with the implicit assumption that the enantiomers behave identically. Whilst their chemical and physical behaviour may be so, it is not surprising, in view of the chiral nature of the enzymes, that their biochemical fates differ. In their preliminary experiments with (1*RS*)-*cis*- and (1*RS*)-*trans*-permethrin (i.e. unresolved enantiomer pairs as used in most studies of pyrethroid metabolism), Devonshire and Moores (1982) detected no hydrolysis of the (1*RS*)-*cis* isomers by E4; however, hydrolysis of the (1*RS*)-*trans* isomers was at rates proportional to the E4 activity in S, R1 and R2 aphids, variants with increasing molar amounts of E4 and pyrethroid resistance. Had they not investigated the hydrolysis of individual enantiomers by E4, they might have concluded that hydrolysis of the physiologically active (1*R*)-*trans* isomer was the main cause of resistance in this species. This demonstrates that considerable caution is needed when interpreting the results of studies on the metabolism of the unresolved enantiomers (Elliott and Janes, 1979).

Sitophilus granarius and Tribolium castaneum

Adults of *S. granarius* strongly resistant to several pyrethroids produced two to three times more demethyl pyrethrin II by oxidation of the isobutenyl moiety of the chrysanthemic acid side chain of pyrethrin I, and a conjugate resulting from the terminal hydroxylation in the pentadienyl side chain of the alcohol, than did the susceptible insects. Both strains also produced six other unidentified metabolites. When pyrethrin I was applied with piperonyl butoxide, hydrolysis was the main metabolic pathway in susceptible (S) and resistant (R) insects, and in R insects it was then not much faster than in the S strain (Rowlands and Lloyd, 1976). Tetramethrin was also metabolized faster by R than susceptible *S. granarius* (Lloyd *et al.*, 1976); hydrolysis played a negligible role in the metabolism of this pyrethroid and oxidation of the isobutenyl moiety was even more pronounced than for pyrethrin I.

In the multiresistant CTC 12 strain of *T. castaneum*, tetramethrin was metabolized to an unidentified metabolite which was excreted by the resistant and susceptible insects (Champ and Dyte, 1976).

Amblyseius fallacis

On the basis of bioassays with permethrin, with and without synergists (piperonyl butoxide and DEF), Scott *et al.* (1983) concluded that pyrethroid resistance in this predatory mite involves '*kdr*' and/or esterase activity. Microelectrophoretic-carboxylesterase banding patterns characteristic in permethrin-resistant strains suggest involvement of one or more esterases in pyrethroid resistance (Croft, 1983).

GENETICS OF RESISTANCE TO PYRETHROIDS

The preliminary examination of the genetics of resistance to pyrethroids in several arthropods indicates that in most, resistance is multifactorial, but not in *Ae. aegypti*, where it is attributed to gene DDT2-py (Malcolm, 1983a,b).

Musca domestica

In houseflies (Plapp and Hoyer, 1967; Tsukamoto, 1969, 1983; Farnham, 1973; Rupes and Pinterova, 1976; Nicholson and Sawicki, 1982; Sawicki *et al.*, 1984; Sawicki *et al.*, in preparation) several resistance mechanisms have been isolated and inbred into a susceptible background, using a standard autosome substitution technique (Tsukamoto, 1964). The majority of the resistance factors reported in pyrethroid-resistant strains by different laboratories have been insufficiently characterized to establish their homology or identity.

Singly, almost all the pyrethroid resistance factors confer weak resistance, even when derived from strains virtually immune to the pyrethroid used for the selection. This implies that the very strong resistance of the parent strain results most probably from strong interactions between these weak factors. Some appear to be selected preferentially and/or confer stronger resistance to the selector than to most other pyrethroids.

In the Orlando DDT strain, pyrethroid resistance was due to *kdr*-0 on autosome 3 (Plapp and Hoyer, 1967, 1968a). Analysis of a Swedish strain (Tsukamoto, 1969) showed that the contribution of homozygous autosomes to total kill resistance by natural pyrethroids was 3 > 2 > 5 > 4. Rupes and Pinterova (1976) also isolated an allele of *kdr* on the third autosome of a DDT-selected strain.

Three distinct factors with different levels of resistance and expression were found in the Danish strain NPR selected with natural pyrethroids (Farnham, 1971, 1973). Factor *Py ex* on autosome 2 gave stronger resistance to the synthetic than to the natural pyrethroids, especially in the presence of the synergist Sesamex. Yet with the factor *Py-ses* on autosome 5, which conferred weak resistance to the natural pyrethroids but none to 5-benzyl-3-furylmethyl esters, synergism to natural pyrethrins was even more pronounced than in the susceptible strain. *Kdr* on autosome 3, which gave the strongest resistance, especially to 5-benzyl-3-furylmethyl esters, was unaffected by the synergists used, and greatly delayed but did not eliminate knock-down. *Pen*, the penetration delaying factor, further delayed the onset of knock-down of flies with *kdr*, but did not increase resistance at kill, although it may have intensified the activity of the other two factors.

None of these factors alone conferred greater resistance than 5-fold to natural pyrethrins, yet a proportion of the parental strain from which these factors were isolated was immune to 50 μg natural pyrethroids per female.

The differential action of the synergist Sesamex with *Py-ex* and *Py-ses* is noteworthy, and should serve as a warning against overhasty attribution and identification of resistance mechanisms based solely on the response to synergists of uncharacterized and heterogeneous resistant populations.

The weak (4-fold) resistance to natural pyrethrins in a substrain of the NPR strain examined by Farnham (1971, 1973), selected with and strongly resistant to resmethrin, indicates that resmethrin did not select *Py-ses* which confers resistance only to the natural pyrethrins in strain NPR.

Recent work (Sawicki *et al.*, in preparation) on several strains of houseflies from the UK has identified additional factors, bringing to seven the number of factors likely to confer resistance to pyrethroids in these flies. A minor factor has been detected on autosome 1, but has not yet been examined in detail. It intensifies resistance of *kdr* to deltamethrin (Farnham *et al.*, 1984). The additional factors on autosome 2 are: the esterase E0.39 (Sawicki *et al.*, 1984), and one or more intensifiers of the *super-kdr* resistance to deltamethrin factor 161, and a very weak uncharacterized factor selected by synergized bioresmethrin.

E0.39 is an esterase controlled by a dominant or semi-dominant gene. It is either closely linked with, or responsible for, resistance to several pyrethroids, but not to the natural compounds, and for 10–15-fold resistance to trichlorfon and malathion. The esterase is very common in British houseflies, and is also present in French, Belgian and Swiss strains, but was not found in any of the Danish strains collected by Keiding since 1947 (Sawicki *et al.*, 1984), even in those resistant to trichlorfon, the pyrethroids, or insecticides of both groups.

Kdr, described in detail earlier, unlike the esterase E0.39, whose distribution appears not to be universal, has been recorded throughout much of the Northern Hemisphere: in the USA (Milani and Franco, 1959; Plapp and Hoyer, 1968a; De Vries and Georghiou, 1981a), Canada (Farnham *et al.*, 1984) and Europe, especially in Scandinavia and North Germany (Keiding, 1981; Keiding and Skovmund, 1983) and the UK (Chapman and Lloyd, 1981; Sawicki *et al.*, 1981).

The *Pen* factor on the same autosome as *kdr* greatly retards penetration of pyrethrin I. Four hours after the topical application of [^{3}H]pyrethrin I, over 95% of the insecticide was washed off the surface of flies with *Pen*, whereas only 30% was washed off flies without *Pen* (Farnham, 1973). It is generally assumed that *Pen* acts as a major intensifier of metabolic resistance; by retarding entry of the insecticide it enables the metabolic resistance mechanisms to detoxify the penetrated pyrethroids sufficiently to prevent their reaching lethal concentrations. Such concentrations are likely to be much stronger in insects with than without nerve insensitivity conferred by *kdr*. Therefore the intensifying activity of *Pen* is effective when *kdr* is present together with metabolic resistance mechanisms, but not when metabolic mechanisms are absent (Farnham, 1973).

Other arthropods

Pyrethroid resistance in *B. germanica* is caused by a single autosomal incompletely dominant gene (Cochran, 1973), but can also be caused by the pleiotropic effect of a recessive gene(s) selected by diazinon (Katz *et al.*, 1973) or propoxur (Collins, 1975).

In *C. pipiens quinquefasciatus* the autosomal resistance is codominant in larvae of the (1*R*)-*trans*-permethrin-selected strain, and partly recessive in the (1*R*)-*cis*-permethrin-selected strain. In both, resistance is probably multifactorial (Priester and Georghiou, 1979). Presumably, additional factor(s) selected by (1*R*)-*trans*-permethrin, which confer stronger resistance in this strain, also cause a difference in overall dominance.

So far, only one semi-dominant gene, DDT2-py responsible for some of the DDT but all pyrethroid resistance, and presumably analogous for *kdr* in the housefly, has been found in *Ae. aegypti* (Malcolm, 1983a,b).

Resistance in *P. xylostella* is partly recessive and is probably controlled by several factors (Liu *et al.*, 1981). In the phytoseiid mites *Amblyseius fallacis* and *Metaseiulus occidentalis* resistance is also polyfactorial (Hoy and Knop, 1981; Croft and Whalon, 1983). In *M. occidentalis* incompatibility was observed between the base strain males and selected strain females. In *A. fallacis*, resistance is believed to be mainly controlled by recessive gene(s); the response of the F_1 hybrid intermediate in susceptibility was a little closer to the susceptible than to the resistant parents (Croft, 1983).

DETECTION AND INTERPRETATION OF PYRETHROID RESISTANCE

Resistance is a dynamic phenomenon in which levels of tolerance, cross-resistance patterns, gene frequencies and the overall response of field populations alter ceaselessly as pests come into contact with more, new or different insecticides. This dynamic aspect is ignored in internationally recommended tests for detecting resistance which rely on the comparison of tolerance of field strains with susceptible standard strains defined as 'populations never subjected to insecticidal pressure, and in which resistant insects are rare' (Busvine, 1980). Since such strains no longer represent the overall response of most pest species in the field, the internationally recommended tests seldom, if ever, give adequate information about the practical effectiveness of the insecticide (Sawicki and Denholm, 1984), and the literature abounds with claims of resistance to pyreth-

roids and other insecticides which for the most part are of no practical importance.

It is often assumed, without much evidence, that a 10-fold resistance
measured in laboratory tests signifies resistance of practical importance. This
convenient assumption is hardly ever based on concrete evidence because the
tests used to determine resistance rarely measure or represent control in the
field, e.g. the topical application of insecticides with microapplicators does not
determine the effectiveness of the antifeedant or repellent effects of synthetic
pyrethroids against lepidopteran larvae (Tan, 1981); this can play an important
part in the damage suffered by the crop. El-Guindy *et al.* (1982b) demonstrated
the considerable disparity in the response of *Spodoptera littoralis* to several synthetic pyrethroids in leaf tests that stimulate field treatments better than topical
application tests. Thus effectiveness, levels of resistance and ranking order can
depend on test methods.

Ultimately however, economically significant levels of resistance depend on
the killing power of the recommended field dose. This dose, usually aimed at
controlling at least 90% of the treated field population, is based on field trials
and not on laboratory tests against standard susceptible strains. Moreover, the
recommended dose has usually built-in 'over-kill' so as to achieve truly effective
control, and this 'over-kill' has successfully dealt even with *H. virescens* field
strains which, on the basis of laboratory tests, were over 50-fold more resistant
to pyrethroids than standard laboratory strains.

There have been very few attempts to establish the relationship between the
level of resistance measured in the laboratory and the efficacy of control in the
field (Micks *et al.*, 1980; Ball, 1981) and fewer still to quantify the relationship
between the frequency of an insecticide-resistant genotype, levels of resistance
as measured by bioassays, and control failure.

A study of the tolerance to several pyrethroids and the frequency of *kdr*
homozygotes in 41 housefly strains collected on British animal farms
demonstrated apparent ubiquitous resistance in laboratory tests using a
susceptible laboratory strain as standard (Farnham *et al.*, 1984). Yet although
all field strains were at least four times more resistant to synthetic pyrethroids at
LD_{50} than the standard susceptible strain, and 26 had detectable levels of *kdr*,
pyrethroids failed to control houseflies on only nine of the farms. On these, 10%
or more of the flies were homozygous for *kdr*, and/or LD_{50}s in bioassays corresponded to resistance factors ranging from $\times$ 20 to bioresmethrin to $\times$ 4 to
the natural pyrethrins. Regression analysis showed a good relationship between
LD_{50}s to pyrethroids, and the frequency of *kdr* homozygotes. With this single
instance of quantification of the relationship between the frequency of an insecticide-resistant genotype, levels of resistance as measured by bioassays and control failure, the authors showed the possibility of establishing accurately when
control failed because of resistance.

A great deal of work on similar lines needs to be done for other pests to obtain these parameters if resistance as the cause of control failure is no longer to remain an area of uncertainty and a contentious issue between the user, applicator and the producer of the insecticide.

COUNTERING PYRETHROID RESISTANCE

Several strategies have recently been investigated in both the laboratory and the field to delay or prevent resistance to pyrethroid insecticides.

Minimal use of pyrethroids

This solution has been adopted by Egyptian authorities to control *S. littoralis* on cotton. By the early 1970s, *S. littoralis* had become resistant to virtually all insecticides then used against cotton pests in Egypt (Salama, 1983). The authorities, faced with this very serious situation, restricted the use of photostable pyrethroids solely to *S. littoralis* on cotton, allowing only a single application of pyrethroids per year throughout the country. This was a radical departure from previous practice which allowed unlimited applications per year of the same insecticide against *S. littoralis*, not only on cotton but other crops as well. It had led to the sequential development of resistance to most available insecticides in *S. littoralis*.

The new strategy has proved extremely effective, for although pyrethroids have been used for 6 years, there has been no resistance in the field, although many laboratory experiments have demonstrated that *S. littoralis* has the potential to develop strong resistance to pyrethroids (El-Guindy *et al.*, 1982b; Riskallah *et al.*, 1983). Similar though less drastic steps have been adopted in Eastern Australia to overcome loss of control through pyrethroid resistance in *H. armigera*, which suddenly erupted in January 1983 at Emerald and has been detected in samples of larvae from other areas of Eastern Australia (Gunning *et al.*, 1984). The strategy adopted enforced a ban on pyrethroid usage at Emerald; elsewhere pyrethroid use has been restricted to a 42-day period between January 10 and February 20, the time needed for the first generation of *H. armigera* to complete its life cycle on cotton. No pyrethroids have been allowed after that date. This voluntary strategy applies to all crops harbouring *H. armigera* (Edge, 1983, personal communication).

During the 1983–84 cotton season there was good adherence to the strategy, and pyrethroid failures were very few. Laboratory assays showed that about 10% of the individuals collected in cotton growing areas were heterozygous for pyrethroid resistance, but 'homozygotes' were rare (R. Gunning and N. Forrester, 1984, personal cummunication).

Early reports (Murray and Cull, 1984) indicated that at Emerald the frequency of resistant individuals had dropped since the ban on pyrethroids. Subsequent tests showed some resurgence of resistance in this area, where endosulfan was applied frequently (R. Gunning, 1984, personal communication).

Rotation of pyrethroids with insecticides of other groups

Georghiou *et al.* (1983) examined the effects of selection on resistance in subpopulations of a 'synthetic' strain of *C. quinquefasciatus* incorporating the genomes of three resistant strains: Temephos-R, Propoxur-R and Permethrin-R by single insecticides (temephos, permethrin, propoxur and diflubenzuron) over several generations followed by the relaxation of pressure, as well as selection by short- and long-term rotation of these chemicals. After selection with single insecticides, resistance persisted longest with propoxur, least with temephos and was intermediate for permethrin. In long-term rotation, temephos and permethrin reciprocally suppressed but did not eliminate resistance to each other, but propoxur prolonged permethrin resistance. In short-term selection, resistance became very strong, especially to permethrin. Georghiou *et al.* concluded that long-term rotation can delay resistance if the chemicals used either confer negative cross-resistance (few such chemicals are available) or produce unstable resistance (this last property is unpredictable). Rotation therefore can work but its outcome is often likely to be unpredictable since use of non-pyrethroids may have unexpected consequences on pyrethroid resistance: thus the outcome of the treatment of a strain of houseflies with DDT and trichlorfon was very strong resistance to pyrethroids, since the former selected *kdr*, the latter E0.39 (Sawicki *et al.*, 1984). Both mechanisms confer cross-resistance to pyrethroids. There is strong circumstantial evidence that the very rapid and widespread resistance to pyrethroids in houseflies on animal farms in the UK resulted from the widespread sequential use of those insecticides on these farms.

MacDonald *et al.* (1983a,b) investigated the effects of spraying permethrin or dichlorvos singly, alternately or together in the laboratory against a 'synthetic' strain produced by crossing a susceptible and a multiresistant field strain which resisted, amongst others, dichlorvos and the pyrethroids. They also used both insecticides singly or alternately to control houseflies on several animal farms.

After eight generations of selection in the laboratory, resistance to permethrin was strongest ($\times$ 73) in the strain selected with permethrin only, less strong ($\times$ 31) in the strain selected with dichlorvos and permethrin alternately, and weakest ($\times$ 4 and $\times$ 6) in the other two strains. In the dichlorvos-selected strain, resistance to this compound was highest ($\times$ 27), it was 10-fold in the strain selected alternately, and 4- and 7-fold in the other two strains. Clearly, the mixture was the most effective in preventing build-up of resistance. However, MacDonald *et al.* (1983a) concluded that the mixture was unlikely to be effective in the field, because the persistence of permethrin and the short residual life of dichlorvos could result in flies emerging 3 to 4 days after dichlorvos treatment, thereby being exposed to and selected by permethrin alone; thus alternation was the next best solution. Field tests during two successive summers on one of the farms confirmed that alternate use of permethrin and dichlorvos did not elicit resistance, whereas the use of permethrin alone resulted on one farm in very strong resistance, but not on another. The authors suggest that frequent sanitation delayed the development of resistance on that farm. In spite of this plausible explanation, this series of experiments has left several interesting points unresolved, notably—why in laboratory tests the permethrin/dichlorvos mixture suppressed the selection of *kdr* which is present in the Guelph strain (Farnham *et al.*, 1984) and why the alternate use of a very persistent and a very transient insecticide on the farm did not result in the selection of resistance to be persistent (pyrethroid) insecticides. These points warrant further investigation in view of the initial successful outcome of the strategy used by MacDonald *et al.* (1983b).

Reliance on synergists or cotoxicity

Following the discovery by Nolan and Bird (1977) that small amounts of pyrethroids increase the effectiveness of organophosphorus acaricides against OP-resistant cattle tick, commercial formulations containing mixtures of the two groups of insecticides have recently been introduced to control OP- and pyrethroid-resistant ticks in Australia. It is not possible to predict the long-term effects of these mixtures, particularly against strains of ticks with pyrethroid resistance, e.g. the Malchi strain (Schnitzerling *et al.*, 1982), but work by Mac-Donald *et al.* (1983a) on houseflies indicates that mixtures may be preferable to rotation. Cotoxicity probably works through the inhibition by the OPs and/or pyrethroids of the same enzyme(s), presumably esterases which degrade (hydrolyse) both groups of chemicals. In 1981 Procida introduced Decis b (25 g/l deltamethrin: 400 g/l heptenophos) for the control of aphids on sugar-beet, claiming that this mixture is far more effective than either compound used separately in controlling aphids and decreasing virus. Work in this laboratory suggests that factors other than cotoxicity, e.g. behavioural factors, might be important in this context (Rice, 1984, personal communication).

El-Guindy *et al.* (1982a) have examined the effect of mixtures of pyrethroids and OPs against *Pectinophora gossypiella* (Saunders), and obtained somewhat surprising results Whereas the cotoxicity factors between chlorpyriphos and deltamethrin or cypermethrin are positive (+50 and +60), the cotoxicity factors between this OP and fenvalerate or deltamethrin/profenofos are negative (−30, and −100 respectively). This suggests that there is no general rule as to the effectiveness of OP/pyrethroid mixtures to a single, let alone several, species.

The recent successful control of pyrethroid-resistant Colorado beetles with pyrethroids synergized with piperonyl butoxide (Glynn-Jones, 1983) has rekindled interest in the use of synergists to decrease or eliminate pyrethroid resistance in agricultural pests. Regrettably most synergists are not suitable for field use and DEF, often used in laboratory work to demonstrate the presence of esterase as resistance mechanisms to pyrethroids, is unsuitable because it is very phytotoxic.

Use of pyrethroids with short persistence

When photostable pyrethroids were first introduced for housefly control in the UK, Sawicki (1979, unpublished, in private correspondence to the parties concerned) predicted that persistent pyrethroids would rapidly cause strong pyrethroid resistance in British houseflies. He advised the use of pyrethroids with short persistence, such as bioresmethrin, arguing that resistance was less likely to develop to these insecticides because of their short-lived selecting power. His predictions were based on observations made on three farms on which flies had become resistant to pyrethroids either through the application of residues of the persistent analogues, or through the excessive and prolonged use of synergized natural pyrethrins, and the experience of Keiding (1978) in Denmark where photostable pyrethroids are banned for the control of houseflies on animal farms. The warning went unheeded. Persistent pyrethroids were introduced for the residual treatment of farms, and strong pyrethroid resistance developed rapidly on many British farms.

The long-term (12 months or more) effects of bioresmethrin were examined against a fly population that had become uncontrollable with permethrin within a month of the first application of this insecticide on the farm (Denholm et al., 1983). Before introducing bioresmethrin space sprays, insecticidal treatments were withheld for a year to enable tolerance to revert to its pre-permethrin treatment level. Bioresmethrin was then sprayed during a 12-month period wherever fly numbers reached or exceeded an arbitrarily fixed level (about once a fortnight). Control and pyrethroid tolerance were monitored by bioassay and, in the laboratory, selection with permethrin was applied to flies that had been treated in the field for several months with bioresmethrin in order to determine if bioresmethrin-treated flies had retained the potential for pyrethroid resistance.

Although bioresmethrin did not elicit any detectable resistance throughout the 12 months' trial, laboratory selection with permethrin for three generations resulted in resistance equal to that recorded when permethrin failed in the field. Thus the bioresmethrin treatment elicited no resistance in spite of the resistance potential present in the population. Resistance failed to develop because bioresmethrin was used at intervals sufficiently long for breeding to dilute any resistance that might have been selected during the spraying treatments of the pig pens, and although kill was usually as high as 90%, these treatments did not appear to select preferentially resistant individuals. Thus, in enclosed areas with little or no migration, pyrethroids with little persistence can be used without eliciting resistance. It must be stressed however that very frequent use of such pyrethroids stimulate continuous selection by the more persistent analogues; both lead to the rapid appearance of resistance (Denholm et al., 1983).

These findings are not in full accord with the predictions of mathematical simulations by Taylor and Georghiou (1982) regarding the effect of pesticide persistence on the rate of build-up of resistance. Their model did not anticipate the non-appearance of resistance when short-lived compounds are applied to control adult insects only, but predicted instead that application of a pesticide with a half-life of one day to a closed population would result in the GF50 (time taken for R gene frequency to attain 50%) being reached within 25 weeks, only 5 weeks later than with compounds with a half-life of up to 10 days. This discrepancy between predictions and results demonstrates the need to exercise extreme caution when contemplating the use of mathematical predictions, models or theoretically devised strategies to prevent or delay the build-up of resistance.

Pyrethroid-resistant phytoseiid mites in integrated pest management programmes

Although photostable pyrethroids are more acaricidal to susceptible strains of A. fallacis than to tetranychid mites (Croft et al., 1982) pyrethroid-resistant strains of A. fallacis, initially selected from field populations sprayed with fenvalerate and permethrin for 3 years (Strickler and Croft, 1981) and further selected in the greenhouse (Croft, 1983) were successfully released in a commercial orchard to control Panonychus ulmi (Koch), and Tetranychus urticae (Koch) (Whalon et al., 1982). Although the orchard was regularly sprayed in 1980 and 1981 with permethrin and fenvalerate to control Cydia pomonella (L.) pyrethroid-resistant mites survived these treatments, and inbred with indigenous populations. This however resulted in considerable reversion towards pyrethroid susceptibility. From experience gained in these experiments, Croft (1983) suggested using photostable pyrethroids twice per season in integrated pest management programmes in areas with pyrethroid-resistant predatory mites: the first treatment to control early season pests of apple trees, the second to control the coddling moth, both sprays to coincide with maxima in the influx of migrant A. fallacis, thus ensuring maximal resistance in predatory mites as well as adequate control of the major orchard pests.

Hoy et al. (1983) obtained similarly promising results on releasing pyrethroid-resistant strains of M. occidentalis in pyrethroid-treated orchards in the north-west of the USA.

CONCLUSIONS

So far the worst predictions on build-up of resistance to pyrethroids, made when the photostable analogues were first introduced for crop protection in the mid-1970s, have fortunately not been fulfilled. A number of major pests now have pyrethroid-resistant strains, but serious control problems have so far been restricted to only a few areas such as Central America and Thailand, where insecticidal usage is often excessive (Vaughan and Leon, 1977). World-wide resistance problems caused by pyrethroids are not on a level comparable to those encountered with insecticides of other groups. However, since pyrethroids have been introduced only recently, and in several major agricultural areas are only being introduced now, it is difficult to forecast what problems massive use of pyrethroids may produce.

The resistance potential is also difficult to evaluate. Where pyrethroids have been used for several years thay have most probably already reselected cross-resistance mechanisms first revealed by previously used insecticides. In such areas, new cases of pyrethroid resistance, if and when they arise, will probably result from continued, and relentless reliance on pyrethroids for pest control; it is doubtful if the massive use of these insecticides against cotton pests in the USA can continue indefinitely without eliciting control problems.

Pyrethroid resistance can reveal itself unexpectedly and become very quickly uncontrollable; the best such example involves *H. armigera* at Emerald, Eastern Australia (Gunning *et al.*, 1984). That is why it is essential to monitor changes in response to detect incipient resistance before control failure. For this it is necessary to have effective and reliable tests, accurate baseline data, and a good knowledge of the population dynamics of the pest or pest complex for working out quickly the most effective alternative control strategies. At present most of these are non-realizable desiderata because the necessary knowledge is lacking. Most of the research on the effects of pyrethroids on insects has so far been restricted to the four original photostable pyrethroids, mainly permethrin, often using isomers with different biochemical and toxicological properties. It is to be hoped that, with the new pyrethroids, further attention will be paid to obtaining more easily interpretable data.

It is now clear that pyrethroids as a group are neither chemically, physiologically nor biochemically homogeneous, and the task of unravelling differences in their effects on susceptible let alone resistant insects is immense. There are not only very serious gaps in our knowledge of the biochemistry of these compounds in resistant insects. Our knowledge of *kdr* is likewise still very limited. The facile assumption that *kdr* is homologous in all arthropods is probably erroneous and cross-resistance to DDT and nerve insensitivity to pyrethroids should be reassessed very critically.

Likewise very little is known about the power of different pyrethroids to select for resistance. Our knowledge of this topic in most insects is restricted to only one stage of the life cycle, even though other stages may play a significant role in the process. We still know very little about the behavioural response of resistant strains (Virgona *et al.*, 1983), and the many other factors that accelerate or delay resistance.

Yet in spite of these wide gaps in our knowledge about pyrethroid resistance, it is very encouraging to note that the greatest progress in controlling, preventing or delaying development of resistance has been achieved in the field of actual pyrethroid usage. Pyrethroids are probably the first insecticides for which nation-wide strategies have not only been formulated and applied, but also proved to work. There is therefore a great deal that can be done and the aim must surely be to build on this promising start. The same major effort as has been directed at ensuring that pyrethroids meet statutory requirements for registration will have to be applied to learn more about the ways arthropods are affected by and resist pyrethroids. The *raison d'être* of insecticides is after all the effective, safe and reliable control of insect pests, and pyrethroids are particularly suited for the task.

ACKNOWLEDGEMENTS

I thank very much A. L. Devonshire, M. Elliott, N. F. Janes, R. A. Nicholson, M. P. Osborne and D. M. Soderlund for their helpful comments.

REFERENCES

Abdel-Aal, Y. A. I., and Soderlund, D. M. (1980). 'Pyrethroid-hydrolysing esterases in Southern armyworm larvae: tissue distribution, kinetic properties and selective inhibition', *Pestic. Biochem. Physiol.*, **14**, 282–289.

Attia, F. I., and Hamilton, J. T. (1978). 'Insecticide resistance in *Myzus persicae* in Australia', *J. Econ. Entomol.*, **71**, 851–853.

Ball, H. J. (1981). 'Insecticide resistance—a practical assessment', *Bull. Ent. Soc. Am.*, **27**, 261–262.

Bashir, N. H. H., and Crowder, L. A. (1983). 'Mechanism of permethrin tolerance in the common green lacewing (Neuroptera: Chrysopidae)', *J. Econ. Entomol.*, **76**, 407–409.

Beeman, R. W. (1982). 'Recent advances in mode of action of insecticides', *Ann. Rev. Entomol.*, **27**, 253–281.

Bills, G. T. (1980). 'Effective housefly control in British piggeries', *International Pest Control*, **22**, 84–85.

Blackith, R. E. (1953). 'Bioassay systems for pyrethrins. V. Experiments with a resistant strain of *Calandra granaria*', *Ann. appl. Biol.*, **40**, 106–112.

Briggs, G. G., Elliott, M., and Janes, N. F. (1983). 'Present status and future prospects for synthetic pyrethroids', in *Pesticide Chemistry. Human Welfare and Environment* (eds J. Miyamoto and P. C. Kearney), Vol. 2, pp. 157–164, Pergamon Press, Oxford.

Brown, T. M., Bryson, K., and Payne, G.T. (1982). 'Pyrethroid susceptibility in methyl parathion-resistant tobacco budworm in South Carolina (USA)', *J. Econ. Entomol.* **75**, 301–303.

Buchi, R. (1981). 'Evidence that resistance against pyrethroids in aphids *Myzus persicae* and *Phorodon humuli* is not correlated with high carboxylesterase activity', *Z. Pflanzenkrankheiten und Pflanzenschutz*, **88**, 631–633.

Busvine, J. R. (1951). 'Mechanism of resistance to insecticide in houseflies', *Nature*, **168**, 193–195.

Busvine, J. R. (1958). 'Insecticide resistance in bed-bugs', *Bull. W.H.O.*, **19**, 1042–1052.

Busvine, J. R. (1980). Recommended methods for measurement of pest resistance to pesticides, Food and Agricultural Organization of the United Nations, Rome, pp. 132.

Carter, S. W., Chadwick, P. R., and Wickham, J. C. (1975). 'Comparative observations on the activity of pyrethroids against some susceptible and resistant stored products beetles', *J. Stored Prod. Res.*, **11**, 135–142.

Casida, J. E., Gammon, D. W., Glickman, A. H., and Lawrence, L. J. (1983). 'Mechanisms of selective action of pyrethroid insecticides', *Ann. Rev. Pharmacol. Toxicol.*', **23**, 413–438.

Casida, J. E., Gaughan, L. C., and Ruzo, L. O. (1979). 'Comparative metabolism of pyrethroids derived from 3-phenoxybenzyl and α-cyano 3-phenoxybenzyl alcohols, in *Advances in Pesticide Science* (ed. H. Geissbuehler), pp. 182–189, Pergamon Press, Oxford.

Casida, J. E., Ueda, K., Gaughan, L. C., Jao, L. T., and Soderlund, D. M. (1976). 'Structure biodegradability relationships in pyrethroid insecticides', *Archs Environ. Contam. Toxicol.*, **3**, 491–500.

Casida, J. E., and Ruzo, L. O. (1980). 'Metabolic chemistry of pyrethroid insecticides', *Pestic. Sci.*, **11**, 257–269.

Chadwick, P. R., Invest, J. F., and Bowron, M. J. (1977). 'An example of cross-resistance to pyrethroids in DDT resistant *Aedes aegypti*', *Pestic Sci.*, **8**, 618–624.

Chambers, J. (1980). 'An introduction to the metabolism of pyrethroids'; *Residue Rev.*, **73**, 101–124.

Champ, B. R., and Dyte, C. E. (1976). 'Report of the FAO global survey of pesticide susceptibility of stored grain pests', *FAO Plant Production and Protection Services* No. 5, Food and Agriculture Organisation of the United Nations, Rome, pp. 297.

Chang, C. K., and Jordan, T. W. (1982). 'Penetration and metabolism of topically applied permethrin and cypermethrin in pyrethroid-tolerant *Wiseana cervinata* larvae', *Pestic. Biochem. Physiol.*, **17**, 196–204.

Chang, C. K., and Jordan, T. W. (1983). 'Inhibition of permethrin in hydrolysing esterases from *Wiseana cervinata* larvae', *Pestic. Biochem. Physiol.*, **19**, 190–195.

Chang, C. P., and Plapp, F. W. Jr. (1983a). 'DDT and pyrethroids: receptor binding and mode of action in the housefly', *Pestic. Biochem. Physiol.*, **20**, 76–85.

Chang, C. P., and Plapp, F. W. Jr. (1983b). 'DDT and pyrethroids: receptor binding in relation to knockdown resistance (*kdr*) in the house fly', *Pestic. Biochem. Physiol.*, **20**, 86–91.

Chang, C. P., and Plapp, F. W. Jr. (1983c). 'DDT and synthetic pyrethroids mode of action, selectivity, and mechanism of synergism into tobacco budworm (*Lepidoptera, Noctuidae*) and a predator, *Chrysopa carnea Stephens* (*Neuroptora, Chrysopidae*)', *J. Econ. Entomol.*, **76**, 1206–1210.

Chapman, P. A., and Lloyd, C. J. (1981). 'The spread of resistance among houseflies in the United Kingdom', *Proceedings 1981 British Crop Protection Conference*, 628–632.

Chapman, R. B., and Penman, D. E. (1979). 'Negatively correlated cross-resistance to a synthetic pyrethroid in organophosphorus resistant *Tetranychus urticae*', *Nature*, **281**, 298–299.

Chapman, R. B., and Penman, D. E. (1980). 'The toxicity of mixtures of a pyrethroid with organophosphorus insecticide to *Tetranychus urticae*. Koch', *Pestic. Sci.*, **11**, 600–604.

Cheng, E. Y. (1981). 'Insecticide resistance study in *Plutella xylostella* L. II. A general survey (1980–81)', *J. Agric. Res. of China*, **30**, 285–293.

Chiang, C., and Devonshire, A. L. (1982). 'Changes in membrane phospholipids, identified by Arrhenius plots of acetylcholinesterase and associated with pyrethroid resistance (*kdr*) in houseflies (*Musca domestica*)', *Pestic. Sci.*, **13**, 156–160.

Chung, T. C., Sun, C. N., and Hung, C. Y. (1981). 'Brown planthopper resistance to several insecticides in Taiwan', *International Rice Research Newsletter*, **6**, 19.

Cichy, D. (1971). 'The role of some ecological factors in the development of pesticide resistance in *Sitophilus oryzae* L. and *Tribolium castaneum* Herbst.', *Ekol. pol.*, **19**, 563–616.

Clements, A. N., and May, T. E. (1977). 'The actions of pyrethroids upon the peripheral nervous system and associated organs in locust', *Pestic. Sci.*, **8**, 661–680.

Cochran, D. G. (1973). 'Inheritance and linkage of pyrethrins resistance in the German cockroach', *J. Econ. Entomol.*, **66**, 27–30.

Cole, M. M., and Clarke, P. H. (1961). 'Development of resistance to synergized pyrethrins in body lice and cross-resistance to DDT', *J. Econ. Entomol.*, **54**, 649–651.

Collins, W. J. (1975). 'Resistance in *Blattella germanica* (L.) (*Orthoptera, Blattidae*): the effect of propoxur selection and non-selection on the resistance spectrum developed by diazinon selection', *Bull. Ent. Res.*, **65**, 399–403.

Collins, W. J. (1976). 'German cockroach resistance: propoxur selection induces the same resistance spectrum as diazinon selection', *Pestic. Sci.*, **7**, 171–174.

Croft, B. A. (1983). 'Status and management of pyrethroid resistance in the predatory mite, *Amblyseius fallacis* (Acarina: Phytoseiidae)'. *The Great Lakes Entomologist*, **16**, 17–32.

Croft, B. A., and Wagner, S, W. (1981). 'Selectivity of acaricidal pyrethroids to permethrin-resistant strains of *Amblyseius fallacis*', *J. Econ. Entomol.*, **74**, 703–706.

Croft, B. A., Wagner, S. W., and Scott, J. G. (1982). 'Multiple and cross-resistance to insecticides in pyrethroid resistant strains of the predatory mite *Amblyseius fallacis*', *Environ. Entomol.*, **11**, 161–164.

Croft, B. A., and Whalon, M. E. (1983). 'Inheritance and persistence of permethrin resistance in the predatory mite *Amblyseius fallacis* (Acarina: Phytoseiidae)', *Environ. Entomol.*, **12**, 215–218.

Crowder, L. A., Tollefson, M. S., and Watson, T. F. (1979). 'Dosage-mortality studies of synthetic pyrethroids and methyl parathion on the tobacco budworm in Central Arizona', *J. Econ. Entomol.*, **72**, 1–3.

Davidson, J., Herath, P. R. J., and Hemingway, J. (1980). 'The present status of research on resistance to pesticides in insects of medical importance (other than house-flies)', World Health Organisation, *VBC/EC/80.21*.

Davies, M. J., Keiding, J., and von Hofsten, C. G. (1958). 'Resistance to pyrethrins and to pyrethrins-piperonyl butoxide in a wild strain of *Musca domestica* L. in Sweden', *Nature*, **182**, 1816–1817.

Davies, J. W., Harding, J.A., and Wolfenbarger, D. A. (1975). 'Activity of a synthetic pyrethroid against cotton insects', *J. Econ. Entomol.*, **68**, 373–374.

Davies, J. W., Wolfenbarger, D. A., and Harding, J. A. (1977). 'Activity of several synthetic pyrethroids against the boll weevil and *Heliothis* spp.', *Southwest. Ent.*, **2**, 164–169.

Denholm, I., Farnham, A. W., O'Dell, K., and Sawicki, R M. (1983). 'Factors affecting resistance to insecticides in house-flies, *Musca domestica* L. (Diptera: Muscidae). I. Long-term control with bioresmethrin of flies with strong pyrethroid-resistance potential', *Bull. Ent. Res.* **73**, 481–489.

De Vries, D. H., and Georghiou, G. P. (1980). 'A wide spectrum of resistance to pyrethroid insecticides in *Musca domestica*', *Experientia*, **36**, 226–227.

De Vries, D. H., and Georghiou, G. P. (1981a). 'Decreased nerve sensitivity and decreased cuticular penetration as mechanisms of resistance to pyrethroids in a (1R)-trans-permethrin-selected strain of the house fly', *Pestic. Biochem. Physiol.*, **15**, 234–241.

De Vries, D. H., and Georghiou, G. P. (1981b). 'Absence of enhanced detoxication of permethrin in pyrethroid-resistant houseflies', *Pestic. Biochem. Physiol*, **15**, 242–252.

Devonshire, A. L. (1977). 'The properties of a carboxylesterase from the peach-potato aphid, *Myzus persicae* (Sulz.) and its role in conferring insecticide resistance', *Biochem. J.*, **167**, 675–683.

Devonshire, A. L., and Moores, G. D. (1982). 'A carboxylesterase with broad substrate specificity causes organophosphorus, carbamate and pyrethroid resistance in peach-potato aphids (*Myzus persicae*)', *Pest. Biochem. Physiol.*, **18**, 235–246.

Du Toit, G. O. G., Townsend, R. J., and Armstrong, S. J. (1978). 'Lack of response in porina (*Wiseana* sp.: Hepialidae) caterpillar to treatment with pyrethroid insecticides', *N.Z. J. exp. Agric.*, **6**, 175–176.

El-Guindy, M. A., Abdel-Sottar, M. M., Dogheim, S. M. A., Madi, S. M., and Issa, Y. H. (1982a). 'The joint action of certain insecticides to a field strain of the pink bollworm *Pectinophora gossypiella*', *International Pest Control*, **24**, 154–156, 160.

El-Guindy, M. A., Madi, S. M., Keddis, M. E., Issa, Y. H., and Abdel-Sattar, M. M. (1982b). 'Development of resistance to pyrethroids in field populations of the Egyptian Cotton Leafworm *Spodoptera littoralis* (Boisd.)', *International Pest control*, **24**, 7–11.

Elhag, G. A., and Horn, D. J. (1983). 'Resistance greenhouse whitefly (*Homoptera: Aleurodidae*) to insecticides in selected Ohio greenhouses'. *J. Econ. Entomol.*, **76**, 945–948.

Elliott, M., Farnham, A. W., Janes, N. F., Needham, P. B., Pulman, D. A., and Stevenson, J. H. (1973). 'A photostable pyrethroid', *Nature*, **248**, 169–170.

Elliott, M., Farnham, A. W., Janes, N. F., Needham, P. H., and Pulman, D. A. (1974). 'Synthetic insecticide with a new order of activity', *Nature*, **248**, 710–711.

Elliott, M., and Janes, N. F. (1979). 'Recent structure-activity correlations in synthetic pyrethroids', in *Advances in Pesticide Science*, (ed. H. Geissbuchler), pp. 166–173. Pergamon Press, Oxford.

Elliott, M., Janes, N. F., and Potter, C. (1978). 'The future of pyrethroids in insect control', *Ann. Rev. Ent.*, **23**, 443–469.

Eto, M., Oshima, V., Kitakata, S., Tanaka, F., and Kojima, K. (1966). 'Studies on saligenic cyclic phosphorus esters with insecticidal activity. Part X. Synergism of malathion against susceptible and resistant insects', *Botyu Kagaku*, **31**, 33–38.

Farnham, A. W. (1971). 'Changes in the cross-resistance patterns of houseflies selected with natural pyrethrins or resmethrin', *Pestic. Sci.*, **2**, 138–143.

Farnham, A. W. (1973). 'Genetics of resistance of pyrethroid selected houseflies *Musca domestica* L.', *Pestic. Sci.*, **4**, 513–520.

Farnham, A. W. (1977). 'Genetics of resistance of houseflies (*Musca domestica* L.) to pyrethroids. I. Knockdown resistance', *Pestic. Sci.*, **8**, 631–636.

Farnham, A. W., and Sawicki, R. M. (1976). 'Development of resistance to pyrethroids in insects resistant to other insecticides', *Pestic. Sci.*, **7**, 278–282.

Farnham, A. W., O'Dell, K. E., Denholm, I., and Sawicki, R. M. (1984). 'Factors affecting resistance to insecticides in house-flies, *Musca domestica* L. (Diptera: Muscidae). III. Relationship between the level of resistance to pyrethroids, control failure, and the frequency of gene *kdr*', *Bull. Ent. Res.*, **74**, 581–589.

Fine, B. C. (1961). 'Pattern of pyrethrin resistance in houseflies', *Nature*, **191**, 884–885.

Frazmann, B. A., Hamilton, J. T., and Attia, F. I. (1980). 'A field and laboratory evaluation of insecticides on a multi-resistant strain of *Myzus persicae* (Sulzer) (Hemiptera: Aphididae)', *Protection Ecology*, **2**, 41–46.

Fullbrook, S. L., and Holden, J. S. (1980). 'Possible mechanisms of resistance to permethrin in cotton pests', in *Insect Neurobiology and Pesticide Action*, pp. 281–287, The Society of the Chemical Industry, Pesticide Group, London.

Gammon, D. W. (1980). 'Pyrethroid resistance in a strain of *Spodoptera littoralis* is correlated with decreased sensitivity of the CNS *in vitro*', *Pestic. Biochem. Physiol.*, **14**, 53–62.

Gammon, D. W., Brown, M. A., and Casida, J. E. (1981). 'Two classes of pyrethroid action in the cockroach', *Pestic. Biochem. Physiol.*, **15**, 181–191.

Gammon, D. W., and Holden, J. S. (1980). 'A neural basis for pyrethroid resistance in larvae of *Spodoptera littoralis*', in *Insect Neurobiology and Pesticide Action*, pp. 481–488, The Society of the Chemical Industry, Pesticide Group, London.

Gammon, D. W., Lawrence, L. J., and Casida, J. E. (1982). 'Pyrethroid toxicology; protective effects of diazepam and phenobarbital in the mouse and the cockroach', *Toxicol. appl. Pharmacol.*, **66**, 280–296.

Georghiou, G. P., Lagunes, A., and Baker, J. D. (1983). 'Effect of insecticide rotations on evolution of resistance', in *Pesticide Chemistry. Human Welfare and the Environment*, (eds J. Miyamoto and P. C. Kearney), pp. 183–189, Pergamon Press, Oxford.

Georghiou, G. P., and Mellon, R. B. (1983). 'Pesticide resistance in time and space' in *Pest Resistance to Pesticides*, (eds G. P. Georghiou and T. Saito), pp. 1–46, Plenum Press, New York.

Gibson, R. W., Rice, A. D., and Sawicki, R. M. (1982). 'Effects of the pyrethroid deltamethrin on the acquisition and inoculation of viruses by the aphid *Myzus persicae*', *Ann. appl. Biol.*, **100**, 55 59.

Glickman, A. H., and Casida, J. E. (1982). 'Species and structural variations affecting pyrethroid neurotoxicity', *Neurobehav. Toxicol. Teratol.*, **4**, 793–799.

Glynn Jones, D. G. (1983). 'The use of piperonyl butoxide to increase the susceptibility of insects which have become resistant to pyrethroids and other insecticides', *International Pest Control*, **25**, 14–15, 21.

Grigolo, A., and Oppenoorth, F. J. (1966). 'The importance of DDT-acting diochlorinase for the effect of the resistance gene *kdr* in the housefly, *Musca domestica* L.' *Genetica*, **37**, 159–170.

Gunning, R. V., Easton, C. S., Greenup, L. R., and Edge, V. E. (1984). 'Synthetic pyrethroid resistance in *Heliothis armiger* (Hübner) in Australia', *J. Econ. Entomol.*, **77**, 1283–1287.

Harding, J. A., Huffman, F. R., Wolfenbarger, D. A., and Davis, J. W. (1977). 'Insecticidal activity of *alpha*-cyano-3-phenoxybenzyl pyrethroids against the boll weevil and tobacco budworm', *Southwest. Ent.* **2**, 42–45.

Harris, C. R., Turnbull, S. A., Whistlecraft, J. W., and Surgedner, G. A. (1982). 'Multiple resistance shown by field strains of housefly *Musca domestica* (Diptera: Muscidae) to organochlorine, organophosphorus, carbamate and pyrethroid insecticides', *Can. Entomol.*, **114**, 447–454.

Hemmingway, J. (1981). 'Genetics and biochemistry of insecticide resistance in anopholenes', PhD Thesis, London University.

Ho, S. H., Lee, B. H., and See, D. (1983). 'Toxicity of deltamethrin and cypermethrin to the larvae of the diamond-back moth, *Plutella xylostella*', *Toxicol. Lett.* **19**, 127–131.

Holborn, J. M. (1957). 'The susceptibility to insecticide of laboratory cultures of an insect species', *J. Sci. Food Agric.*, **8**, 182–188.

Holden, J. S. (1979). 'Absorption and metabolism of permethrin and cypermethrin in the cockroach and the cotton-leaf worm larvae', *Pestic. Sci.*, **10**, 295–307.

Hoy, M. A., and Knop, N. F. (1981). 'Selection for and genetic analysis of permethrin resistance in *Metaseiulus occidentalis*: genetic improvement of a biological control agent', *Entomol. Exp. Appl.*, **30**, 10–18.

Hoy, M. A., Knop, N. F., and Joos, J. L. (1980). 'Pyrethroid resistance persists in spider mite predator', *California Agriculture*, **34**, 11–12.

Hoy, M. A., Westigard, P. H., and Hoyt, S. C. (1983). 'Release and evaluation of a laboratory-selected, pyrethroid-resistant strain of the predaceous mite *Metaseiulus occidentalis* (Acari: Phytoseiidae) in Southern Oregon pear orchards and a Washington apple orchard', *J. Econ. Entomol.*, **76**, 383–388.

Hutson, D. H. (1979). 'The metabolic fate of synthetic pyrethroid insecticides in mammals', in *Progress in Drug Metabolism* (eds J. W. Bridges and L. F. Chassaud), Vol. 3, pp. 215–252, Wiley, Chichester.

Hussey, N. W. (1964). Report of the Glasshouse Crops Research Institute, 1964, p. 77.

Ishaaya, I., and Casida, J. E. (1980). 'Properties and toxicological significance of esterases hydrolyzing permethrin and cypermethrin in *Trichoplusia ni* larval gut and integument', *Pestic. Biochem. Physiol.*, **14**, 178–184.

Ishaaya, I., and Casida, J. E. (1981). 'Pyrethroid esterase(s) may contribute to natural pyrethroid tolerance of larvae of the common lacewing', *Environ. Entomol.*, **10** 681–684.

Jao, L. T., and Casida, J. E. (1974). 'Insect pyrethroid-hydrolyzing esterases', *Pestic. Biochem. Physiol.*, **4**, 465–472.

Katz, A. J., Collins, W. J., and Skavaril, R. V. (1973). 'Resistance in the German cockroach (Orthoptera: Blattidae): the inheritance of diazinon resistance and cross-resistance'. *J. Med. Entomol.*, **10**, 599–604.

Kao, H. L., Liu, M. Y., and Sun, C. N. (1981). 'Green rice leafhopper resistance to malathion, methyl parathion, carbaryl, permethrin and fenvalerate in Taiwan', *International Rice Research Newsletter*, **6**, 19.

Keiding, J. (1976). 'Development of resistance to pyrethroids in field populations of Danish houseflies', *Pestic. Sci.*, **7**, 283–291.

Keiding, J. (1977). 'Resistance in the housefly in Denmark and elsewhere' in *Pesticide Management and Insecticide Resistance*. (eds D. L. Watson and A. W. A. Brown), Academic Press, New York, San Francisco and London.

Keiding, J. (1978). 'Consequences of pyrethroid-resistance on Danish farms', in *Danish Pest Infestation Laboratory Annual Report, 1977*, p. 42.

Keiding, J. (1981). 'Tests of resistance in fly populations on Danish farms', in *Danish Pest Infestation Laboratory Annual Report, 1980*, pp. 40–44.

Keiding, J., and Skovmand, O. (1983). 'Insecticide resistance in *Musca domestica*' in *Danish Pest Infestation Laboratory Annual Report, 1982*, pp. 46–58.

Keller, J. C., Clark, P. H., and Lofgren, C. S. (1956). 'Susceptibility of insecticide resistant cockroaches to pyrethrins', *Pest Control*, **24**, 14–15, 30.

Künast, C. (1979). 'Die entwicklung der Permethrinresistenz bei der Stubenfliege (*Musca domestica* L.) in sudden schen Raum', *Zeitschrift fur Angewandte Zoologie*, **66**, 385–390.

Künast, C., Lamina, J., Mederer, H., and Rommertz, J. (1981). 'The present availability of dimethoate and permethrin in housefly control in stables in the south of the Federal Republic of Germany', *Duet. Tier. W.*, **88**(S), 187–192.

Lawrence, L. J., and Casida, J. E. (1983). 'Stereospecific action of pyrethroid insecticides on the gamma-aminobutyric acid receptor ionophore complex', *Science*, **221**, 1399–1401.

Lee, P. W., Stearns, S. M., and Sanborn, J. R. (1982). 'Penetration and metabolism of Pydrin[R] insecticide in susceptible and resistant houseflies', Paper No. 69, Pesticide Chemistry Division, *American Chemical Society, 183rd National Meeting, Las Vegas, Nev.*

Lhoste, J. (1977). 'Pyréthrines naturelles et pyrethrinoides de synthèse', *Travaux de la Société de Pharmacie de Montpellier*, **37**, 307–327.

Liu, M. Y., Tzeng, Y. J., and Sun, C. N. (1981). 'Diamondback moth resistance to several synthetic pyrethroids', *J. Econ. Entomol.*, **74**, 393–396.

Liu, M. Y., Tzeng, Y. J., and Sun, C. N. (1982). 'Insecticide resistance in the diamond-back moth', *J. Econ. Entomol.*, **75**, 153–155.

Lloyd, C. J. (1969). 'Studies on the cross-tolerance to DDT-related compounds of a pyrethrin-resistant strain of *Sitophilus granarius* (L.) (*Coleoptera, Curculionidae*)', *J. Stored Prod. Res.*, **5**, 337–356.

Lloyd, C. J. (1973). 'The toxicity of pyrethrins and five synthetic pyrethroids, to *Tribolium castaneum* (Herbst) and susceptible and pyrethrin resistant *Sitophilus granarius* (L.)', *J. Stored Prod. Res.*, **9**, 77–92.

Lloyd, C. J., and Parkin, E. A. (1963). 'Further studies on a pyrethrum resistant strain of the granary weevil *Sitophilus granarius* (L.)', *J. Sci. Food Agric.*, **14**, 655–663.

Lloyd, C. J., Rowlands, D. G., and Ruczkowski, G. E. (1976). 'Incidence of resistance in the United Kingdom. The rust-red flour beetle. Resistance to synthetic pyrethroids', *Pest Infest. Control 1971–73*, 81–82.

Lloyd, C. J., and Ruczkowski, G. E. (1980). 'The cross-resistance to pyrethrins and eight synthetic pyrethroids of an organophosphorus-resistant strain of the rust-red flour beetle *Tribolium castaneum* (Herbst.)', *Pestic. Sci.*, **11**, 331–340.

Lund, A. E., and Narahashi, T. (1983). 'Kinetics of sodium channel modifications as the basis for the variation in the nerve membrane effects of pyrethroids and DDT analogues', *Pestic. Biochem. Physiol.*, **20**, 203–216.

McClanahan, R. J., and Founk, J. (1983). 'Toxicity of insecticides to green peach aphid (Homoptera Aphididae) in the laboratory and field tests, 1971–1982', *J. Econ. Entomol.*, **76**, 899–905.

MacDonald, R. S., Surgeoner, G. A., Solomon, K. R., and Harris, C. R. (1983a).'Effect of four spray regimes on the development of permethrin and dichlorvos resistance in the laboratory, by the housefly (Diptera: Muscidae)', *J. Econ. Entomol.*, **76**, 417–422.

MacDonald, R. S., Surgeoner, G. A., Solomon, K. R., and Harris, C. R. (1983b). 'Development of resistance to permethrin and dichlorvos by the housefly (Diptera: Muscidae) following continuous and alternative insecticide use on four farms', *Canad. Entomol.*, **115**, 1555–1561.

Malcolm, C. A. (1981). 'A laboratory investigation of the genetics of pyrethroid resistance in the mosquito *Aedes aegypti*', PhD Thesis, University of Manchester.

Malcolm, C. A. (1983a). 'The genetic basis of pyrethroid and DDT resistance inter-relationships in *Aedes aegypti*, I. Isolation of DDT and pyrethroid resistance factors'. *Genetica*, **60**, 213–219.

Malcolm, C. A. (1983b). 'The genetic basis of pyrethroid and DDT resistance inter-relationships in *Aedes aegypti*, II. Allelism of R^{DDT2} and R^{PY}. *Genetica*, **60**, 221–229.

Malcolm, C. A., and Wood, R. J. (1982). 'Location of a gene conferring resistance to knockdown by permethrin, and bioresmethrin in adults of the BKPM 3 strain of *Aedes aegypti*', *Genetica*, **59**, 233–237.

Malinowski, H. (1980). 'The resistance spectrum of DDT-selected houseflies, *Musca domestica* L. (Diptera, Muscidae)', *Pol. Pismo Entomol.*, **50**, 559–567.

Martinez-Carvillo, J. L., and Reynolds, H. J. (1983). 'Dosage-mortality studies with pyrethroids and other insecticides on the tobacco budworm (Lepidoptera: Noctuidae) from the Imperial Valley, California'. *J. Econ. Entomol.*, **76**, 983–986.

Micks, D. W., Moon, W. B., and McNeill, J. C. (1980). 'Malathion tolerance vs. resistance in *Culex quinquefasciatus*', *Mosq. News.*, **40**, 520–523.

Milani, R. (1954). 'Comportamento mendeliano della resistenza alla azione abbattente del DDT: correlazione tra abbattimento e mortalita in *Musca domestica* L. ', *Riv. Parassit.*, **15**, 513–542.

Milani, R. (1960). 'Genetic studies on insecticide resistant insects', *Misc. Publ. Ent. Soc. Amer.*, **2**, 75–83.

Milani, R., and Franco, M. G. (1959). 'Comportamento ereditario della resistenza al DDT in incroci tra il ceppo Orlando-R e ceppi *kdr* e *kdr*$^+$ di *Musca domestica* L.', *Symposia Genetica*, **6**, 269–303.

Miller, T. A., and Adams, M. E. (1982). 'Mode of action of pyrethroids', in *Insecticide Mode of Action* (ed. J. R. Coats), pp. 3–27, Academic Press, New York.

Miller, T. A., Kennedy, J. M., and Collins, C. (1979). 'Central nervous system insensitivity to pyrethroids in the resistant *kdr* strain of housefly *Musca domestica*', *Pestic. Biochem. Physiol.*, **12**, 224–230.

Miller, T. A., Saldago, V. L., and Irving, S. N. (1983). 'The *kdr* factor in pyrethroid resistance in pest resistance to pesticides. *Proceedings of the US Japanese Cooperation Science Program Seminar, 1979* (eds J. P. Georghiou and T. Saito), pp. 353–366, Plenum, New York.

Miyamoto, J. (1976). 'Degradation, metabolism and toxicity of synthetic pyrethroids'; *Environ. Hlth Perspect.*, **14**, 15–26.

Miyamoto, J., and Suzuki, T. (1973). 'Metabolism of tetramethrin in houseflies *in vivo*', *Pestic. Biochem. Physiol.*, **3**, 30–41.

Miyata, T. (1980). 'Negatively correlated cross-resistance to a synthetic pyrethroid in malathion resistant *Laodelphax striatellus* Fallen and *Nilaparvata lugens* Stal.', *XVI International Congress of Entomology, Kyoto*.

Miyata, T., Saito, T., Kassai, T., and Ozaki, K. (1982). 'Negative cross resistance between malathion and fenvalerate in rice leaf hopper', *5th International Congress of Pesticide Chemistry, Kyoto*.

Murray, D., and Cull, P. (1984). 'Resistance strategy works at Emerald', *The Australian Cotton Grower*, 22–25.

Narahashi, T. (1980). 'Site and types of action of pyrethroids', in Pyrethroid Insecticides: Chemistry and Action, pp. 15–17, Tables Rondes Roussel Uclaf No. 37.

Narahashi, T. (1983). 'Nerve membrane sodium channels as the major target site of pyrethroids and DDT', in *Pesticide Chemistry, Human Welfare and the Environment* (eds. J. Miyamoto and P. C. Kearney), vol. 3, pp. 109–114, Pergamon, Oxford.

Nassif, M., Brooke, J. P., Hutchinson, D. B. A., Kamel, O. M., and Savage, E. A. (1980). 'Studies with permethrin against body lice in Egypt', *Pestic. Sci.*, **11**, 679–684.

Nelson, J. O., and Wood, F. E. (1982). 'Multiple and cross-resistance in a field-collected strain of the German cockroach (Orthopthra: Blattellidae). *J. Econ. Entomol.*, **75**, 1052–1054.

Nicholson, R. A., Botham, R. P., and Collins, C. (1983). 'The use of [^{3}H] permethrin to investigate the mechanisms underlying its differential toxicity to adult and larval stages of the sheep blowfly *Lucilia sericata*'. *Pestic. Sci.*, **14**, 57–63.

Nicholson, R. A., Chalmers, A. E., Hart, R. J., and Wilson, R. G. (1980a). 'Pyrethroid action and degradation in the cattle tick (*Boophilus microplus*)', in *Insect Neurobiology and Pesticide Action*, pp. 289–295, The Society of Chemical Industry, Pesticide Group, London.

Nicholson, R. A., Hart, R. J., and Osborne, M. P. (1980b). 'Mechanisms involved in the development of resistance to pyrethroids with particular reference to knockdown resistance in houseflies', in *Insect Neurobiology and Pesticide Action*, pp. 465–471, The Society of Chemical Industry, Pesticide Group, London.

Nicholson, R. A., and Miller, T. A. (1985). Multifactorial resistance to *trans*-permethrin in field collected strains of the tobacco budworm (*Heliothis virescens* F.)', *Pestic. Sci.*, in the press.

Nicholson, R. A., and Sawicki, R. M. (1982). 'Genetic and biochemical studies of resistance to permethrin in a pyrethroid-resistant strain of the housefly (*Musca domestica* L.)', *Pestic. Sci.*, **13**, 357–366.

Nicoli, R. M., and Sautet, J. (1955). 'Rapport sur la fréquence et la sensibilité aux insecticides de *Pediculus humanus humanus* L. dans le sud-est de la France', *Monogr. Inst. nat. Hyg.*, **8**, 78 pp.

Nolan, J., and Bird, P. E. (1977). 'Co-toxicity of synthetic pyrethroids and organophosphorus compounds against the cattle tick (*Boophilus microplus*)', *J. Aust. Ent. Soc.*, **16**, 252.

Nolan, J., Roulston, W. J., and Schnitzerling, H. J. (1979). 'The potential of some synthetic pyrethroids for control of the cattle tick (*Boophilus microplus*)', *Aust. Vet. J.*, **55**, 463–466.

Nolan, J., Roulston, W. J., and Wharton, R. H. (1977). 'Resistance to synthetic pyrethroids in a DDT resistant strain of (*Boophilus microplus*)', *Pestic. Sci.*, **8**, 484–486.

Omer, S. M., Georghiou, G. P., and Irving, S. N. (1980). 'DDT/pyrethroid resistance inter-relationships in *Anopheles stephensi*', *Mosq. News*, **40**, 200–209.

Osborne, M. P., and Hart, R. J. (1979). 'Neurophysiological studies of the effects of permethrin upon pyrethroid resistant (*kdr*) and susceptible strains of dipterous larvae', *Pestic. Sci.*, **10**, 407–413.

Osborne, M. P., and Smallcombe, A. (1983). 'Site of action of pyrethroid insecticides in neural membranes as revealed by the kdr resistance factor', in *Pesticide Chemistry. Human Welfare and the Environment* (eds. J. Miyamoto and P. C. Kearney), vol. 3, pp. 103–107, Pergamon, Oxford.

Ozaki, K. (1980). 'Resistance of rice insect pests to insecticides in Japan', *XVI International Congress of Entomology, Kyoto*.

Parella, M. P. (1983). 'Evaluations of selected insecticides for control of permethrin-resistant *Liriomyza trifolii* (Diptera: Agromyzidae) on chrysanthemum'. *J. Econ. Entomol.*, **76**, 1460–1464.

Parkin, E. A., and Lloyd, C. J. (1960). 'Selection of a pyrethrum resistant strain of grain weevil *Calandra granaria* L. ', *J. Sci. Food Agric.*, **11**, 471–477.

Plapp, F. W., and Bull, D. L. (1978). 'Toxicity and selectivity of some insecticides to *Chrysopa carnea*, a predator of the tobacco budworm', *Environ. Entomol.*, **7**, 431–434.

Plapp, F. W., Jr., and Hoyer, R. F. (1967). 'Insecticide resistance in the housefly; resistance spectra and preliminary genetics of resistance in eight strains', *J. Econ. Entomol.*, **60**, 768–774.

Plapp, F. W., and Hoyer, R. F. (1968a). 'Possible pleiotropism of a gene conferring resistance to DDT, DDT analogues and pyrethrins in the housefly and *Culex tarsalis.*', *J. Econ. Entomol.*, **61**, 761–765.

Plapp, F. W., and Hoyer, R. F. (1968b). 'Insecticide resistance in the house fly: decreased rate of absorption as the mechanism of action of a gene that acts as an intensifier of resistance', *J. Econ. Entomol.*, **61**, 1298–1303.

Prasittisuk, C., and Busvine, J. R. (1977). 'DDT resistant mosquito strains with cross-resistance to pyrethroids', *Pestic. Sci.*, **8**, 527–533.

Priester, T. M., and Georghiou, G. P. (1978). 'Introduction of high resistance to permethrin in *Culex pipiens quinquefasciatus*', *J. Econ. Entomol.*, **71**, 197–200.

Priester, T. M., and Georghiou, G. P. (1979). 'The inheritance of resistance to permethrin in *Culex pipiens quinquefasciatus*', *J. Econ. Entomol.*, **72**, 124–127.

Priester, T. M., and Georghiou, G. P. (1980). 'Cross-resistance spectrum in pyrethroid resistant *Culex pipiens fatigans*', *Pestic. Sci.*, **11**, 617–624.

Procida (1982). 'Decis b jaunisse de la betterave', (pamphlet publicitaire) OK R.C. Marseille B 054 805 346.

Rice, A. D., Gibson, R. W., and Stribley, M. F. (1983). 'Effects of deltamethrin on walking, flight and potato virus Y-transmission by pyrethroid resistant *Myzus persicae*'. *Ann. appl. Biol.*, **102**, 229–236.

Riddles, P. W., Davey, P. A., and Nolan, J. (1983a). 'Carboxylesterases from *Boophilus microplus* hydrolyze *trans*-permethrin'. *Pestic. Biochem. Physiol.*, **20**, 133–140.

Riddles, P. W., Schnitzerling, H. J. and Davey, P. A. (1983b). 'Application of *trans* and *cis* isomers of *p*-nitrophenyl-(1*R,S*)-3-(2,2-dichlorovinyl)-2,2-dimethylcyclopropane-carboxylate to the assay of pyrethroid-hydrolyzing esterases'. *Anal. Biochem.*, **132**, 105–109.

Riskallah, M. R. (1983). 'Esterases and resistance to synthetic pyrethroids in the Egyptian cotton leafworm', *Pestic. Biochem. Physiol.*, **19**, 184–189.

Riskallah, M. R., Abd-Elghafar, S. F., Abd-Elghar, M R., and Nasser, M. E. (1983). 'Development of resistance and cross-resistance in fenvalerate and deltamethrin selected strains of *Spodoptera littoralis* (Boisd.)', *Pestic Sci.*, **14**, 508–512.

Rongsriyam, Y., and Busvine, J. R. (1976). 'Cross resistance in DDT-resistant strains of various mosquitoes (*Diptera, Culicidae*)', *Bull. Ent. Res.*, **65**, 459–471.

Roslatseva, S. A., Zolotova, J. B., Agashkova, T. M., and Polyakova, V. K. (1979). 'Study of the development of resistance to Neopynamin in houseflies, *Musca domestica*, under laboratory conditions', *Med. Parasit. Bol.*, **48**, 65–71.

Rowlands, D. J., and Lloyd, C. J. (1976). 'Incidence of resistance in the United Kingdom. The grain weevil. Metabolism and synergism of pyrethroids in susceptible and resistant strains', *Pest Infest. Control. Lab. Rep.*, **1971–1973**, pp. 75–77.

Rupes, V., and Pinterova, J. (1976). 'Factor of resistance to DDT, methoxychlor and pyrethrin on chromosome III in two strains of housefly, *Musca domestica*.' *Acta entomologica bohemoslovaca*, **73**, 293–301.

Rupes, V., Pinterova, J., Ledvinka, J., and Chmela, J. (1983). 'Insecticide resistance in houseflies *Musca domestica* (L) in Czechoslovakia 1976–1980', *Int. Pest Control*, **25**, 106–108.

Ruscoe, C. N. E. (1977). 'The new NRDC pyrethroids as agricultural insecticides', *Pestic. Sci.*, **8**, 236–242.

Salama, H. S. (1983). 'Cotton pest management in Egypt' *Crop Protection*, **2**, 183–191.

Saldago, V. L., Irving, S. N., and Miller, T. A. (1983a). 'Depolarization of motor nerve terminals by pyrethroids in susceptible and *kdr*-resistant houseflies', *Pestic. Biochem. Physiol.*, **20**, 100–114.

Saldago, V. L., Irving, S. N., and Miller, T. A. (1983b). 'The importance of nerve terminal depolarization in pyrethroid poisoning of insects', *Pestic. Biochem. Physiol.*, **20**, 169–182.

Sautet, J., Aldighieri, J., and Aldighieri, R. (1959). 'Nouvelles études sur la sensibilité aux insecticides de *Pediculus humanus humanus* K. Linnaeus 1758 en France metropolitaine (sud-est, Paris, nord)', *Bull. Soc. Path. exot.*, **52**, 87–97.

Sawicki, R. M. (1975). 'Effects of sequential resistance on pesticide management', *Proceedings 8th British Insecticide and Fungicide Conference*, pp. 799–811.

Sawicki, R. M. (1978). 'Unusual response of DDT-resistant houseflies to carbinol analogues of DDT', *Nature*, **875**, 433–444.

Sawicki, R. M., and Denholm, I. (1984). 'Adaptation of insects to insecticides' in *Origins and Development of adaptation Ciba Foundation Symposium 102* (eds. D. Evered and J. M. Collins), pp. 152–162, Pitman, London.

Sawicki, R. M., Devonshire, A. L., Farnham, A. W., O'Dell, K., Moores, G. D., and Denholm, I. (1984). 'Factors affecting resistance to insecticides in houseflies, *Musca domestica* L. (Diptera: Muscidae). II. Close linkage on autosome 2 between an esterase and resistance to trichlorphon and pyrethroids'. *Bull. Ent. Res.*, **74**, 197–206.

Sawicki, R. M., and Farnham, A. W. (1967). 'Genetics of resistance to insecticides in the SKA strain of *Musca domestica* I. Location of the main factors responsible for the maintenance of high DDT-resistance in diazinon-selected SKA flies'. *Entomol. Exp. Appl.* **10**, 253–262.

Sawicki, R. M., and Farnham, A. W. (1968). 'Genetics of resistance to insecticides in the SKA strain of *Musca domestica* III. Location and isolation of the factors of resistance to dieldrin'. *Entomol. Exp. Appl.*, **11**, 133–142.

Sawicki, R. M., and Rice, A. D. (1978). 'Response of susceptible and resistant peach-potato aphids *Myzus persicae* (Sulz.)', *Pestic. Sci.*, **9**, 513–516.

Sawicki, R. M., Farnham, A. W., Denholm, I., and O'Dell, K. (1981). 'Housefly resistance to pyrethroids in the vicinity of Harpenden', *Proceedings 1981 British Crop Protection Conference*, pp. 609–613.

Schnitzerling, H. J., Noble, P. J., Macqueen, A., and Dunham, R. J. (1982). 'Resistance of the buffalo fly, *Haematobia irritans exigua* (De Meijera) to two synthetic pyrethroids', *J. Aust. Ent. Soc.*, **21**, 77–80.

Schnitzerling, H. J., Nolan, J., and Hughes, S. (1983). 'Toxicology and metabolism of some synthetic pyrethroids in larvae of susceptible and resistant strains of the cattle tick, *Boophilus microplus* (Can.)', *Pestic. Sci.*, **14**, 64–72.

Scott, J. G., Croft, B. A., and Wagner, S. W. (1983). 'Studies in the mechanism of permethrin resistance in *Amblyseius fallacis* (Acarina: Phytoseiidae) in relation to previous insecticide use on apple', *J. Econ. Entomol.*, **76**, 6–10.

Scott, J. G., and Matsumura, F. (1981). 'Characteristics of a DDT-induced case of cross-resistance to permethrin in *Blatella germanica*', *Pestic. Biochem. Physiol.*, **16**, 21–27.

Scott, J. G., and Matsumura, F. (1983). 'Evidence for two types of toxic action of pyrethroids on susceptible and DDT-resistant German cockroaches', *Pestic. Biochem. Physiol.*, **19**, 141–150.

Shelton, A. M., and Soderlund, D. M. (1983). Varying susceptibility to methomyl and permethrin in widely separated cabbage looper (Lepidoptera: Nocturidae) population within Eastern North America. *J. Econ. Entomol.*, **76**, 987–989.

Shono, T., Unai, T., and Casida, J. E. (1978). 'Metabolism of permethrin in American cockroach adults, house fly adults and cabbage looper larvae', *Pestic. Biochem. Physiol.*, **9**, 96–106.

Soderlund, D. M. (1979). 'Pharmacokinetic behaviour of enantiomeric pyrethroid esters in the cockroach, *Periplaneta americana* L.', *Pestic. Biochem. Physiol.*, **12**, 38–48.

Soderlund, D. M., Ghiasuddin, S. M., and Helmuth, D. W. (1983a). 'Receptor-like stereospecific binding of a pyrethroid insecticide to mouse brain membranes'. *Life Sci.*, **33**, 261–267.

Soderlund, D. M., Heissney, C. W., and Jiang, M. (1983b). 'Metabolism of fenvalerate by resistant Colorado Beetle adults', Paper No. 28, Pesticide Chemistry Division, *American Chemical Society, 186th National Meeting*.

Soderlund, D . M., Sanborn, J. B., and Lee, P. W. (1983c). 'Metabolism of pyrethrins and pyrethroids in insects', in *Progress in Pesticide Biochemistry* (eds D. H. Hutson and T. R. Roberts), vol. 3 pp. 402–435, John Wiley & Sons, Chichester.

Sparks, T. (1981). 'Development of insecticide resistance in *Heliothis zea* and *Heliothis virescens* in North America', *Bull. Ent. Soc. Am.*, **27**, 186–192.

Spiers, R. D., and Zettler, J. L. (1969). 'Toxicity of three organophorus compounds and pyrethrins to malathion resistant *Tribolium castaneum*', *J. Stored Prod. Res.*, **4**, 279–283.

Strickler, K., and Croft, B. A. (1981). 'Selection for permethrin resistance in the predatory mite *Amblyseius fallacis*', *Entomol. Exp. Appl.*, **31**, 339–345.

Sun, Y. P., and Johnson, E. R. (1960). 'Analysis of joint action of insecticides against house flies', *J. Econ. Entomol.*, **53**, 887–892.

Tan, K. H. (1981). 'Antifeeding effect of cypermethrin and permethrin at sub-lethal levels against *Pieris brassicae* larvae', *Pestic. Sci.*, **12**, 619–626.

Taylor, C. E., and Georghiou, G. P. (1982). 'Influence of pesticide persistence in the evolution of resistance', *Environ. Entomol.*, **11**, 746–750.

Taylor, R. N. (1982). 'Insecticide resistance in houseflies from the Middle East and North Africa with notes on the use of various bioassay techniques', *Pestic. Sci.*, **13**, 415–425.

Taylor, R. N., Pope, A. R. J., and Hirst, R. G. (1981). 'Insecticide resistance in houseflies *Musca domestica* from two sites in Southern Africa', *Phytophylactica*, **13**, 1–4.

Tsukamoto, M. (1964). 'Methods for the linkage-group determination of insecticide-resistance factors in the housefly', *Botyu Kagaku*, **29**, 51–60.

Tsukamoto, M. (1969). 'Biochemical genetics of insecticide resistance in the housefly'. *Residue Rev.*, **25**, 289–314.

Tsukamoto, M. (1983). 'Methods of genetic analysis of insecticide resistance', in *Pest Resistance to Pesticides* (eds G. P. Georghiou and T. Saito), pp. 71–98, Plenum Press, New York.

Tsukamoto, M., Narahashi, T., and Yamasaki, T. (1965). 'Genetic control of low nerve sensitivity to DDT in insecticide resistant houseflies', *Botyu Kagaku*, **30**, 128–132.

Tsukamoto, M., and Suzuki, R. (1964). 'Genetic analyses of DDT-resistance in two strains of the housefly', *Botyu Kagaku*, **29**, 76–89.

Twine, P. H., and Reynolds, H. T. (1980). 'Relative susceptibility and resistance of the tobacco budworm to methyl parathion and synthetic pyrethroids in Southern California', *J. Econ. Entomol.*, **73**, 239–242.

Ueda, K., Gaughan, L. C., and Casida, J. E. (1975). 'Metabolism of four resmethrin isomers by liver microsomes', *Pestic. Biochem. Physiol.*, **5**, 280–294.

Vaughan, M. A., and Leon, Q. G. (1977). 'Pesticide management on a major crop with severe resistance problems', in *Proceedings of XV International Congress of Entomology* (eds J. S. Packer and D. White), pp. 812–815, The Entomological Society of America.

Vijverberg, H. P. M., Van der Zalm, J. M., and Van der Bercken, J. (1982). 'Similar mode of action of pyrethroids and DDT on sodium channel gating in myelinated nerves', *Nature*, **295**, 601–603.

Virgona, C. T., Holan, G., and Shipp, E. (1983). 'Repellency of insecticides to resistant strains of housefly', *Entomol. Exp. Appl.*, **34**, 287–290.

Wang Boonkong, S. (1981). 'Chemical control of cotton insect pests in Thailand', *Trop. Pest. Management*, **27**, 495–500.

Wardlow, L. R., Ludlam, A. B., and Bradley, L. F. (1976). 'Pesticide resistance in glasshouse whitefly *Trialeurodes vaporariorum* West', *Pestic. Sci.*, **7**, 320–324.

Whalon, M. E., Croft, B. A., and Mowry, T. M. (1982). 'Introduction and survival of susceptible and pyrethroid-resistant strains of *Amblyseius fallacis* (Acarina: Phytoseiidae) in a Michigan apple orchard'. *Environ. Entomol.*, **11**, 1096–1099.

Whitehead, G. B. (1959). 'Pyrethrum resistance conferred by resistance to DDT in the blue tick', *Nature*, **184**, 378–379.

Wolfenbarger, D. A., Bodegas, P. R. V., and Flores, R. G. (1981). 'Development of resistance in *Heliothis* spp. in the Americas, Australia, Africa and Asia', *Bull. Ent. Soc. Am.*, **27**, 181–185.

Wolfenbarger, D. A., Harding, J. A., and Davis, J. W. (1977). 'Isomers of (3-phenoxyphenyl)methyl-(+)-*cis,trans*-3(2,2-dichloroethyl)-2,2-dimethylcyclopropanecarboxylate against boll weevils and tobacco budworm', *J. Econ. Entomol.*, **70**, 226–228.

Wright, J. W., and Brown, A. W. A. (1957). 'Survey of possible insecticide resistance in body lice', *Bull. W.H.O.*, **16**, 9–31.

Wright, J. W., and Pal, R. (1965). 'Second survey of insecticide resistance in body lice 1958–63', *Bull. W.H.O.*, **33**, 485–501.

Zettler, J. L., McDonald, L. L., Redlinger, L. M., and Jones, R. D. (1973). '*Plodia interpunctella* and *Cadra cautella*, resistance in strains to malathion and synergised pyrethrins', *J. Econ. Entomol.*, **66**, 1049–1050.

Insecticides
Edited by D. H. Hutson and T. R. Roberts
© 1985 John Wiley & Sons Ltd

CHAPTER 5

Mammalian toxicology of pyrethroids

A. J. Gray and D. M. Soderlund

Abbreviations:
i.p., intraperitoneal; i.v., intravenous; p.o., *per os*; i.c.v., intracerebroventricular; CNS, central
nervous system.

INTRODUCTION

The pyrethroids are now broadly recognized as the fourth major class of synthetic organic insecticides. Since the commercial production of the first photostable pyrethroid in 1976, this group of compounds has achieved world-wide use with widespread agricultural and environmental health applications. Commercial sales of pyrethroids for agricultural usage exceeded $450 million in 1981 and are anticipated to take a 15% share of the world insecticide market by 1990. Of the major insecticide classes, the pyrethroids as a group are among the most potent as insecticides but exhibit comparatively low toxicity to mammals (Elliott, 1977a). The combination of these properties gives the pyrethroids a degree of selectivity that is almost unique among conventional insecticides. In this context, studies on the mammalian toxicology of pyrethroids not only establish the safety of these compounds prior to their introduction and use, but also attempt to understand physiological and pharmacological events that confer selectivity.

Scope and relationship to other reviews

In this review, we summarize the toxicological consequences following administration of pyrethrins and pyrethroids to mammals. We discuss the impact of pyrethroid structure, metabolism, distribution, and route of administration on the toxicity of these compounds and explore the existing information on the consequences of subacute and chronic pyrethroid intoxication. We also summarize progress to date in defining the pharmacological basis of the neurotoxic action of pyrethroids. The metabolic fate of individual pyrethroids is discussed only in those instances where it has particular relevance to toxicity; detailed reviews of the metabolism of the pyrethrins (Casida, 1973) and pyrethroids (Hutson, 1979; Casida et al., 1979; Casida and Ruzo, 1980; Chambers, 1980; Casida, 1983) are available elsewhere. We have also included a limited appraisal of the literature on the toxicology of pyrethroids in fish and birds even though this topic is not strictly encompassed by the title of this review. Four recent reviews on the mechanism of action of pyrethroids include some discussion of their mammalian toxicology (Glickman and Casida, 1982; Miller and Adams, 1982; Casida et al., 1983; Ruigt, 1984). However, this review, which includes papers published in early 1984, is the first that provides a comprehensive view of all aspects of the mammalian toxicology of pyrethroid insecticides.

Pyrethroid structures and nomenclature

The pyrethroids examined in this review are esters of 3-substituted 2,2-dimethylcyclopropanecarboxylic acids, or analogous 2-aryl-3-methyl-

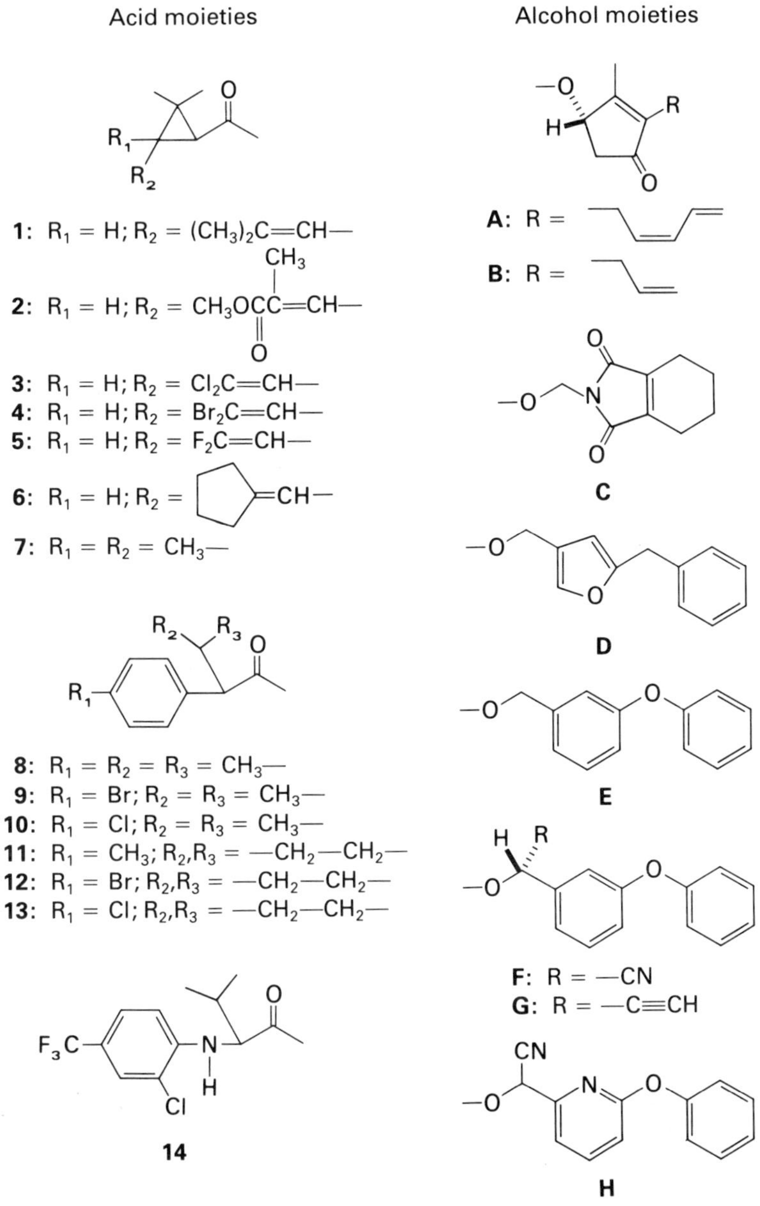

Acid moieties   Alcohol moieties

1: R_1 = H; R_2 = $(CH_3)_2C$=CH—

2: R_1 = H; R_2 = CH$_3$OCC=CH—

A: R =

B: R =

3: R_1 = H; R_2 = Cl$_2$C=CH—
4: R_1 = H; R_2 = Br$_2$C=CH—
5: R_1 = H; R_2 = F$_2$C=CH—

6: R_1 = H; R_2 =   =CH—

7: R_1 = R_2 = CH$_3$—

C

D

8: R_1 = R_2 = R_3 = CH$_3$—
9: R_1 = Br; R_2 = R_3 = CH$_3$—
10: R_1 = Cl; R_2 = R_3 = CH$_3$—
11: R_1 = CH$_3$; R_2,R_3 = —CH$_2$—CH$_2$—
12: R_1 = Br; R_2,R_3 = —CH$_2$—CH$_2$—
13: R_1 = Cl; R_2,R_3 = —CH$_2$—CH$_2$—

E

F: R = —CN
G: R = —C≡CH

14

H

Figure 1 Structures of the acid and alcohol moieties of pyrethrins and pyrethroids considered in this review. Alcohol moieties **A**, **B**, **F**, and **G** are shown in the configuration giving esters of highest insecticidal activity and mammalian toxicity

butyric acids lacking the cyclopropane ring, and an appropriate alcohol containing one or more unsaturated moieties. The structures of the separate acid and alcohol moieties are shown in Figure 1. Specific pyrethroids or their component acid and alcohol moieties that are named or discussed in the test are identified by number and letter as shown in Figure 1.

Most pyrethroids have two or three chiral centres, resulting in four or eight optical and geometrical isomers that differ in insecticidal activity, mammalian toxicity and metabolic fate. It is therefore imperative that the pyrethroid structure is defined as specifically as possible in order to avoid confusion when discussing the toxicology of each compound. The nomenclature proposed by Elliott *et al.,* (1974c), which is used throughout this review, describes the configuration of cyclopropane C-1 in the acid moiety as *R* or *S* according to the sequence rule. The relative configuration of the substituent at cyclopropane C-3 is then designated *cis* or *trans* with reference to the plane of the ring and the substituent at cyclopropane C-1 (Figure 2). The isomers of the non-cyclopropane acids are designated *R* or *S* at C-2 (Figure 2) and the absolute configuration of chiral alcohol moieties is defined *R* or *S* as appropriate. Thus one of the most insecticidal isomers of cypermethrin (Figure 1, **3F**) is designated as [1*R,cis,αS*]-cypermethrin.

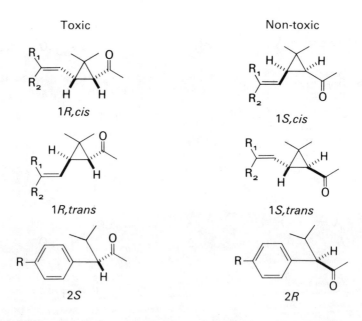

Figure 2 Stereochemical configurations of 3-substituted-2,2-dimethyl-cyclopropanecarboxylic acids and 2-substituted 3-methylbutyric acids that give toxic and non-toxic pyrethroid esters

Development of the pyrethroids

The history of the development of pyrethroids has been summarized in several reviews (Elliott, 1971; 1974; 1977a,b; 1980; Elliott *et al.*, 1974a) and it seems superfluous to duplicate these efforts here. Nevertheless, a brief description of the key events in the optimization of pyrethroid chemistry seems appropriate to provide the reader with a context in which to appreciate the diversity of structures for which information on mammalian toxicology is available.

Synthetic pyrethroids have been developed over a period of more than 40 years through the synthesis and evaluation of analogues of the pyrethrins, the naturally occurring insecticidal constituents of the flowers of *Tanacetum cinerariaefolium*. Although the pyrethrins are adequate insecticides, their extreme photolability has precluded their use in agriculture. Pyrethrin I (**1A**), the most potent and most abundant of the pyrethrins, has served as the model in efforts to synthesize analogues that retain the desirable insecticidal and toxicological properties of the pyrethrins, but are more stable in the environment. By 1965 a wide variety of synthetic pyrethroids had been prepared, but none of these was superior to pyrethrin I in insecticidal activity or photostability. Nevertheless, several of these compounds, such as allethrin (**1B**; Schechter *et al.*, 1949) and tetramethrin (**1C**; Kato *et al.*, 1964) have achieved widespread use in applications where photostability is either not necessary or undesirable. In 1967, Elliott and co-workers announced the discovery of resmethrin (**1D**; Elliott *et al.*, 1967), the first synthetic ester with insecticidal activity clearly superior to pyrethrin I. Although resmethrin itself was still photolabile, the discovery of this compound stimulated research to design photostabilized esters that retained these desirable insecticidal properties.

The first significant success in achieving a photostabilized pyrethroid with high insecticidal activity was realized through use of the 3-phenoxybenzyl alcohol moiety. Phenothrin (**1E**; Elliott, 1971; Fujimoto *et al.*, 1973), the chrysanthemate ester of this alcohol, is significantly more photostable than resmethrin yet retains nearly the same level of insecticidal activity. Replacement of the isobutenyl moiety of phenothrin with a dichlorovinyl substituent produced permethrin (**3E**; Elliott *et al.*, 1973), the first highly insecticidal pyrethroid sufficiently photostable for agricultural use. Further increases in insecticidal potency were achieved through introduction of a cyano substituent at the benzylic carbon of the 3-phenoxybenzyl moiety of phenothrin and related dihalovinyl esters to give compounds such as cyphenothrin (**1F**), cypermethrin (**3F**) and bromocyphenothrin (**4F**; Elliott *et al.*, 1974b, 1975; Matsuo *et al.*, 1976). Deltamethrin, which is prepared as the resolved 1*R,cis,αS* isomer of bromocyphenothrin, is one of the most active insecticides of any class, having a potency of 600 times that of DDT against *Anopheles stephensi* (Barlow and Hadaway, 1975) and 34 times that of [1*R,trans*]-resmethrin (bioresmethrin) against *Musca domestica* (Elliott *et al.*, 1974b; Elliott, 1977a). Moreover, both cypermethrin and deltamethrin have been shown to retain residual activity for at least six weeks in field conditions (Elliott, 1977b).

A further major step in pyrethroid development was the finding that 2-aryl-3-methylbutyric acid esters of pyrethroid alcohols possessed significant pyrethroid-like insecticidal activity and good photostability (Ohno *et al.*, 1974). This discovery demonstrated that the cyclopropane ring was not necessary for high insecticidal activity and defined an entirely new series of potent pyrethroid esters. The first of this group to reach commercial use was fenvalerate (**10F**).

The compounds mentioned above exemplify critical steps in the evolution of pyrethroids, but they represent only a small proportion of the known active insecticides in this class. Current development efforts are directed at obtaining photostable esters of the highest potency. As a result, most of the compounds presently under development fall into two structural groups, both of which share the α-cyano-3-phenoxybenzyl alcohol moiety: cyclopropanecarboxylate esters similar to cypermethrin and deltamethrin; and 2-aryl-3-methylbutyrate esters related to fenvalerate. Because of this convergence, the structural differences between pyrethroids may appear to be subtle to those unfamiliar with these compounds; however, small changes in substituents and stereochemistry are sufficient to produce compounds differing in their insecticidal potency, spectrum of activity, and mammalian toxicology.

ACUTE NEUROTOXICITY

Structure–toxicity relationships

Synthetic pyrethroids are generally viewed as being among the safest insecticides available on the basis of their low acute toxicity to mammals (Table 1). However, this is not due to a lack of intrinsic neurotoxicity, since several pyrethroids are highly potent neurotoxins when administered by routes (e.g. intravenous or intracerebral) that minimize metabolism and permit more efficient delivery to the nervous system (Table 1). The involvement of complex pharmacokinetic and pharmacodynamic factors complicates interpretation of pyrethroid toxicity data following oral or intraperitoneal administration and limits the usefulness of these data in determining intrinsic structure–toxicity relationships. However, recent surveys of the intravenous (Verschoyle and Aldridge, 1980) and intracerebral (Lawrence and Casida, 1982) toxicities of pyrethroids to rats and mice describe structure–toxicity relationships that may reflect more closely the intrinsic neurotoxicity of pyrethroids in mammals. The following discussion of structure–toxicity relationships is based primarily on these studies.

The neurotoxic action of pyrethroids is highly stereospecific. Only esters of 1*R* cyclopropanecarboxylates and isosteric 2*S* isomers of non-cyclopropane acids are toxic (Figure 2), whereas the corresponding 1*S* cyclopropanecarboxylates and their equivalent 2R acyclic analogues are completely non-toxic and have no synergistic or antagonistic effect on the toxicity of active enantiomers *in vivo*. The absolute configuration at cyclopropane C-3, which determines whether esters have the *cis* or *trans* configuration across the ring, also strongly influences mammalian toxicity. With the exception of [1R,*cis*]-phenothrin, insecticidal esters having the 1R,*cis* configuration are potent mammalian toxicants, but several of the corresponding 1R,*trans* isomers (e.g. tetramethrin, resmethrin, permethrin) exhibit no detectable toxicity following the intracerebral administration of large doses (Lawrence and Casida, 1982). The unequivocal demonstration of the absence of intrinsic neurotoxicity in such cases requires samples of high isomeric purity. Thus, the apparent neurotoxicity of [1R,*trans*]-resmethrin following intravenous administration to rats (White *et al.*, 1976) was later shown to result from the *cis* isomer (1.3–5.5%) content of the sample used (Gray and Connors, 1980). The more recent report of measurable intracerebral toxicity of [1RS,*trans*]-permethrin in mice (Staatz *et al.*, 1982a), which has not been confirmed in other studies, may also reflect *cis* isomer contamination.

Changes in the nature of the 3-substituent of cyclopropanecarboxylates also influence mammalian toxicity. Replacement of the isobutenyl moiety of the non-toxic 1R,*trans* isomers of resmethrin or phenothrin with a dihalovinyl group (Figure 1, **3–5**) or a cyclopentylidenemethyl group (Figure 1, **6**) frequently results in compounds which are potent neurotoxicants (Table 1). These substituents also increase toxicity in the corresponding *cis*-substituted esters (Table 1). Although the methyl groups of the isobutenyl moiety are important sites of oxidative attack in chrysanthemate esters, the increases in mammalian toxicity afforded by substitution at this position probably result from alterations in intrinsic neurotoxicity rather than from increased stability (see below).

In mammals, as in insects, the presence of an α-cyano substituent in the S configuration at the benzylic position of 3-phenoxybenzyl alcohol greatly enhances neurotoxic activity. This effect is highly stereospecific, since the corresponding α-R epimers, which have the appropriate configuration in the acid moiety, are completely non-toxic even when administered intracerebrally (Lawrence and Casida, 1982). The α-cyano substituent also alters structure–toxicity relationships in the acid moiety. The most dramatic effect involves the non-toxic *trans*-substituted cyclopropanecarboxylates of 3-phenoxybenzyl alcohol, such as *trans*-phenothrin and *trans*-permethrin; addition of the α-cyano substituent to esters, giving *trans*-cyphenothrin and *trans*-cypermethrin, results in intrinsically more potent neurotoxicants.

It is of interest to compare mammalian structure–toxicity relationships with the extensive information on pyrethroid structure–activity relationships in insects (for reviews, see: Elliott *et al.*, 1974a, 1978b; Elliott and Janes, 1979). In general, only those compounds having useful insecticidal activity have been

Table 1 Pyrethroid structures, toxicity to rats and mice by various routes of administration and the signs of toxicity produced

Pyrethroid[a]	Structure[b] acid	alcohol	Specific isomer[a]	Configuration[c]	Toxicity to Rats[d] (mg/kg) Oral	i.v.	Toxicity to Mice[e] Oral (mg/kg)	i.p. (mg/kg)	i.v. (mg/kg)	i.c.v. (μg/g)	Poisoning[f] syndrome
Pyrethrin I	1	A			260–420 (M)	5	350	240		} >2500	T
Pyrethrin II	2	A			>600 (M)	1	250				T
Allethrin	1	B			1030	4	2000	30		290	T
			Bioallethrin	1R,trans,αRS	430					>860	T
			S-bioallethrin	1R,trans,αS							T
Tetramethrin	1	C		1RS,cis,trans	>5000						T
Resmethrin	1	D		1R,cis	1995(1441–2765)	160–170	490	320		41(24–64)	T
			Cismethrin	1R,cis	32(23–44)	6–7	160			>8600	T
			Bioresmethrin	1R,trans	>8000	340	800	>1500			T
Ethanomethrin	6	D		1RS,cis,trans							T
				1R,cis		2–5				6.8(3.8–15)	T
				1R,trans						14(7.8–31)	T
NRDC 108	7	D	NRDC 141	1R,cis	147(115–187)	2–5		10			T
Chlororesmethrin	3	D	NRDC 174	1R,cis	18(12.5–25)	1.4–2.8					T
Fluororesmethrin	5	D	NRDC 173	1R,trans	14(9.9–20)	0.35–0.65					T
				1RS,cis,trans	126	1.4–2.8		>1500			T
Phenothrin	1	E		1R,cis	>10,000	>600	>5000	>1500		>4300	T
				1R,trans			>2500	>1500		>8600	T
							>5000				T
Permethrin	3	E	NRDC 148	1RS,cis		>270	85	1000	19.8(16.8–22.3)	11(6.8–40)	T
			NRDC 146	1RS,trans		>270	3200	1000	93.2(83–104.5)		T
			NRDC 147	1R,cis		>270					T
			NRDC 157	1R,trans		>30					T
Bromophenothrin	4	E		1R,cis,αS		5				3.9(1.5–9.4)	CS/T (II)
Cyphenothrin	1	F	RU 29209	1R,trans,αRS		17–25				12(2.5–21)	T/CS (II)
			RU 29208	1R,cis,αRS		6–8					CS
Bromocyphenothrin	4	F	NRDC 156	1R,trans,αRS		114–170		10	2.36(1.92–2.89)	1.2(0.8–1.7)	CS
			NRDC 158	1R,cis,αS	20(14–28)	2.0–2.6					CS
			Deltamethrin	1R,trans,αRS		17–25					CS
			RU 26979	1R,trans,αS		50–60					CS
Cypermethrin	3	F		1RS,cis,trans,αRS	500	6–9		28		0.6(0.3–1.3)	CS
			NRDC 160	1RS,cis,αRS		170–180				1.6(1.1–4.3)	CS
			NRDC 159	1RS,trans,αRS		0.52–0.75		>500			CS
Fluorocyphenothrin	5	F	RU 26607	1RS,cis,αRS		5–7					CS
			RU 26941	1R,cis,αS							T
Fenpropathrin	7	F	S-3206	1R,trans,αS	28.3(17–46)	2.5		15		6.1(3.9–9.4)	T(I/II)
RES 2637	8	F				225–280					TS
RES 2664	9	F				150–225					CS
Fenvalerate	10	F	S-5602		450	50–100		>500		1.0(0.4–2.2)	TS
S-5655	10	G				200–250					T
NRDC 179	11	F				400–500					CS
NRDC 180	12	F				250–450					CS
NRDC 181	13	F				300–450					CS

[a] Approved common names are italicized; other compounds containing an altered substituent at cyclopropane C-3 are designated according to the substituent modification of the parent ester (e.g. bromocyphenothrin, the dibromo analogue of cyphenothrin); identification numbers of some unnamed compounds have been retained.

[b] The numbers and letters refer to structures given in Figure 1.

[c] The pyrethroid was a mixture of all isomers unless otherwise specified. The stereochemistry of the acid moieties are illustrated in Figure 2.

[d] The rat i.v. toxicity data were taken from Verschoyle and Aldridge (1980). The rat oral toxicity data were taken from a number of sources cited in the text but were mainly from Verschoyle and Barnes (1972) or from a personal communication of Verschoyle. The values given are the LD$_{50}$ (95% confidence limits), or the approximate lethal dose, in mg/kg. The vehicle for i.v. administration was always glycerol formal but varied for oral dosing (see later section). The values are for female animals unless indicated otherwise by (M).

[e] The i.p. toxicity data for mice were compiled from a number of sources cited in the text but were mainly derived from Soderlund and Casida (1977a) and Casida et al. (1983). The test compounds were administered in methoxytriglycerol to male mice. The oral LD$_{50}$ values for female mice were taken from Miyamoto (1976), the i.v. data from Staatz et al. (1982a). The intracerebroventricular (i.c.v.) data from Lawrence and Casida (1982) were obtained in male mice using methoxytriglycerol as solvent and are given in μg/g brain weight. The i.c.v. toxicities of the cypermethrin isomers are for the pure 1R,αS isomers.

[f] The signs of toxicity are classified according to Verschoyle and Aldridge (1980) and are described in detail in a later section. The T and CS syndromes generally correspond to the type I and II syndromes, respectively, used by Lawrence and Casida (1982); exceptions are indicated in the table. Cyphenothrin isomers did not produce signs of toxicity that were exclusively of one class in rats; the predominant syndrome is indicated first.

examined in mammals. Consequently, structure–toxicity relationships drawn from mammalian data do not reflect the overall relationship of pyrethroid-like structure and toxicity, but rather describe the mammalian toxicity of a restricted group of potent insecticidal pyrethroids. In this context, available data suggest that the site of pyrethroid action in mammals possesses even greater specificity than that in the insect nervous system. [1R,trans]-Cyclopropanecarboxylate esters of primary alcohols are potent insecticides, but most of these compounds are relatively non-toxic to mammals. Similarly, esters of [R]-α-cyano-3-phenoxybenzyl alcohol possess measurable insecticidal activity (similar to, or somewhat less than, their non-cyano analogues; Elliott et al., 1978a), but these esters are apparently non-toxic in mammals. In addition to these differences, the comparison of brain levels of deltamethrin found in mice at death following a twice LD_{50} dose (0.55–1.47 nmol/g brain; Ruzo et al., 1979) with the mean steady-state levels of its less insecticidal non-cyano analogue in the ventral nerve cord of cockroaches following a twice LD_{50} dose (0.11 nmol/g nerve cord; Soderlund, 1979) suggests that the mammalian site of pyrethroid action may be generally less sensitive than that in the insect nervous system. Confirmation of this latter relationship requires a well characterized biochemical or receptor assay for the pyrethroid site of action that can be applied to both mammalian and insect nervous tissue.

Correlation of structure with signs of toxicity

The signs of toxicity produced by administration of the pyrethrins, bioallethrin, the resmethrins and NRDC 108 to rats were first described by Verschoyle and Barnes (1972) and included tremors not unlike those produced by DDT. In contrast, the signs of toxicity produced by deltamethrin were found to be distinctly different (Barnes and Verschoyle, 1974). This led to the examination of a number of pyrethroid isomers in order to determine whether any relationship existed between structure and signs of toxicity (Verschoyle and Aldridge, 1980). The signs of toxicity produced following i.v. injection of toxic doses of these pyrethroids to rats generally fell into two distinct syndromes, defined as T and CS (Table 1). An alternative classification of pyrethroid action into two types (type I and type II) based on both their effects on cockroach cercal nerve preparations (Gammon et al., 1981) and the signs of toxicity observed following intracerebral administration to mice (Lawrence and Casida, 1982) has been proposed. A third classification scheme has been used by the World Health Organization (1979) in which the syndromes produced have been defined as class 1 or class 2. In most cases pyrethroids categorized as Type I or Class 1 correspond to the T syndrome and type II or class 2 to the CS syndrome although some inconsistencies do exist. We have chosen the T and CS classification scheme in the context of this review because it permits the classification of pyrethroid action in terms clearly related to the mammalian toxicology of these compounds.

The T syndrome is characterized by rapid onset of aggressive sparring behaviour and an increased sensitivity to external stimuli. These symptoms are quickly followed by the appearance of a fine tremor, gradually becoming more severe until the animals become prostate with coarse whole-body tremors. The core temperature of these animals increases, paralleling the development of tremors, probably due to the increased muscle activity (White *et al.*, 1976). The onset of rigor often precedes death (Verschoyle and Aldridge, 1980). The CS syndrome is characterized by initial hyperactive behaviour (grooming, chewing, pawing and burrowing or 'bulldozing'), facial tremor, hunched-back posture and profuse salivation. A coarse whole-body tremor develops, together with a hypersensitivity to auditory and tactile stimuli and motor incoordination ('splayed hind legs'), progressing to induced and finally spontaneous choreathetosis (sinuous writhing movements). Clonic (alternate muscular contraction and relaxation) and tonic (rigid extension of limbs) seizures are usually observed in the terminal stages (Barnes and Verschoyle, 1974; Ray and Cremer, 1979). The core temperature may decrease, possibly in part due to evaporation of saliva covering the animal (Ray and Cremer, 1979).

The time of onset and duration of particular signs varies considerably according to the route of administration, vehicle, and the compound under investigation. However, in general, following i.v. administration of an approximate LD_{50} dose, pyrethroids that cause the T syndrome produce tremors within 5 minutes whilst those causing the CS syndrome usually take 10–20 minutes to produce coarse whole-body tremors and writhing. Death may occur from 10 minutes to 2 hours after an approximate i.v. LD_{50} dose: survivors appear near normal after 2–6 hours (Gray, 1980; Verschoyle and Aldridge, 1980). Cismethrin and deltamethrin are frequently used in detailed studies into the mechanism of action of pyrethroids since they produce signs of toxicity typical of the T and CS syndromes, respectively, and are readily available as single optically pure isomers. The signs of poisoning are clearly correlated with pyrethroid structure. Pyrethroids containing both the S-α-cyano-3-phenoxybenzyl alcohol and a halogenated $1R$-acid moiety usually produce the CS syndrome whereas those not containing both of these constituents generally produce the T syndrome. Pyrethrins and pyrethroids containing the 5-benzyl-3-furylmethyl (**D**), 3-benzylbenzyl or 3-phenoxybenzyl (**E**) alcohols almost always cause the T syndrome of poisoning regardless of the acid component (Table 1) (Verschoyle and Aldridge, 1980). Compounds containing the S-α-cyano-3-phenoxybenzyl alcohol moiety (**F**) and a dihalovinyl derivative of chrysanthemic acid (**3, 4, 5**), a halogenated 2-phenylisovaleric acid (**9, 10**) or a halogenated cyclopropylphenylacetic acid (**12, 13**) generally produce the CS syndrome (Verschoyle and Aldridge, 1980). Esters of α-cyano-3-phenoxybenzyl alcohol and nonhalogenated acids such as the cyclopropylphenylacetic acid (**11**) or tetramethylcyclopropanecarboxylic acid (fenpropathrin, S-3206, **7**) both produce the T syndrome according to Verschoyle and Aldridge (1980). However, the latter compound is reported to cause the class 2 syndrome (WHO, 1979) or is classified as type II, but with some tremors (Lawrence and Casida, 1982), and was intermediate (type I/II) in the cockroach nerve preparation (Gammon et al., 1981). There is, therefore, some controversy about which poisoning syndrome is produced by fenpropathrin. Two compounds not included in the Lawrence and Casida (1982) study, RES 2637 (**8F**) and S-5655 (**10G**), were classified by Verschoyle and Aldridge (1980) as producing both tremor and salivation (TS).

Although 1R,cis and 1R,trans isomers usually produce the same poisoning syndrome, there are some exceptions. The 1R,cis isomer of the pyrethroid consisting of 3-(2,2-difluorovinyl)-2,2-dimethylcyclopropanecarboxylic acid (**5**) and S-α-cyano-3-phenoxybenzyl alcohol (**F**) (fluorocyphenothrin, Table 1) produces the CS syndrome whilst its 1R,trans isomer produces the T syndrome. A similar anomaly occurs with the 1R,cis and 1R,trans isomers of chrysanthemic acid (**1**) when esterified with the S-α-cyano-3-phenoxybenzyl alcohol to produce cyphenothrin (Table 1).

Toxicity by direct injection into the central nervous system

The signs of toxicity produced after intracerebroventricular (i.c.v.) injection were similar to those produced following per os (p.o.) intraperitoneal (i.p.) or intravenous (i.v.) administration of much higher doses of these compounds (Gray and Rickard, 1982b; Lawrence and Casida, 1982; Staatz et al., 1982a). The CNS concentrations of pyrethroids following i.c.v. injection of severely toxic or lethal doses closely correspond to those found after peripheral administration of toxic doses, assuming uniform distribution of the complete i.c.v. dose throughout the central nervous system (Table 2) (White et al., 1976; Ruzo et al., 1979; Gray et al., 1980; Glickman and Lech, 1982; Gray and Rickard, 1982a,b; Lawrence and Casida, 1982; Staatz et al., 1982a). There are, however, some anomalies in the signs of toxicity produced after i.c.v. injection. Salivation was not observed following i.c.v. administration of deltamethrin to rats (Gray and Rickard, 1982b) or mice (Staatz et al., 1982a) and was not reported after i.c.v. injection of other typical CS syndrome pyrethroids (cypermethrin, deltamethrin and fenvalerate) to mice (Lawrence and Casida, 1982). However, salivation has been observed following i.c.v. administration of some T syndrome pyrethroids (e.g. cismethrin, Gray, 1980; Lawrence and Casida, 1982). Fenpropathrin was found to produce both tremors and choreoathetosis when given by the i.c.v. route (Lawrence and Casida, 1982) but is reported to cause only tremors when administered i.v. to rats (Verschoyle and Aldridge, 1980).

Although the signs of toxicity following i.c.v. administration are similar to those observed following parenteral or oral administration, the cause of death is not necessarily the same. Death after i.c.v. injection of deltamethrin to rats was associated with severe respiratory distress; the lungs of these animals appeared congested at post-mortem examination and histological studies indicated severe oedema which had probably resulted in loss of respiratory function and death (Gray and Rickard, 1982b). A number of other convulsant compounds have also been shown to cause hypertension and effects on respiration by their action on the cardiovascular and respiratory centres of the CNS. In contrast, in this same study, no deaths were observed following i.c.v. administration of cismethrin despite the incidence of severe tremors that persisted for up to 5 hours. Even doses of cismethrin equivalent to the i.v. LD_{50} did not cause death when given by the i.c.v. route. Hence peripheral actions, such as on the cardiovascular or respiratory systems, may also contribute to the lethality of some pyrethroids (Gray and Rickard, 1982b). The onset of motor signs of toxicity after i.c.v. administration of deltamethrin, cis-permethrin, and cismethrin to mice and rats was delayed when compared with i.v. injection (Gray and Rickard, 1982b; Staatz et al., 1982a). The onset of motor signs was more rapid when deltamethrin and cismethrin were injected directly into the spinal cord, although signs of toxicity were only observed caudal to the site of injection (Gray and Rickard, 1982b).

Using radiolabelled cismethrin and deltamethrin, it was demonstrated that the concentrations of pyrethroid around the site of injection following i.c.v. injection were considerably higher than those found in the brain after peripheral administration and did not correlate with the onset of signs. It was not until the concentration of cismethrin or deltamethrin in the pons/medulla, cerebellum and upper spinal cord approached the critical threshold determined by peripheral administration of these pyrethroids (Table 2) that the tremor or writhing movements started (Gray and Rickard, 1982b). Injection of labelled cismethrin or deltamethrin into the lumbar region of the spinal cord (Gray and Rickard, 1982b) indicated that very little flow of the pyrethroid to the brainstem and higher centres occurred but the concentrations around the site of injection were much higher than found after i.v. or oral administration (White et al., 1976; Gray et al., 1980; Gray and Rickard, 1982a) or during the i.c.v. studies (Gray and Rickard, 1982b). It was concluded that, although an action at the spinal level alone was sufficient to account for the motor signs of toxicity, higher pyrethroid concentrations may be required if only the spinal cord is affected. When other areas of the CNS, such as the cerebellum, which acts as a modulator of spinal reflexes, are also affected, the toxic events in the spinal cord may become amplified. These studies have demonstrated that the majority of signs of toxicity of both CS and T syndromes can be reproduced by i.c.v. or spinal injection of pyrethroids. The two different poisoning syndromes therefore cannot be distinguished by their major site of action on the mammalian nervous system (i.e. not peripheral vs central) as was previously suggested (Barnes and Verschoyle, 1974; Verschoyle and Aldridge, 1980) by interpretation of the observed signs of toxicity.

Table 2 Comparison of the concentration of pyrethroid found in the CNS following i.v. or p.o. administration with that required to produce toxicity by i.c.v. injection[a][b]

		Concentration (nmol/g) in the CNS associated with toxicity	
Pyrethroid	Species	after i.v. or p.o. administration	after i.c.v. administration
Deltamethrin	rat	0.5 (toxic)	1.8 (death)
	mouse	1.0 (toxic)	0.5 (toxic), 0.6 (LD_{50})
Cismethrin	rat	1.5–5.5 (toxic), 10–15 (death)	20 (toxic)
	mouse		120 (LD_{50})
cis-Permethrin	mouse	15.6 (death)	5.8 (toxic), 28 (LD_{50})

[a] In order to facilitate comparison, the concentrations of the pyrethroids in the CNS have been converted from the various units used by the original authors to nmol/g tissue. These calculations were based on the assumptions that all the i.c.v. administered dose was uniformly distributed throughout the CNS, that the mice weighed 25 g and had a total CNS weight of 1 g, and that the rat CNS weight was 2 g.

[b] The approximate severity of the toxicity produced by the indicated dose is given in parentheses. The data presented in this table were compiled from a number of sources which are cited in the test of this section.

The use of direct injection of pyrethroids into the CNS may prove to be a useful technique to study the mechanism of action and the intrinsic toxicity of given pyrethroid isomers since it minimizes both the amount of compound required and the contribution of absorption, distribution and metabolism. However, care must be taken when extrapolating findings from i.c.v. administration studies to the toxicity produced by the more usual routes of administration as they do not take into consideration peripheral effects which may contribute to the lethality of the pyrethroid. Additionally, i.c.v. or spinal injection may cause signs of toxicity resulting in death that have not been observed after peripheral administration.

Role of metabolism in determining toxicity

Pyrethroids are rapidly metabolized in mammals *in vivo* by hydrolysis of the central ester bond and oxidative attack at several sites to give complex arrays of primary and secondary metabolites (for reviews, see: Casida *et al.*, 1975/76, 1979; Miyamoto, 1976; Ruzo and Casida, 1977; Hutson, 1979; Casida and Ruzo, 1980; Chambers, 1980). Since these metabolites have little or no demonstrable neurotoxicity (White *et al.*, 1976; Casida *et al.*, 1983), it is likely that the first metabolic attack on the parent ester results in detoxication. The critical role of these detoxication processes in limiting access of pyrethroids to sites of neurotoxic action is evident in the synergism of pyrethroid toxicity to rats and mice by organophosphate esterase inhibitors (Abernathy *et al.*, 1973; Jao and Casida, 1974; Soderlund and Casida, 1977a; Ruzo *et al.*, 1979; Gaughan *et al.*, 1980) and monooxygenase inhibitors (Jao and Casida, 1974;

Table 3 Synergism of the acute intraperitoneal toxicity of pyrethroids to mice by S,S,S-tributyl phosphorotrithioate (DEF) and piperonyl butoxide (PB)[a]

Pyrethroid	Acid moiety		Factor of synergism[b]	
	Structure	Configuration	DEF	PB
5-Benzyl-3-furylmethyl (**D**) esters				
Bioresmethrin	1	1R,trans	None	None
Cismethrin	1	1R,cis	11	25
trans-Ethanomethrin	6	1R,trans	> 188	> 60
cis-Ethanomethrin	6	1R,cis	4	5
α-Cyano-3-phenoxybenzyl (**F**) esters[c]				
trans-Cypermethrin	3	1R,trans	> 20	> 10
cis-Cypermethrin	3	1R,cis	5	6
Deltamethrin	4	1R,cis	2	3.5
Fenvalerate	10	2R,S	> 32	> 45

[a] Data from: Abernathy et al., (1973); Jao and Casida (1974); Soderlund and Casida (1977a).
[b] LD_{50} alone/LD_{50} plus synergist; see Table 1 for unsynergized LD_{50} values.
[c] Racemic except in deltamethrin.

Crawford and Hutson, 1977; Soderlund and Casida, 1977a; Ruzo et al., 1979). Both the esterase inhibitor S,S,S-tributyl phosphorotrithioate (DEF) and the oxidase inhibitor piperonyl butoxide (PB) increase the intraperitoneal toxicity of a variety of pyrethroid esters to mice (Table 3). The factors of synergism are small for compounds such as cis-ethanomethrin, cis-cypermethrin, and deltamethrin, all of which are relatively potent intraperitoneal toxicants in the absence of synergists (Table 1). In contrast, these synergists dramatically increase the toxicity of compounds such as trans-ethanomethrin, trans-cypermethrin, and fenvalerate (Table 3), which are weak intraperitoneal toxicants when administered alone (Table 1). Neither DEF nor PB synergizes trans-resmethrin, thus confirming that the low mammalian toxicity of this compound is due to its lack of intrinsic neurotoxicity.

The synergism data suggest that the relative importance of hydrolytic and oxidative attack varies from pyrethroid to pyrethroid and from isomer to isomer. Thus, effects of pyrethroid structure on biodegradability may complicate intrinsic structure–toxicity relationships. A survey of the in vitro degradation of more than 40 pyrethroids by mouse liver microsomal esterases and monooxygenases (Soderlund and Casida, 1977b) attempted to establish intrinsic structure–biodegradability relationships. Selected data from this study (Table 4) illustrate the general structure–biodegradability relationships for pyrethroids in this system: trans-substituted esters of primary alcohols (Figure 1, **D, E**) are rapidly hydrolysed, whereas the corresponding cis-substituted isomers and all esters of secondary alcohols (Figure 1, **A, B, F, G**) were more resistant to hydrolysis. In contrast, the rates of oxidative metabolism were much

Table 4 Rates of hydrolysis and oxidation of selected pyrethroids by mouse liver microsomal enzymes[a]

Substrate (acid moiety)	Relative rate[b]	
	Hydrolysis	Oxidation
5-Benzyl-3-furylmethyl (D) esters		
[1R,*trans*]-Resmethrin (1)	79	20
[1R,*cis*]-Resmethrin (1)	<3	29
[1R,*trans*]-Ethanomethrin (6)	19	12
[1R,*cis*]-Ethanomethrin (6)	<3	19
3-Phenoxybenzyl (E) esters		
[1R,*trans*]-Phenothrin (1)	59	27
[1R,*cis*]-Phenothrin (1)	<4	37
[1R,*trans*]-Permethrin (3)	77	30
[1R,*cis*]-Permethrin (3)	<2	26
[2S]-S-5439 (10)	<3	23
α-Cyano-3-phenoxybenzyl (F) esters[c]		
[1R,*trans*]-Cyphenothrin (1)	3	5
[1R,*cis*]-Cyphenothrin (1)	<3	8
[1RS,*trans*]-Cypermethrin (3)	17	4
[1RS,*cis*]-Cypermethrin (3)	<2	5
Deltamethrin (4)	<3	<4
[2S]-Fenvalerate (10)	<2	11

[a] Data from: Soderlund and Casida (1977b).
[b] The rate of [1R,*trans*]-resmethrin metabolism by combined oxidases and esterases is set equal to 100.
[c] Racemic except in deltamethrin.

less variable, except that α-cyano-3-phenoxybenzyl esters were consistently more resistant to oxidative attack than their non-cyano analogues. These studies also demonstrated that modification of the isobutenyl substituent of the acid moiety, a known site of oxidative attack in chrysanthemate esters, had little influence on rates of oxidative detoxication. Thus, the increased toxicity of some of these chrysanthemate analogues is not likely to be a result of their increased metabolic stability.

These findings suggest that the greatest impact of metabolic detoxication on observed toxicity is expected to be found among good esterase substrates that are also potent neurotoxicants. A clear example is *trans*-ethanomethrin, which is synergized dramatically in mice by DEF (Table 3). The effects of the structure of pyrethroids on their metabolic stability in general, and on their relative susceptibility to hydrolysis in particular, were thought at one time to influence structure–toxicity relationships. However, recent studies of the intracerebral toxicity of pyrethroids have shown that the structural features that permit rapid metabolism in many cases also confer low intrinsic neurotoxicity (e.g. *trans*-substitution, presence of a primary alcohol moiety), and those that give metabolically stable compounds also confer high intrinsic toxicity (e.g. *cis*-substitution, presence of an *S*-α-cyano substituent). Consequently, whereas the toxicity of almost all pyrethroids is modified by metabolic detoxication, a critical impact of metabolism on structure–toxicity relationships is evident in only a few instances.

An underlying assumption in the correlations of structure with biodegradability summarized above is that the liver is the principal site of pyrethroid detoxication in mammals. Although it is likely that the liver is the predominant site of pyrethroid metabolism, monooxygenase activity is known to be present in a number of other tissues (Burke and Orrenius, 1982) and pyrethroid-hydrolysing activity has been demonstrated in rat plasma, and in kidney, stomach, and brain preparations of several mammals (Abernathy *et al.*, 1973; White *et al.*, 1976; Ruzo *et al.*, 1979; Gray *et al.*, 1980; Glickman and Lech, 1981; Ghiasuddin and Soderlund, 1984). In two instances, the esterase activities of these other tissues appear to differ from the liver esterases in their substrate specificity. First, rat plasma esterases do not preferentially hydrolyse *trans*-substituted esters, but instead hydrolyse *trans* and *cis* isomers at equal rates (White *et al.*, 1976; Gray *et al.*, 1980). Second, soluble esterases in mouse brain hydrolyse the non-cyclopropane esters fenvalerate and fluvalinate (**14F**) at unusually rapid rates relative to those for other pyrethroids (Ghiasuddin and Soderlund, 1984). Hydrolysis by non-hepatic esterases may be particularly important in determining the circulating levels of esters that are relatively resistant to detoxication in the liver. For example the rapid hydrolysis of fenvalerate by brain esterases may explain why this compound, a relatively poor substrate for hepatic esterases (Table 4), is synergized more than 30-fold by DEF in mice (Table 3).

Toxicity to fish and birds

Pyrethroids are particularly toxic to fish (Mauck *et al.*, 1976; Miyamoto, 1976; Zitko *et al.*, 1977). Unlike mammals, fish are almost equally susceptible to both the $1R,cis$ and $1R,trans$ isomers, but as in insects, the $1S$ isomers are essentially non-toxic (Miyamoto, 1976). The toxicity of the non-cyano pyrethroids (pyrethrum, allethrin, resmethrin and permethrin) increases with octanol/water partition coefficient (Zitko *et al.*, 1977). Permethrin, which had a lethal threshold of approximately 0.023 μM in Atlantic salmon, was the most persistent and accumulated in the fish with a coefficient of 55 with respect to water concentration (Zitko *et al.*, 1977). Some tissues bioaccumulate pyrethroids more effectively than others. Fat of rainbow trout exposed to *cis*- or *trans*-permethrin contained radiolabelled pyrethroid concentrations of up to 400 times that of the water (Glickman *et al.*, 1981). Studies with pyrethroids containing an α-cyano group (deltamethrin, cypermethrin and fenvalerate) indicated that they were more toxic to salmon, lobsters and rainbow trout than would be predicted from their octanol/water partition coefficients when compared to compounds without the α-cyano group (Coats and O'Donnell-Jeffery, 1979; Zitko et al., 1979). This suggests that the intrinsic toxicity of the α-cyano compounds to fish may be greater than that of the non-cyano pyrethroids.

When compared on a whole-body basis, mice hydrolysed trans-permethrin 184 times faster than did rainbow trout (Glickman and Lech, 1981). However, trans-permethrin was still 35 times more toxic to trout than to mice even when both oxidative and hydrolytic detoxication routes were inhibited (Glickman et al., 1982). Brain concentrations of trans-permethrin that were associated with toxicity in trout were 4 nmol/g whether administered i.v. or i.p. and lethality occurred when concentrations of 5–15 nmol/g were achieved (Glickman and Lech, 1982). In the same study, signs of toxicity were not observed in mice until brain levels of 64–77 nmol/g were achieved. Similar brain concentrations of both cis and trans isomers (5 nmol/g) were associated with toxicity in trout but in mice cis isomer levels of only 15 nmol/g were required for toxicity compared to 77 nmol/g of the trans isomer. The brain concentrations of cis- and trans-cypermethrin associated with toxicity in trout were 0.5 and 0.4 nmol/g respectively (Edwards and Millburn, 1984), but in the same study the mouse brain concentration of cis-cypermethrin associated with signs of toxicity was 4.2 nmol/g. The cis and trans isomers of permethrin and cypermethrin are therefore equipotent in the fish brain, but the mouse nervous system appears to be more isomer-specific, the cis isomers being intrinsically more potent than the trans isomers. Hence a number of factors contribute to the sensitivity of fish to pyrethroids including bioaccumulation, slow detoxication and the sensitivity of the fish nervous system to both 1R,cis and 1R,trans isomers.

Despite the high toxicity of pyrethroids to fish, few environmental problems have arisen. This is presumably a result of the very low application levels required for insect control and the adsorption of residues onto soil and organic matter where rapid decomposition minimizes the danger of aquatic concentrations achieving toxic levels (Ruscoe, 1977; Elliott et al., 1978b; Crossland et al., 1982).

There is very little information available on the toxicity of pyrethroids to birds but all the data that have been published indicate that birds are fairly tolerant to this class of insecticide. Daily oral administration of 52 mg/kg pyrethrum to common house sparrows (Passer domesticus) produced only some weight loss associated with hyperactivity and excitability, but no mortality (Saxena and Saxena, 1973). Similar results were observed after intramuscular administration of up to 50 mg/kg of a 20% pyrethrum extract to blue rock pigeon (Saxena and Bakre, 1976). A more detailed study (Miyamoto, 1976) reported that lovebirds were not killed by oral administration of 2000 mg/kg of racemic allethrin, tetramethrin, mixed $1R,trans$- and $1R,cis$-furamethrin, resmethrin or phenothrin, although $1R,trans$-allethrin had an LD_{50} of 1600 mg/kg (male) and 840 mg/kg (female). Mortalities were observed in mallard ducks fed for 5 days on a diet containing 4640 or 10,000 mg/kg deltamethrin; some ataxia and incoordination were also apparent at the high dose level (FAO, 1981). A dose-related decrease in food intake and weight gain was also observed. At the same doses, bobwhite quail showed similar signs of toxicity although no mortalities occurred (FAO, 1981). The low toxicity of *cis*-cypermethrin to quail was partially due to the tolerance of high brain concentrations (10 nmol/g) before onset of toxicity, coupled with rapid metabolism and excretion (Edwards and Millburn, 1984). The acute LD_{50} of deltamethrin to wild duck exceeds 4000 mg/kg and that of permethrin is greater than 13,500 mg/kg (Elliott *et al.*, 1978b). Since the doses required to cause toxicity in birds far exceed those that are likely to be ingested from foodstuffs contaminated with pyrethroids, no significant environmental hazard is anticipated in this context.

Hazard to man

The extremely low application levels required for insecticidal activity, together with the generally low mammalian toxicity of the pyrethroids when administered by oral or dermal routes, indicate that the prospect of systemic poisoning with pyrethroids during spraying or by consumption of contaminated foodstuffs is unlikely. Systemic poisoning by inhalation is unlikely to occur in practice since the droplets produced by recommended spray applicators are too large to be inhaled under normal spray conditions. However, as many of the pyrethroids are potent neurotoxins, misuse or abuse of some of the concentrated commercial preparations could result in poisoning. Such a case was recently reported in the literature when a man died 3 hours after ingesting beans that had been cooked in a preparation containing 10% cypermethrin instead of oil (Poulos *et al.*, 1982). The only tissue analysed *post mortem* was the stomach which was found to contain 0.7 g of cypermethrin.

The *in vitro* percutaneous penetration of permethrin through human skin was slower than that measured through rabbit, mouse, rat and guinea pig skin, but approximately equal to that of the pig (Wang *et al.*, 1981). Additionally, these workers demonstrated that permethrin radiolabel penetrated skin at a rate slower than the other pesticides examined and only 0.62% of the applied permethrin radioactivity (200 μg/cm²) penetrated the human skin over an 8-hour period. Unfortunately no assessment of the degree of deposition of the pyrethroid into skin was made.

Allergic reactions (dermatitis, asthmatic symptoms) have been observed in humans after prolonged exposure to pyrethrins or pyrethrum extract (Mitchell *et al.,* 1972; Baer *et al.,* 1973; Carlson and Villaveces, 1977). However, experiments in guinea pigs have demonstrated that the contact allergic reactions are due to glycopeptides or glycoproteins in the pyrethrum extract and are probably not associated with the pyrethrins themselves (Rickett and Tyszkiewicz, 1973). These allergens are now removed during the refining processes (Barthel, 1973; Griffin, 1973).

The major hazard to workers handling pyrethroids appears to be local irritative effects particularly of the face and upper respiratory tract. These effects occur predominantly following handling of the α-cyano-substituted pyrethroids cypermethrin, fenpropathrin, fenvalerate and deltamethrin (Le Quesne *et al.,* 1980; Tucker and Flannigan, 1983). One report has also implicated permethrin (Kolmodin-Hedman *et al.,* 1982) but this has not been confirmed (FAO, 1982) and no report of any irritation was made when this pyrethroid was applied at up to 0.25 g per person, diluted in kieselguhr and talc, in an assessment of its usefulness in the control of body lice (Nassif *et al.,* 1980). The irritative signs usually occur following direct contact with the pyrethroid as a solution, a fine powder (Le Quesne *et al.,* 1980) or a dried wettable powder residue on crops (Kolmodin-Hedman *et al.,* 1982). The areas of the face most frequently affected include the lips, around the eyes and mouth, cheeks and occasionally the nose (Tucker and Flannigan, 1983). Symptoms usually commence 30 minutes to 3 hours after exposure and persist for 30 minutes to 8 hours, depending on the extent of the exposure (Le Quesne *et al.,* 1980), although there is one report of symptoms occurring up to 2 days following deltamethrin exposure (FAO, 1981).

The signs of toxicity have been described in a variety of ways, as episodes of a hot and cold irritating sensation, a prickling, tingling or burning sensation and are often referred to as paraeshtesia. These effects may become more painful when the skin is in contact with water or when it is perspiring (FAO, 1981; Kolmodin-Hedman *et al.,* 1982; Tucker and Flannigan, 1983) and may become intense at lower ambient temperatures (Kolmodin-Hedman *et al.,* 1982). These local irritant effects may be alleviated by application of local anaesthetics such as procaine (Gray, unpublished observations; Glomot, 1982). Other effects noted particularly after exposure to deltamethrin or fenvalerate include rhinorrhoea or lacrimation (FAO, 1981; Kolmodin-Hedman *et al.,* 1982). It has been concluded that these effects are due to a direct action of the pyrethroids on sensory nerves in the face, since no abnormal neurological signs were observed and the electrophysiological responses of the arms and legs were normal (Le Quesne *et al.,* 1980). These irritant effects are probably the result of the repetitive firing of sensory nerve terminals in the skin in a similar manner to that reported in frogs (Vijverberg *et al.,* 1982). Although paraesthesia will cause considerable discomfort, this is a transient effect and is unlikely to result in any permanent damage. It is probably more significant as an indicator of over-exposure to the pyrethroids concerned.

OTHER CONSEQUENCES OF ACUTE AND CHRONIC PYRETHROID POISONING

Neuropathology and neurochemistry

Since many pyrethroids are potent neurotoxicants the possibility that they may cause permanent nerve damage has been considered. There is some histological evidence that axonal swelling and degeneration and myelin fragmentation of sciatic nerve occur following administration of acutely toxic doses of fenvalerate or permethrin (presumably to rats) although the clinical significance of these findings to man was not clear (FAO, 1980). The usual test species for the delayed neuropathy produced by organophosphate compounds is the hen, but high doses of fenvalerate (FAO, 1980) or deltamethrin (FAO, 1981; Glomot, 1982) failed to produce any signs of peripheral neuropathy.

A transient functional impairment was determined in rats following oral administration of toxic, and in some cases lethal, doses of permethrin (400 mg/kg per day), deltamethrin (5 mg/kg per day) or cypermethrin (50–100 mg/kg per day) for 7 days or resmethrin for 7 days at 500 mg/kg per day followed by three daily doses of 1000 mg/kg (Rose and Dewar, 1983). The decrease in mean slip angle (i.e. the angle of elevation of an inclined plane at which the animals slide down the slope) was usually greatest toward the end of the treatment regimen and persisted for up to 10 days after the last dose. In contrast resmethrin administration produced a significant decrease in mean slip angle only on the ninth day after the last dose. Moreover, unlike the other three pyrethroids, resmethrin administration did not cause signs of toxicity during the treatment period, but produced signs between 6 and 10 days after the last dose.

Two lysosomal enzymes, β-glucuronidase and β-galactosidase, are sensitive indicators of peripheral nerve damage (Dewar, 1981). Administration of permethrin (400–1200 mg/kg per day, males; 800–1200 mg/kg per day, females), cypermethrin (1000–1400 mg/kg per day, females; 150–1400 mg/kg per day, males) or deltamethrin (20 mg/kg per day) for 7 days, produced a significant dose-related increase in β-galactosidase and β-glucuronidase activity of sciatic/ posterior tibal nerve preparations 3–4 weeks later, particularly in the distal section of the nerve (Rose and Dewar, 1983). No such changes were found after oral administration of resmethrin for 7 days at 1000 mg/kg per day followed by three daily doses of 2000 mg/kg. These biochemical changes occurred after all clinical signs of toxicity had ceased and never exceeded 150% of control values. These changes are minor compared to the increases in β-glucuronidase activity produced by compounds known to cause peripheral neuropathy: acrylamide (200%); methylmercuric chloride (580%); and a substituted phosphine (1000%) (Rose and Dewar, 1983). The lesion also appears to be reversible since, after 7 daily oral treatments with 150 mg/kg cypermethrin, a maximal increase in β-glucuronidase activity was found at 22 days but was not significantly different from that of control rats 38 days after the initial dose (Rose and Dewar, 1983). There-fore, since these effects on peripheral nerves are slight when compared to known neuropathic agents, are reversible and do not correlate with any clinical signs, they do not appear to be cause for undue concern. This conclusion is sup-ported by the lack of evidence of functional impairment of peripheral nerve con-duction velocity in workers exposed to various pyrethroids (Le Quesne *et al.*, 1980).

Mutagenicity, carcinogenicity and teratogenicity

The mutagenic, carcinogenic and teratogenic activity of the pyrethrins has been reviewed by Williams (1973) who concluded that there was no evidence to sug-gest that these toxic effects could be produced by pyrethrins. There is also very little evidence to implicate the pyrethroids as mutagens, carcinogens or terato-gens. Since the data obtained with pyrethroids in such investigations is generally negative, it is usually only published in summary form in evaluation reports of the EPA and WHO.

Fenvalerate has been found to cause no oncogenic, reproductive, mutagenic or teratogenic effects in a number of test species, including 2-year feeding studies with rats and mice (250–300 ppm in diet), 3-generation reproduction studies (250 ppm) and a number of short-term mutagenicity assays, bone mar-row cytogenicity assay (up to 150 mg/kg), the dominant lethal bioassay (100 mg/kg) and a host-mediated bioassay in mice (50 mg/kg) (Anon, 1980; FAO, 1980; Campt, 1983). Some chromosomal aberrations and alterations in the mitotic index were reported in bone marrow and testis cells taken from rats treated orally with fenvalerate at 100 mg/kg, a dose that resulted in the death of 15 out of 21 animals (Chatterjee *et al.*, 1982).

Equally thorough tests with deltamethrin have also proved negative despite the oral administration of toxic or sometimes lethal doses of 50–100 ppm in the diet (Kavlock *et al.*, 1979; FAO, 1981; Glomot, 1982; Poláková and Vargová, 1983). Teratogenicity studies with deltamethrin revealed possible embryo-toxicity at the highest doses (decreased foetal weight and, in one study, delayed ossification in mice and rats dosed 10 mg/kg per day foetal losses in rabbits dosed 16 mg/kg per day) but no teratogenic effects were attributed to the com-pound (Kavlock *et al.*, 1979; FAO, 1981; Glomot, 1982).

Cypermethrin administered in a single oral dose of 259 mg/kg, or five daily doses of 77.9 mg/kg, to mice caused a 3–4% incidence of chromosomal aberra-tions in bone marrow cells in contrast to the 2% incidence in controls (Páldy, 1981). The incidence was the same for deltamethrin (11 mg/kg or 3.3 mg/kg per day for 5 days) and slightly higher (5% incidence) for permethrin (150 mg/kg or 45.2 mg/kg per day for 5 days) although none of these was significantly different from control values.

Administration of permethrin to rats, mice and rabbits at 5000 ppm caused no reproductive effects (Anon, 1979; FAO, 1982). One report has suggested an effect of permethrin on placental biochemistry in rats fed 2000 or 4000 ppm in the diet, resulting in a slight, but significant, increase in the resorption of foetuses (Spencer and Berhane, 1982). Permethrin is negative in the Ames, mouse lymphoma cell transformation and mouse dominant lethal mutagenicity assays (Anon, 1979) and there was no definite evidence of oncogenicity in three 2-year rat feeding studies (Johnson, 1982). However there has been some controversy regarding the findings of similar studies in mice. In some of these studies a statistically significant increase in the incidence of lung and liver tumours in female mice treated with 750 or 375 mg/kg per day for 2 years, and a suggestion of permethrin-induced lung tumours at 250 mg/kg per day (Johnson, 1982). Since the relevance of such tumours in mice to the prediction of oncogenic potential in humans is uncertain, and the dose levels at which it occurs far exceed those likely to be encountered during use of permethrin, the EPA concluded that 'even if permethrin is a human oncogen (which is unlikely), it is highly unlikely that it would present a significant risk to humans at the levels to which they will be exposed' (Johnson, 1982).

Miyamoto (1976) concluded that the racemic mixtures and pure isomeric forms of allethrin, phenothrin, furamethrin, permethrin, tetramethrin and resmethrin were not mutagenic or radiomimetic in Ames assays and mouse host-mediated bioassays and also reported that no teratogenic effects had been observed in rats, mice, and, in some cases, rabbits dosed with many of these pyrethroids. Ames assays with fenpyrithrin, a pyrethroid containing the dichlorovinyl-substituted cyclopropanecarboxylic acid moiety (3) esterified with the novel α-cyano-(6-phenoxy-2-pyridinyl)methyl alcohol (H), indicated that this pyrethroid and its metabolites also lacked mutagenic activity (Malhotra et al., 1981).

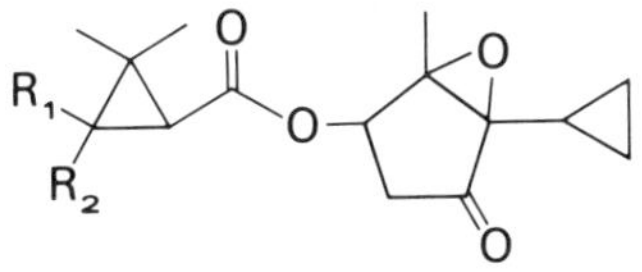

Figure 3 Mutagenic cyclopropyl-epoxy derivatives of allethrin and related esters

The only commercial pyrethroid for which there is any evidence of mutagenic activity is allethrin which, in the presence of metabolic enzymes, caused a 77% incidence of chromosome aberrations in Chinese hamster cells incubated with pyrethroid at 1.9 μg/ml (Matsuoka et al., 1979). This was confirmed in an Ames assay (Kimmel et al., 1982) which demonstrated that photo-oxidation products of allethrin, or its analogues terallethrin (7B) and dichloroallethrin (3B), may be converted to a cyclopropylepoxy product (Figure 3) which is mutagenic. S-Bioallethrin and terallethrin were not mutagenic even when a microsomal metabolism system was included although their cyclopropyl derivatives, which may occur as impurities, did increase the number of revertants (Kimmel *et al.*, 1982). Therefore it appears that the mutagenic activity of allethrin is due to photodegradation products and their metabolites.

Pharmacological and physiological consequences

Many pyrethroids and pyrethrins cause liver enlargement when large oral or i.p. doses are administered over a period of a few weeks or more (Springfield *et al.*, 1973; Miyamoto, 1976). The major cause for the increased liver weight following pyrethrum administration was shown to be hypertrophy with some increase in liver total lipid (Springfield *et al.*, 1973). The animals had a decreased hexabarbitone (hexabarbital) sleeping time without change in barbitone (barbital) sleeping time and had an increased rate of metabolism of *p*-nitroanisole and EPN to *p*-nitrophenol, although *N*-demethylation of aminopyrine was not affected (Springfield *et al.*, 1973). It was concluded that the increased metabolic activity was a result of pyrethrin-induced increases in NADPH cytochrome *c* reductase activity and cytochrome P450 content of the liver. Oral administration of permethrin to rats for 4 days (80% *cis* isomer) or 8 days (40% *cis* isomer) increased liver NADPH cytochrome *c* reductase and cytochrome P450 activities although a similar dose of cypermethrin had no effect (Carlson and Schoenig, 1980). Technical permethrin, [1*R,cis*]-permethrin and [1*R,trans*]-permethrin were all approximately equipotent in stimulating ethoxycoumarin-*O*-de-ethylase (1.5–2.5-fold), ethoxyresorufin-*O*-de-ethylase (1.2–2.5-fold) and benzphetamine-*N*-demethylase activity and caused significant increases in liver weight and cytochrome P450 content when orally administered to rats for 4 days at 100 mg/kg (Glickman *et al.*, 1985). The induction pattern was similar to that of phenobarbitone, but dissimilar to that of 3-methylcholanthrene, and the soret maxima of the hepatic cytochrome P450 were not different from control. Liver hypertrophy and induction of enzyme activity has also been demonstrated following feeding permethrin to rats at 2500 ppm for 4 weeks, but was no longer apparent 4 weeks after treatment had been discontinued (Litchfield, 1983). Treatment with cypermethrin for 13 weeks with a diet containing 150 ppm did not result in an increased liver weight in rats although a proliferation of the smooth endoplasmic reticulum and increased aminopyrine-*N*-demethylase activity were found. These effects were no longer apparent after a 4-week recovery period (Litchfield, 1983). Since these hepatic changes are apparently reversible, usually require the administration of a high dose of pyrethroid and produce an enzyme induction pattern similar to that of phenobarbitone, they are probably of minimal toxicological significance.

Single oral administration of cypermethrin to mice (48 mg/kg) caused a

15.7% and 42.8% increase in the activity of B6-dependent kynurenine amino-transferase (KAT) and kynurenine hydrolase (KH) respectively, although no significant changes were determined when the pyrethroid was given at 12 mg/kg per day for 6 consecutive days (El-Sewedy *et al.*, 1982b). In contrast, fenvalerate (1 × 60 mg/kg or 6 × 20 mg/kg per day) decreased KH activity by 35–45%. The multiple dose increased KAT activity by 26% and acid ribonuclease activity by 14%. The activity of hepatic β-glucuronidase was increased by 31% and 47%, respectively, following the single or multiple treatments with fenvalerate (El-Sewedy *et al.*, 1982a). These authors suggest that such changes in kynurenine metabolism may eventually result in the accumulation of endogenous carcinogens, although there is no evidence to suggest that fenvalerate or cypermethrin increase the incidence of tumours (see previous section). Allethrin (82 μM) increased mouse liver ATPase activity *in vitro* (Payne *et al.*, 1973), altered the light scattering properties of mouse liver mitochondria at 0.22 μM, and caused an increase in mitochondrial respiration (Settlemire *et al.*, 1974). The latter authors proposed that these effects reflected an increase in membrane permeability or in the activity of the respiratory chain.

Protein binding of cismethrin and bioresmethrin radiolabel occurs after incubation with liver homogenates *in vitro* (Ueda *et al.*, 1975a) and in subcellular fractions of liver taken from animals administered these compounds (Graillot and Hoellinger, 1982; Hoellinger *et al.*, 1983). Although the bound species is probably an oxidation product of the alcohol moiety it has not been identified and its toxicological significance remains obscure.

The cardiovascular effects of pyrethroids investigated in rats following the administration of toxic doses of cismethrin or deltamethrin, which produce signs that are representative of the T and CS poisoning syndromes respectively (Table 1, Verschoyle and Aldridge, 1980), have been summarized elsewhere (Ray, 1982a). Administration of a lethal dose of deltamethrin caused a transient decrease in arterial blood pressure (BP) which then became elevated once writhing became established (Ray and Cremer, 1979; Ray, 1982b; Bradbury *et al.*, 1983). In contrast, in the pithed animal, in which the CNS structures responsible for the homoeostasis of the cardiovascular system are no longer functional, i.v. injection of deltamethrin resulted in a pronounced rise in mean and differential arterial BP and slight increase in heart rate (HR) which were not abolished by predosing with *l*-propranolol and phentolamine (Forshaw and Bradbury, 1983). Pulse pressure was also increased but prior to the onset of any motor signs. These cardiovascular changes persisted even when the spinal cord was pithed (Bradbury *et al.*, 1983). It has been demonstrated that, in the intact animal, a toxic dose of deltamethrin causes a marked (> 100 fold) increase in plasma noradrenaline and adrenaline, particularly during choreoathetosis (Cremer and Seville, 1982) and this massive increase in circulatory catecholamine may be partially responsible for some of the cardiovascular changes observed. However, in an isolated working heart preparation, deltamethrin (0.25 μmol/heart) increased aortic output and mean systolic pressure; these effects were only partially abolished by pretreatment of the animals with reserpine (Forshaw and Bradbury, 1983). This finding indicates that deltamethrin also has a positive inotropic effect directly on the myocardium, similar to that of veratridine (Honerjäger, 1980; Forshaw and Bradbury, 1983). These cardiovascular changes observed following administration of deltamethrin are not restricted to the rat since an increase in pulse pressure and cardiac output has also been observed in anaesthetized dogs (Navarro-Delmasure, 1981). Other CS syndrome pyrethroids may also cause cardiovascular changes; cypermethrin was found to increase the pressor response to isoprenaline, but only at lethal or near lethal doses (Holland and Smith, 1980). However, direct effects of pyrethroids on the cardiovascular system of intact animals may be masked by the compensatory action of the homoeostatic centres of the CNS which are maintained until the onset of severe neurotoxic signs. Unlike deltamethrin, cismethrin caused no cardiovascular changes in the intact or pithed animal nor did this pyrethroid have any effect on the isolated working heart preparation (Ray 1982b; Cremer and Ray, 1983; Forshaw and Bradbury, 1983) despite causing a 10-fold increase in plasma noradrenaline and a 7-fold increase in plasma adrenaline during tremors (Cremer and Seville, 1982). However, these catecholamine increases corresponded to a slight rise in blood glucose and lactate (6-fold), increased oxygen consumption and rectal temperature (up to 40 °C) and a fall in P_{CO_2} due to hyperventilation during tremors (Cremer and Seville, 1982; Ray, 1982b). These changes are consistent with physiological responses following a period of sustained exercise (Cremer and Seville, 1982) and are therefore probably the result of the increased muscle activity associated with the tremor.

Following administration of a lethal dose of deltamethrin, blood glucose and lactate concentrations increased. These increases were most pronounced once writhing had begun, although normal blood gases were maintained until a few minutes prior to death (Ray and Cremer, 1979; Cremer and Seville, 1982; Ray, 1982b). Cypermethrin has also been shown to increase blood glucose and lactate prior to the onset of choreoathetosis, although a significant increase in ammonia was not observed until after the first seizure had occurred (Lock and Berry, 1980, 1981). Blood glucose concentration was also increased in pithed rats administered deltamethrin, but did not return to normal upon the cessation of writhing (Bradbury et al., 1983). This increase was not observed in spinal rats pretreated with mephenesin, which protected against deltamethrin-induced choreoathetosis, but was only slightly reduced by similar treatment in the intact animals. This indicates that the rise in blood glucose was partially a result of the signs of toxicity but also involved an action of the pyrethroid on supraspinal centres of the CNS (Bradbury et al., 1983). Brain glucose concentration was also significantly increased during choreoathetosis, except in the cerebellum, following i.p. administration of a lethal dose of deltamethrin. However, glucose utilization was only significantly increased in the hypothalamus, colliculi, and, most markedly, in the cerebellum (Cremer et al., 1980, 1983). Deltamethrin also increased brain blood flow at this time although the regional distribution of these two parameters was different, indicating dissociation of these two normally highly integrated processes (Cremer et al., 1980; Ray, 1982b). Cypermethrin caused a slight increase in cerebellar glucose concentration during seizures although it was not elevated once writhing had begun (Lock and Berry, 1980, 1981). In contrast, these animals had elevated cerebellar lactate levels at all stages of toxicity, including salivation.

Cismethrin administration resulted in an up to 8-fold increase in blood flow to several brain regions. These increases were abolished in the cortical regions by pretreatment with atropine sulphate, thereby indicating that a cholinergic mechanism was involved in the cortical vasodilation (Cremer and Ray, 1983). However, atropine pretreatment only produced a 50% reduction in brain blood flow in the cerebellum. Since cismethrin administration has been also shown to cause a 2-fold increase in cerebellar glucose utilization (Cremer et al., 1983), the increase in blood flow to this region is probably due to increased cerebellar neuronal activity.

Oral administration of large doses of deltamethrin (50 mg/kg) or cismethrin (100 mg/kg) to rats produced decreased cerebral and cerebellar acetylcholine concentrations with little or no change in their cyclic adenosine monophosphate (cAMP) levels during the peak signs of toxicity (Aldridge et al., 1978). These two pyrethroids also produced slight increases in cyclic guanosine monophosphate (cGMP) in the brain (minus cerebellum), but up to 10-fold increases in cerebellar cGMP levels. DDT, which was also included in this study, gave less marked changes in the cGMP with little or no change in acetylcholine and cAMP. Cypermethrin has also been shown to cause a progressive increase in cerebellar cGMP concentration of up to 10-fold control values during the writhing without producing a significant change in cAMP (Lock and Berry, 1980, 1981). These workers could not reproduce the increase in cGMP by incubation of cypermethrin with cerebellar slices in vitro. The increase in brain cGMP was investigated in detail following deltamethrin administration at several dose levels (Brodie and Aldridge, 1982) and was found to be time-dependent, but not dose-dependent, and tended to reflect the increase in motor activity as the signs of toxicity became more severe. In these same experiments, no change in cortical cGMP was determined, suggesting that deltamethrin acted more as a locomotor stimulant than as a convulsant.

Deltamethrin administration caused a slight increase in brain gamma-aminobutyric acid (GABA) concentration and a complementary, but generally non-significant, decrease in glutamate levels during choreoathetosis (Cremer et al., 1980). There was no significant change in cerebellar concentration of glutamate, glutamine, GABA, ATP, creatine phosphate or ammonia following administration of a severely toxic dose of cypermethrin, although a significant decrease in aspartate concentration was observed during writhing (Lock and Berry, 1980, 1981).

The biochemical changes observed in the brain indicate that there is a considerable increase in cerebellar neural activity following pyrethroid administration. However, the decrease in acetylcholine and increase in cGMP, blood flow and glucose utilization are probably not due to direct action of pyrethroids in this region but are a result of the increased motor activity associated with the signs of toxicity. It is generally acknowledged that administration of convulsants, electroshock or an increase in motor activity, all result in an increase in cerebellar cGMP (see Brodie and Aldridge, 1982, for references) and that such increases tend to correlate with the nature and quantity of movement. The neurochemical changes therefore tend to be a response to the toxic effects caused by pyrethroids and are not indicative of a primary event leading to the manifestation of motor signs. The contribution of the biochemical and physiological changes observed to the toxicity and lethality of pyrethroids, particularly deltamethrin and cypermethrin, therefore remains obscure.

TOXICOKINETICS

Effects of route of administration and vehicle on toxicity

In order to assess the true toxic potential of a compound it must be administered as a preparation from which it is readily available for absorption and subsequent distribution to, and interaction with, the site of action. The vehicle and route of administration are both important determinants of such parameters. For example, if a pyrethroid is administered orally in a vehicle which reduces the extent or the rate of absorption, an underestimate of the potential toxicity of the compound will result. Such tests are only valid when an assessment of the toxicity of the formulated commercial product is required.

A comparison of the toxicity of pyrethroids is often complicated by the different species, sexes, routes of administration and vehicles used by various workers. Mice are usually more susceptible than rats and the females of these species are more sensitive than males to pyrethroid toxicity (Miyamoto, 1976; Chanh et al., 1981). The limited information available indicates that dogs, sheep and cattle are no more, and often less, susceptible to pyrethroids than rats although cats may be more sensitive (James, 1980). It is difficult to produce toxicity by the dermal route, even when pyrethroids are applied at the g/kg level (James, 1980). Deltamethrin, possibly the most toxic commercial pyrethroid, failed to produce signs of neurotoxicity when dermally applied in xylene to female rats for 5 days at 800 mg/kg (Kavlock et al., 1979). Inhalation toxicities are difficult to compare with other routes since they are usually quoted as the toxic concentration in the exposure chamber and no assessment of absorption is made. However, the

Table 5 Effects of vehicle on the oral toxicity of pyrethroids to female rats

Pyrethroid	Vehicle	LD_{50} (95% confidence limits) (mg/kg)	Reference
Resmethrin	Glycerol formal	> 3000	Verschoyle and Barnes, 1972
	Polyethylene glycol 400	1995 (1441–2765)	
	Dimethylsulphoxide	1347 (1129–1605)	
Cismethrin	Glycerol formal	168 (115–245)	
	Arachis oil	119 (78–181)	
	Polyethylene glycol 400	86 (53–139)	
	Dimethylsulphoxide	63 (41–95)	
Deltamethrin	Glycerol formal	40% mortality at 60 mg/kg[a]	Chanh et al., 1981
	Arachis oil	31 (29–34)	Kavlock et al., 1979
	Arachis oil	20 (14–28)	Gray, 1980
	Corn oil	52	Crofton and Reiter, 1982
	Sesame oil	139 (105–157)	
	Polyethylene glycol 200	86 (71–106)	Glomot, 1982

[a] Deltamethrin in male rats, all other values for females.

Table 6 Comparison of oral and i.v. toxicities of cismethrin and deltamethrin to female rats determined by various research groups

| Pyrethroids | Route of administration | | | | | Oral/i.v. ratio | Source |
	Vehicles[a]	Oral LD$_{50}$[b]	Vehicles[a]	i.v. LD$_{50}$[b]			
Cismethrin	AO	119(78–181)	GF	6–7		20	Verschoyle and Barnes, 1972
Cismethrin	AO	32(23–44)	GF	5.0(4.1–6.3)		6.4	Gray, 1980
Deltamethrin	AO	25–63	GF	2.25		20	Barnes and Verschoyle, 1974
Deltamethrin	AO	20(14–28)	GF	3.0(2.2–4.1)		6.7	Gray, 1980
Deltamethrin	AO	31(29–34)	Acetone	4.0(2.9–5.3)		7.8	Kavlock et al. 1979
Deltamethrin	PEG 200	86(71–106)	PEG 200	3.3(2.9–3.8)		26	Glomot, 1972

[a] AO, arachis oil; GF, glycerol formal; PEG 200, polyethylene glycol 200.
[b] LD$_{50}$ values (and 95% confidence limits) given in most cases, other values without limits are estimates or the range in which 50% mortality would be expected.

inhalation toxicities of deltamethrin and resmethrin have been reported in mg/kg using calculations based on assumptions of minute volume, absorption from the lung and pyrethroid concentration in the inspired material. The reported inhalation LD_{50} values were 99–243 mg/kg for resmethrin in mice (Berteau and Dean, 1978) and 36 (male) and 28 (female) mg/kg for deltamethrin in rats (Kavlock et al., 1979), compared to oral LD_{50} values of 1390 mg/kg and 52 (male) and 31 (female), respectively, determined in these same studies. The particle size of inhaled aerosols is an important factor in determining whether a compound will reach lung alveoli for absorption into the blood stream. Since this will largely depend on the method of spraying and the formulation, inhalation exposure to the commercial formulation under similar conditions to those used in the field is probably the most valid assessment of potential hazard.

The oral toxicity of pyrethroids is highly dependent on the solvent vehicle used to administer the compound (Table 5). The oral toxicity may be slightly increased if the animals had been fasted overnight prior to treatment (most of the values in Tables 1 and 5 were obtained with fasted animals) but since no published report comparing oral toxicity in fasted and fed animals is available, the importance of this factor cannot be determined. The toxicity of pyrethroids by the i.p. route would be expected to be greater than by the oral route, although again the vehicle is an important determinant. When cismethrin was administered i.p. to female rats as a solution in glycerol formal, no toxicity was observed at doses of 40 mg/kg (Gray, 1980) indicating that this pyrethroid was more toxic by the oral route (Table 1). In contrast, the i.p. LD_{50} of deltamethrin to rats was half of its oral LD_{50} when administered as a solution in glycerol formal (9.5 mg/kg, 95% confidence limits 7.5–12.1) (Gray, 1980).

Some pyrethroids and pyrethrins are extremely toxic when administered by i.v. injection (Table 1). In a study of the oral and i.v. toxicities of pyrethrins and pyrethroids to rats (Verschoyle and Barnes, 1972), the pyrethrins, bioallethrin, NRDC 108 and cismethrin were 10–200 times more toxic by the i.v. route than when administered orally (Tables 1 and 6). A number of authors have shown an oral/i.v. LD_{50} ratio of 8 : 20 for deltamethrin (Barnes and Verschoyle, 1974; Kavlock et al., 1979; Gray, 1980; Glomot, 1982). An oral/i.v. LD_{50} ratio of 5 : 20 also exists for a number of other pyrethroids (Tables 1 and 6). Pyrethroids that have a low inherent toxicity (e.g. bioresmethrin) pose a particular problem as it is necessary to inject concentrated solutions i.v. in order to maintain an acceptable volume of vehicle. Since pyrethroids have a very low aqueous solubility, a precipitate forms when these concentrated dosing solutions are mixed with an aqueous medium such as blood. It has been shown that a large proportion of an i.v. administered dose of a concentrated bioresmethrin solution in glycerol formal (500 mg/ml) accumulated in the lungs of rats (Gray and Connors, 1980). In addition, the toxicity of cismethrin appeared to be decreased, and the onset of signs delayed, when the compound was administered in a concentrated form (300–500 mg/ml in glycerol formal). This

was due to the initial accumulation of the precipitated material in the fine capill-
ary network of the lung and subsequent slow release into the blood stream for
redistribution to other tissues, including the nervous system (Gray and Connors, 1980). However, despite the possibility of dosing artefacts when concentrated solutions are injected, the i.v. toxicity data determined by various
research groups is more consistent than that determined via other routes of
administrtation (Table 6).

The time to onset of signs of toxicity and their duration are also dependent on
the route and vehicle as well as the pyrethroid administered. However, in
general, signs of toxicity occur more rapidly when the pyrethroid is
administered by routes or in vehicles which result in lower LD_{50} values, e.g. by
i.v. injection. The duration of these signs tends to be longer when an equitoxic
dose of the pyrethroid is given orally or i.p. compared to i.v. particularly in
vehicles that may limit the rate of absorption of the compound.

Correlation between the tissue concentration of pyrethroids and their metabolites with signs of toxicity

There are few reports of the kinetics of absorption and distribution following
administration of toxic doses of pyrethroids and those that are available are
mainly restricted to deltamethrin and the isomers of resmethrin. Since most
metabolism studies use non-toxic doses of the pyrethroid and usually do not
involve measurement of pyrethroid concentration in tissue at times when toxicity would have occurred, they are of limited value in the assessment of the relationship between pyrethroid tissue concentrations and toxicity. However, in a
study of the metabolism of ^{14}C-alcohol [1RS,$trans$]-resmethrin following oral
(500 mg/kg) or i.v. (50 mg/kg) administration to rats, some assessment of the
concentrations of parent compound and radioactive metabolites in tissue was
made soon after dosing (Miyamoto et al., 1972). Three hours after oral
administration, approximately 40% of the dose remained in the gastrointestinal
tract although only 5% was intact pyrethroid. High concentrations of radiolabel were found in lung, heart, kidney, liver and blood 3 hours after the oral
dose, but less than 0.3 nmol/g was present as parent compound. After i.v.
administration, the concentration of pyrethroid in blood and liver decreased
rapidly, and was barely detectable after 60 minutes. Within 2.5 minutes of dosing, the alcohol metabolites, 5-benzyl-3-furylcarboxylic acid (BFCA) and
5-benzyl-3-furyl alcohol (BFA) were demonstrated in blood and liver. Although
the levels of BFA decreased to undetectable levels after 15–60 minutes, the
BFCA was maintained at high concentrations, particularly in blood
(240 nmol/g), for the duration of the study (120 minutes). In contrast, the
majority of the radioactivity in the brain was identified as parent pyrethroid,

with considerably lower concentrations of the alcohol metabolites. High concentrations (~ 75 nmol/g) of resmethrin isomers were achieved in the brain within 2.5 minutes, and peak concentrations of 90 nmol/g were found at 15 minutes. Unlike all of the other tissues examined, the levels of parent pyrethroid in the brain decreased slowly, and 120 minutes after dosing approximately 30 nmol/g remained. No mention of any signs of toxicity was made in this report and it is unlikely that any occurred, since these authors reported the oral and i.v. LD$_{50}$ for [1$RS,trans$]-resmethrin as > 5g/kg and 90 mg/kg, respectively (Miyamoto *et al.*, 1972).

In another study (White *et al.*, 1976) the tissue distribution of parent cismethrin and bioresmethrin were compared at equitoxic doses after i.v. administration to female rats. In addition, the blood and brain concentrations of cismethrin and bioresmethrin were compared after oral administration and cismethrin tissue concentrations after oral administration at environmental temperatures of 4, 20 and 30 °C. It was discovered that, irrespective of the route of administration or the environmental temperature, brain concentrations of cismethrin always correlated with the toxicity observed in rats, concentrations of ~ 3 nmol/g being associated with an increase in core temperature and onset of tremor, ~ 6–9 nmol/g with a marked increase in core temperature (39–40° C), and severe tremors and death occurred when brain levels exceeded 12 nmol/g. Much higher brain levels of bioresmethrin were tolerated before signs of toxicity (~ 10 nmol/g) or death (> 90 nmol/g) occurred. The delay of 30 minutes before the onset of tremors after i.v. administration of bioresmethrin (200 mg/kg) corresponded to a similar delay in the accumulation of this pyrethroid in the brain. This delay in the onset of toxicity after i.v. injection of the concentrated solution required to produce toxicity with bioresmethrin has been shown to be due to a dosing artefact (Gray and Connors, 1980). Due to the low aqueous solubility of pyrethroids, when injected by the i.v. route at high concentrations, the compound precipitates and is trapped in the lung from where it is slowly released back into the blood stream. This same study demonstrated that the bioresmethrin samples analysed contained 1.3–5.5% cismethrin and it was probably this that was responsible for the toxicity of the bioresmethrin reported in previous studies (Verschoyle and Barnes, 1972; White *et al.*, 1976).

Cismethrin was considerably less toxic when orally administered than when injected i.v. and this was suggested to be due to the slow absorption of the pyrethroid from the intestine allowing detoxication by liver and plasma enzymes (White *et al.*, 1976). However, after signs of toxicity appeared, they were more prolonged after oral than after i.v. administration due to the continuous infusion of the toxic compound into the circulation and, presumably, subsequent transport to the site of action. The lower oral toxicity of cismethrin at high environmental temperatures was not due to slower absorption from the gastrointestinal tract although entry into the brain was delayed. However, as the i.v. toxicity of cismethrin did not vary with changes in environmental temperature, the negative temperature–toxicity correlation observed following oral administration is unlikely to reflect a change in the sensitivity of the nervous system (White et al., 1976).

In a third study (Gray et al., 1980), cismethrin and bioresmethrin, radiolabelled in the alcohol moiety, were administered to rats at 2.5 mg/kg by i.v. injection. The levels of both parent compounds in blood decreased rapidly after i.v. injection, with less than 1% of the initial dose remaining in the blood after 2–3 minutes. Within 2.5 minutes of dosing, metabolites of both compounds could be detected in blood, particularly BFCA after bioresmethrin administration. Retention of this metabolite in blood probably contributed to higher levels of bioresmethrin radiolabel, than of cismethrin, in many tissues. The concentrations of both cismethrin and bioresmethrin in liver decreased rapidly and were barely detectable after 60 minutes. As found in the earlier studies, peak brain levels of both parent compounds were achieved rapidly (1–2.5 minutes) and were maintained at substantially higher concentrations than those found in liver and blood. In agreement with White et al., (1976), a cismethrin threshold of approximately 3.5 nmol/g in the CNS was associated with tremors, although a similar concentration of bioresmethrin failed to produce toxic effects. However, the clearance of bioresmethrin and its associated radioactivity from the brain was more rapid than that of cismethrin, indicating that minor, but selective, metabolism may occur in nervous tissue.

Radioactivity from ^{14}C-acid, ^{14}C-alcohol and ^{14}C-cyano labelled deltamethrin was distributed to all tissues examined within 1 minute of i.v. injection of a toxic, but non-lethal, dose of 1.75 mg/kg to rats, and initial tissue concentrations for all radiolabelled preparations were similar (Gray and Rickard, 1981). This dose of deltamethrin (520 nmol/150 g rat) was approximately half that of an equitoxic dose of cismethrin (1109 nmol/150 g rat) used in a previous study (Gray et al., 1980) and resulted in tissue concentrations of radiolabel that were consistently 50% those of animals administered cismethrin. The major exception was the CNS, in which a peak deltamethrin radiolabel concentration of only 0.5–1.0 nmol/g was achieved within 1 minute of injection (Gray and Rickard, 1981). In a continuation of this study (Gray and Rickard, 1982a), peak concentrations of extractable deltamethrin in spinal cord and brain were determined within 1 minute of i.v. injection. However, these peak concentrations did not correspond with the onset of motor signs of toxicity since there was a delay of 5–10 minutes before onset of coarse whole-body tremor and choreoathetosis did not develop until 20 minutes (Gray and Rickard, 1981, 1982a). This delay was not due to redistribution within the CNS, as at all times examined there was a uniform distribution of compound throughout the brain and spinal cord (Gray and Rickard, 1981; Rickard and Brodie, 1985). These initially high CNS levels of deltamethrin would have been partially due to the contribution of deltamethrin in blood (20–26 nmol/g at 1 minute) which may account for the poor correlation between CNS deltamethrin concentrations and motor signs of toxicity. However, the concentration of pentane-unextractable radiolabel appeared to be the best correlate with the severity of toxicity, reaching peak levels of 0.2 nmol/g at 15 minutes, which persisted until 60 minutes when animals showed signs of recovery (Gray and Rickard, 1982a). The majority of this radiolabel was identified as deltamethrin metabolites, particularly 3-phenoxybenzoic acid (PBacid) (Gray and Rickard 1982a; Rickard and Brodie, 1985), and could be removed from the brain by perfusion with saline (Rickard and Brodie, 1985). The PBacid in the brain was in equilibrium with that in blood and may be partially due to deltamethrin metabolism in the CNS, although the majority was probably the result of PBacid entering the CNS from blood (Rickard and Brodie, 1985). Since these metabolites are considerably less toxic than deltamethrin when given i.p. to mice (Casida *et al.*, 1979) and PBacid did not produce toxicity when injected into the brain (Rickard and Brodie, 1985) they are unlikely to contribute to the observed toxicity of deltamethrin. However, the toxicological significance of the 0.066 nmol/g of alcohol-labelled material which remained after ethyl acetate extraction (Gray and Rickard, 1982a), even when the brain was perfused with saline (Rickard and Brodie, 1985), possibly warrants further investigation. The threshold of deltamethrin in brain and spinal cord after i.p. administration of different dose levels (Rickard and Brodie, 1985) confirmed that the levels of parent compound associated with choreoathetosis were in the range of 0.4–1.0 nmol/g and lower concentrations of 0.1–0.2 nmol/g and 0.2–0.35 nmol/g correlated with salivation and tremor respectively. A similar brain concentration of approximately 1 nmol/g has also been proposed as the toxic threshold level for deltamethrin in mice (Ruzo *et al.*, 1979).

Following i.v. administration of 1.75 mg/kg, blood concentrations of deltamethrin were in the region of 20–26 nmol/g at 1 minute and decreased with an approximate half-life of 5 minutes. Trace amounts (0.03 nmol/g) could still be detected at 240 minutes (Gray and Rickard, 1982a), indicating the greater *in vivo* stability of deltamethrin when compared to cismethrin (Gray *et al.*, 1980). Peak blood concentrations of deltamethrin (3–5 nmol/g) were achieved 85 minutes after i.p. administration of 8 mg/kg and corresponded with the onset of choreoathetosis (Rickard and Brodie, 1985). Lower blood concentrations, in the range of 2.5–3.8 nmol/g and 1.5–2.4 nmol/g were associated with tremor and salivation, respectively (Rickard and Brodie, 1985). There is, therefore, only a poor correlation between the blood concentrations of deltamethrin and the severity of signs of toxicity when the deltamethrin was given by different routes of administration. Peak liver levels of deltamethrin (9 nmol/g) were reached 5 minutes after i.v. injection and then decreased rapidly to 1.5 nmol/g at 15 minutes (Gray and Rickard, 1982a) but were much lower (peak of 3.3 nmol/g at 85 minutes) and appeared to be relatively stable over the time course after i.p. administration (Rickard and Brodie, 1985). The majority of the radioactivity in liver was not extracted in either study and probably represented various polar metabolites of deltamethrin.

In an investigation of the brain concentrations of several pyrethroids follow-ing i.p. administration of non-toxic (2.5 mg/kg) doses to mice, peak mouse brain concentrations of *cis*-permethrin were approximately 10-fold higher than those of the cyano-substituted pyrethroids: deltamethrin, cypermethrin and fenvaler-ate (Marei *et al.*, 1982). Since a greater proportion of an i.v. administered cis-methrin dose was determined in the CNS than that of deltamethrin (Gray *et al.*, 1980; Gray and Rickard, 1982a), these findings indicate that pyrethroids which do not contain the α-cyano moiety may have greater access to the CNS than those with this group. The lower brain levels of *trans*-permethrin in these exper-iments (Marei *et al.*, 1982) were probably due to its more rapid ester hydrolysis, since pretreatment with esterase inhibitors increased brain concentrations of the *trans* isomer to values similar to those obtained with *cis*-permethrin.

These toxicokinetic studies indicate that although there may be some correla-tion between peak blood concentrations of pyrethroid and the severity of tox-icity, the concentrations of pyrethroid in blood decreased more rapidly than in the CNS, particularly after i.v. injection. There appears to be a threshold con-centration in the CNS for each pyrethroid studied, which must be exceeded before a particular sign of toxicity occurs. This sign of toxicity then persists for as long as this concentration is maintained. The non-cyano pyrethroids *cis*-permethrin and cismethrin appear to achieve higher CNS concentrations than many of the α-cyano pyrethroids studied. However, approximately 10-fold higher CNS concentrations of cismethrin than deltamethrin were required before toxicity occurred. The 1*R,trans* analogue of cismethrin, bioresmethrin, achieved similar CNS concentrations after i.v. injection of an equimolar dose, but failed to produce toxicity even when considerably higher CNS concentra-tions were achieved. This indicated that the lower mammalian toxicity of bio-resmethrin is not due to a more rapid or different metabolism than cismethrin, but that this isomer has little or no intrinsic mammalian toxicity. The blood kinetic studies indicate that, although still rapidly metabolized, deltamethrin is considerably more stable than cismethrin. The greater stability of deltamethrin and related compounds is predicted by the *in vitro* structure–biodegradability studies (Table 4) and may partially account for both their higher mammalian toxicity and the greater persistence of the motor signs when compared to the more biodegradable esters.

Elimination and tissue retention

Metabolic studies, conducted mainly in rats and mice, have demonstrated that pyrethroids are rapidly excreted as a variety of metabolites. Generally more than 50% of an oral or i.p. dose is excreted within the first 24 hours and only a small proportion, if any, of this has been demonstrated to be parent compound. The acid- and alcohol-radiolabelled preparations of most pyrethroids are usually eliminated at similar rates. However, following administration of bio-resmethrin to rats, the alcohol radiolabel was eliminated more gradually than the radioactivity derived from the acid moiety (Ueda *et al.*, 1975b). The alcohol component of bioresmethrin was retained in blood, liver and kidney as BFCA (Miyamoto *et al.*, 1972; Gray *et al.*, 1980), possibly due to enterohepatic circulation of this metabolite and its conjugates (Miyamoto *et al.*, 1972). However, since BFA and BFCA are not particularly toxic (White *et al.*, 1976; Rickard and Brodie, 1985) and the toxicity of bioresmethrin has been accounted for by small amounts of cismethrin impurity found in bioresmethrin samples (Gray and Connors, 1980) it seems unlikely that the retention of this metabolite has any toxicological significance. The alcohol moiety of many of the photostable pyrethroids, 3-phenoxybenzyl alcohol, is apparently not retained to the same extent. Although it is rapidly oxidized and eliminated as a glucuronide conjugate in the bile which is subsequently hydrolysed by gut flora and reabsorbed, it is then rapidly hydroxylated and excreted as a sulphate conjugate in the urine (Huckle *et al.*, 1981).

The rate of elimination of radioactivity following administration of deltamethrin, cypermethrin, fenvalerate and other pyrethroids radiolabelled in the α-cyano group is the slowest of all preparations. This is apparently due to the release of the cyano moiety, during the ester cleavage, which is then incorporated into the body thiocyanate pool and retained in the skin and stomach (Casida *et al.*, 1983).

Determination of tissue concentrations of radioactivity at the termination of metabolism experiments has indicated that the long-term retention of pyrethroids and their metabolites is unlikely. However, up to 3 days after dosing, kidney, liver and blood contain higher levels of radioactivity than most other tissues. Within 6–8 days the amounts of radiolabel in these had decreased to barely detectable levels. Retention of many pyrethroids has been demonstrated in body fat, presumably due to the lipophilic nature of these compounds. Following the administration of *cis*-cypermethrin (2.5 mg/kg) to rats, fat residues of almost 2.4 nmol/g were still present at 8 days and the majority was identified as parent compound (Crawford *et al.*, 1981). These concentrations had decreased to 0.12 nmol/g 42 days after treatment, and a half-life of approximately 12 days was determined. The *trans* isomer of cypermethrin was not retained in fat to the same extent and was present at particularly low concentrations in male rats (Crawford *et al.*, 1981). A similar finding has also been reported following administration of permethrin isomers to rats (Elliott *et al.*, 1976; Gaughan *et al.*, 1977); the *cis* isomers were maintained at higher concentrations in fat than were the *trans* isomers and radioactivity from the alcohol-labelled material persisted for longer than that of the acid moiety. The fat of rats analysed 8 days after oral administration of 1.5 mg/kg alcohol-labelled fenpropathrin (**7F**), a pyrethroid which does not have *cis* and *trans* isomers (but possesses two *cis*- and two *trans*-methyl groups on the cyclopropane ring) was also found to contain considerably higher levels of radioactivity (~ 5 nmol/g) than the other tissues examined (Crawford and Hutson, 1977). The persistence of several pyrethroids in fat have been determined simultaneously by gas chromatography (Marei *et al.*, 1982). These studies demonstrated that even though only small oral doses were administered (3 mg/kg), the pyrethroids could still be detected in fat after 21 days. The half-lives of these residues were 5–6 days for *trans*-permethrin and deltamethrin and 7–10 days for fenvalerate, *cis*-permethrin and cypermethrin.

Retention of pyrethroids in fat has also been demonstrated in a number of other mammals, including goats (Hunt and Gilbert, 1977; Ivie and Hunt, 1980), cows (Gaughan *et al.*, 1978a), lamb (Wszolek *et al.*, 1981) and also in chickens (Gaughan *et al.*, 1978b). Trace amounts of pyrethroid have also been detected in milk collected from goats and cows and in the eggs of chickens administered permethrin, particularly the *cis* isomer (Hunt and Gilbert, 1977; Gaughan *et al.*, 1978a,b; Ivie and Hunt, 1980).

ELECTROPHYSIOLOGY AND NEUROPHARMACOLOGY

Electrophysiological studies

The actions of pyrethroids (predominantly those now classified as producing the T syndrome) in several non-mammalian nerve preparations have been described in detail (for reviews, see Narahashi, 1971, 1976, 1983; Wouters and van den Bercken, 1978; Vijverberg and van den Bercken, 1982; Ruigt, 1984). Elicited action potentials in pyrethroid-treated nerves show a negative after-potential that leads to long trains of repetitive discharges from single stimulus. After prolonged exposure or treatment with high concentrations, nerve block is achieved. Voltage clamp studies have demonstrated that these neurotoxic effects result from a marked prolongation of the sodium inactivation component of the action potential.

The first direct comparison of insecticidal α-cyano and non-cyano esters, otherwise identical in structure, revealed two distinct types of action on locust leg muscle-nerve preparations (Clements and May, 1977). Subsequent studies have shown that α-cyano esters do not produce effects typical of T syndrome compounds (e.g. repetitive discharge) on axonal preparations but instead produce stimulation frequency-dependent blockage of action potentials (Duclohier and Georgescauld, 1979; Gammon *et al.*, 1981; Vijverberg *et al.*, 1982, 1983). Nevertheless, CS syndrome compounds produce a slowing of sodium inactivation under voltage clamp conditions that is qualitatively similar to, but much more prolonged than that produced by T syndrome compounds (Duclohier and Georgescauld, 1979; Vijverberg *et al.*, 1982, 1983; Lund and Narahashi, 1983). The extension of electrophysiological studies to examine the effects of pyrethroids on the mammaliam CNS has been limited by its relative inaccessibility to single-cell recording and voltage clamp techniques. Consequently, information on electrophysiological effects of pyrethroids in mammals is limited to descriptions of the effects on sensory nerves and on more complex CNS responses.

Carlton (1977) described two effects of intravenous cismethrin (multiple doses of 10 mg every 2–3 min) on the sensory nervous system of the spinal rabbit. Cismethrin lowered the threshold of some sensory (hair follicle) receptors and also produced spontaneous activity in some sensory cells. Cismethrin also caused repetitive firing of an afferent nerve preparation following electrical stimulation. Both cismethrin and deltamethrin increased the excitability of rat tail nerve preparations following intravenous administration of 0.5 mg/kg (Parkin and Le Quesne, 1982; Takahashi and Le Quesne, 1982). Deltamethrin gave a prolonged (up to 400 ms) period of excitability following a nerve impulse, whereas cismethrin increased excitability for a brief period (2–4 ms) following stimulation (Takahashi and Le Quesne, 1982).

Pyrethroid-dependent changes in CNS electrophysiology have also been noted. Intravenous cismethrin increased the excitability of both monosynaptic and polysynaptic reflex arcs in the spinal rabbit (Carlton, 1977) and enhanced dorsal root potential amplitude in rats (Smith, 1980). In contrast, the recognized GABA antagonist bicuculline reduced dorsal root potential amplitude. A recent study (Staatz and Hosko, 1985) compared the action of cis-permethrin and deltamethrin on rat and cat spinal neurons. Both deltamethrin (1–2 mg/kg i.v.) and cis-permethrin (10 mg/kg i.v.) increased the level of spontaneous activity in rat lumbar ventral root preparations and increased the amplitude of both monosynaptic and polysynaptic reflex responses. Deltamethrin (2 mg/kg i.v.) and cis-permethrin (20 mg/kg i.v.) also caused sustained increases in the spontaneous firing of cat spinal interneurons; subsequent treatment with diazepam (0.5 mg/kg i.v.) reduced the pyrethroid-enhanced firing by approximately 50% in each case. Ray (1980) reported the effect of deltamethrin on electroencephalographic (EEG) recordings in the rat following intravenous or intraperitoneal administration. In these studies, the dominant EEG feature of the CS poisoning syndrome was the development of sharp waves evoked by sensory stimuli. These waves developed rapidly into spontaneous spike sequences of cortical origin following individual spasms.

These studies confirm that the responses of mammalian sensory and CNS preparations to pyrethroids are qualitatively similar to those described previously for invertebrate and non-mammalian vertebrate preparations. Although they provide little insight into the precise mechanism of pyrethroid neurotoxicity in mammals they indicate that pyrethroid action is not restricted to any particular region of the nervous system. The signs of toxicity are probably due to a combination of the effects on both the peripheral and central nervous systems although some regions (e.g. spinal interneurons and peripheral sense organs) may appear more sensitive due to their higher intrinsic activity.

The resulting poisoning syndrome is therefore possibly more dependent on the range of pharmacological consequences that result from the pyrethroid–receptor interaction rather than several interactions at different anatomical sites.

Pyrethroid–receptor interactions

Although electrophysiological studies have provided information on neuronal responses to pyrethroid intoxication, they have not adequately described the specific interactions that occur between pyrethroids and their site(s) of action. The stringent steric requirements for pyrethroid-like activity and the reversibility of pyrethroid intoxication following sublethal doses both suggest that poisoning results from a specific and reversible interaction of a pyrethroid with a chiral receptor. Recently, several laboratories have attempted to clarify this pyrethroid–receptor interaction.

On the basis of electrophysiological studies, the voltage-dependent sodium channel of the nerve is anticipated to be involved in the action of some if not all pyrethroids. Sodium channel function can be observed directly in cultured neuroblastoma cells and brain synaptosomes by measuring sodium channel-dependent radiosodium flux across cellular or vesicular membranes (Catterall, 1980). Using this approach, Jacques et al. (1980) demonstrated that pyrethroids (exemplifying both T and CS syndromes) had no direct effect on sodium channel-dependent ^{22}Na uptake by mouse neuroblastoma cells, but that they potentiated the activation of ^{22}Na uptake by both alkaloid (veratridine, batrachotoxin) and polypeptide (scorpion, sea anemone) toxins. The pyrethroid effect was inhibited by tetrodotoxin. These studies imply that pyrethroid effects on the sodium channel may involve a binding site different from those for other known sodium channel-active drugs and toxins. However, interpretation of these results is complicated by the finding that potent mammalian neurotoxicants such as cis-permethrin and cismethrin apparently bind to the sodium channel without potentiating activation-dependent sodium flux, thereby acting as antagonists of the action of other pyrethroids. The significance of these effects on sodium channel function in the neuroblastoma model in explaining pyrethroid neurotoxicity in vivo remains to be established. A recent study (Ghiasuddin and Soderlund, 1985) used the radiosodium flux technique to examine effects of pyrethroids on mouse brain synaptosomal sodium channels. NRDC 157 (which produced the T syndrome) and deltamethrin had no effect on radiosodium flux alone, but both compounds enhanced veratridine-dependent sodium channel activation. This effect was stereospecific; the corresponding 1S enantiomers were inactive in this assay at concentrations up to 10 μM. Half-maximal enhancement was achieved at 25 nM (deltamethrin) or 220 nM (NRDC 157). Using the activity of deltamethrin for comparison, this is the most potent pharmacological effect reported to date for pyrethroids in mammalian CNS preparations.

In another study of pyrethroid effects on sodium channels, Nicholson et al. (1983) examined the effects of pyrethroids on sodium flux-dependent release of tritiated GABA from preloaded guinea pig brain synaptosomes. Both delta-methrin and permethrin, as well as veratridine and DDT, stimulated the release of labelled GABA from superfused guinea pig synaptosomes. Of the insecticides examined, deltamethrin was the most potent and produced the greatest stimulation. The deltamethrin effect was completely blocked by tetrodotoxin, thus demonstrating a specific effect of deltamethrin on the sodium channel. These findings differ from those obtained by measuring radiosodium flux in that the pyrethroids produced an apparent sodium channel-specific effect in the absence of veratridine activation. Because pyrethroids are highly lipophilic, it is likely that the effective pyrethroid concentration achieved in superfused synaptosomes was much higher than the concentrations used in radiosodium flux studies.

Although the findings summarized above implicate the sodium channel as the site of pyrethroid action, other types of pyrethroid–nerve interaction have also been reported. The action of diazepam as a pyrethroid antidote, with apparently greater potency against the signs of toxicity produced by CS syndrome compounds (Gammon et al., 1982; Staatz et al., 1982a), has led to investigations of pyrethroid interactions with the GABA receptor–ionophore complex (Leeb-Lundberg and Olsen, 1980; Lawrence and Casida, 1983; Squires et al., 1983). Using a radioreceptor assay, Lawrence and Casida (1983) demonstrated that toxic pyrethroids that produce the CS syndrome inhibited the binding of [^{35}S]t-butylbicyclophosphorothionate (TBPS), a potent cage convulsant and specific GABA receptor ligand, by up to 37% when incubated at 5 µM in rat brain subcellular fractions. In contrast, non-toxic isomers of these pyrethroids and all compounds that produce the T syndrome were less effective. The relative inhibitory potencies of these CS type pyrethroids correlated with their relative neurotoxicities by intracerebral injection. Interpretation of these findings is complicated by the high concentrations of pyrethroid required to inhibit TBPS binding. In particular, the IC_{50} for deltamethrin in this assay (5.6 µM; Squires et al., 1983) is more than 200-fold higher than the concentration of deltamethrin required to produce half-maximal enhancement of sodium channel activation (25 nm; Ghiasuddin and Soderlund, 1985). Other neuro-pharmacological effects reported recently for pyrethroids include inhibition of squid neural Ca-ATPases (Clark and Matsumura, 1982), inhibition of perhydrohistrionicotoxinin binding to the Torpedo electroplax acetylcholine receptor/ionophore (Abbassy et al., 1983), and partial inhibition of kainic acid binding to mouse brain preparations (Staatz et al., 1982b). Although these studies suggest other possible sites of pyrethroid action, the relevance of these findings to the description of pyrethroid neurotoxicity in mammals is limited either in the scope of compounds tested or in the applicability of the model system employed.

Another approach to defining the pyrethroid receptor is to use the pyrethroid itself as the radioligand. Recently, Soderlund *et al.,* (1983) used the experimental T syndrome pyrethroid NRDC 157, labelled with tritium at a high specific activity, to demonstrate a high-affinity, saturable, stereospecific binding site for this ligand in mouse brain membrane preparations. Receptor-like stereospecific binding represented only a small proportion (2.8%) of total binding; the remainder was predominantly non-specific and unsaturable, probably reflecting the extensive association of this lipophilic ligand with nerve membrane lipids. The stereospecific site was half-saturated at 40 nM and fully saturated at concentrations in excess of 100 nM. These indices of binding affinity agree well with the potency of NRDC 157 as an enhancer of sodium channel activation in mouse brain synaptosomes (Ghiasuddin and Soderlund, 1985). The stereospecific binding capacity of this preparation was 2.3–2.8 pmol/mg protein, a value similar to the specific saxitoxin binding capacity of sodium channels in rat brain synaptosomes (4.9 pmol/mg protein; Catterall *et al.,* 1979). These findings represent the first demonstration of receptor-like binding of a pyrethroid radioligand. However, further studies of stereospecific binding using a broad range of pyrethroid ligands are required to establish whether this binding site is the receptor involved in pyrethroid intoxication.

From the foregoing discussion, it is clear that there is no consensus on the precise molecular mechanism of pyrethroid neurotoxicity. The common stereochemical requirements for toxic action among all pyrethroids suggest interaction with a single type of binding site, yet the sharp and consistent differentiation of T and CS biological activities in both insects and mammals implies the existence of two types or subtypes of neurotoxic action. One hypothesis, based primarily on electrophysiological studies of pyrethroid action and supported in part by pharmacological studies, reconciles these two apparently conflicting observations by suggesting that T and CS syndrome compounds act at a common site, the nerve membrane sodium channel, but produce qualitatively different biological activities depending on the gating kinetics of the pyrethroid-modified channel (Vijverberg *et al.,* 1982, 1983; Lund and Narahashi, 1983). An alternative hypothesis, based on the antagonism of the onset of the CS syndrome by diazepam and the specific inhibition of TBPS binding by CS syndrome compounds, suggests two sites of action: the nerve membrane sodium channel for compounds that produce the T poisoning syndrome, and the GABA receptor–ionophore complex for those that cause the CS syndrome (Gammon *et al.,* 1981, 1982; Lawrence and Casida, 1983). At present neither hypothesis alone sufficiently explains all of the observed actions of pyrethroids *in vivo* and in isolated systems *in vitro*. In particular, further work is clearly needed to define the relationship between the pyrethroid–GABA receptor interactions and the well documented effects of these compounds on the sodium channel.

Pharmacological antagonists and potentiators of pyrethroid toxicity

It is generally accepted that the signs of toxicity produced by pyrethroids in mammals are due to an action of these compounds on the nervous system. Therefore, a number of workers have investigated the effect of selected drugs, chosen for their specific sites and modes of action, on pyrethroid-induced signs of toxicity in an attempt to elucidate which neurotransmitter systems and neuronal pathways are involved in the poisoning syndromes. Staatz et al. (1982a) demonstrated that drugs which caused a reduction in noradrenaline synthesis or release, such as clonidine, diethyldithiocarbamic acid and reserpine, potentiated the toxicity of i.v. administered permethrin in mice. However, the adrenergic receptor blocker, phenoxybenzamine, was less effective and phenylephrine and pargyline, which increase adrenergic tone, had no effect on toxicity. Pretreatment with compounds that selectively block dopamine receptors caused little change in permethrin toxicity, with the exception of chlorpromazine which increased toxicity, possibly because this compound is less selective and also causes blockade of noradrenaline receptors. Chlorpromazine was found to be ineffective in protecting against deltamethrin toxicity in rats although it did prevent salivation (Bradbury et al., 1983). Atropine pretreatment, which inhibits salivation induced by both deltamethrin (Ray and Cremer, 1979) and cypermethrin (Lock and Berry, 1981) had no effect on permethrin-induced signs of toxicity (Staatz et al., 1982a). The muscarinic blocker, mecamylamine, and hemicholinium (administered i.c.v.), a drug which depletes the acetylcholine content of nerve terminals, both significantly enhanced the toxicity of permethrin to mice.

The GABA antagonist, picrotoxin, did not significantly potentiate permethrin toxicity, while amino-oxyacetic acid, which raises brain GABA concentrations, and diazepam (at high doses, 10 mg/kg,) which may increase the availability of GABA to its receptors, both caused a decrease in permethrin toxicity (Staatz et al., 1982a). Diazepam pretreatment at 0.3 mg/kg i.p. has also been shown to delay the onset of signs following i.c.v. administration of deltamethrin and, at 3 mg/kg, to be effective in protecting against the lethality of this pyrethroid (Gammon et al., 1982). The LD_{50} of i.c.v. administered permethrin was increased almost 10-fold in mice pretreated with 3 mg/kg diazepam, although the time course of the signs of toxicity was not affected (Gammon et al., 1982). This study also demonstrated that diazepam was capable of causing a slight delay in the onset of signs of toxicity following [2S, αS]-fenvalerate, but not S-bioallethrin, although no data were given for the effect of diazepam pretreatment on the lethality of these two pyrethroids. In contrast, diazepam and clonazepam (Cremer et al., 1980) and fletazepam (Gray and Gray, 1984) were found to be ineffective in protecting against deltamethrin administered i.p. or i.v. to rats. This may represent a species difference, but is more probably due to the variation in route of administration; the i.c.v. route used by Gammon et al., (1982) would have resulted in greater concentrations of pyrethroid in the higher

Table 7 Effective doses of compounds which antagonise pyrethroid toxicity

Pyrethroid	Route of administration of pyrethroid	Species	Protectant[a]	Effective dose or ED$_{50}$ of protectant
Cismethrin	i.v.	Rat	Mephenesin	5 mg/kg/min + 10 mg/kg/min[b,c,d]
	i.v.	Rat	Propranolol	4.5 mg/kg/min + 1.8 mg/kg/min[b,c,d]
cis-Permethrin	i.v.	Mouse	Diazepam	10 mg/kg[e]
	i.v.	Mouse	Amino-oxyacetic acid	50 mg/kg[e]
	i.v.	Mouse	Cyclohexamide	1 mg/kg[e]
	i.c.v.	Mouse	Diazepam	3 mg/kg[d]
Deltamethrin	i.c.v.	Mouse	Diazepam	3 mg/kg[d], 1 mg/kg[f]
	i.v.	Rat	Mephenesin	4 mg/kg/min + 2.5 mg/kg/min[b,c,d]
	i.v.	Rat	Propranolol	4.5 mg/kg/min + 1.8 mg/kg/min[b,c,d]
	i.v.	Rat	Procainamide	4 mg/kg/min + 1 mg/kg/min[b,c,d]
	i.v.	Rat	Mephenesin	1.3 mg/kg/min[b,d,g]
	i.v.	Rat	Chlorphenesin carbamate	2.2 mg/kg/min[b,d,g]
	i.v.	Rat	Methocarbamol	3.7 mg/kg/min[b,d,g]
	i.v.	Rat	Methocarbamol	159 mg/kg[d,g]
	i.v.	Rat	Chlorzoxazone	67 mg/kg[d,g]
	i.v.	Rat	Metaxalone	84 mg/kg[d,g]

[a] All protectants were given as i.p. pretreatments unless indicated otherwise.
[b] Administered by continuous i.v. infusion.
[c] Pretreatment by i.v. infusion at first rate shown and continued at the second rate after pyrethroid administration. Data from Bradbury et al., 1981, 1983.
[d] Reduced incidence of mortality or prevented mortality. Diazepam data from Gammon et al., 1982.
[c] Reduced incidence or delayed onset of loss of righting reflex. Data from Staatz et al., 1982a.
[f] Delayed onset of 'convulsions'. Data from Gammon et al., 1982.
[g] Dose of protectant which resulted in 50% survival (ED$_{50}$) of animals given a twice LD$_{50}$ dose of the pyrethroid which normally results in 100% mortality. Data from Gray and Gray, 1985.

brain centres and a different distribution than would occur after parenteral administration. Although it has never been proven to be effective if given after ingestion of, or exposure to, a pyrethroid, diazepam is still the recommended therapy for the treatment of pyrethroid poisoning (DHSS, 1983). Propranolol reduced the severity of toxicity and the incidence of mortality due to cismethrin and deltamethrin, but only if administered prior to the pyrethroid. Both isomers of propranolol were equally effective, indicating that the membrane-stabilizing, local anaesthetic, properties were responsible for the reduction in severity of pyrethroid motor symptoms, rather than β-receptor blockade (Bradbury et al., 1983) (Table 7). In this study the local anaesthetic procainamide also reduced the severity of deltamethrin-induced signs of toxicity if given as a pretreatment, but was ineffective against cismethrin (Table 7). General anaesthetics such as phenobarbitone (Gammon et al., 1982) and ether (Bradbury et al., 1981) also protect against pyrethroid toxicity but only prevent mortality when administered at near-lethal doses. These studies have demonstrated that many compounds which cause selective alterations in the activity of neurotransmitter systems in the CNS can modify the signs of toxicity or lethality of pyrethroids. However, there is insufficient evidence at present to implicate any one transmitter system as being primarily involved in the pyrethroid poisoning syndromes.

The spinally acting muscle relaxant, mephenesin, antagonized all motor signs of toxicity in rats following i.v. administration of deltamethrin or cismethrin, although higher doses are required with cismethrin (Bradbury et al., 1981, 1983) (Table 7). Mephenesin was effective whether given as a pretreatment or after the onset of signs of toxicity, but must be given as a continuous i.v. infusion because of its rapid metabolism and clearance (Bradbury et al., 1981, 1983; Gray and Gray, 1985). Infusion rates of mephenesin that were effective in protecting against the signs of toxicity of cismethrin and deltamethrin in intact rats (salivation not affected) also prevented the characteristic motor sign of toxicity produced by these two pyrethroids in spinal rats (Bradbury et al., 1983).

The potency of ten muscle relaxant compounds in delaying or reducing the severity of the signs of toxicity or protecting against death following the i.v. administration of a lethal dose of deltamethrin, has been reported recently (Gray and Gray, 1985). This study indicated that pretreatment with compounds that have a primarily spinal locus of action (metaxalone, mephenesin, methocarbamol, chlorphenesin carbamate and chlorzoxazone) afforded the greatest protection against deltamethrin lethality (Table 7) and that methocarbamol, mephenesin and chlorphenesin carbamate were also effective if infused after the pyrethroid. Some muscle relaxant compounds delayed the onset, or modified the signs, of deltamethrin toxicity (meprobamate, carisoprodal, fletazepam, orphenadrine citrate and the peripheral muscle relaxant dantrolene sodium) without protecting against the lethality of this pyrethroid (Gray and Gray, 1985). Another, spinally acting muscle relaxant, baclofen, was found to be ineffective in protecting against pyrethroid toxicity (Bradbury et al., 1983). This was presumably due to this compound acting by decreasing the release of excitatory neurotransmitter from presynaptic nerve terminals rather than by an enhancement of post-synaptic inhibition. The compounds which were found to be effective against deltamethrin lethality were also capable of modifying strychnine toxicity and lethality. Since strychnine has a primarily spinal locus of action, and only muscle relaxants with membrane-stabilizing effects at the spinal interneuron level antagonized the toxicity of both strychnine and deltamethrin, it was concluded that the pyrethroid lethality must be associated with an action at the spinal level of the mammalian CNS (Gray and Gray, 1985).

CONCLUSIONS

The literature encompassed in this review supports the contention that pyrethroid insecticides are generally safe when administered by routes commonly used to assess acute or chronic hazard. However, when administered by routes that minimize detoxication and increase delivery of the pyrethroid to the CNS, neurotoxicity often results at extremely low doses, demonstrating that many pyrethroids are intrinsically potent neurotoxicants in mammals. The signs of toxicity produced can be classified into two distinct types; T and CS, the latter is generally produced by pyrethroids containing both a halogenated acid and the S-α-cyano-3-phenoxybenzyl alcohol. Both of these poisoning syndromes appear to result predominantly from interactions between pyrethroids and sites in the CNS. It is, therefore, not possible to distinguish between the two syndromes on the basis of different target regions. Although considerable information exists on the electrophysiological and neurochemical consequences of pyrethroid poisoning, a coherent picture relating these effects to the specific production of the T and CS poisoning syndromes has not yet emerged. The electrophysiological changes that occur in the mammalian nervous system following exposure to toxic pyrethroids appear to be similar to those of amphibia and insects, although mammalian sites of action are apparently less sensitive and more stereospecific in their response. The high toxicity of pyrethroids to fish may result in part from the efficient bioaccumulation of these lipophilic compounds from an aqueous environment, but lower rates of detoxication and greater intrinsic sensitivity of the CNS, when compared to mammals, are also contributing factors. In contrast, low sensitivity of the CNS to pyrethroids contributes significantly to the very low acute toxicity of these compounds to birds.

The low oral toxicity of pyrethroids has permitted their administration in the diet for long-term feeding studies at very high doses. These studies have revealed no severe chronic effects and indicate that, at the low concentrations required for insect control, pyrethroid residues in foodstuffs and in water do not constitute a significant hazard to man or wildlife. The low oral and dermal toxicities of these compounds suggests that acute systemic poisoning is only likely to occur if large quantities of the concentrated pyrethroid formulation are ingested. The hazard most commonly encountered in the manufacture and use of some pyrethroids is paraesthesia, a transient irritative effect involving the skin and upper respiratory tract.

The wide variety of neuroactive compounds have been examined for their impact on pyrethroid toxicity and symptomology. Several of these antagonize pyrethroid toxicity, but none has been clinically proven to be a useful antidote for pyrethroid intoxication. Nevertheless, these compounds may be of further value both as leads in the identification of useful antidotes and as chemical probes in studies of the mechanisms and consequences of pyrethroid-induced neurotoxicity.

Recently there has been considerable progress in defining pyrethroid–receptor interactions in mammalian nervous tissue. Most of the evidence to date implicates the nerve membrane sodium channel as the site of pyrethroid action, but other studies have implicated a specific interaction of CS syndrome compounds with the GABA receptor. Further investigation is clearly needed to identify the site(s) of toxic action and establish the causal relationship between the pyrethroid–receptor interaction and the observed syndromes of intoxication.

ACKNOWLEDGEMENTS

The authors wish to thank the following for permission to include unpublished data or cite papers in press in this review: Mr Richard Verschoyle, Ms Christina Staatz, Ms Janet Martin and Ms Julie Gray. A.J.G. is grateful to the latter for her advice on the section on pharmacological antagonists and for her encouragement during the preparation of this review. We are indebted to Ms Rose McMillen-Sticht for her painstaking preparation of the figures and Ms Suzanne Burnett for her patience and skill in typing the numerous draft manuscripts. A.J.G. is grateful to colleagues at the ICI Central Toxicology Laboratory, in particular Dr E. A. Lock, for their constructive criticism of the draft manuscript.

REFERENCES

Abbassy, M. A., Eldefrawi, M. E., and Eldefrawi, A. T. (1983). 'Pyrethroid action on the nicotinic acetylcholine receptor/channel', *Pestic. Biochem. Physiol.*, **19**, 299–308.

Abernathy, C. O., Ueda, K., Engel, J. L., Gaughan, L. C., and Casida, J. E. (1973). 'Substrate-specificity and toxicological significance of pyrethroid-hydrolyzing esterases of mouse liver microsomes', *Pestic. Biochem. Physiol.*, **3**, 300–311.

Aldridge, W. N., Clothier, B., Forshaw, P., Johnson, M. K., Parker, V. H., Price, R. J., Skilleter, D. N., Verschoyle, R. D., and Stevens, C. (1978). 'The effect of DDT and the pyrethroids cismethrin and deltamethrin on the acetyl choline and cyclic nucleotide content of rat brain', *Biochem. Pharmacol.*, **27**, 1703–1706.

Anon. (1979). '"Conditional" tolerances established for permethrin', *Pestic. Tox. Chem. News*, **7**, 9–11.

Anon. (1980). 'Tolerances set for mesurol, fenvalerate, oxamyl', *Pestic. Tox. Chem. News*, **8**, 22–24.

Baer, R. L., Ramsey, D. L., and Biondi, E. (1973). 'The most common contact allergens' *Archs Dermatol.*, **108**, 74–78.

Barlow, F., and Hadaway, A. B. (1975). 'The insecticidal activity of some synthetic pyrethroids against mosquitoes and flies', *PANS*, **21**, 233–238.

Barnes, J. M., and Verschoyle, R. D. (1974). 'Toxicity of new pyrethroid insecticide', *Nature*, **248**, 711.

Barthel, W. F. (1973). 'Toxicity of pyrethrum and its constituents to mammals', in *Pyrethrum the Natural Insecticide* (ed. J. E. Casida), pp. 123–142, Academic Press, New York.

Berteau, P. E., and Dean, W. A. (1978). 'A comparison of oral and inhalation toxicities of four insecticides to mice and rats', *Bull. Environ. Contam. Toxicol.*, **19**, 113–120.

Bradbury, J. E., Forshaw, P. J., Gray, A. J., and Ray, D. E. (1983). 'The action of mephenesin and other agents on the effects produced by two neurotoxic pyrethroids in the intact and spinal rat', *Neuropharmacol.*, **22**, 907–914.

Bradbury, J. E., Gray, A. J., and Forshaw, P. (1981). 'Protection against pyrethroid toxicity in rats with mephenesin', *Toxicol. appl. Pharmacol.*, **60**, 382–384.

Brodie, M. E., and Aldridge, W. N. (1982). 'Elevated cerebellar cyclic GMP levels during the deltamethrin-induced motor syndrome', *Neurobehav. Toxicol. Teratol.*, **4**, 109–113.

Burke, M. D., and Orrenius, S. (1982). 'Isolation and comparison of endoplasmic reticulum membranes and their mixed function oxidase activities from mammalian extrahepatic tissues', in *Hepatic Cytochrome P-450 Monooxygenase System* (eds J. B. Schenkman and D. Kupfer), pp. 47–98, Pergamon Press, Oxford.

Campt, D. D. (1983). 'Cyano (3-phenoxyphenyl)methyl 4-chloro-alpha-(1-methyl) benzeneacetate; proposed tolerance', *Fed. Reg.* **48**, 484–485.

Carlson, G. P., and Schoenig, G. P. (1980). 'Induction of liver microsomal NADPH cytochrome *c* reductase and cytochrome P-450 by some new synthetic pyrethroids', *Toxicol. appl. Pharmacol.*, **52**, 507–512.

Carlson, J. E., and Villaveces, J. W. (1977). 'Hypersensitivity pneumonitis due to pyrethrum', *J. Am. med. Ass.*, **237**, 1718–1719.

Carlton, M. (1977). 'Some effects of cismethrin on the rabbit nervous system', *Pestic. Sci.*, **8**, 700–712.

Casida, J. E. (1973). 'Biochemistry of the pyrethrins', in *Pyrethrum the Natural Insecticide* (ed. J. E. Casida), pp. 101–120, Academic Press, New York.

Casida, J. E. (1983). 'Novel aspects of metabolism of pyrethroids', in *Pesticide Chemistry: Human Welfare and Environment* (eds J. Miyamoto and P. C. Kearney), Vol. 2, pp. 187–192, Pergamon, Oxford.

Casida, J. E., Gammon, D. W., Glickman, A. H., and Lawrence, L. J. (1983). 'Mechanisms of selective action of pyrethroid insecticides', *Ann. Rev. Pharmacol. Toxicol.*, **23**, 413–438.

Casida, J. E., Gaughan, L. C., and Ruzo, L. O. (1979). 'Comparative metabolism of pyrethroids derived from 3-phenoxybenzyl and α-cyano-3-phenoxybenzyl alcohols', in *Advances in Pesticide Science* (ed. H. Geissbuehler), pp. 182–189, Pergamon, Oxford.

Casida, J. E., and Ruzo, L. O. (1980). 'Metabolic chemistry of pyrethroid insecticides', *Pestic. Sci.*, **11**, 257–269.

Casida, J. E., Ueda, K., Gaughan, L. C., Jao, L. T., and Soderlund, D. M. (1975/76). 'Structure-biodegradability relationships in pyrethroid insecticides', *Archs Environ. Contam. Toxicol.*, **3**, 491–500.

Catterall, W. A. (1980). 'Neurotoxins that act on voltage-sensitive sodium channels in excitable membranes', *Ann. Rev. Pharmacol. Toxicol.*, **20**, 15–43.

Catterall, W. A., Morrow, C. S., and Hartshorne, R. P. (1979). 'Neurotoxin binding to receptor sites associated with voltage-sensitive sodium channels in intact, lysed, and detergent-solubilized brain membranes', *J. Biol. Chem.*, **254**, 11379–11387.

Chambers, J. (1980). 'An introduction to the metabolism of pyrethroids', *Residue Rev.*, **73**, 101–124.

Chanh, P. H., Navarro-Delmasure, C., Chanh, A. P. H., and Martinez, C. (1981). 'Toxicity of decamethrin depends on the solvent, the administration routes, the sex and the animal species', *IRCS Med. Sci. Libr. Comp.*, **9**, 565.

Chatterjee, K. K., Talukder, G., and Sharma, A. (1982). 'Effects of synthetic pyrethroids on mammalian chromosomes. I. Sumicidin', *Mut. Res.*, **105**, 101–106.

Chen, Y.-L., and Casida, J. E. (1969). 'Photodecomposition of pyrethrin I, allethrin, phthathrin and dimethrin. Modifications in the acid moiety', *J. Agric. Food Chem.*, **17**, 208–215.

Clark, J. M., and Matsumura, F. (1982). 'Two different types of inhibitory effects of pyrethroids on nerve Ca- and Ca^+ Mg-ATPase activity in the squid, *Loligo pealei'*, *Pestic. Biochem. Physiol.*, **18**, 180–190.

Clements, A. N., and May, T. E. (1977). 'The actions of pyrethroids upon the peripheral nervous system and associated organs in the locust', *Pestic. Sci.*, **8**, 661–680.

Coats, J. R., and O'Donnell-Jeffery, N. L. (1979). 'Toxicity of four synthetic pyrethroid insecticides to rainbow trout', *Bull. Environ. Contam. Toxicol.*, **23**, 250–255.

Crawford, M. J., Croucher, A., and Hutson, D. H. (1981). 'Metabolism of cis- and trans-cypermethrin in rats. Balance and tissue retention study', *J. Agric. Food Chem.*, **29**, 130–135.

Crawford, M. J., and Hutson, D. H. (1977). 'The metabolism of the pyrethroid insecticide (±)-α-cyano-3-phenoxy-benzyl 2,2,3,3-tetramethylcyclopropanecarboxylate, WL41706, in the rat', *Pestic. Sci.*, **8**, 579–599.

Cremer, J. E., Cunningham, V. J., Ray, D. E., and Sarna, G. S. (1980). 'Regional changes in brain glucose utilization in rats given a pyrethroid insecticide', *Brain Res.*, **194**, 278–282.

Cremer, J. E., Cunningham, V. J., and Seville, M. P. (1983). 'Relationships between extraction and metabolism of glucose, blood flow, and tissue blood volume in regions of rat brain', *J. Cereb. Blood Flow Metab.*, **3**, 291–302.

Cremer, J. E., and Ray, D. E. (1983). 'The influence of atropine on regional brain blood flow', Communication to the Physiological Society, November, 1983.

Cremer, J. E., and Seville, M. P. (1982). 'Comparative effects of two pyrethroids, deltamethrin and cismethrin, on plasma catecholamines and on blood glucose and lactate', *Toxicol. appl. Pharmacol.*, **66**, 124–133.

Crofton, K. M., and Reiter, L. W. (1982). 'The effects of deltamethrin, a synthetic pyrethroid insecticide on locomotor activity in rats', *The Toxicologist*, **2**, Abstract 339.

Crossland, N. O., Shires, S. W., and Bennett, D. (1982). 'Aquatic toxicology of cypermethrin. III. Fate and biological effects of spray drift deposits in fresh water adjacent to agricultural land', *Aquatic Toxicol.*, **2**, 253–270.

DHSS (1983). Pesticide Poisoning—Notes for the Guidance of Medical Practitioners, pp. 57–58, HMSO, London.

Dewar, A. J. (1981). 'Neurotoxicity testing with particular reference to biochemical methods', in *Testing for Toxicity* (ed. J. Gorrod), pp. 119–217, Taylor Francis Ltd, London.

Duclohier, H., and Georgescauld, D. (1979). 'The effects of the insecticide decamethrin on action potential and voltage-clamp currents of *Myxicola* giant axons', *Comp. Biochem. Physiol.*, **62C**, 217–223.

Edwards, R., and Milburn, P. (1984). 'Toxicity and metabolism of cypermethrin in fish compared with other vertebrates', Presented at the *Symposium on Pyrethroid Insecticides in the Environment*, 12th and 13th April 1984, Southampton University, England.

Elliott, M. (1969). 'Structural requirements for pyrethrin-like activity', *Chem. Ind. (London)*, 776–781.

Elliott, M. (1971). 'The relationship between structure and the activity of pyrethroids', *Bull. W.H.O.*, **44**, 315–324.

Elliott, M. (1974). 'Future use of natural and synthetic pyrethroids', in *Future for Insecticides: Needs and Prospects. Advances in Environmental Science and Technology*, **6**, (eds R. L. Metcalf and J. J. McKelvey, Jr.), pp. 163–193, John Wiley and Sons, New York.

Elliott, M. (1977a). 'Synthetic pyrethroids', in *Synthetic Pyrethroids* (ed. M. Elliott), pp. 1–28, American Chemical Society, Washington, DC.

Elliott, M. (1977b). 'Synthetic insecticides designed from natural pyrethrins', *Pontifae Academiae Scientiarum Scripta Varia*, **41**, 157–184.

Elliott, M. (1980). 'Established pyrethroid insecticides', *Pestic. Sci.*, **11**, 119–128.

Elliott, M., Farnham, A. W., Janes, N. F., Needham, P. H., and Pearson, B. C. (1967). '5-Benzyl-3-furylmethyl chrysanthemate: a new potent insecticide', *Nature*, **213**, 493–494.

Elliott, M., Farnham, A. W., Janes, N. F., Needham, P. H., and Pulman, D. A. (1974a). 'Insecticidally active conformations of pyrethroids', in *Mechanism of Pesticide Action* (ed. G. K. Kohn), pp. 80–91, American Chemical Society, Washington, DC.

Elliott, M., Farnham, A. W., Janes, N. F., Needham, P. H., and Pulman, D. A. (1974b). Synthetic insecticide with a new order of activity', *Nature*, **248**, 710–711.

Elliott, M., Farnham, A. W., Janes, N. F., Needham, P. H., and Pulman, D. A. (1975). 'Insecticidal activity of the pyrethrins and related compounds. VII. Insecticidal dihalovinyl analogues of *cis* and *trans* chrysanthemates', *Pestic. Sci.*, **6**, 537–542.

Elliott, M., Farnham, A. W., Janes, N. F., Needham, P. H., Pulman, D. A., and Stevenson, J. H. (1973). 'A photostable pyrethroid', *Nature*, **246**, 169–170.

Elliott, M., Farnham, A. W., Janes, N. F., and Soderlund, D. M. (1978a). 'Insecticidal activity of the pyrethrins and related compounds. Part XI. Relative potencies of isomeric cyano-substituted 3-phenoxybenzyl esters', *Pestic. Sci.*, **9**, 112–116.

Elliott, M., and Janes, N. F. (1979). 'Recent structure-activity correlations in synthetic pyrethroids', In *Advances in Pesticide Science* (ed. H. Geissbuehler), pp. 166–173, Pergamon, Oxford.

Elliott, M., Janes, N. F., and Potter, C. (1978b). The future of pyrethroids in insect control', *Ann. Rev. Entomol.*, **23**, 443–469.

Elliott, M., Janes, N. F., and Pulman, D. A. (1974c). 'The pyrethrins and related compounds. Part XVIII. Insecticidal 2,2-dimethylcyclopropanecarboxylates with new unsaturated 3-substituents', *J. Chem. Soc. Perkin Trans.*, **1**, 2470–2474.

Elliott, M., Janes, N. F., Pulman, D. A., Gaughan, L. C., Unai, T., and Casida, J. E. (1976). 'Radiosynthesis and metabolism in rats of the 1*R* isomers of the insecticide permethrin', *J. Agric. Food Chem.*, **24**, 270–276.

El-Sewedy, S. M., Mostafa, M. H., El-Bassiouni, E. A., Abdel-Rafee, A., and El-Sebae, A. H. (1982a). 'Effect of fenvalerate on kynurenine metabolizing enzymes and acid ribonuclease of mouse liver', *J. Environ. Sci. Hlth*, **B17**(5), 571–579.

El-Sewedy, S. M., Zahran, M. A., Zeidan, M. A., Mostafa, M. H., and El-Bassiouni, E. A. (1982b). 'Effect and mechanism of action of methomyl and cypermethrin insecticides on kynurenine metabolizing enzymes of mouse liver', *J. Environ. Sci. Hlth*, **B17**(5), 527–537.

FAO (1980). 'Pesticide residues in food—1979', *FAO Plant Production and Protection Paper 20*, Report, Food and Agriculture Organisation of the United Nations, Rome.

FAO (1981). 'Pesticide residues in food—1980', *FAO Plant Production and Protection Paper 26,* Evaluations, Food and Agriculture Organisation of the United Nations, Rome.

FAO (1982). 'Pesticide residues in food—1981', *FAO Plant Production and Protection Paper 37,* Report, Food and Agriculture Organisation of the United Nations, Rome.

Forshaw, P. J., and Bradbury, J. E. (1983). 'Pharmacological effects of pyrethroids on the cardiovascular system of the rats', *Eur. J. Pharmacol.,* **91,** 207–213.

Fujimoto, K., Itaya, N., Okuno, Y., Kadota, T., and Yamaguchi, T. (1973). 'A new insecticidal pyrethroid ester', *Agr. Biol. Chem.,* **37,** 2681–2682.

Gammon, D. W., Brown, M. A., and Casida, J. E. (1981). 'Two classes of pyrethroid action in the cockroach', *Pestic. Biochem. Physiol.,* **15,** 181–191.

Gammon, D. W., Lawrence, L. J., and Casida, J. E. (1982). 'Pyrethroid toxicology: protective effects of diazepam and phenobarbital in the mouse and the cockroach', *Toxicol. appl. Pharmacol.,* **66,** 290–296.

Gaughan, L. C., Ackerman, M. E., Unai, T., and Casida, J. E. (1978a). 'Distribution and metabolism of *trans-* and *cis*-permethrin in lacting Jersey cows', *J. Agric. Food Chem.,* **26,** 613–618.

Gaughan, L. C., Engel, J. L., and Casida, J. E. (1980). 'Pesticide interactions: effects of organophosphorus pesticides on the metabolism, toxicity, and persistence of selected pyrethroid insecticides', *Pestic. Biochem. Physiol.,* **14,** 81–85.

Gaughan, L. C., Robinson, R. A., and Casida, J. E. (1978b). 'Distribution and metabolic fate of *trans-* and *cis*-permethrin in laying hens', *J. Agric. Food Chem.,* **26,** 1374–1380.

Gaughan, L. C., Unai, T., and Casida, J. E. (1977). 'Permethrin metabolism in rats', *J. Agric. Food Chem.,* **25,** 9–17.

Ghiasuddin, S. M., and Soderlund, D. M. (1984). 'Hydrolysis of pyrethroid insecticides by soluble mouse brain esterases', *Toxicol. appl. Pharmacol.,* **74,** 390–396.

Ghiasuddin, S. M., and Soderlund, D. M. (1985). 'Pyrethroid insecticides: potent, stereospecific enhancers of mouse brain sodium channel activation', *Pestic. Biochem. Physiol.,* in press.

Glickman, A. H., and Casida, J. E. (1982). 'Species and structural variations affecting pyrethroid neurotoxicity', *Neurobehav. Toxicol. Teratol.,* **4,** 793–799.

Glickman, A. H., Elcombe, C. R., and Lech, J. J. (1985). 'Induction of hepatic monooxygenase activity in rats by technical and pure permethrin isomers', submitted to *Archs Toxicol.*

Glickman, A. H., Hamid, A. A. R., Ricket, D. E., and Lech, J. J. (1981). 'Elimination and metabolism of permethrin isomers in rainbow trout', *Toxicol. appl. Pharmcol.,* **57,** 88–99.

Glickman, A. H., and Lech, J. J. (1981). 'Hydrolysis of permethrin, a pyrethroid insecticide, by rainbow trout and mouse tissues *in vitro*: a comparative study, *Toxicol. appl. Pharamcol.,* **60,** 186–192.

Glickman, A. H., and Lech, J. J. (1982). 'Differential toxicity of *trans*-permethrin in rainbow trout and mice. II. Role of target organ sensitivity', *Toxicol. appl. Pharmacol.,* **66,** 162–171.

Glickman, A. H., Weitman, S. D., and Lech, J. J. (1982). 'Differential toxicity of *trans*-permethrin in rainbow trout and mice. I. Role of biotransformation', *Toxicol. appl. Pharmacol.,* **66** 153–161.

Glomot, R. (1982). 'Toxicity of deltamethrin to higher vertebrates', *Deltamethrin monograph,* pp. 109–137, Roussel Uclaf, Paris, France.

Graillot, C., and Hoellinger, H. (1982). 'Binding of two pyrethroid isomers, cismethrin and bioresmethrin, to liver proteins', *Toxicol. appl. Pharmacol.,* **66,** 313–318.

Gray, A. J. (1980). 'Toxicity and comparative pharmacokinetics of some pyrethroid insecticides in rats', PhD thesis, University of London, England.

Gray, A. J., and Connors, T. A. (1980). 'Delayed toxicity after intravenous administration of bioresmethrin to rats, *Pestic. Sci.*, **11**, 361–366.

Gray, A. J., Connors, T. A., Hoellinger, H., and Nguyen-Hoang-Nam (1980). 'The relationship between the pharmacokinetics of intravenous cismethrin and bioresmethrin and their mammalian toxicity', *Pestic. Biochem. Physiol.*, **13**, 281–293.

Gray, J. A., and Gray, A. J. (1985). 'A pharmacological comparison of muscle relaxant effects on deltamethrin toxicity in rats', Submitted to *Toxicol. appl. Pharmacol.*

Gray, A. J., and Rickard, J. (1981). 'Distribution of radiolabel in rats after intravenous injection of a toxic dose of ^{14}C-acid-, ^{14}C-alcohol- or ^{14}C-cyano-labelled deltamethrin, *Pestic. Biochem. Physiol.*, **16**, 79–85.

Gray, A. J., and Rickard, J. (1982a). 'The toxicokinetics of deltamethrin in rats after intravenous administration of a toxic dose', *Pestic. Biochem. Physiol.*, **18**, 205–215.

Gray, A. J., and Rickard, J. (1982b). 'Toxicity of pyrethroids to rats after direct injection into the central nervous system', *Neurotoxicol.*, **3**, 25–35.

Griffin, C. S. (1973). 'Mammalian toxicology of pyrethrin', *Pyrethrum Post*, **12**, 50–58.

Hoellinger, H., Sonnier, M., Pichon, J., Lecorsier, A., Do-Cao-Thang, and Nguyen-Hoang-Nam (1983). '*In vitro* covalent binding of cismethrin and bioresmethrin to hepatic proteins', *Toxicol. Lett.*, **19**, 179–187.

Holland, J. P., and Smith, I. K. (1980). 'The effect of synthetic pyrethroids on the cardiovascular system of the rat', *Toxicol. Lett.*, **5** (Sp. 1) 50, Abs. 0.70.

Honerjäger, P. (1980). 'A neurally-mediated inotropic effect of veratridine and cevadine on the isolated guineapig papillary muscle, Naunyn-Schmiedeb', *Archs Pharmacol.*, **314**, 157.

Huckle, K. R., Chipman, J. K., Hutson, D. H., and Milburn, P. (1981). 'Metabolism of 3-phenoxybenzoic acid and the enterohepatorenal disposition of its metabolites in the rat', *Drug. Metab. Dispos.*, **9**, 360–368.

Hunt, L. M., and Gilbert, B. N. (1977). 'Distribution and excretion rates of ^{14}C-labeled permethrin isomers administered orally to four lactating goats for 10 days', *J. Agric. Food Chem.*, **25**, 673–676.

Hutson, D. H. (1979). 'The metabolic fate of synthetic pyrethroid insecticides in mammals' in *Progress in Drug Metabolism* (eds J. W. Bridges and L. F. Chasseaud), Vol. 3, pp. 215–252, Wiley, Chichester.

Ivie, G. W., and Hunt, L. M. (1980). 'Metabolites of *cis*- and *trans*-permethrin in lactating goats', *J. Agric. Food Chem.*, **28**, 1131–1138.

Jacques, Y., Romey, G., Cavey, M. T., Kartalovski, B., and Lazdunski, M. (1980). Interaction of pyrethroids with the Na$^+$ channel in mammalian neuronal cells in culture', *Biochem. Biophys. Acta*, **600**, 882–897.

James, J. A. (1980). 'The toxicity of synthetic pyrethroids to mammals', in *Trends Vet. Pharmacol. Toxicol., Dev. Anim. Vet. Sci.*, **6**, 249–255.

Jao, L. T., and Casida, J. E. (1974). 'Esterase inhibitors as synergists for $(+)$-*trans*-chrysanthemate insecticide chemicals', *Pestic. Biochem. Physiol.*, **4**, 456–464.

Johnson, E. L. (1982). 'Tolerances and exemptions from tolerances for pesticide chemicals in or on raw agricultural commodities; permethrin', *Fed. Reg.*, **147**, 45008–45010.

Kato, T., Ueda, K., and Fujimoto, K. (1964). 'New insecticidally active chrysanthemates', *Agric. Biol. Chem.*, **28**, 914–915.

Kavlock, R., Chernoff, N., Baron, R., Linder, R., Rogers, E., and Carver, B. (1979). 'Toxicity studies with decamethrin, a synthetic pyrethroid insecticide', *J. Environ. Path. Toxicol.*, **2**, 751–765.

Kimmel, E. C., Casida, J. E., and Ruzo, L. O. (1982). 'Identification of mutagenic photoproducts of the pyrethroids allethrin and terallethrin', *J. Agric. Food Chem.*, **30**, 623–626.

Kolmodin-Hedman, B., Swensson, Å., and Åkerblom, M. (1982). 'Occupational exposure to some synthetic pyrethroids (permethrin and fenvalerate)', *Archs Toxicol.*, **50**, 27–33.

Lawrence, L. J., and Casida, J. E. (1982). 'Pyrethroid toxicology: mouse intracerebral structure–toxicity relationships', *Pestic. Biochem. Physiol.*, **18**, 9–14.

Lawrence, L. J., and Casida, J. E. (1983). 'Stereospecific action of pyrethroid insecticides on the γ-aminobutyric acid receptor-ionophore complex', *Science*, **221**, 1399–1401.

Leeb-Lundberg, F., and Olsen R. W. (1980). 'Picrotoxinin binding as a probe of the GABA postsynaptic membrane receptor–ionophore complex', in *Psychopharmacology and Biochemistry of Neurotransmitter Receptors* (eds H. I. Yamamura, R. W. Olsen, and E. Usdin), pp. 593–606, Elsevier, New York.

Le Quesne, P. M., Maxwell, I. C., and Butterworth, S. T. G. (1980). 'Transient facial sensory symptoms following exposure to synthetic pyrethroids: a clinical and electrophysiological assessment', *Neurotoxicol.*, **2**, 1-11.

Litchfield, M. H. (1983). 'Characterisation of the principal mammalian toxicological and biological actions of synthetic pyrethroids', in *Pesticide Chemistry: Human Welfare and the Environment* (eds J. Miyamoto and P. C. Kearney), Vol. 2, pp. 207–211, Pergamon Press, Oxford.

Lock, E. A., and Berry, P. N. (1980). 'Biochemical changes in the rat cerebellum following cypermethrin administration', in *Mechanisms of Toxicity and Hazard Evaluation* (eds B. Holmstedt, R. Lauwerys, M. Mercier, and M. Roberfroid), pp. 623–626, Elsevier/North-Holland Biomedical Press, Amsterdam.

Lock, E. A., and Berry, P. N. (1981). 'Biochemical changes in the rat cerebellum following cypermethrin administration', *Toxicol. appl. Pharmacol.*, **59**, 508–515.

Lund, A. E., and Narahashi, T. (1983). 'Kinetics of sodium channel modification as the basis for the variation in nerve membrane effects of pyrethroids and DDT analogs', *Pestic. Biochem. Physiol.*, **20**, 203–216.

McLaughlin, G. A. (1973). 'History of pyrethrum', in *Pyrethrum the Natural Insecticide* (ed. J. E. Casida), pp. 3–16, Academic Press, New York.

Malhotra, S. K., Van Heertum, J. C., Larson, L. L., and Ricks, M. J. (1981). 'Dowco 417: a potent synthetic pyrethroid insecticide', *J. Agric. Food Chem.*, **29**, 1287–1289.

Marei, A. E.-S. M., Ruzo, L. O., and Casida, J. E. (1982). 'Analysis and persistence of permethrin, cypermethrin, deltamethrin, and fenvalerate in the fat and brain of treated rats', *J. Agric. Food Chem.*, **30**, 558–562.

Matsuo, T., Itaya, N., Mizutomi, T., Ohno, N., Fujimoto, K. Okuno, Y., and Yoshioka, H. (1976). '3-Phenoxy-α-cyano-benzyl esters, the most potent synthetic pyrethroids', *Agric. Biol. Chem.*, **40**, 247–249.

Matsuoka, A., Hayashi, M., and Ishidate, M. (1979). 'Chromosomal aberration tests on 29 chemicals combined with S9 mix *in vitro*', *Mut. Res.*, **66**, 277–290.

Mauck, W. L., Olson, L. E., and Marking, L. L. (1976). 'Toxicity of natural pyrethroids and five pyrethroids to fish', *Archs Environ. Contam. Toxicol.*, **4**, 18–29.

Miller, T. A., and Adams, M. E. (1982). 'Mode of action of pyrethroids' in *Insecticide Mode of Action* (ed. J. R. Coats), pp. 3–27, Academic Press, New York.

Mitchell, J. C., Dupuis, G., and Towers, G. H. N. (1972). 'Allergic contact dermatitis from pyrethrum (*Chrysanthemum* spp). The roles of pyrethrosin, a sesquiterpene lactone, and of pyrethrin II, *Br J. Dermatol.*, **86**, 568–573.

Miyamoto, J. (1976). 'Degradation, metabolism and toxicity of synthetic pyrethroids', *Environ. Hlth Persp.*, **14**, 15–28.

Miyamoto, J., Nishida, T., and Ueda, K. (1972). 'Metabolic fate of resmethrin, 5-benzyl-3-furymethyl *dl-trans*-chrysanthemate in the rat', *Pestic. Biochem. Physiol.*, **1**, 293–306.

Narahashi, T. (1971). 'Mode of action of pyrethroids', *Bull. W.H.O.*, **44**, 337–345.

Narahashi, T. (1976). 'Nerve membrane as a target of pyrethroids', *Pestic. Sci.*, **7**, 267–272.

Narahashi, T. (1983). 'Neurophysiological study of pyrethroids: molecular and membrane mechanism of action', In *Pesticide Chemistry: Human Welfare and the Environment* (eds J. Miyamoto and P. C. Kearney), Vol. 2, pp. 179–186, Pergamon, Oxford.

Nassif, M., Brooke, J. P., Hutchinson, D. B. A., Kamel, O. M., and Savage, E. A. (1980). 'Studies with permethrin against bodylice in Egypt', *Pestic. Sci.*, **11**, 679–684.

Navarro-Delmasure, C. (1981). 'Recherches sur le méchanisme d'action pharmacologique de la decamethrine: Role de la neurotranmission adrenergique et des prostaglandines', Thesis, Université Paul Sabatier, Toulouse, France.

Nicholson, R. A., Wilson, R. C., Potter, C., and Black, M. H. (1983). 'Pyrethroid- and DDT-evoked release of GABA from the nervous system *in vitro*', in *Pesticide Chemistry: Human Welfare and the Environment* (eds J. Miyamoto and P. C. Kearney), Vol. 3, pp. 75–78, Pergamon, Oxford.

O'Brien, R. D. (1967). 'Pyrethroids', in *Insecticides—Action and Metabolism* (ed. R. D. O'Brien), pp. 164–172, Academic Press, New York.

Ohno, N., Fujimoto, K., Okuno, Y., Mizutoni, T., Hirano, M., Itaya, N., Honda, T., and Yoshioka, H. (1974). 'A new class of pyrethroid insecticides; α-substituted phenylacetic acid esters', *Agric. Biol. Chem.*, **38**, 881–883.

Páldy, A. (1981). 'Examination of the mutagenic effect of synthetic pyrethroids on mouse bone-marrow cells', *Proc. Hung, Ann. Meet. Biochem.*, **21**, 227–228.

Parkin, P. L., and Le Quesne, P. M. (1982). 'Effect of a synthetic pyrethroid deltamethrin on excitability changes following a nerve impulse', *J. Neurol. Neurosurg. Psychiat.*, **45**, 337–342.

Payne, N. B., Herzberg, G. R., and Howland, J. L. (1973). 'Influence of some insecticides on the ATPase of mouse liver mitochondria', *Bull. Environ. Contam. Toxicol.*, **10**, 365–367.

Poláková, H., and Vargová, M. (1983). 'Evaluation of the mutagenic effects of deltamethrin: cytogenetic analysis of bone marrow', *Mut. Res.*, **120**, 167–171.

Poulos, L., Athanaselis, S., and Coutselinis, A. (1982). 'Acute intoxication with cypermethrin (NRDC 149)', *J. Toxicol. Clin. Toxicol.*, **19**, 519–520.

Ray, D. E. (1980). 'An EEG investigation of decamethrin-induced choreoathetosis in the rat', *Exp. Brain Res.*, **38**, 221–227.

Ray, D. E. (1982a). 'The contrasting actions of two pyrethroids (deltamethrin and cismethrin) in the rat', *Neurobehav. Toxicol. Teratol.*, **4**, 801–804.

Ray, D. E. (1982b). 'Changes in brain blood flow associaed with deltamethrin-induced choreoathetosis in the rat', *Exp. Brain Res.*, **45**, 269–276.

Ray, D. E., and Cremer, J. E. (1979). 'The action of decamethrin (a synthetic pyrethroid) on the rat', *Pestic. Biochem. Physiol.*, **10**, 333–340.

Rickard, J., and Brodie, M. E. (1985). 'Correlation of blood and brain levels of the neurotoxic pyrethroid deltamethrin with the onset of symptoms in rats', *Pestic. Biochem. Physiol.*, **23**, 143–156.

Rickett, F. E., and Tyszkiewicz, K. (1973). Pyrethrum dermatitis. II. The allergenicity of pyrethrum oleorsin and its cross-reactions with the saline extract of pyrethrum flowers. *Pestic. Sci.*, **4**, 801–810.

Rose, G. P., and Dewar, A. J. (1983). 'Intoxication with four synthetic pyrethroids fails to show any correlation between neuromuscular dysfunction and neurobiochemical abnormalities in rats', *Archs Toxicol.*, **53**, 297–316.

Ruigt, Gé S. F. (1984). 'Pyrethroids', *Comp. Insect Physiol. Biochem. Pharmacol.* pp. 10–75.

Ruscoe, C. N. E. (1977). 'The new NRDC pyrethroids as agricultural insecticides', *Pestic. Sci.*, **8**, 236–242.

Ruzo, L. O., and Casida, J. E. (1977). 'Metabolism and toxicology of pyrethroids with dihalovinyl substituents', *Environ. Hlth Persp.*, **21**, 285–292.

Ruzo, L. O., Engel, J. L., and Casida, J. E. (1979). 'Decamethrin metabolites from oxidative, hydrolytic, and conjugative reactions in mice', *J. Agric. Food Chem.*, **27**, 725–731.

Saxena, S. C., and Bakre, P. P. (1976). 'Toxicity of pyrethrum to blue rock pigeon', *Pyrethrum Post*, **14**, 47–48.

Saxena, P., and Saxena, S. C. (1973). 'Effect of pyrethrum on body and organ weights, food consumption and faeces production of the house sparrow, *Passer domesticus (L)*', *Pyrethrum Post*, **12**, 76.

Schechter, M. S., Green, N., and La Forge, F. B. (1949). 'Constituents of pyrethrum flowers. XXIII. Cinerolone and the synthesis of related cyclopentenolones', *J. Am. Chem. Soc.*, **71**, 3165–3173.

Settlemire, C. T., Hutson, A. S., Jacobs, L. S., Harvey, J. C., and Howland, J. L. (1974). 'Action of some insecticides on membranes of mouse liver mitochondria', *Bull. Environ. Contam. Toxicol.*, **11**, 169–173.

Smith, P. R. (1980). 'The effect of cismethrin on rat dorsal root potentials,' *Eur. J. Pharmacol.*, **66**, 125–128.

Soderlund, D. M. (1979). 'Pharmacokinetic behaviour of enantiomeric pyrethroid esters in the cockroach, *Periplaneta americana (L)*', *Pestic. Biochem. Physiol.*, **12**, 38–48.

Soderlund, D. M., and Casida, J. E. (1977a). 'Substrate specificity of mouse-liver microsomal enzymes in pyrethroid metabolism', in *Synthetic Pyrethroids* (ed. M. Elliott), pp. 162–172, American Chemical Society, Washington, DC.

Soderlund, D. M., and Casida, J. E. (1977b). 'Effects of pyrethroid structure on rates of hydrolysis by mouse liver microsomal enzymes', *Pestic. Biochem. Physiol.*, **7**, 391–401.

Soderlund, D. M., Ghiasuddin, S. M., and Helmuth, D. W. (1983). 'Receptor-like stereospecific binding of pyrethroid insecticide to mouse brain membranes', *Life. Sci.*, **33**, 261–267.

Spencer, F., and Berhane, Z. (1982). 'Uterine and fetal characteristics in rats following a post-implantational exposure to permethrin', *Bull. Environ. Contam. Toxicol.*, **29**, 84–88.

Springfield, A. C., Carlson, G. P., and DeFeo, J. J. (1973). 'Liver enlargement and modification of hepatic microsomal drug metabolism in rats by pyrethrum', *Toxicol. appl. Pharmacol.*, **24**, 298–308.

Squires, R. F., Casida, J. E., Richardson, M., and Saederup, E. (1983). [35S]*t*-Butylbicyclophosphorothionate binds with high affinity to brain-specific sites coupled to γ-aminobutyric acid A and ion recognition sites', *Mol. Pharmacol.*, **23**, 326–336.

Staatz, C. G., Bloom, A. S., and Lech, J. J. (1982a). 'A pharmacological study of pyrethroid neurotoxicity in mice', *Pestic. Biochem. Physiol.*, **17**, 287–292.

Staatz, C. G., Bloom, A. S., and Lech, J. J. (1982b). 'Effect of pyrethroids on [³H]kainic acid binding to mouse forebrain membranes', *Toxicol. appl. Pharmacol.* **64**, 566–569.

Staatz, C. G., and Hosko, M. J. (1985). 'Interaction of pyrethroids with mammalian spinal neurones', submitted to *Pestic. Biochem. Physiol.*

Takahashi, M., and Le Quesne, P. M. (1982). 'The effects of the pyrethroids deltamethrin and cismethrin on nerve excitability in rats', *J. Neurol. Neurosurg. Psychiat.*, **45**, 1005–1011.

Tucker, S. B., and Flannigan, S. A. (1983). 'Cutaneous effects from occupational exposure to fenvalerate', *Archs Toxicol.*, **54**, 195–202.

Ueda, K., Gaughan, L. C., and Casida, J. E. (1975a). 'Metabolism of four resmethrin isomers by liver microsomes', *Pestic. Biochem. Physiol.*, **5**, 280–294.

Ueda, K., Gaughan, L. C., and Casida, J. E. (1975b). 'Metabolism of (+)-*trans*- and (+)-*cis*-resmethrin in rats', *J. Agric. Food Chem.*, **23**, 106–115.

Verschoyle, R. D., and Aldridge, W. N. (1980). 'Structure-activity relationships of some pyrethroids in rats', *Archs Toxicol.*, **45**, 325–329.

Verschoyle, R. D., and Barnes, J. M. (1972). 'Toxicity of natural and synthetic pyrethrins to rats', *Pestic. Biochem. Physiol.*, **2**, 308–311.

Vijverberg, H. P. M., Ruigt, G. S. F., and van den Bercken, J. (1982). 'Structure-related effects of pyrethroid insecticides on the lateral-line sense organ and on peripheral nerves of the clawed frog, *Xenopus laevis*', *Pestic. Biochem. Physiol.*, **18**, 315–324.

Vijverberg, H. P. M., and van den Bercken, J. (1982). 'Action of pyrethroid insecticides on the vertebrate nervous system', *Neuropathol. appl. Neurobiol.*, **8**, 421–440.

Vijverberg, H. P. M., van der Zalm, J. M., van Kleef, R. G. D. M., and van den Bercken, J. (1983). 'Temperature- and structure-dependent interaction of pyrethroids with the sodium channels in frog node of Ranvier', *Biochem. Biophys. Acta*, **728**, 73–82.

Wang, Y.-L., Jin, X.-P., Jiang, X.-Z., Lin, H.-F., Li, F., Jin, P.-H., Yang, X., and Geng, J.-B. (1981). 'Study on the percutaneous absorption of four radioactively labelled agrochemicals', *Acta Acad. Med. Prim. Shanghai*, **8**, 365–370.

White, I. N. H., Verschoyle, R. D., Moradian, M. H. and Barnes, J. M. (1976). 'The relationship between brain levels of cismethrin and bioresmethrin in female rats and neurotoxic effects', *Pestic Biochem. Physiol.*, **6**, 491–500.

Williams, C. H. (1973). 'Tests for possible teratogenic, carcinogenic, mutagenic, and allergenic effects of pyrethrum', in *Pyrethrum the Natural Insecticide* (ed. J. E. Casida), pp. 167–176, Academic Press, New York.

World Health Organization (1979). 'Safe use of pesticides. Third report of the WHO expert committee on vector biology and control', *Technical Report Series No 634*, WHO, Geneva.

Wouters, W., and van den Bercken, J. (1978). 'Action of pyrethroids', *Gen. Pharmacol.*, **9**, 387–398.

Wszolek, P. C., Hogue, D. E., and Lisk, D. J. (1981). 'Accumulation of fenvalerate insecticide in lamb tissues', *Bull. Environ. Contam. Toxicol.*, **27**, 869–871.

Zitko, V., Carson, W. G., and Metcalfe, C. D. (1977). 'Toxicity of pyrethroids to juvenile atlantic salmon', *Bull. Environ. Contam. Toxicol.*, **18**, 35–41.

Zitko, V., Mcleese, D. W., Metcalfe, C. D., and Carson, W. G. (1979). 'Toxicity of permethrin, decamethrin, and related pyrethroids to salmon and lobster', *Bull. Environ. Contam. Toxicol.*, **21**, 338–343.

Insecticides
Edited by D. H. Hutson and T. R. Roberts
© 1985 John Wiley & Sons Ltd

CHAPTER 6

The metabolism and toxicity of insecticides in fish

R. Edwards and P. Millburn

INTRODUCTION

Many pesticides and environmental pollutants are highly toxic to both freshwater and marine fish. In particular, these aquatic vertebrates are extremely sensitive to very low concentrations of dissolved insecticides (see Table 1). In the case of the natural insecticide rotenone, a toxic plant product which inhibits mitochondrial electron transport and is used by South American Indians as a fish poison, this property has been exploited for many years as a registered piscicide (Dawson et al., 1983).

It was thought that the sensitivity of fish to xenobiotics was due to the absence of any detectable capacity for detoxification (Brodie and Maickel, 1962). However, subsequent studies, both in vitro and in vivo, have shown that fish can carry out a range of biotransformation reactions, though generally with lower overall activities than comparable mammalian systems (see Sieber and Adamson, 1977). This subject has been reviewed by Adamson (1967), Buhler and Rasmusson (1968), Chambers and Yarbrough (1976), Bend and James (1978) and Allen et al. (1979), and the relationship between the metabolism and toxicity of several xenobiotics in fish assessed by Lech and Bend (1980). Thus,

Table 1 Toxicity of representative insecticides to four species of fish, expressed as the 96 h LC_{50} in μg/l, in comparison with their lipophilic character expressed as log of the n-octanol/water partition coefficient (log $P_{o/w}$)

		LC_{50}	(96 h)	μg/l(μmol/l)	
Family	Species	DDT[a]	Methyl-[a] parathion	Carbaryl[a]	Cypermethrin[b]
Cyprinidae	Carp	10(0.028)	7130(27)	5280(26)	0.9(0.0022)
Centrarchidae	Bluegill	8(0.023)	5720(22)	6760(36)	—
Salmonidae	Rainbow trout	7(0.020)	2750(10)	4340(22)	0.5(0.0012)
Percidae	Perch	9(0.025)	3060(12)	745(3.7)	—
Log $P_{o/w}$		6.12	1.38	2.39	6.606

[a] Macek and McAllister, 1970.
[b] Stephenson, 1982.

several piscine species are known to hydrolyse esters, to carry out mixed-function oxidase reactions including oxidation and dealkylation and reduction reactions, and subsequently to conjugate the resulting phase I products with glycine, taurine, sulphate, glucuronic acid and glutathione. Mercapturic acid biosynthesis in fish using the winter flounder (*Pseudopleuronectes americanus*), a marine teleost, has been investigated recently (Yagen *et al.*, 1984). All the reactions of this pathway occur in flounder *in vivo*. The substrate specificities and subunit compositions of the hepatic glutathione *S*-transferases of rainbow trout have been reported (Ramage and Nimmo, 1984). This review describes the routes and rates of metabolism of insecticides in fish and discusses the relationship between biotransformation and toxicity.

XENOBIOCHEMISTRY OF FISH

Routes of exposure to xenobiotics

In many cases, the potential bioaccumulation of dissolved foreign compounds into fish can be predicted from the relative lipophilicity of the compound as measured by its partition coefficient (*P*-value) between *n*-octanol and water (Neely *et al.*, 1974). Many synthetic insecticides are apolar (see Table 1) and when dissolved in the surrounding water rapidly accumulate in fish. Uptake may occur across the membranes of the gills, through the intestine and by dermal absorption. In studies with the northern pike (*Esox lucius*), Balk *et al.* (1984) demonstrated that the lipophilic environmental pollutant benzo(*a*)pyrene is absorbed primarily through the gills when dissolved in the water, but if this polycyclic aromatic hydrocarbon is absorbed onto food, uptake is largely intestinal. Similarly, Addison (1976) has shown that ingestion of organochlorine compounds is a significant route of exposure in fish.

Difficulties in determining the relative importance of gill, dermal and intestinal uptake, together with the susceptibility of fish to many xenobiotics and problems encountered in the collection of excreted metabolites, prevent conventional *in vivo* metabolism studies from being carried out in aquatic organisms. However, Crosby *et al.*, (1979) have suggested some ways in which this may be overcome, and McKim and Goeden (1982) have described a technique for measuring the efficiency of uptake of endrin (6; Figure 2, Chapter 1) in brook trout (*Salvelinus fontinalis*). This procedure uses a partitioned chamber in which the fish is immobilized, and this allows the dissolved xenobiotic to pass through the gills under controlled conditions. Thus, the delivered dose can be accurately measured, and the rates of uptake and elimination of the compound monitored by collecting blood and urine samples following catheterization. Factors influencing absorption across the blood–water barrier, such as the degree of water-oxygenation, can also be examined. In their study with endrin, McKim and Goeden (1982) showed that the efficiency of insecticide uptake was decreased with the reduction in oxygen utilization by the trout.

Routes of elimination of metabolites

Biotransformation in fish appears to be carried out largely in the liver (Buhler and Rasmusson, 1968). Hepatic mixed-function oxidases and UDP-glucuronyltransferases have been studied in some detail (see Chambers and Yarbrough, 1976). In addition, detoxification of insecticides takes place in the plasma (Glickman and Lech, 1981), kidneys (Fukami *et al.*, 1969; Glickman and Lech, 1981), intestine (Fukami *et al.*, 1969; Edwards, 1984) and intestinal microflora (Wedemeyer, 1968). For some compounds extrahepatic metabolism may exceed that of the liver (see the pyrethroids section below).

Following biotransformation, the metabolites are eliminated in the urine, or *via* the bile into the faeces, or by diffusion through the gills into the surrounding water. The gill is considered to be a poor excretory unit (Chambers and Yarbrough, 1976) and would certainly not eliminate lipophilic compounds, as originally suggested by Brodie and Maickel (1962). Elimination of polar metabolites in the urine has been observed in catheterized rainbow trout, *Salmo gairdneri* (Glickman *et al.*, 1981), and has been inferred in other studies from the presence of metabolic products in the aquarium water. The renal excretion of xenobiotics in fish has been reviewed by Pritchard and Bend (1984). The bile is the major route of elimination for the metabolites of a number of lipophilic xenobiotics and it has been proposed as a sensitive monitor of aquatic pollution (Statham *et al.*, 1976; Lech and Bend, 1980), since the accumulated products in the gall bladder can be present at very high concentrations compared with that in the surrounding water or other fish tissues. For example, in a study with radiolabelled carbaryl, DDT and the selective sea lamprey larvicide 3-trifluoromethyl-4-nitrophenol, which is excreted in fish bile as a glucuronide (Lech, 1973), the ratio of the concentration of radioactivity in the bile of trout to that in the surrounding water was 947, 124 and 1061, respectively (Lech *et al.*, 1973). Subsequent voiding of the gall bladder contents is dependent upon the ingestion of food (Balk *et al.*, 1984) and thus considerable quantities of metabolites are found in this organ in unfed fish.

As in mammalian species, the relative proportions of the dose eliminated in the bile and urine appear to be dependent on the amphipathic nature of the metabolites, as determined by their molecular weight and the presence of polar anionic groups (see Millburn, 1970, 1976; Hirom *et al.*, 1976). Thus, the glucuronic acid conjugates of *m*-cresol, *o*-chlorophenol and 1-naphthol (mol. wts 284, 305 and 320, respectively) were present at much higher concentrations than the corresponding sulphates (mol. wts 188, 209 and 234) in the bile of 12 freshwater fish species (Layiwola *et al.*, 1983). Furthermore, phenyl sulphate (mol. wt 270) was not identified as a biliary product in goldfish, though it was present in the surrounding water (Nagel and Urich, 1983). In the rat, compounds with molecular weights below about 300 are excreted predominantly in the urine, since the threshold molecular weight for appreciable biliary excretion (i.e. more than 10% of the dose) is 325 ± 50 in this species (Millburn *et al.*, 1967). This threshold varies with mammalian species, being 400 ± 50 for the guinea pig and 475 ± 50 for the rabbit (Hirom *et al.*, 1972). The value for the molecular weight threshold in fish is unknown.

In mammals, compounds excreted in bile can undergo enterohepatic circulation (Parker *et al.*, 1980). Thus, metabolites of the environmental procarcinogen benzo(*a*)pyrene persist *in vivo* in both the rat (Chipman *et al.*, 1981) and the rabbit (Chipman *et al.*, 1982) via enterohepatic recycling. Such metabolites are found in the gall bladder bile of the adult English sole (*Parophrys vetulus*) 24 hours after an oral dose of benzo(*a*)pyrene (Varanasi *et al.*, 1982), and the gall bladder is a major storage site for polycyclic aromatic hydrocarbons and their metabolites in mudsuckers (*Gillichthys mirabilis*), tidepool sculpins (*Oligocottus maculosus*) and sand dabs (*Citharichthys stigmaeus*) (Lee *et al.*, 1972). Such observations raise the possibility that enterohepatic circulation of foreign organic chemicals also takes place in fish, but the extent to which this may occur is unknown at present.

Xenobiotics excreted in bile can undergo enterohepatorenal disposition instead of enterohepatic circulation, e.g. 3-phenoxybenzoic acid, a major metabolite of the photostable pyrethroid insecticides (Huckle *et al.*, 1981). In the rat, 3-phenoxybenzoic acid is hydroxylated and then conjugated to form the ester and ether glucuronides of 3-(4-hydroxyphenoxy)benzoic acid. These glucuronic acid conjugates are excreted in bile and deconjugated in the intestine releasing 3-(4-hydroxyphenoxy)benzoic acid. This phenol is absorbed, conjugated with sulphate in the liver and the sulphate conjugate is eliminated in the urine. In this example, a urinary metabolite (a sulphate) is derived by enterohepatorenal disposition from the primary metabolites (glucuronides) excreted in bile. This sequence of events may occur with certain compounds in other vertebrates apart from mammals. When fish are exposed to phenol, for example, phenyl glucuronide is found in the gall bladder (in bile) but most of the metabolite found in the aquarium water is phenyl sulphate (see Nagel and Urich, 1983).

In fish, bile may also function as an *excretory fluid* for certain lipid-soluble insecticides in the following way. The bile acid conjugates are amphipathic molecules with detergent properties; they make cholesterol and phospholipids soluble in bile by incorporating these lipids into micelles, and they emulsify dietary fats in the intestine during digestion. Bile salts may also emulsify lipophilic organic compounds ingested by fish. Thus, the pyrethroid insecticide cypermethrin, following administration to rainbow trout in their food pellets, is readily eliminated from the intestines of the fish dispersed in secreted bile (Edwards, 1984). This results in a lowered toxicity of the pyrethroid to the fish compared with that following exposure of trout to cypermethrin present in aquarium water. In the latter situation the insecticide rapidly penetrates the fish via the gills.

Factors influencing biotransformation and toxicity

Both the rates and routes of biotransformation are, to a large extent, dependent upon the species (Nagel, 1983) and strain (Pederson *et al.*, 1976) of fish used, as well as on the size, age and sex of the individual animal under study (Forlin, 1980; Stegeman and Chevion, 1980). In this respect fish are similar to mammals (see Williams and Millburn, 1975). Strain-dependent differences may arise from the previous exposure of particular populations of fish to environmental pollutants which induce mixed-function oxidase activity (Pederson *et al.*, 1976). The importance of hepatic enzyme induction in regulating xenobiotic biotransformation and toxicity in fish has been stressed by Lech and Bend (1980), and the environmental induction of cytochrome P450 systems in fish is reviewed by Binder *et al.* (1984). The route of administration of the xenobiotic has also been shown to influence the rate of detoxification (Nagel, 1983). Thus, in goldfish (*Carassius auratus*) orally administered phenol was metabolized more slowly to polar conjugates than when it was dissolved in the surrounding water. Furthermore, increasing the concentration of phenol in the aquarium water decreased the proportion of sulphate conjugates formed while increasing that of phenyl glucuronide. This change in conjugation pattern on increasing the dose of phenol administered to the fish, is similar to that often seen with phenolic substrates in mammals (e.g. phenol itself in non-human primates; Mehta *et al.*, 1978). Sulphation and glucuronidation are competing conjugation reactions for phenols *in vivo*. The former is a readily saturable process with zero-order kinetics at high doses (Bray *et al.*, 1952b), whereas the latter exhibits first-order kinetics and is not readily saturable (Bray *et al.*, 1952a).

The ambient temperature of the water would also be expected to influence the rates of xenobiotic biotransformation, since fish are poikilothermic. Zinck and Addison (1975) investigated the effect of temperature on the rate of conversion of the nerve poison DDT (**1**, Chapter 1) into inactive DDE (**2**, Chapter 1) in brook trout, and found that fish maintained at 18 °C formed twice as much DDE as trout kept at 2 °C. However, recent studies suggest that seasonal temperature variations may not influence hepatic biotransformation so markedly in acclimatized fish. Thus, Curtis (1983) has shown that the biliary excretion of phenolphthalein as its glucuronic acid conjugate is not significantly affected by variations in water temperature in preacclimated steelhead trout (*Salmo gairdneri irideus*). Similarly, monooxygenase and UDP-glucuronyltransferase activities show a compensatory response to environmental changes in temperature, and this appears to be due to changes in the molecular structure of these enzymes and may be mediated by the sex hormones (Koivusaari and Andersson, 1984). Differential temperature effects on hepatic biotransformation are dependent upon the species of fish under study. Thus, the goldfish appears to have a higher optimal temperature for mixed-function oxidase activity than other commonly studied fish species such as rainbow trout (Maemura and Omura, 1983).

Another environmental factor which has been shown to affect hepatic biotransformation in fish is the pH of the surrounding water (Laitinen *et al.*, 1982). Thus, when the pH was lowered from 6.7 to 3, aryl hydrocarbon hydroxylase activity was significantly reduced in laveret (*Coregonus lavaretus*) and splake (a hybrid offspring of two trout species; *Salvelinus fontinalis* × *Salvelinus namaycush*). This is of particular relevance to pesticide usage in areas where aquatic acidification is occurring due to pollution brought about by acid rain.

The newest class of widely used insecticides, the synthetic pyrethroids (see Chapter 1) are currently undergoing hazard assessment in relation to their effects on aquatic environments (Kingsbury and Kreutzweiser, 1985; Stephenson, 1985). Pyrethroids are highly lipophilic, have low water solubility, and in most aquatic environments are rapidly absorbed onto bottom sediments, aquatic vegetation, and other surfaces. They are also readily biodegradable. Thus, it is unlikely that these insecticides will pose a hazard to fish when they are used at commercial application rates (5–50 g/ha) despite the fact that pyrethroids are very toxic *per se* to fish, acting on the central nervous system (see Pyrethroid section below). Therefore, environmental factors such as those given above must be known in assessing the biohazard of pesticides under field conditions.

METABOLISM AND TOXICITY OF INSECTICIDES IN FISH

Some synthetic insecticides are highly toxic to fish (Table 1). In general, the Salmonidae are the most sensitive to the toxic effects of these compounds

(Macek and McAllister, 1970). Due to the experimental difficulties of working with fish *in vivo*, most metabolism studies, which have attempted to correlate biotransformation with toxicity, have been carried out using liver preparations. In addition, some workers have investigated the synergistic effects on fish of insecticides and metabolic inhibitors, which have been dissolved in aquarium water. Piperonyl butoxide and β-diethylaminoethyldiphenyl-*n*-propylacetate-HCl (SKF 525A) are the inhibitors most commonly used to examine the role of monooxygenase enzymes in detoxification (e.g. piperonyl butoxide inhibited the metabolism of pesticides in green sunfish, *Lepomis cyanellus*; it inhibited the epoxidation of aldrin to dieldrin, the *O*-dealkylation of methoxychlor, and the *N*-dealkylation of trifluralin; Reinbold and Metcalf, 1976). Additionally, organophosphate compounds such as tri-*o*-tolyl phosphate have been used to inhibit esterases so as to assess their role in metabolism. In other cases, investigations have been limited to the identification of metabolites which have accumulated in fish tissues during exposure to the dissolved insecticide. The available metabolic and toxicological data are now reviewed in the approximate chronological order of the major usage of each class of pesticide.

Rotenone

The natural insecticide rotenone (**1**), which is isolated from derris root, inhibits mitochondrial respiration by blocking the NADH–dehydrogenase complex of the respiratory chain. As such, it exhibits no species-dependent variations in activity at its site of action, but it is selectively toxic to insects and fish. To examine the role of phase I detoxification enzymes in determining this selectivity, Fukami *et al.* (1969) incubated ^{14}C-labelled rotenone with microsomes prepared from the brain, kidney, liver and intestine of carp (*Cyprinus carpio*) and roach (*Leuciscus rutilus*), and compared the metabolites formed with those obtained from rodent and insect tissues. The livers of the fish species were more active than their other tissues, although oxidative activity was detected in all preparations. The major metabolite in carp liver microsomal incubations was 6′,7′-dihydro-6′,7′-dihydroxyrotenone (**2**) and rotenolone II (**3**) derivatives. Addition of the cytosolic fraction to the incubations led to the extensive formation of water-soluble products, which were presumably conjugates. The extent of inhibition of metabolite formation produced by piperonyl butoxide and SKF 525A was similar in carp and rat liver. When rotenone was administered orally to young carp, the metabolites produced after 22 hours were qualitatively similar to those formed *in vitro*, and showed negligible activity as inhibitors of mitochondrial respiration. Fukami *et al.* (1969) concluded that the fish were capable of carrying out a similar range of biotransformation reactions to the mammals, but that selective toxicity arose because of the low levels of mixed-function oxidase activity in the piscine species. This observation was subsequently supported by a case of rotenone tolerance in mosquitofish (*Gambusia*

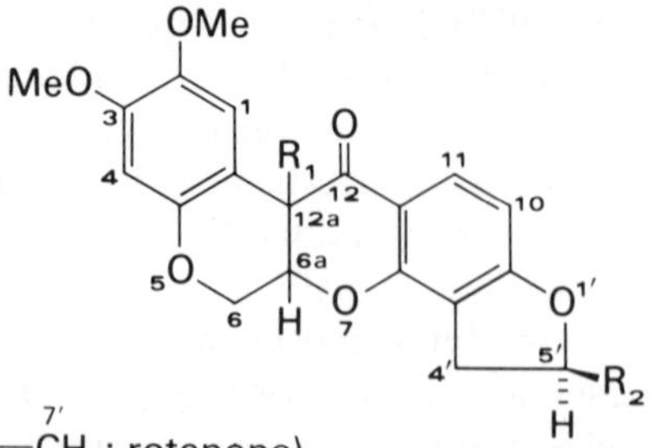

(**1**, R_1 = H; R_2 = C(Me)=CH$_2$; rotenone)

(**2**, R_1 = H; R_2 = C(Me)OH—CH$_2$OH; 6',7'-dihydro-6',7'-dihydroxy-rotenone)

(**3**, R_1 = ---OH; R_2 = C(Me)=CH$_2$; rotenolone II)

affinis). This appeared to be due largely to increased levels of microsomal enzymes (Fabacher and Chambers, 1972a).

Organophosphates

The organophosphorothionate insecticides (see Chapter 1) can undergo bio-transformation reactions which lead to activated products, e.g. the oxidation of the P=S group to the P=O derivative [for example the conversion of parathion (**10**, Chapter 1) to paraoxon (**9**, Chapter 1)], and also to metabolites which possess reduced or negligible activity as acetylcholinesterase inhibitors (see Figure 5, Chapter 1). In a comparative study with piscine and terrestrial vertebrates, Potter and O'Brien (1964) showed that liver slices from brook trout and mud puppies were less active in forming paraoxon from parathion than were liver slices from mammalian and avian species. However, this reduced ability to form an activated product is counterbalanced by a low detoxification rate. Thus, low levels of hydrolytic activity towards paraoxon and malathion were found in sunfish (*Lepomis gibbosus*) and other fish, in comparison with rodents (Murphy, 1966). However, the activity of parathion nitroreductase was similar in the piscine and mammalian species (Hitchcock and Murphy, 1967).

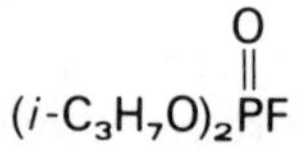

$$(i\text{-}C_3H_7O)_2\overset{\displaystyle O}{\overset{\|}{P}}F$$

(**4**, DFP)

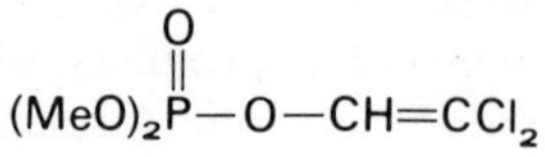

$$(MeO)_2\overset{\displaystyle O}{\overset{\|}{P}}-O-CH=CCl_2$$

(**5**, dichlorvos)

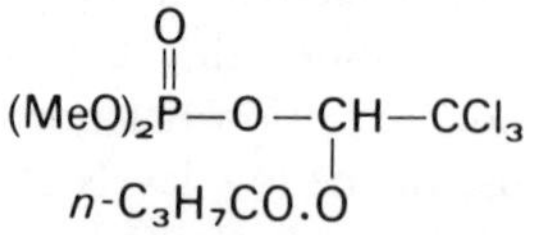

$$(MeO)_2\overset{\displaystyle O}{\overset{\|}{P}}-O-\underset{\underset{\displaystyle n\text{-}C_3H_7CO.O}{|}}{CH}-CCl_3$$

(**6**, butonate)

$$(MeO)_2\overset{\displaystyle O}{\overset{\|}{P}}-O-\underset{\underset{\displaystyle Me}{|}}{C}=CHCOOMe$$

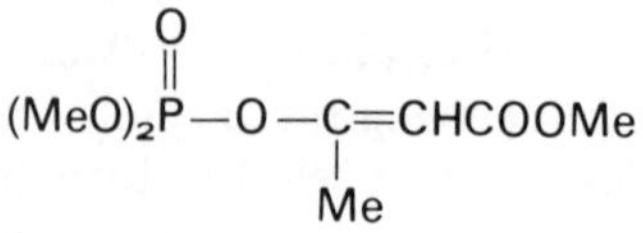

(**7**, mevinphos)

Hogan and Knowles (1968) examined the organophosphate-hydrolysing phosphatases in bluegills (*Lepomis macrochirus*) and channel catfish (*Ictalurus punctatus*) and found that they have similar properties to the corresponding enzymes in mammalian species (i.e. stimulated by Mn^{2+} and inhibited by sulphydryl reagents). There were, however, considerable species-dependent differences in the activities of the liver enzymes towards various organophosphates, although in both fish species the rates of hydrolysis followed the order: di-isopropyl phosphorofluoridate (DFP; **4**) > dichlorvos (**5**) > butonate (**6**) > paraoxon > mevinphos (**7**) (paraoxon and mevinphos were hydrolysed only by channel catfish). In a subsequent study (Hogan and Knowles, 1972), the oxidative metabolism of diazinon (**12**; Chapter 1) was investigated in hepatic subcellular preparations from catfish and bluegills. The microsomes were the most active in forming both diazoxon (the P=O analogue of diazinon) and water-soluble metabolites, namely diethyl phosphorothioic acid and diethyl phosphoric acid (ratio 3:1) resulting from the cleavage of diazinon and diazoxon, respectively.

Few *in vivo* studies have been reported with organophosphate compounds, though Smith *et al.* (1966) have studied the fate in goldfish of *O,O*-diethyl-*O*-(3,5,6-trichloro-2-pyridyl) phosphorothionate (chlorpyrifos **1.1**) labelled with ^{14}C in the 2 and 6 positions of the pyridinol ring. The dissolved organophosphate was taken up rapidly by the fish over an initial 10-hour period and then, following a period of equilibrium (1.5 days), the concentrations of radioactivity in the water began to increase, indicating that water-soluble, detoxification products were being excreted. Following exposure to the insecticide in a model ecosystem containing plants and soil, the fish were killed, their tissues extracted

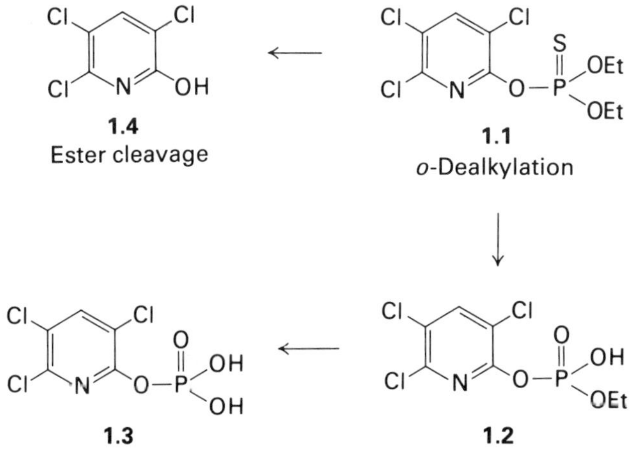

Figure 1 Metabolism of *O,O*-diethyl *O*-(3,5,6-trichloro-2-pyridyl)phosphoro-thionate (chlorpyrifos) in goldfish

with acetone and water and the metabolites analysed by paper and thin-layer chromatography. The majority of the tissue radioactivity was identified as parent compound (**1.1**), though 3,5,6-trichloro-2-pyridyl phosphate (**1.3**), 3,5,6-trichloro-2-pyridinol (**1.4**) and ethyl *O*-3,5,6-trichloro-2-pyridyl phosphorothionate (**1.2**) were also identified (Figure 1). None of the metabolites isolated from fish tissues were cholinesterase inhibitors. Smith *et al.* (1966) compared the uptake of ^{14}C-labelled chlorpyrifos by fish living in tanks containing clear water or water with plants and soil, since the compound was being tested for the control of mosquitoes by spraying various water supplies in which the insect can breed. The radioactivity was rapidly taken up from the water by plants and soil particles, and they concluded that under natural conditions this will limit the amount of chlorpyrifos which can be absorbed by fish.

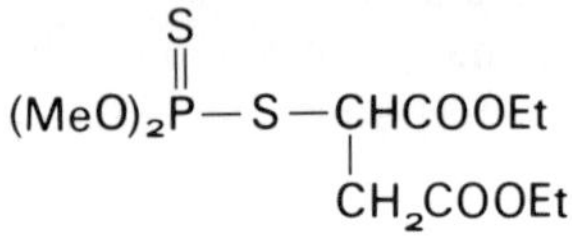

$$
\underset{\textstyle \text{CH}_2\text{COOEt}}{(\text{MeO})_2\overset{\textstyle \overset{\textstyle S}{\|}}{P}-\text{S}-\text{CHCOOEt}}
$$

(**8**, malathion)

In a bioaccumulation study with malathion (**8**) pinfish were shown to form the mono- and dicarboxylic acid metabolites rapidly, such that no parent insecticide was identified in the tissues after a 24-hour exposure to 75 µg/l (Cook and Moore, 1976). This finding illustrates the value of analysing tissues for metabolites, as well as parent insecticide, in monitoring the exposure of fish to pesticides, following field applications.

The relationship between oxidative metabolism and the toxicity of organophosphates to fish has been studied by using inhibitors of oxidative enzymes. Thus, SKF 525A and sesamex antagonize the activation of parathion to paraoxon in mosquitofish (Ludke *et al.*, 1972). In another study with parathion, pretreatment with SKF 525A reduced the inhibition of acetylcholinesterase in the brains of golden shiners and sunfish, but no protective effect was observed in bluegill fish (Gibson and Ludke, 1973). Interestingly, when mosquitofish were exposed to aldrin (**3**, Chapter 1) and parathion simultaneously, the level of acetylcholinesterase inhibition was reduced, as compared to that brought about by parathion treatment alone (Ludke *et al.*, 1972). This appeared to be due to competition between oxidative activation of parathion and aldrin epoxidation. Similarly, mosquitofish which were tolerant to both organochlorines and organophosphates showed an increased capacity to carry out the dearlyation of parathion and methyl parathion (Chambers and Yarbrough, 1973).

An unusual case of tolerance of sunfish to parathion and parathion-methyl (i.p. LD_{50} 110 and > 2,500 mg/kg, respectively) was reported in a comparative study with mice (i.p. LD_{50} 13.5 and 11 mg/kg, respectively) (Benke *et al.*, 1974).

These species differences in toxicity were partly explained by the reduced sensitivity of the sunfish acetylcholinesterases present in brain and muscle to the inhibition by paraoxon and methyl paraoxon. The metabolism of the organophosphates in liver homogenates showed that the sunfish were less active than the mouse in forming the activated oxygen analogue. However, the piscine species also had a reduced hydrolytic deoxyarylation (A-esterase) capacity and was less active in carrying out glutathione-dependent demethylation. The importance of this latter detoxification reaction in the metabolism of parathion-methyl *in vivo* was reflected in the depletion in hepatic glutathione observed after exposure to the insecticide. However, the *in vitro* detoxification of organophosphates by the fish compared with that of the mice may have been underestimated in this study, because assays were carried out at 37 °C; the temperature optima of many biotransformation reactions in fish is more commonly around 25 °C (Chambers and Yarbrough, 1976).

Organochlorines

Sudershan and Khan (1979, 1980a,b, 1981) have investigated the uptake, metabolism and elimination of the cyclodiene insecticides dieldrin (**4**, Chapter 1), endrin (**6**, Chapter 1) and chlordane, and of the photoconversion products photodieldrin and photochlordane which can be present as residues in the environment, in bluegill fish (*Lepomis machrochirus*), a freshwater species. The photoisomers are generally more toxic than the parent cyclodienes to both mammals and fish (see Sudershan and Khan, 1980b). Following a 48-hour exposure of bluegills to either dieldrin (50 µg/l), or endrin (1 µg/l), or chlordane (5 µg/l), or photodieldrin (20 µg/l), or photochlordane (5 µg/l), the ^{14}C-labelled cyclodienes were rapidly absorbed by the fish from the aquarium water. After the transfer of the fish to clean water, the rate of elimination of radioactivity was measured for several weeks. Dieldrin-exposed bluegills eliminated ^{14}C more slowly (16% in 3.3 weeks) than did fish exposed to photodieldrin (50% in 3 weeks), endrin (50% in 4 weeks) or photochlordane (biphasic elimination; 30% in the first 3 weeks of the rapid linear phase). Chlordane was eliminated very slowly: 20% of the absorbed radioactivity over a period of about 6 weeks. The majority of the absorbed radioactivity recovered from fish tissues following exposure to ^{14}C-labelled dieldrin and endrin was in the form of the unchanged insecticide. Thus, ether extracts of eviscerated bluegills contained 53% of the absorbed ^{14}C as intact endrin, pointing to the potential health hazard of the edible part of the fish.

The routes of metabolism of dieldrin in fish appear to be similar to those in mammals (see Chapter 7, Volume 1 of this series; Hutson, 1981). In bluegills, the major metabolites of dieldrin are the caged structure known as dieldrin pentachloroketone (Figure 7.1, 1.4, Volume 1), a product of oxidative dehydrochlorination catalysed by the hepatic monooxygenase enzyme system, and

4,5-*trans*-dihydroaldrindiol (Figure 7.1, 1.3, Volume 1). Dieldrin penta-chloroketone is also a product of photodieldrin metabolism in bluegill fish, and in goldfish (Sudershan and Khan, 1980c). The possible biochemical mechanisms involved in the formation of this intramolecularly bridged metabolite from both dieldrin and photodieldrin are reviewed by Bedford (1975). In mammals, dieldrin pentachloroketone is a minor metabolite of diel-drin and is probably formed as an alternative to the major product, *syn*-12-hydroxydieldrin (Figure 7.1, 1.2, Volume 1). The latter may be present as a metabolite of dieldrin in bluegills based on the chromatographic behaviour of an unidentified radiolabelled product. 12-Hydroxydieldrin has been demonstrated in a fish using a model ecosystem (Sanborn and Yu, 1973). Endrin is hydroxylated in bluegills to give *anti*-12-hydroxyendrin (Figure 7.2, 2.2, Volume 1) together with lesser amounts of *syn*-12-hydroxyendrin (Figure 7.2, 2.3, Volume 1, see also Figure 3, Chapter 1, this volume).

The above metabolites of the cyclodienes are those that have been identified in fish. However, the overall biotransformation of these organochlorines in fish appears to be complex. Thus, in bluegills exposed to [^{14}C]dieldrin, eight radio-active metabolites were separated by thin-layer chromatography. Seven metabolites were found with endrin, and 11 with photodieldrin (13 in goldfish). Most of these metabolites remain unidentified because of the small amounts present in fish. The biotransformation of *cis*-[^{14}C]chlordane and of *cis*-[^{14}C]-photochlordane is similarly diverse in bluegill fish. Both cyclodienes appear to be converted into hydroxylated products, which are then conjugated. The epox-ide metabolite oxychlordane, which is a major metabolite in mammals (see Bedford, 1979), was not found in bluegills.

The epoxidation of aldrin (3, Chapter 1) to dieldrin, which is a persistent insecticide (see Chapter 1), occurs in several freshwater species of fishes, namely mosquitofish (*Gambusia affinis*), golden shiners (*Notemigonus chrysoleucas*), bluegills, green sunfish (*Lepomis cyanellus*), yellow bullhead catfish (*Ictalurus natalis*) (Ludke *et al.*, 1972), Atlantic salmon (*Salmo salar*) fry (Addison *et al.*, 1976) and brook trout (*Salvelinus fontinalis*) (Addison *et al.*, 1977). Insecti-cide-resistant mosquitofish collected from the Mississippi Delta converted aldrin to dieldrin and/or water-soluble products at a greater rate than did susceptible mosquitofish (Wells *et al.*, 1973). Isodrin (5, Chapter 1) is metabolized to its epoxide (endrin) by bluegill liver microsomes (Stanton and Khan, 1973).

The contamination of aquatic environments with persistent insecticides and their bioconcentration through food chains can be an ecological hazard to fish and birds. The epoxidation of aldrin occurs in a variety of aquatic food chain organisms including the following invertebrates: protozoa, coelenterates, worms, arthropods and molluscs (Khan *et al.*, 1972). Lobsters (*Homarus ameri-canus*), which are an important economic species used by man for food, can also convert aldrin to dieldrin (Carlson, 1974). The site of metabolism in these crustaceans is the hepatopancreas and the enzyme system has a greater activity at 20 °C than at 37 °C.

Stanton and Khan (1973) have compared the hepatic mixed-function oxidase activity towards aldrin in bass (*Micropterus dolomieux*) and bluegills with that in mice. Aldrin epoxidase activity was lower in fish, i.e. 35%, 73% and 13% of the activity in mouse for bass fry, bluegill fry and bluegill adults, respectively. The maximum activity for bass fry was found in the temperature range of 22–26 °C, above which the monooxygenase enzyme system appeared to degrade rapidly.

The persistence of dieldrin in different tissues of goldfish (including brain, nerve, muscle, liver, kidney, intestine, skin and ovary) has been studied by Grzenda et al., (1972). The half-life of [^{14}C]dieldrin for individual tissues was around 20 days. The tissue half-life for [^{14}C]DDT, measured under the same conditions, was around 30 days (Grzenda et al., 1970). In fish (see Young et al., 1971; Zinck and Addison, 1975), as in mammals (see Chapter 7, Volume 1 of this series: Hutson, 1981), DDT(1, Chapter 1, this volume) is metabolized by dehydrochlorination to the very persistent DDE (2, Chapter 1) and by reductive dechlorination to DDD, which can undergo further biotransformations to the polar excretory product DDA, 1,1-*bis*-(4-chlorophenyl)acetic acid. According to Wedemeyer (1968), in rainbow trout, the intestinal microflora play a significant role in DDT detoxication.

Carbamates

The carbamate insecticides illustrate some of the species-dependent biotransformations of which fish are capable, namely oxidation, *N*-demethylation, ester hydrolysis and conjugation of the resulting phase I metabolites with sulphate and glucuronic acid.

In a comparative study with the freshwater fish Matsugo (*Pseudorasbora parva*), carbamate insecticides were absorbed less readily from the surrounding tank water than were the organophosphorus insecticides diazinon, fenitrothion, or malathion (Kanazawa, 1975). The rates of carbamate bioaccumulation followed the order 1-naphthyl-*N*-methylcarbamate (carbaryl, **28**, Chapter 1) > 2-*sec*-butylphenyl-*N*-methylcarbamate > 3,5-dimethylphenyl-*N*-methyl-carbamate. A similar pattern was observed when the rates of insecticide elimination from the fish were compared, presumably due to variations in the metabolic activity of Matsugo fish towards these substrates. When the uptake and persistence of [^{14}C]carbaryl was measured in channel catfish (Korn, 1973), a more rapid accumulation of radioactivity in the carcass occurred following oral administration of the insecticide, than when the fish were exposed to carbaryl dissolved in the aquarium water. However, in the latter case, radioactive residues continued to rise over a 56-day period, whereas concentrations of ^{14}C in the tissues of orally dosed catfish remained constant. Korn (1973) suggested that this was due to the absorption from the tank water of 1-naphthol, which is the principal hydrolytic degradation product of carbaryl.

Chin *et al.* (1979) studied the biotransformation of [^{14}C-*naphthyl*]- and [^{14}C-*N-methyl*]-labelled carbaryl in liver explants of five fish species. Analysis of the surrounding growth medium, after incubation of the insecticide with the fish livers at 22 °C, showed that kissing gourami (*Helostoma temmincki*), bluegills, channel catfish, goldfish and yellow perch (*Perca flavescens*) metabolized carbaryl in the order of 8:2:1:1:1 (relative activities of the respective liver tissue). The major polar metabolite in each case had the chromatographic characteristics of the glucuronic acid conjugate of 5,6-dihydro-5,6-dihydroxycarbaryl (see Figure 8, Chapter 1). In addition, hydroxycarbaryl glucuronide, hydroxycarbaryl sulphate (not gourami), 1-naphthyl glucuronide and 1-naphthyl sulphate were also identified (cf. the metabolism of carbaryl in rats and man; see Chapter 7, Volume 1 of this series; Hutson, 1981). Unconjugated dihydrodihydroxycarbaryl was found in the medium, and this appeared to be a mixture of the 3,4 and 5,6 isomers. The relative proportions of these two isomers varied with the fish species. Other neutral metabolites included 4-hydroxycarbaryl, an important component in catfish, and 1-naphthol, which was a minor product in all the species studied. The metabolism of [*naphthyl*-1-^{14}C]-carbaryl was studied in rainbow trout, by analysis of the biliary metabolites (Statham *et al.*, 1975). Following a 24-hour exposure to the insecticide dissolved in the water (0.25 mg/l), the final bile to water ^{14}C ratio was nearly 1000 : 1. Four compounds were purified from bile, namely, unchanged carbaryl, 1-naphthyl glucuronide, 5,6-dihydro-5,6-dihydroxycarbaryl, and an unknown, stable, non-conjugated polar metabolite. The major radioactive biliary component was the glucuronic acid conjugate of 1-naphthol. This contrasts with the five fish species examined *in vitro* by Chin *et al.* (1979), in which 1-naphthyl glucuronide was a minor product, the major hepatic metabolite being the glucuronic acid conjugate of 5,6-dihydro-5,6-dihydroxycarbaryl.

Following exposure of rainbow trout to 4-methylamino-*m*-tolyl *N*-methylcarbamate (aminocarb, **9**) dissolved in the aquarium water, the bioaccumulation ratio (total residue in fish/concentration of aminocarb in water) ranged from 1.7 to 3.3 at different concentrations of the carbamate (Szeto and Holmes, 1982). When the fish were subsequently transferred to fresh water, 90% of the tissue residue was eliminated in 72 hours. Analysis of whole fish showed that the major metabolite was the *N*-demethylated derivative, 4-methylamino-*m*-tolyl

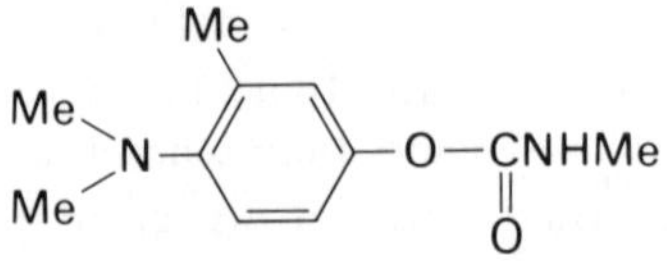

(**9**, aminocarb)

N-methylcarbamate. The 4-amino analogue was not found in any of the fish analysed. The authors concluded that the low toxicity of aminocarb to trout was due to a combination of its poor bioaccumulation and rapid elimination.

Due to the increasing use of insecticides in the paddyfields of West Malaysia there is a decline in the numbers of paddyfield fish, which are an important source of protein for the rural population. Gill (1980) has investigated the *in vitro* metabolism of carbofuran (R=H, Figure 9, Chapter 1) in the paddyfield fish *Trichogaster pectoralis*. When liver microsomes were incubated with [ring-^{14}C]carbofuran at 37 °C in the presence of NADPH, the major metabolites were the *N*-hydroxymethyl derivative of the insecticide and 3-hydroxycarbofuran in the ratio 2 : 1. In addition, 3-ketocarbofuran and the hydrolysis product 3-ketocarbofuranphenol were present as minor metabolic products. Biotransformation in the absence of reduced cofactors was negligible. Similarly, oxidative metabolism of carbofuran is an important metabolic pathway in rat liver microsomes. However, in the mammalian system the major product is 3-hydroxycarbofuran (Dorough, 1968) with *N*-hydroxymethylcarbofuran being formed in lesser amounts than is the case with the liver microsomal-NADPH system of *T. pectoralis*. Furthermore, the rate of carbofuran biotransformation is comparatively lower in fish liver than in rodent liver. Gill (1980) concluded, therefore, that care must be taken in extrapolating insecticide degradation rates obtained from laboratory animals such as rodents to give detoxication rates in fish.

Pyrethroids

The relationship between the biotransformation and toxicity of the synthetic pyrethroids in rainbow trout, a salmonid species known to be very sensitive to this class of insecticide (see Table 1; Mauck *et al.*, 1976; Stephenson, 1982), has been studied recently. The metabolism of pyrethroids in most vertebrate species is largely dependent upon the *cis/trans* conformation of the substituent groups at C-1 and C-3 of the cyclopropane ring (Figure 2). Thus, the *cis*-isomers (in which the C-1 carboxyl group and the C-3 dichlorovinyl group are on the same side of the cyclopropane ring) are preferentially metabolized by aromatic hydroxylation, whereas the *trans*-isomers (C-1 and C-3 substituents on opposite sides of the ring) are better substrates for ester cleavage (Hutson, 1979). The hydrolysis products of permethrin, namely the cyclopropane carboxylic acid derivative and 3-phenoxybenzyl alcohol, are not lethal to juvenile Atlantic salmon (*Salmo salar*; highest tested concentration, 5 mg/l), by contrast the pyrethroid itself (*cis* + *trans* isomers) is extremely toxic (lethal threshold, approximately 9 µg/l, Zitko *et al.*, 1977).

When trout were exposed to [^{14}C]-labelled *cis*- and *trans*-permethrin (**2.1**) dissolved in the aquarium water (5 µg/l), radioactivity rapidly accumulated in the fat and liver, though after 48 hours' exposure 35% and 50% respectively of

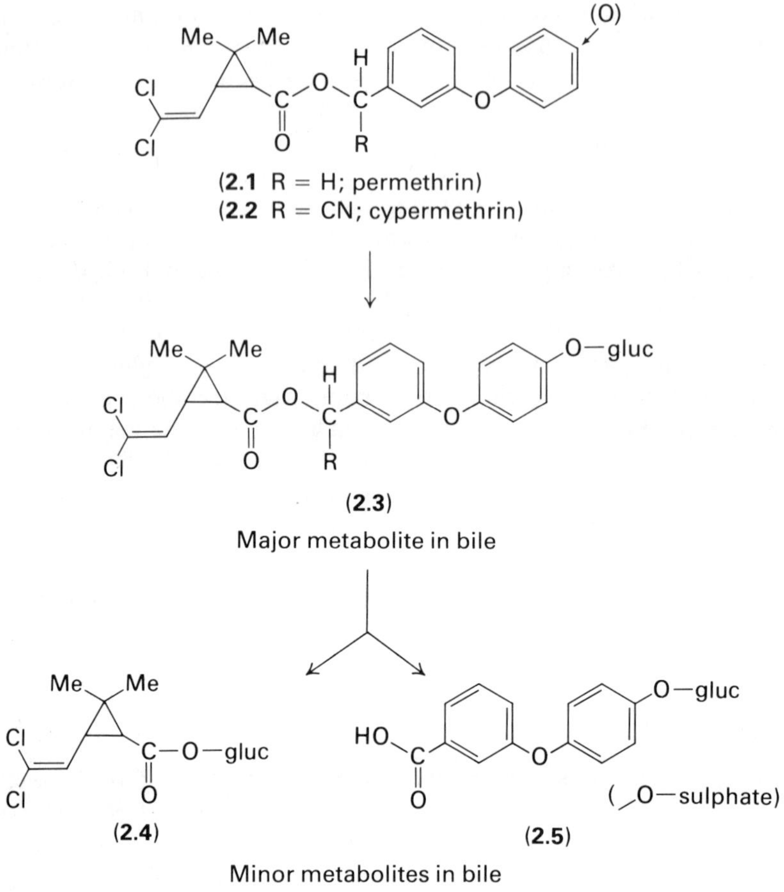

(2.1 R = H; permethrin)
(2.2 R = CN; cypermethrin)

(2.3)
Major metabolite in bile

(2.4)   (2.5)
Minor metabolites in bile

Figure 2 Metabolism and biliary excretion of synthetic pyrethroids in rainbow trout

the dose had been excreted, largely in the bile (Glickman *et al.*, 1981). The major biliary metabolite was 4′-HO-permethrin glucuronide (**2.3**, R=H). The polar urinary metabolites were not characterized, but enzymic hydrolysis suggested that they were sulphate conjugates (enterohepatorenal disposition of metabolites?—see section on routes of elimination of metabolites). The fate of the α-cyano analogue of permethrin, cypermethrin (**2.2**), has also been investigated in trout (Edwards and Millburn, 1985). Following exposure to *cis*-cypermethrin (5 μg/l), labelled with ^{14}C in either the acid or alcohol moiety, 27% of the dose was recovered in the bile within 24 hours. In common with permethrin, the major biliary metabolite was the glucuronide of the 4′-hydroxylated pyrethroid (**2.3**, R=CN). In addition, the glucuronide of the

Table 2 Biotransformation of cypermethrin isomers (10^{-5} M) by liver, brain and intestine 10,000 g supernatant fractions and plasma. (data are given as nmol metabolized/min per kg body weight)

Species	Cis-				Trans-			
	liver	brain	intestine	plasma	liver	brain	intestine	plasma
Trout	10	0.03	4.6	1.2	3	0.08	4.4	7.0
Frog	24	0.16	—	—	19	0.31	—	—
Mouse	263	0.46	22	22	297	1.6	64	77
Quail	170	0.08	13	1.5	186	0.39	39	2.7

In all four species the liver was a major site of detoxification, though intestine and plasma were also relatively important (esteratic). Metabolism in brain was low compared to other tissues, but may be important by acting at the target organ. Lower biotransformation rates *in vitro* in quail compared to mice reflect the smaller proportional organ weights (Table 3). This was not the case *in vivo* (quail » mouse).

dichlorovinyl-dimethyl-cyclopropanecarboxylic acid (**2.4**) and the ether glu-curonide and sulphate of 4′-hydroxy-3-phenoxybenzoic acid (**2.5**) were identi-fied as minor metabolites in bile and aquarium water (Edwards, 1984).

Trout liver microsomes metabolize the *cis*- and *trans*-isomers of both permethrin (Glickman *et al.*, 1979) and cypermethrin (Edwards, 1984) to the 4′-hydroxy derivatives of the intact esters and their corresponding ether glu-curonic acid conjugates. This suggests that, compared with other vertebrates, trout liver is deficient in pyrethroid-hydrolysing esterases. However, hydrolysis of *trans*-permethrin in trout kidney and plasma (Glickman and Lech, 1981), and of *trans*-cypermethrin in plasma and intestinal preparations (Table 2; Edwards, 1984), occurs to the extent that the extrahepatic metabolism of the *trans*-isomers of these pyrethroids exceeds that of the liver.

A comparison of the synergistic effects of piperonyl butoxide and tri-*o*-tolyl phosphate (a carboxylesterase inhibitor) on the toxicity of *trans*-permethrin in trout and mice, confirm that the detoxification of pyrethroids in the fish species is largely oxidative (Glickman *et al.*, 1982). Thus, tri-*o*-tolyl phosphate pretreat-ment potentiates the toxicity of *trans*-permethrin in mice, but has no effect in trout. Furthermore, some synergism was observed in the fish with the *trans* isomer and piperonyl butoxide. Overall rates of ester hydrolysis of permethrin in tissue preparations were lower in trout than in mice (Glickman and Lech, 1981; Glickman *et al.*, 1982). Similarly, the combined hydrolytic, oxidative and conjugative activity towards *cis*- and *trans*-cypermethrin was considerably lower in this piscine species compared with that in frogs, mice or quail (Tables 2 and 3, Edwards, 1984). This, in conjunction with the higher intrinsic toxicity of pyrethroids to the CNS of trout compared with other vertebrates (Glickman and Lech, 1982; Table 4, Edwards and Millburn, 1985), accounts for the sen-sitivity of salmonids to pyrethroids. These points are summarized for cypermethrin in Table 5.

Other fish species appear to be rather less sensitive to these insecticides and this may be due to differences in the rates and routes of detoxification used. Thus, carp liver microsomes (Family: Cyprinidae) are more active than trout liver microsomes in hydrolysing *trans*-permethrin (Glickman *et al.*, 1979) and in studies with fenvalerate (**33**, Chapter 1) in a model aquatic ecosystem, the residues in carp were largely polar products derived from ester cleavage (Ohkawa *et al.*, 1980). A greater capacity for the hydrolysis of pyrethroids in carp may account for their greater tolerance to the dissolved insecticide (see Table 1). Interestingly, Fabacher and Chambers (1972b) have reported that mosquitofish, which were tolerant to organochlorines, were also resistant to pyrethrum extracts and allethrin. This appeared to be due to elevated levels of mixed-function oxidases.

The relationship between metabolism, environmental exposure and the tox-icity of pesticides is well illustrated by considering the potential hazard to fish of two of the current generation of synthetic pyrethroid insecticides, cypermethrin

Table 3   Relative weights of liver, brain, intestine and plasma in four vertebrate species

Species	% Body weight			
	liver	brain	intestine	plasma
Trout	1.1	0.1	7.4	2.1
Frog	2.1	0.4	—	—
Mouse	4.7	1.2	12.6	3.0
Quail	1.8	0.4	6.4	6.3

Table 4   Brain concentration of cypermethrin associated with acute toxic signs

Species	Concentration of *cis*-cypermethrin (μg/g) $\pm$ s.e.m. ($n \geqslant 7$)
Frog	0.08 $\pm$ 0.03
Trout	0.23 $\pm$ 0.05*
Mouse	1.71 $\pm$ 0.33
Quail	3.94 $\pm$ 0.88

*trans = 0.15 $\pm$ 0.02.
Animals were exposed to toxic doses of [^{14}C-*cis*-cypermethrin] and their brains were removed on the manifestation of toxic signs. Following solvent extraction and clean-up, brain samples were analysed by h.p.l.c., the recoveries of [^{14}C-*cis*-cypermethrin] being monitored by the addition of [^{14}C-*trans*-cypermethrin] as an internal standard.

Table 5   Cypermethrin: selective toxicity

1.   Brain sensitivity   frog $\geqslant$ trout $>>$ mouse $>$ quail

2.   Biotransformation

   (A). Rates
                             quail $>$ mouse $>$ frog $\geqslant$ trout
   (B). Routes

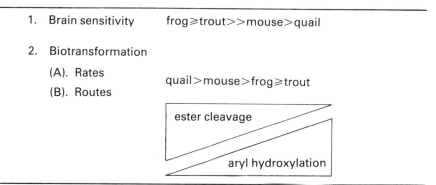

and WL85871 (FASTAC, registered Shell trademark). The latter is a new insecticide, which is very active against the major pests of cotton, vegetables, fruit and soya beans. WL85871 consists of a pair of enantiomers, namely the (1*R,cis*)*S* and $(1S,cis)R$ isomers, from among the eight isomers that constitute cypermethrin. In particular, the $1R$ isomer is highly active as a neurotoxicant in insects and mammals (Gray and Soderlund, Chapter 5, this volume). Cypermethrin and WL85871 preparations have similar properties in aquatic toxicology studies performed in the laboratory (Stephenson, 1982, 1985). Thus, in clean water tests they are both highly toxic to rainbow trout with effects on the fish being observed at concentrations of less than 1 µg/l. However, in field experiments in which the insecticides are sprayed onto the surface of the ponds, the toxicity is much reduced, mainly due to the binding of the chemicals to various adsorbing surfaces (LD_{50} values, rainbow trout, are 90 and 70 g/ha for cypermethrin and WL85871, respectively). Therefore, these pyrethroids are unlikely to be toxic to fish as a result of their normal agricultural use (recommended application rate of cypermethrin is 25–75 g/ha; see Stephenson, 1985, and WL85871 is expected to be applied to crops at the lower rate of 10–30 g/ha; see Le Blanc, 1985). The use in aquatic ecosystems of the resolved active *cis* isomers of cypermethrin, as contained in WL85871, will reduce the potential biohazard of the pyrethroid to fish for a further reason. In rainbow trout the *trans* isomers contained in cypermethrin are as toxic as the *cis* isomers because the *cis* and *trans* isomers are detoxified at comparable rates (Table 2) and are equally active as neurotoxicants (Table 4). Therefore, with WL85871, the fish are exposed to considerably less active ingredient than is the case for cypermethrin.

CONCLUSIONS

Fish generally possess most of the phase I and phase II (conjugation) pathways necessary for the biotransformation of insecticides and other xenobiotics, but the rate of matabolism in piscine species is often slower than that in mammals. However, in spite of low enzyme activities, fish appear to be able to metabolize agrochemicals reasonably effectively *in vivo*, provided that exposures are low. The major route of exposure, i.e. chemicals present in the aqueous environment is, of course, unique to fish and other aquatic organisms. This must be considered in comparative metabolism and toxicological studies with terrestrial, arboreal and aerial species. Several factors influence the aquatic exposure levels of insecticides. Thus, the formulation and commercial methods of application of a chemical can affect its solubility in water. Also, availability is partly determined by the binding of compounds to aquatic plants, bottom sediments and other surfaces. The extent to which a chemical is readily biodegradable will also influence its exposure levels to fish. Therefore, for some insecticides there is a lack of correlation between their toxicity to fish in clean water tests performed in the laboratory compared with tests carried out under field conditions. For example, cypermethrin and WL85871 (see Pyrethroids section) are unlikely to be hazardous to fish when applied to crops at the recommended levels.

REFERENCES

Adamson, R. H. (1967). 'Drug metabolism in marine vertebrates', *Fedn Proc. Fedn Am. Socs exp. Biol.*, **26**, 1047–1055.

Addison, R. F. (1976). 'Organochlorine compounds in aquatic organisms: their distribution, transport and physiological significance', in *Effects of Pollutants on Aquatic Organisms* (ed. A. P. M. Lockwood), Cambridge Univ. Press, Cambridge, pp. 127–143.

Addison, R. F., Zinck, M. E., and Leahy, J. R. (1976). 'Metabolism of single and combined doses of ^{14}C-aldrin and ^{3}H-*p,p'*-DDT by Atlantic salmon (*Salmo salar*) fry', *J. Fish. Res. Bd Can.*, **33**, 2073–2076.

Addison, R. F., Zinck, M. E. and Willis, D. E. (1977). 'Mixed function oxidase enzymes in trout (*Salvelinus fontinalis*) liver: absence of induction following feeding of *p,p'*-DDT or *p,p'*-DDE', *Comp. Biochem. Physiol.*, **57C**, 39–43.

Allen, J. L., Dawson, V. K., and Hunn, J. B. (1979). 'Biotransformation of selected chemicals by fish', in *Pesticide and Xenobiotic Metabolism in Aquatic Organisms* (eds M. A. Q. Khan, J. J. Lech and J. J. Menn), Chapter 7, pp. 121–129, ACS symposium series; 99.

Balk, L., Meijer, J., DePierre, J. W., and Appelgren, L-E. (1984). 'The uptake and distribution of [^{3}H] Benzo[*a*]pyrene in the northern pike (*Esox lucius*). Examination by whole-body autoradiography and scintillation counting', *Toxicol. appl. Pharmacol.*, **74**, 430–449.

Bedford, C. T. (1975). In *Foreign Compound Metabolism in Mammals*, Specialist Periodical Reports, Vol. 3, pp. 406–407, The Chemical Society, London.

Bedford, C. T. (1979). In *Foreign Compound Metabolism in Mammals*, Specialist Periodical Reports, Vol. 5, pp. 466–468, The Chemical Society, London.

Bend, J. R., and James, M. O. (1978). 'Xenobiotic metabolism in marine and freshwater species', in *Biochemical and Biophysical Perspectives in Marine Biology* (eds D. C. Malins and J. R. Sargent), Vol. 4, pp. 125–188, Academic Press, New York.

Benke, G. M., Cheever, K. L., Mirer, F. E., and Murphy, S. D. (1974). 'Comparative toxicity, anticholinesterase action and metabolism of methyl parathion and parathion in sunfish and mice', *Toxicol. appl. Pharmacol.*, **28**, 97–109.

Binder, R. L., Melancon, M. J., and Lech, J. J. (1984). 'Factors influencing the persistence and metabolism of chemicals in fish', *Drug Metab. Rev.*, **15**, 697–724.

Bray, H. G., Humphris, B. G., Thorpe, W. V., White, K., and Wood, P. B. (1952a). 'Kinetic studies on the metabolism of foreign organic compounds. 3. The conjugation of phenols with glucuronic acid', *Biochem. J.*, **52**, 416–419.

Bray, H. G., Humphris, B. G., Thorpe, W. V., White, K., and Wood, P. B. (1952b). 'Kinetic studies on the metabolism of foreign organic compounds. 4. The conjugation of phenols with sulphuric acid', *Biochem. J.*, **52**, 419–423.

Brodie, B. B., and Maickel, R. P. (1962). 'Comparative biochemistry of drug metabolism', in *Proc. 1st Int. Pharmac. Mtg* (eds. B. B. Brodie and E. G. Erdos), Vol. 6, pp. 299–324, Macmillian Press, New York.

Buhler, D. R., and Rasmusson, M. E. (1968). 'The oxidation of drugs by fishes', *Comp. Biochem. Physiol.*, **25**, 223–239.

Carlson, G. P. (1974). 'Epoxidation of aldrin to dieldrin by lobsters', *Bull. Environ. Contam. Toxicol.*, **11**, 577–582.

Chambers, J. E., and Yarbrough, J. D. (1973). 'Organophosphate degradation by insecticide-resistant and susceptible populations of mosquitofish (*Gambusia affinis*)', *Pestic. Biochem. Physiol.*, **3**, 312–316.

Chambers, J. E., and Yarbrough, J. D. (1976). 'Xenobiotic biotransformation systems in fishes', *Comp. Biochem. Physiol.*, **55C**, 77–84.

Chin, B. H., Sullivan, L. J., and Eldridge, J. E. (1979). '*In vitro* metabolism of carbaryl by liver explants of bluegill, catfish, perch, goldfish and kissing gourami', *J. Agric. Food Chem.*, **27**, 1395–1398.

Chipman, J. K., Bhave, N. A., Hirom, P. C., and Millburn, P. (1982). 'Metabolism and excretion of benzo(*a*)pyrene in the rabbit', *Xenobiotica*, **12**, 397–404.

Chipman, J. K., Hirom, P. C., Frost, G. S., and Millburn, P. (1981). 'The biliary excretion and enterohepatic circulation of benzo(*a*)pyrene and its metabolites in the rat', *Biochem. Pharmacol.*, **30**, 937–944.

Cook, G. H., and Moore, J. C. (1976). 'Determination of malathion, malaoxon, and mono- and dicarboxylic acids of malathion in fish, oyster and shrimp tissue', *J. Agric. Food Chem.*, **24**, 631–634.

Crosby, D. G., Landrum, P. F., and Fischer, C. C. (1979). 'Investigation of xenobiotic metabolism in intact aquatic animals', in *Pesticide and Xenobiotic Metabolism in Aquatic Organisms* (eds M. A. Q. Khan, J. J. Lech and J. J. Menn), Chapter 14, pp. 217–231, ACS symposium series; 99.

Curtis, L. R. (1983). 'Glucuronidation and biliary excretion of phenolphthalein in temperature-acclimated steelhead trout (*Salmo gairdneri*)', *Comp. Biochem. Physiol.*, **76C**, 107–111.

Dawson, V. K., Harman, P. D., Schultz, D. P., and Allen, J. L. (1983). 'Rapid method for measuring rotenone in water at piscicidal concentrations', *Trans. Am. Fish Soc.*, **112**, 725–727.

Dorough, H. W. (1968). 'Metabolism of Furadan (NIA-10242) in rats and houseflies', *J. Agric. Food Chem.*, **16**, 319–325.

Edwards, R. (1984). 'The comparative metabolism and toxicity of the pyrethroid insecticide cypermethrin in vertebrates', PhD Thesis, Univ. of London.

Edwards, R., and Millburn, P. (1985). 'Toxicity and metabolism of cypermethrin in fish compared with other vertebrates', *Society of Chemical Industry Symposium. Pyrethroid Insecticides in the Environment.* (Southampton, April, 1984), *Pestic. Sci.*, **16**, 201–202.

Fabacher, D. L., and Chambers, H. (1972a). 'Rotenone tolerance in mosquito-fish', *Environ. Pollut.*, **3**, 139–141.

Fabacher, D. L., and Chambers, H. (1972b). 'Apparent resistance to pyrethroids in organochlorine-resistant mosquitofish', *Proc. 26th Ann. Conf. S.E. Assoc. Game Fish Commis.*, pp. 461–464.

Forlin, L. (1980). 'Effects of Clophen A50, 3-methylcholanthrene, pregnenalone 16 α-carbonitrile and phenobarbital on the hepatic microsomal cytochrome P-450 dependent monooxygenase system of rainbow trout (*Salmo gairdneri*) of different age and sex', *Toxicol. appl. Pharmacol.*, **54**, 420–430.

Fukami, J-I., Shishido, T., Fukunaga, K., and Casida, J. E. (1969). 'Oxidative metabolism of rotenone in mammals, fish and insects and its relation to selective toxicity', *J. Agric. Food Chem.*, **17**, 1217–1226.

Gibson, J. R., and Ludke, J. L. (1973). 'Effect of SKF 525A on brain acetylcholinesterase inhibition by parathion in fishes', *Bull. Environ. Contam. Toxicol.*, **9**, 140–142.

Gill, S. S. (1980). '*In vitro* metabolism of carbofuran by liver microsomes of the padifield fish *Trichogaster pectoralis*', *Bull. Environ. Contam. Toxicol.*, **25**, 697–701.

Glickman, A. H., Hamid, A. A. R., Rickert, D. E., and Lech, J. J. (1981). 'Elimination and metabolism of permethrin isomers in rainbow trout', *Toxicol. appl. Pharmacol.*, **57**, 88–98.

Glickman, A. H., and Lech, J. J. (1981). 'Hydrolysis of permethrin, a pyrethroid insecticide, by rainbow trout and mouse tissues *in vitro*: A comparative study', *Toxicol. appl. Pharmacol.*, **60**, 186–192.

Glickman, A. H., and Lech, J. J. (1982). 'Differential toxicity of *trans*-permethrin in rainbow trout and mice. II. Role of target organ sensitivity', *Toxicol. appl. Pharmacol.*, **66**, 162–171.

Glickman, A. H., Shono, T., Casida, J. E., and Lech, J. J. (1979). '*In vitro* metabolism of permethrin isomers by carp and rainbow trout liver microsomes', *J. Agric. Food Chem.* **27**, 1038–1041.

Glickman, A. H., Weitman, S. D., and Lech, J. J. (1982). 'Differential toxicity of *trans*-permethrin in rainbow trout and mice. I. Role of biotransformation', *Toxicol. appl. Pharmacol.*, **66**, 153–161.

Grzenda, A. R., Paris, D. F., and Taylor, W. J. (1970). 'The uptake, metabolism and elimination of chlorinated residues by goldfish (*Carassius auratus*) fed a ^{14}C-DDT contaminated diet', *Trans. Am. Fish. Soc.*, **99**, 385–396.

Grzenda, A. R., Taylor, W. J., and Paris, D. F. (1972). 'The elimination and turnover of ^{14}C-dieldrin by different goldfish tissues', *Trans. Am. Fish. Soc.*, **101**, 686–690.

Hirom, P. C., Millburn, P., and Smith, R. L. (1976). 'Bile and urine as complementary pathways for the excretion of foreign organic compounds', *Xenobiotica*, **6**, 55–64.

Hirom, P. C., Millburn, P., Smith, R. L., and Williams, R. T. (1972). 'Species variations in the threshold molecular-weight factor for the biliary excretion of organic anions', *Biochem. J.*, **129**, 1071–1077.

Hitchcock, M., and Murphy, S. D. (1967). 'Enzymatic reduction of O,O-(4-nitrophenyl) phosphorothioate, O,O-diethyl O-(4-nitrophenyl) phosphate, and O-ethyl O-(4-nitrophenyl) benzene thiophosphonate by tissues from mammals, birds and fishes', *Biochem. Pharmacol.*, **16**, 1801–1811.

Hogan, J. W., and Knowles, C. O. (1968). 'Degradation of organophosphates by fish liver phosphatases', *J. Fish. Res. Bd Can.*, **25**, 1571–1579.

Hogan, J. W., and Knowles, C. O. (1972). 'Metabolism of diazinon by fish liver microsomes', *Bull. Environ. Contam. Toxicol.*, **8**, 61–64.

Huckle, K. R., Chipman, J. K., Hutson, D. H., and Millburn, P. (1981). 'Metabolism of 3-phenoxybenzoic acid and the enterohepatorenal disposition of its metabolites in the rat', *Drug Metab. Dispos.*, **9**, 360–368.

Hutson, D. H. (1979). 'The metabolic fate of synthetic pyrethroid insecticides in mammals', in *Progress in Drug Metabolism* (eds J. W. Bridges and L. F. Chasseaud), Vol. 3, pp. 215–252, Wiley, Chichester.

Hutson, D. H. (1981). 'The metabolism of insecticides in man', in *Progress in Pesticide Biochemistry* (eds D. H. Hutson and T. R. Roberts), Vol. 1, pp. 287–333, Wiley, Chichester.

Kanazawa, J. (1975). 'Uptake and excretion of organophosphorus and carbamate insecticides by freshwater fish, motsugo, *Pseudorasbora parva*', *Bull. Environ. Contam. Toxicol.*, **14**, 346–352.

Khan, M. A. Q., Kamal, A., Wolin, R. J., and Runnels, J. (1972). '*In vivo* and *in vitro* epoxidation of aldrin by aquatic food chain organisms', *Bull. Environ. Contam. Toxicol.*, **8**, 219–228.

Kingsbury, P. D., and Kreutzweiser, D. P. (1985). 'The impact of permethrin on aquatic ecosystems in Canadian forests', *Society of Chemical Industry Symposium. Pyrethroid Insecticides in the Environment.* (Southampton, April, 1984), *Pestic. Sci.*, **16**, 202–203.

Koivusaari, U., and Andersson, T. (1984). 'Partial temperature compensation of hepatic biotransformation enzymes in juvenile rainbow trout (*Salmo gairdneri*) during the warming of water in spring', *Comp. Biochem. Physiol.*, **78B**, 223–226.

Korn, S. (1973). 'The uptake and persistence of carbaryl in channel catfish', *Trans. Am. Fish. Soc.*, **102**, 137–139.

Laitinen, M., Nieminen, M., and Hietanen, E. (1982). 'The effect of pH changes of water on the hepatic metabolism of xenobiotics in fish', *Acta Pharmacol. Toxicol.*, **51**, 24–29.

Layiwola, P. J., Linnecar, D. F. C., and Knights, B. (1983). 'The biotransformation of three ^{14}C-labelled phenolic compounds in twelve species of freshwater fish', *Xenobiotica*, **13**, 107–113.

Le Blanc, J. (1985). 'Field experiments on the effects of a new pyrethroid insecticide, FASTAC, on bees foraging artificial aphid honeydew on winter wheat', *Society of Chemical Industry Symposium. Pyrethroid Insecticides in the Environment.* (Southampton, April, 1984), *Pestic. Sci.,* **16**, 206.

Lech, J. J. (1973). 'Isolation and identification of 3-trifluoromethyl-4-nitrophenyl glucuronide from bile of rainbow trout exposed to 3-trifluromethyl-4-nitrophenol', *Toxicol. appl. Pharmacol.*, **24**, 114–124.

Lech, J. J., and Bend, J. R. (1980). 'Relationship between biotransformation and the toxicity and fate of xenobiotic chemicals in fish', *Environ. Hlth Persp.*, **34**, 115–131.

Lech, J. J., Pepple, S. K., and Statham, C. N. (1973). 'Fish bile analysis: a possible aid in monitoring water quality', *Toxicol. appl. Pharmacol.*, **25**, 430–434.

Lee, R. F., Sauerheber, R., and Dobbs, G. H. (1972). 'Uptake metabolism and discharge of polycyclic aromatic hydrocarbons by marine fish', *Marine Biol.*, **17**, 201–208.

Ludke, J. L. L., Gibson, J. R., and Lusk, C. I. (1972). 'Mixed function oxidase activity in freshwater fishes: aldrin epoxidation and parathion activation', *Toxicol. appl. Pharmacol.*, **21**, 89–97.

Macek, K. J., and McAllister, W. A. (1970). 'Insecticide susceptibility of some common fish family representatives', *Trans. Am. Fish. Soc.*, **1**, 20–27.

Maemura, S., and Omura, T. (1983). 'Drug-oxidising mono-oxygenase system in liver microsomes of goldfish (*Carassius auratus*)', *Comp. Biochem. Physiol.*, **76C**, 45–51.

Mauck, W. L., Olson, L. E., and Marking, L. L. (1976). 'Toxicity of natural pyrethrins and five pyrethroids to fish', *Archs Environ. Contam. Toxicol.*, **4**, 18–29.

McKim, J. M., and Goeden, H. M. (1982). 'A direct measure of the uptake efficiency of a xenobiotic chemical across the gills of brook trout (*Salvelinus fontinalis*) under normoxic and hypoxic conditions', *Comp. Biochem. Physiol.*, **72C**, 65–74.

Mehta, R., Hirom, P. C., and Millburn, P. (1978). 'The influence of dose on the pattern of conjugation of phenol and 1-naphthol in non-human primates', *Xenobiotica*, **8**, 445–452.

Millburn, P. (1970). 'Factors in the biliary excretion of organic compounds', in *Metabolic Conjugation and Metabolic Hydrolysis* (ed. W. H. Fishman), Vol. 2, pp. 1–74, Academic Press, New York.

Millburn, P. (1976). 'Excretion of xenobiotic compounds in bile', in *The Hepato-biliary System* (ed. W. Taylor), pp. 109–129, Plenum Press, New York.

Millburn, P., Smith, R. L., and Williams, R. T. (1967). 'Biliary excretion of foreign compounds. Biphenyl, stilboestrol and phenolphthalein in the rat: molecular weight, polarity and metabolism as factors in biliary excretion', *Biochem. J.*, **105**, 1275–1281.

Murphy, S. D. (1966). 'Liver metabolism and toxicity of thiophosphate insecticides in mammalian, avian and piscine species', *Pharm. Soc. exp Biol. Med.*, **123**, 392–398.

Nagel, R. (1983). 'Species differences, influence of dose and application on biotransformation of phenol in fish', *Xenobiotica*, **13**, 101–106.

Nagel, R., and Urich, K. (1983). 'Quinol sulphate, a new conjugate of phenol in goldfish', *Xenobiotica*, **13**, 97–100.

Neely, W. B., Branson, D. R., and Blau, G. E. (1974). 'Partition coefficient to measure bioconcentration potential of organic chemicals in fish', *Environ. Sci. Tech.*, **8**, 1113–1115.

Ohkawa, H., Kikuchi, R., and Miyamoto, J. (1980). 'Bioaccumulation and biodegradation of the (*S*)-acid isomer of fenvalerate (Sumicidin) in an aquatic model ecosystem',
J. Pestic. Sci., **5**, 11–22.

Parker, R. J., Hirom, P. C., and Millburn, P. (1980). 'Enterohepatic recycling of
phenolphthalein, morphine, lysergic acid diethylamide (LSD) and diphenylacetic acid
in the rat. Hydrolysis of glucuronic acid conjugates in the gut lumen', *Xenobiotica*, **10**,
689–703.

Pedersen, M. G., Hershberger, W. K., Zachariah, P. K., and Juchau, M. R. (1976). 'Hepatic biotransformation of environmental xenobiotics in six strains of rainbow trout
(*Salmo gairdneri*)', *J. Fish. Res. Bd Can.*, **33**, 666–675.

Potter, J. L., and O'Brien, R. D. (1964). 'Parathion activation by livers of aquatic and
terrestrial vertebrates', *Science*, **144**, 55–56.

Pritchard, J. B., and Bend, J. R. (1984). 'Mechanisms controlling the renal excretion of
xenobiotics in fish: effects of chemical structure', *Drug. Metab. Rev.*, **15**, 655–671.

Ramage, P. I. N., and Nimmo, I. A. (1984). 'The substrate specificities and subunit compositions of the hepatic glutathione *S*-transferases of rainbow trout (*Salmo gairdneri*)', *Comp. Biochem. Physiol.*, **78B**, 189–194.

Reinbold, K. A., and Metcalf, R. L. (1976). 'Effects of the synergist piperonyl butoxide
on metabolism of pesticides in green sunfish', *Pestic. Biochem. Physiol.*, **6**, 401–412.

Sanborn, J. R., and Yu, C. C. (1973). 'The fate of dieldrin in a model ecosystem', *Bull.
Environ. Contam. Toxicol.*, **10**, 340–346.

Sieber, S. M., and Adamson, R. H. (1977). 'The metabolism of xenobiotics by fish', in
Drug metabolism—from Microbe to Man (eds D. V. Parke and R. L. Smith),
pp. 233–245, Taylor and Francis Ltd, London.

Smith, G. N., Watson B. S., and Fisher, F. S. (1966). 'The metabolism of [14C]
O,O-diethyl *O*-(3,5,6-trichloro-2-pyridyl) phosphorothioate (Dursban) in fish', *J.
Econ. Entomol.*, **59**, 1464–1475.

Stanton, R. H., and Khan, M. A. Q. (1973). 'Mixed-function oxidase activity toward
cyclodiene insecticides in bass and bluegill sunfish', *Pestic. Biochem. Physiol.*, **3**,
351–357.

Statham, C. N., Melancon, Jr. M. J., and Lech, J. J. (1976). 'Bioconcentration of xenobiotics in trout bile: a proposed monitoring aid for some waterborne chemicals', *Science*, **193**, 680–681.

Statham, C. N., Pepple, S. K., and Lech, J. J. (1975). 'Biliary excretion products of
1-[1-14C]naphthyl-*N*-methylcarbamate (carbaryl) in rainbow trout (*Salmo gairdneri*)',
Drug Metab. Dispos., **3**, 400–406.

Stegeman, J. J., and Chevion, M. (1980). 'Sex differences in cytochrome P-450 and mixed
function oxygenase activity in gonadally mature trout', *Biochem. Pharmacol.*, **29**,
553–558.

Stephenson, R. R. (1982). 'Aquatic toxicology of cypermethrin. I. Acute toxicity to some
freshwater fish and invertebrates in laboratory tests', *Aq. Toxicol.*, **2**, 175–185.

Stephenson, R. R. (1985). 'Assessing the hazard of pyrethroid insecticides in the aquatic
environment', *Society of Chemical Industry Symposium. Pyrethroid Insecticides in the
Environment* (Southampton, April, 1984), *Pestic. Sci.*, **16**, 199–201.

Sudershan, P., and Khan, M. A. Q. (1979). 'Metabolic and elimination products of
[14C]photodieldrin from bluegill fish', *Pestic. Biochem. Physiol.*, **12**, 216–223.

Sudershan, P., and Khan, M. A. Q. (1980a). 'Metabolic fate of [14C]endrin in bluegill
fish', *Pestic. Biochem. Physiol.*, **14**, 5–12.

Sudershan, P., and Khan, M. A. Q. (1980b). 'Metabolism of *cis*-[14C]chlordane and
cis-[14C]photochlordane in bluegill fish', *J. Agric. Food Chem.*, **28**, 291–296.

Sudershan, P., and Khan, M. A. Q. (1980c). 'Metabolic fate of [^{14}C]photodieldrin in goldfish', *Pestic. Biochem. Physiol.,* **13**, 148–157.
Sudershan, P., and Khan, M. A. Q. (1981). 'Metabolism of [^{14}C]dieldrin in bluegill fish', *Pestic. Biochem. Physiol.,* **15**, 192–199.
Szeto, S. Y., and Holmes, S. B. (1982). 'The lethal toxicity of Maticil^R 1.8F to rainbow trout and its *in vivo* metabolism', *J. Environ. Sci. Hlth,* **B17**, 51–61.
Varanasi, U., Stein, J. E., Nishimoto, M., and Hom, T. (1982). 'Benzo(a)pyrene metabolites in liver, muscle, gonads and bile of adult English sole (*Parophrys vetulus*)', in *Polynuclear Aromatic Hydrocarbons: Seventh International Symposium on Formation, Metabolism and Measurement* (eds M. W. Cooke and A. J. Dennis), pp 1221–1234, Battelle Press, Columbus, Ohio.
Wedemeyer, G. (1968). 'Role of intestinal microflora in the degradation of DDT by rainbow trout (*Salmo gairdneri*)', *Life Sci.,* **7**, 219–223.
Wells, M. R., Ludke, J. L., and Yarbrough, J. D. (1973). 'Epoxidation and fate of [^{14}C]aldrin in insecticide-resistant and susceptible populations of mosquitofish (*Gambusia affinis*)', *J. Agric. Food Chem.,* **21**, 428–429.
Williams, R. T., and Millburn, P. (1975). 'Detoxication mechanisms—the biochemistry of foreign compounds', in *MTP International Review of Science. Biochemistry Series One, Vol 12, Physiological and Pharmacological Biochemistry* (ed. H. K. F. Blaschko), pp. 211–266, Butterworths, London.
Yagen, B., Foureman, G. L., Ben-Zvi, Z., Ryan, A. J., Hernandez, O., Cox, R. H., and Bend, J. R. (1984). 'The metabolism and excretion of ^{14}C-styrene oxide-glutathione adducts administered to the winter flounder, *Pseudopleuronectes americanus*, a marine teleost. Identification of the corresponding S-cysteine derivatives as major urinary metabolites', *Drug Metab. Dispos.,* **12**, 389–395.
Young, R. G., St. John, L., and Lisk, D. J. (1971). 'Degradation of DDT by goldfish', *Bull. Environ. Contam. Toxicol.,* **6**, 351–354.
Zinck, M. E., and Addison, R. F. (1975). 'The effect of temperature on the rate of conversion of p,p'-DDT to p,p'-DDE in brook trout', *Can. J. Biochem.,* **53**, 636–639.
Zitko, V., Carson, W. G., and Metcalfe, C. D. (1977). 'Toxicity of pyrethroids to juvenile Atlantic salmon', *Bull. Environ. Contam. Toxicol.,* **18**, 35–41.

Insecticides
Edited by D. H. Hutson and T. R. Roberts
© 1985 John Wiley & Sons Ltd

CHAPTER 7

Chitin synthesis inhibitors as insecticides

Nicolas P. Hajjar

INTRODUCTION

Chitin synthesis inhibitors with potent insecticidal properties were accidentally discovered in the 1970s by investigators at Philips-Duphar, The Netherlands, in the course of preparing and examining derivatives of the herbicides dichlobenil and fenuron (van Daalen *et al.*, 1972; Wellinga *et al.*, 1973a,b). These derivatives and analogues, known as the benzoylphenyl ureas (BPUs), did not have any herbicidal activity but were surprisingly very potent toxicants to insects upon feeding (Verloop and Ferrell, 1977). The insecticidal properties of these chemicals were unusual in that toxicity signs were observed in immatures only, and were limited to the moulting process. Initial histological and biochemical radiotracer studies demonstrated that these chemicals caused defects in the process of cuticle deposition, and inhibited chitin biosynthesis in fifth instar larvae of *Pieris brassicae* L. (Mulder and Gijswijt, 1973; Post and Vincent, 1973; Post *et al.*, 1974). Other groups of chemicals found to inhibit chitin synthesis include the pyrimidine nucleoside antibiotics, the polyoxins (Endo and Misato, 1969; Endo *et al.*, 1970; Ohta *et al.*, 1970) and the nikkomycins (Dähn *et al.*, 1976; Brillinger, 1979; Kobinata *et al.*, 1980). These chemicals have good fungicidal activities (Endo *et al.*, 1970; Dähn *et al.*, 1976; Kobinata *et al.*, 1980), but their use as insecticides may be limited.

The discovery of insect chitin synthesis inhibitors led to extensive research in the field of insect cuticle biochemistry and chitin synthesis, and the development of a variety of *in vitro* chitin synthesizing systems, tissue cultures and cell-free enzyme preparations. Hundreds of chemicals have been screened in these systems and many have been found to be potent inhibitors. The importance of this unique insecticidal action is the presence of chitin biosynthesis in insects but not in animals or plants. Thus, chitin synthesis inhibitors have selectivity advantages over broad spectrum insecticides that exert their action on the nervous system or on bioenergetic mechanisms present in both mammals and a variety of insect pests. They are also the most promising insect growth regulators for further research and development.

Recently the effects of various insecticides on the insect cuticle have been reviewed by Chen and Mayer (1984). Marks *et al.* (1982) reviewed and compared the modes of action of chitin synthesis inhibitors in fungi and *in vitro* cockroach leg regenerates. This chapter focuses on chemicals inhibiting chitin synthesis in insects and recent progress in the elucidation of their modes of action. To appreciate fully the mode of action of insect chitin synthesis inhibitors, it is essential to highlight the significant role of chitin in the insect cuticle. A brief review of the biochemical structure and function of the insect cuticle and chitin biosynthesis is, therefore, warranted.

CHITIN AND THE INSECT CUTICLE

Chitin is the most abundant organic component of insect and other invertebrate exoskeletons, and of the cell walls of most fungi (Neville, 1975; Muzzarelli, 1977). Its distribution in the animal phyla has been reviewed by Jeuniaux (1963) and Muzzarelli (1977). Chitin (poly-β-(1-4)-N-acetyl-D-glucosamine) is essentially a long, unbranched polysaccharide chain of N-acetylglucosamine

---

Abbreviations:
Glucosamine, GA; N-acetylglucosamine, NAGA; uridine diphosphate-N-acetyl-glucosamine, UDP-NAGA; benzoylphenyl ureas, BPUs; chitin synthetase, CSase; diflubenzuron, DFB.

(NAGA) monomers joined by β-1,4-glucosidic linkages (Rudall and Kenchington, 1973; Hackman, 1974). However, a small percentage of D-glucosamine monomers (GA) may be present (Neville, 1975). Chitin is not known to occur in any conformation other than the two residue per turn form (**1**), referred to as the biose unit (Rudall and Kenchington, 1973; Hackman, 1974).

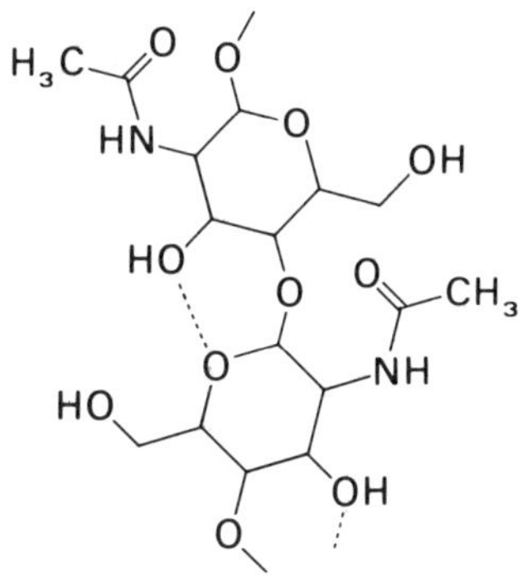

(1)

The insect cuticle is a rigid and complex structure that affords the insect its integrity, skeletal muscle support, and protection against physical damage and pathogens; it also restricts water loss from the body surface (Neville, 1975). The cuticle is composed of several distinct layers, namely, the basement membrane and a single layer of epidermal cells, the procuticle and the epicuticle. The procuticle constitutes the major part of the cuticle.

Chitin occurs only in the procuticle and it usually accounts for about 20–50% of the dry weight of the insect integument (Hackman, 1974; Neville, 1975). In addition to its presence in the cuticle, chitin is also a major component of the peritrophic membranes. Three distinct crystalline forms of chitin known as α, β and γ have been characterized by X-ray diffraction studies (Rudall, 1963; Rudall and Kenchington, 1973). The forms differ from one another in the direction of neighbouring chains and the presence of bound molecules of water (Hackman, 1974). Only α-chitin, which is comprised of adjacent chains with opposite direction to each other, is known to occur in the insect integument. Adjacent chains are present as long microfibrils in the integument, and are usually about 2.8 nm thick and of indefinite length. Chitin microfibrils originate on the external plasma membrane of the epidermal cells, and may be organized in various patterns. The organization appears to be of importance to the mechanical properties of the cuticle (Neville, 1975). α-Chitin has also been found in the peritrophic membrane of honeybee, bumblebee and wasp larvae. β-Chitin, on the other hand, has adjacent chains parallel to each other, and is formed by the midgut where the material is to be used for making cocoons. γ-Chitin, in which adjacent chains are thought of as two in one direction and one in the opposite direction, is predominantly found in the peritrophic membrane of several insect species, e.g., locust and cockroach adults, and silkworm and sawfly larvae (Rudall and Kenchington, 1973).

In addition to chitin, another major component of insect cuticle is protein; together they make up more than 95% of the cuticle (Neville, 1975; Sridhara, 1983). Electron microscopy studies indicate that chitin microfibrils are more or less regularly spaced within a protein matrix. The relative amounts of chitin and protein vary considerably between species, and either may account for one-half or more of the dry weight of the integument. Cuticles of *Calliphora* larvae contain 44.2% chitin and 55.3% protein (Hackman and Goldberg, 1977). Although it is widely recognized that the cuticle contains many different proteins, little is known about their nature. Sridhara (1983) reported the presence of 30–40 polypeptides, separable on SDS-PAGE electrophoresis, from both pupal and adult untanned cuticles of the oak silkmoth *Antheraea polyphemus*. However, only about 10 major proteins with molecular weights ranging from 15,000 to 45,000 Daltons were present in each stage, and the protein compositions of pupal and adult cuticles were different from one another. Similar findings on the number and molecular weight range of proteins and stage specificity have been reported for larvae, pupae and adult *Tenebrio molitor* (Roberts and Willis, 1980) and *Drosophila melanogaster* (Chihara et al., 1982), although the molecular weights of most larval proteins are well below 27,000 Daltons. Twenty-one soluble protein fractions were also separated from the untanned cuticle of *Sarcophaga bullata* maggots with molecular weights of 16,000–24,000 Daltons (Lipke et al., 1981).

The interactions between chitin and protein are complex and not well understood, but are important in maintaining the structure and material properties of the cuticle. Chitin microfibrils are bound to protein by both covalent and non-covalent bonds and this complex is immersed in a matrix of loosely bound protein (Hackman, 1976). Proteins covalently bind to chitin by means of aspartyl and/or histidyl residues to form N-acylglucosamine glycoproteins (Hackman, 1974). In *S. bullata* larvae, about 36.2% of the cuticle protein was found to be covalently bound to chitin (Lipke et al., 1981). Non-covalent binding between chitin and protein chains, chitin chains and proteins also plays a major role in maintaining the structure of the cuticle. Hackman and Goldberg (1978) suggested that non-covalent binding between chitin and proteins is non-specific.

Intermolecular covalent cross-linking of proteins to N-acetyldopamine during sclerotization also results in pronounced changes in the properties of the cuticle. One theory is that proteins covalently bind to the β-carbon of the side chain of N-acetyldopamine and/or to the quinone ring following oxidation of N-acetyldopamine by phenoloxidase (Andersen, 1974, 1976; Karlson and Sekeris, 1976). Others believe that the mechanism of sclerotization is a result of cuticular dehydration, and that cross-linking of cuticular components is of little importance (Vincent and Hillerton, 1979; Hillerton and Vincent, 1979).

In addition to chitin and protein, the insect integument contains many other minor components such as lipids, pigments, and hydrocarbons. However, a discussion of these components is beyond the scope of this chapter. For further details on the various biochemical and physiological aspects of chitin and the insect integument, the reader is referred to reviews by Hackman, 1974; Neville, 1975; Hepburn, 1976; Muzzarelli, 1977; and Andersen, 1979.

CHITIN SYNTHESIS AND HORMONAL CONTROL

Chitin synthesis is associated with the moulting process and development of insects. *De novo* synthesis of chitin occurs during the formation of new cuticle at the moult and for varying periods following moulting (Surholt, 1975). The moulting process allows immature insects and other arthropods to shed the old cuticle, which is rigid and limits growth, and to replace it with a new one which is sufficiently flexible to permit expansion and growth. The moulting cycle and chitin synthesis are controlled by the moulting hormone, β-ecdysone or 20-hydroxy ecdysterone (Schneiderman and Gilbert, 1964; Kimura, 1973; Karlson et al., 1975; Quennedey et al., 1983). β-Ecdysone initiates a multiplicity of events during the moult in both the epidermal cells and fat body. These include DNA and RNA synthesis, organelle formation, massive intermoult synthesis and differentiation in the epidermis and activation of chitinase activity (Neville, 1975; Dean et al., 1980; Quennedey et al., 1983).

The biosynthetic pathway for chitin in insects is based on initial studies conducted by Candy and Kilby (1962) with the desert locust *Schistocerca gregaria*. Using homogenates from the wing tissue of 2- to 3-day-old adults, they demonstrated the presence of enzymes necessary to convert glucose, GA, and NAGA into uridine diphosphate *N*-acetylglucosamine (UDP-NAGA). However, attempts to demonstrate the formation of chitin from UDP-NAGA in locust wing cell-free extracts were unsuccessful. The final polymerization step is catalysed by chitin synthetase (CSase), UDP-2-acetamido-2-deoxy-D-glucose: chitin 4-β-acetamidodeoxyglycosyltransferase (EC. 2.4.1.16). The utilization of CSase was first reported in cell-free extracts of the fungus *Neurospora crassa* (Glaser and Brown, 1957). Porter and Jaworski (1965) isolated a cell-free preparation from *Prodenia eridania* larvae capable of forming chitin from UDP-NAGA. However, other attempts to demonstrate CSase activity in cell-free extracts from insects have been until recently unsuccessful (Fristrom, 1968; Ishaaya and Casida, 1974). During the same period, several *in vitro* systems were developed that were capable of synthesizing chitin in reasonable yields from convenient precursors such as glucose, GA, NAGA or UDP-NAGA. These systems were used extensively in studies on the mode of action of chitin synthesis inhibitors and screening for potential inhibitors. Hence, the properties of these systems will be briefly discussed.

In vitro abdomen systems and tissue cultures

Various tissues from a number of holometabolous (Lepidoptera, Diptera) and hemimetabolous (Orthoptera, Hemiptera) insect species were found to be suitable *in vitro* systems for chitin synthesis (Table 1). These systems can be divided into two groups: those that utilize whole or part of the intact abdominal integuments, can synthesize chitin within a few hours and do not require activation by β-ecdysone; and those that utilize specific tissue cultures requiring assay periods of up to 2 weeks and activation by exogenous β-ecdysone. In tissue cultures, the cuticle is apparently formed *de novo* rather than by adding onto an already existing cuticle.

The tissues were usually incubated with buffered solutions or culture media together with radiolabelled precursors, and the chitin synthesized was identified by various methods. The rates of [^{14}C] incorporation into chitin ranged from about 0.1% to 5.0% for each system, and varied with the [^{14}C] precursor used. These differences may be associated with the enzymatic reactions involved at the point of entry of each substrate into the biosynthetic pathway of chitin, as well as with other metabolic processes occurring simultaneously (Hajjar and Casida, 1979; Mayer et al., 1980b). The enzymatic reactions involved during chitin biosynthesis in insects are well known (Candy and Kilby, 1962; Chen and Mayer, 1984).

Cell-free preparations

Recently, cell-free enzyme preparations have been isolated from the larvae or pupae of numerous holometabolous insects. These insects include four species from each of the following orders: Diptera (Mayer et al., 1980a; Turnbull and Howells, 1983), Coleoptera (Cohen and Casida, 1980a) and Lepidoptera (Cohen and Casida, 1980a, 1982a; Mitsui et al., 1981). In addition to these insect species, a CSase system was isolated from larvae of the brine shrimp *Artemia salina* (Horst, 1981). These cell-free enzymes are capable of incorporating NAGA from UDP-NAGA into either cuticular or peritrophic membrane chitin. This product has been found in all cases to be insoluble in aqueous organic solvents, resistant to alkaline hydrolysis or protease, susceptible to acid hydrolysis, and susceptible in varying degrees to chitinases. Acid hydrolysis is complete and the product is invariably identified as GA; however, degradation by chitinase is usually 70–90% complete following prolonged incubations of 20- to 48-hour periods (Cohen and Casida, 1980a; Mayer et al., 1980a; Horst, 1981; Turnbull and Howells, 1983). The products of chitinase digestion depend on the enzyme source and purity. These have been identified as NAGA (Mayer et al.,

Table 1 *In vitro* systems capable of chitin biosynthesis from glucose, glucosamine, or *N*-acetylglucosamine precursors

Insect species	Tissue/organ	Developmental stage	Inhibition (I_{50})[a]		Reference
			Polyoxin D	Benzoylphenyl ureas	
Intact abdomen tissue systems					
Locusta migratoria	Integument	Nymph	Polyoxin A *ca.* 10^{-6} M	—	Surholt (1975)
Melanoplus sanguinipes	Integument	Adult	1.5×10^{-5} M	—	Vardanis (1976)
Oncopeltus fasciatus	Integument	Adult		DFB, 5.5×10^{-7} M; BAY SIR 8514, 69% at 3×10^{-5} M	Hajjar and Casida (1979)
Pieris brassicae	Integument	Larva	80% at 8×10^{-5} M	DFB, 80% at 5×10^{-5} M	Gijswijt *et al.* (1979)
Heliothis zea	Integument	Pharate adult	79% *ca.* 1.92×10^{-5} M	DFB, 82% at 6.4×10^{-9} M	Kaska *et al.* (1980)
Spodoptera littoralis	Integument	Larva		DFB, 4.0×10^{-9} M; CGA 112 913, 2.3×10^{-7} M	Neumann and Guyer (1983)
Musca domestica	Integument	Larva	*ca.* 5.3×10^{-8} M	DFB, *ca.* 5.3×10^{-9} M	van Eck (1979)
Lucilia cuprina	Integument	Larva	6×10^{-7} M	DFB, 7×10^{-7} M	Turnbull and Howells (1982)
Stomoxys calcitrans	Imaginal epidermis	Pupa	1.3×10^{-5} M	DFB, 5.2×10^{-8} M; BAY SIR 8514, 4.4×10^{-7} M; EL 494, 8.6×10^{-6} M	Mayer *et al.* (1980b)
Tissue cultures					
Drosophila melanogaster	Wing imaginal discs	Larva			Mandaron (1976)
Plodia interpunctella	Wing imaginal discs	Larva	*ca.* 1.92×10^{-3} M	DFB, no inhibition at 3.2×10^{-4} M	Oberlander and Leach (1974)
Chilo suppressalis	Integument	Larva			Agui *et al.* (1969)
Chilo suppressalis	Integument	Larva	9.8×10^{-6} M	DFB, 8.3×10^{-8} M; TH 6038, 5.8×10^{-7} M	Nishioka *et al.* (1979); Kitahara *et al.* (1983); Nakagawa *et al.* (1984)
Manduca sexta	Abdominal epidermis	Larva		DFB, 1.1×10^{-9} M	Mitsui *et al.* (1980)
Mamestra brassicae	Abdominal epidermis	Larva	8.8×10^{-8} M	8.9×10^{-9} M	Mitsui *et al.* (1981)
Leucophaea maderae	Leg regenerate	Nymph	3.4–750×10^{-9} M	DFB, 9.4–61.1×10^{-11} M; BAY SIR 8514, 2.1×10^{-11} M	Sowa and Marks (1975); Marks and Sowa (1976); Leighton *et al.* (1981)

[a] Unless otherwise indicated.

Table 2 Source, properties and inhibition of chitin synthetase from insects and brine shrimp

Enzyme source/ species	Subcellular distribution	Chitin synthetase assay	Rate of incorporation	Requirements/ stimulation	Inhibition	
					Nucleoside antibiotics	Benzoylphenyl ureas/others
Integument						
L. cuprina[a,b] (early 3rd instar larvae)	Crude homogenate (cell membrane 1000 g pellet)	80 µl crude homogenate (ca. 11.5 integuments) in cell culture media with 20% fetal serum, 15 µl [14C]UDP-NAGA (10 mM); 30 min at 30 °C	K_m ca. 0.8 mM, ca. 200 pmol/h/ incubation	Mg^{2+} (10 mM) and GA, NAGA (20 mM), no effect; KCN (20 mM) ca. 100%	No preincubation; polyoxin D; $K_i = 4.0 \times 10^{-8}$ M	DFB; $K_i = 5\text{-}8 \times 10^{-6}$ M
M. domestica						
C. erythrocephala						
T. ni[c] (5th instar larvae)	Cell membrane 12,000 g pellet	100 µl enzyme (0.5–1.3 mg protein), 200 µl 25 mM Tris-HCl pH 7.2, MgCl₂ (10 mM) NAGA (18 mM), DTT (1 mM), 5 µl [3H]UDP-NAGA (0.135 µM); 30 min at 30 °C	Close to that of T. castaneum	Mg^{2+}, NAGA; BSA (10 mg/ml) and digitonin (4 mg/ml) during enzyme extraction	10 min preincubation; polyoxin B, $I_{50} = 5 \times 10^{-7}$ M; polyoxin D, $I_{50} = 1.4 \times 10^{-5}$ M; nikkomycin, $I_{50} = 6 \times 10^{-9}$ M; UPT, $I_{50} = 7 \times 10^{-5}$ M	DFB no inhibition at 3×10^{-4} M; captan $I_{50} = 7 \times 10^{-5}$ M
Wing tissues						
H. cecropia[c] (1-wk old pupae)	Cell membrane 12,000 g pellet	Same as T. ni	Slightly lower than that reported for T. ni	Mg^{2+}; NAGA (18 mM) no effect	10 min. preincubation; polyoxin D 14% at 3×10^{-4} M; nikkomycin, $I_{50} = 10^{-3}$ M	BAY SIR 8514 & DFB no inhibition at 3×10^{-3} M; captan, $I_{50} = 4 \times 10^{-5}$ M
Whole insects						
S. calcitrans[d] (4-day old pupae)	Cell membrane 10,000 g pellet	100 µl enzyme (0.7 mg protein) in MOPS pH 7.0, DTE (0.5 mM), [3H]UDP-NAGA (0.1 mM); 2 h at 30 °C	K_m 31.7 ± 0.7 µM, V_{max} 135 ± 10.3 pmol/h/mg protein	Mg^{2+}, NAGA (1 mM) inhibit activity; trypsin ca. 30% at 40 µg/ml	Polyoxin D, 28% at 2×10^{-4} M, 40% at 10^{-3} M	DFB, BAY SIR 8514, EL 494 no inhibition at 2×10^{-5} M; tunicamycin no inhibition at 1.2×10^{-4} M

M. brassicae[e] (early last instar larvae or 8-day old pupae)	Microsomes 78,000 g pellet	1.1 ml enzyme (3 g of larvae) 25 mM Tris-HCl pH 7.5; MgCl$_2$(5 mM), mercaptoethanol (5 mM); EDTA(0.5 mM), NAGA (50 mM), glycerin 25%; [^{14}C]UDP-NAGA(2 μM), 2.5 mg chitodextrin/ml; 2 h at 25 °C	*ca.* 3000 dpm/h	Mg^{2+}, NAGA	Polyoxin D 65% at 1.6 × 10^{-6} M	DFB no inhibition at 2.7 × 10^{-4} M
Gut (peritrophic membrane) *T. castaneum*[f,b] *T. confusum* *T. brevicornis*	Microsomes 105,000 g pellet	350 μl 25 mM Tris-HCl pH 7.2, MgCl$_2$(10mM), NAGA (17 mM), DTT (1 mM), 5 μl [^{3}H]UDP-NAGA (0.2 μM); 30 min at 22 °C	3300–5000 dpm/ min/mg protein for *Triboleum* species	Mg^{2+}, NAGA; trypsin *ca.* 40% at 13 μg/ml	10 min preincubation; polyoxin D, I$_{50}$ = 4 × 10^{-6} M nikkomycin, I$_{50}$ = 2 .. 10^{-8} M	DFB, BAY SIR 8514, tunicamycin, no inhibition at 5 × 10^{-4} M; captan, 95% at 6 × 10^{-5} M
G. melonella			3800 dpm/min/mg protein			
T. molior (**last instar larvae**)			6300 dpm/min/mg protein			
Whole brine shrimp *A. salina*[g]	Microsomes 30,000 g pellet	1 ml enzyme (0.1–2 mg protein), 50 mM Hepes pH 7.1, NaCl (0.4 M), MgCl$_2$(30 mM), glycerol 5‰ DTT (0.65 mM); [^{14}C]-UDP-NAGA (0.168 μM), 60–120 min at 37 °C	0.484 pmol/h/mg protein	Mg^{2+} NAGA (100mM) 38%; trypsin *ca.* 70% at 100 μg/ml	Polyoxin D, 17% at *ca.* 10^{-5} M	60 min preincubation, DFB 52–92% inhibition at 3.2 × 10^{-6} M

[a] Turnbull and Howells (1983).
[b] Chitin synthetase properties and inhibition determined primarily in the first insect listed.
[c] Cohen and Casida (1982a).
[d] Mayer *et al.*, (1980a, 1981).
[e] Mitsui *et al.*, (1981).
[f] Cohen and Casida (1980a, b).
[g] Horst (198.).

1980a), GA, NAGA, chitobiose (Turnbull and Howells, 1983) and N-acetyl-chitobiose (Horst, 1981).

The properties of these enzymes are listed in Table 2. Although several similarities exist among the enzymes, there also are major differences with regard to enzyme requirements, kinetics and inhibition by nucleoside antibiotics and BPUs. The enzymes are particulate, i.e. membrane-bound, although it is unclear with what membrane the CSases are associated (Chen and Mayer, 1984). The subcellular distribution of the enzyme from $A.$ $salina$ may be associated with the plasma membrane (Horst, 1981). Studies with yeasts and fungi indicate that the enzymes from $Streptomyces$ $cerevisiae$ (Duran et $al.$, 1975) and $Phycomyces$ $blakesleeanus$ (Jan, 1974) are also localized in the plasma membrane.

The enzyme preparation from $Lucilia$ $cuprina$ is not activated by NAGA at concentrations up to 20 mM (Turnbull and Howells, 1983). The enzyme may also have no requirements for Mg^{2+}, since increasing the Mg^{2+} concentration in the incubation from 1 mM to 10 mM had no effect on activity. Similar findings have been reported for $Stomoxys$ $calcitrans$ L. enzyme preparation; in fact, at 1 mM concentrations both the Mg^{2+} and NAGA inhibited enzyme activity (Mayer et $al.$, 1980a). Findings from both studies strongly suggest that dipterous cell-free enzymes are not activated by either Mg^{2+} or NAGA, and may differ from CSases of insects from other orders. The enzyme preparations from $Hyalophora$ $cecropia$ (Cohen and Casida, 1982a) and $A.$ $salina$ (Horst, 1981) do not require NAGA for activation. The enzymes from the other species require both Mg^{2+} and NAGA for activation, and are similar in that respect to fungal enzymes (Jan, 1974, de Rousset-Hall and Gooday, 1975; Leighton et $al.$, 1981). NAGA probably serves as an allosteric affector.

The activities of the gut enzyme from $Triboleum$ $castaneum$ (Cohen and Casida, 1980a) and the enzymes from $S.$ $calcitrans$ (Mayer et $al.$, 1980a) and $A.$ $salina$ (Horst, 1981) are slightly activated (about 30–70%) by the addition of trypsin at 10–100 µg/ml concentrations. Although these findings may indicate the presence of an endogenous activation factor, such as unmasking of catalytic or allosteric sites of the enzymes, it is unclear whether activation is associated with a zymogenic form of the enzyme (Cohen and Casida, 1982b). Proteolytic activation of CSase zymogen has been demonstrated for $S.$ $cerevisiae$ (Cabib, 1972) and $Candida$ $albicans$ (Hardy and Gooday, 1983).

The enzymes from $T.$ $castaneum$ and $S.$ $calcitrans$ are also insensitive to tunicamycin (Cohen and Casida, 1980b; Mayer et $al.$, 1981). Cohen and Casida (1980b, 1982b) concluded that the enzymatic reaction of $T.$ $castaneum$ CSase is not mediated by a lipid carrier essential for the synthesis of many glycoproteins. However, the role, if any, of lipid-linked oligosaccharides in chitin biosynthesis in insects is not yet clear.

Attempts to obtain active enzyme preparations from the following insects or insect organs were unsuccessful: $Myzus$ $persicae$ (Sulzer), whole homogenates of nymphs or adults; *Oncopeltus fasciatus* (Dallas), whole homogenates of embryos, last instar nymphs, and newly emerged adults; *Teleogryllus commodus* (Walker) gut from eighth instar female nymphs; *Byrsotria fumigata* (Guérin) guts from females; *Spodoptera exigua* (Hübner) gut from last instar larvae; *D. melanogaster* imaginal disks treated with β-ecdysone (Cohen and Casida, 1980a); and *Boarmia selenaria, Earias insulana, Heliothis virescens, O. fasciatus, S. exigua* and *T. castaneum* integuments from last instar larvae (Cohen and Casida, 1982a).

CHITIN SYNTHESIS INHIBITORS

Several chemicals have been found to inhibit *in vitro* and/or *in vivo* chitin synthesis in insects, crustaceans or fungi. Chitin synthesis inhibition has been demonstrated for the BPUs, polyoxins and nikkomycins. Inhibition of chitin synthesis has also been suggested for other chemicals, such as for a triazine insect growth regulator, an organophosphate fungicide, tunicamycins, avermectins and plumbagin. In addition, several groups of chemicals, including known sulphenimide fungicides, triazine herbicides and organochlorine insecticides, are good inhibitors of *in vitro* chitin synthesis in abdomen systems and tissue cultures or cell-free preparations. Inhibition, in most cases, has been associated with the final polymerization step catalysed by CSase.

$$(NAGA)_n + UDP\text{-}NAGA \rightarrow (NAGA)_{n+1} + UDP$$

The effects of these chemicals on chitin synthesis, and the significance of these findings in relation to their modes of action, are discussed in the following sections.

Benzoylphenyl ureas (BPUs)

Following the discovery of BPUs, extensive research was conducted with diflubenzuron (DFB), 1-(2,6-difluorobenzoyl)-3-(4-chlorophenyl) urea (**2**), to determine its insecticidal properties and biochemical effects in various insects, crustaceans and fungi. Several analogues have since been synthesized and are being developed for commercial use (Table 3). These analogues have similar ovicidal and larvicidal effects as DFB but their toxicities vary with different insects. However, none is apparently as effective as DFB against dipterous larvae except for EL 494. The primary effects of BPUs (**2–7**) involve the inhibition of chitin synthesis in immature insects. The ovicidal activity of DFB, and most probably other BPUs, has also been associated with the disruption of cuticle formation in the developing embryo (Grosscurt, 1978). Other biochemical and histological effects thought to be either primary or secondary have also been reported and will be discussed in relation to proposed mode of action of BPUs in a subsequent section.

Table 3   Benzoylphenyl urea insecticides

Name	Chemical nomenclature	Reference
BAY SIR 8514 (3)	2-chloro-N-[[[4-(trifluoromethoxy)-phenyl]amino]-carbonyl]benzamide	Zoebelein *et al.* (1980)
EL 494 (4)	N-[[[5-(4-bromophenyl)-6-methyl-2-pyrazinyl]amino]-carbonyl]-2,6-dichlorobenzamide	Eli Lilly (1977)
CME 134 (5)	1-(3,5-dichloro-2,4-difluoro-phenyl)-3-(2,6-difluorobenzoyl)-urea	Becher *et al.* (1983)
CGA 112913 (6)	1-[3,5-dichloro-4-(3-chloro-5-trifluoromethyl-2-pyridyloxy)-phenyl]-3-(2,6-difluorobenzoyl) urea	Neumann and Guyer (1983)
XRD 473 (7)	N-[[[3,5-dichloro-4-(1,1,2,2-tetrafluoroethoxy)phenyl]amino]-carbonyl]-2,6-difluorobenzamide	Sbragia *et al.* (1983)

In vivo inhibition of chitin formation by DFB and DU 19111, 1-(2,6-dichlorobenzoyl)-3-(3,4-dichlorophenyl) urea (**8**), was first demonstrated by means of histological and microautoradiographical procedures in *P. brassicae* fifth instar larvae (Mulder and Gijswijt, 1973). Practically no radioactivity was present in the endocuticle of DU 19111-treated larvae injected with [^{3}H]glucose (Post *et al.*, 1974). Similar histological findings were observed in the integuments of DFB-treated locusts (Ker, 1977) and *T. molitor* (Soltani *et al.*, 1984). Biochemical studies also indicated reduced incorporation of [^{14}C]glucose into chitin extracted from integuments of DU 19111-treated *P. brassicae* larvae (Post and Vincent, 1973). A dose-related decrease in cuticle chitin content was reported in DFB-treated housefly larvae (Ishaaya and Casida, 1974), *O. fasciatus* nymph (Hajjar and Casida, 1978, 1979) and *Culex pipiens* larvae (Hajjar, 1979). Inhibition of chitin biosynthesis by DFB was also reported for adult locusts (Hunter and Vincent, 1974), *Manduca sexta* (Mitsui *et al.*, 1980) and *Spodoptera littoralis* larvae (Neumann and Guyer, 1983).

Although BPUs inhibit chitin synthesis, no effects on protein biosynthesis have been observed in several insect species. These include *P. brassicae* (Post and Vincent, 1973; Post *et al.*, 1974), *Musca domestica* (Ishaaya and Casida, 1974) and *S. gregaria* (Ker, 1977) immatures and *Anthonomus grandis* adults (Mitlin *et al.*, 1977).

The effects of DFB on chitin biosynthesis in the peritrophic membrane are not as dramatic and clear as those observed in the integuments. Mulder and Gijswijt (1973) did not find any histological differences between the peritrophic membranes of the DFB-treated and untreated *P. brassicae* larvae. Clarke *et al.* (1977) reported that DFB partially blocked chitin synthesis during the production of peritrophic membrane in locusts. Similarly, the *in vivo* and *in vitro* rates of peritrophic membrane formation in adult *Calliphora erythrocephala* were

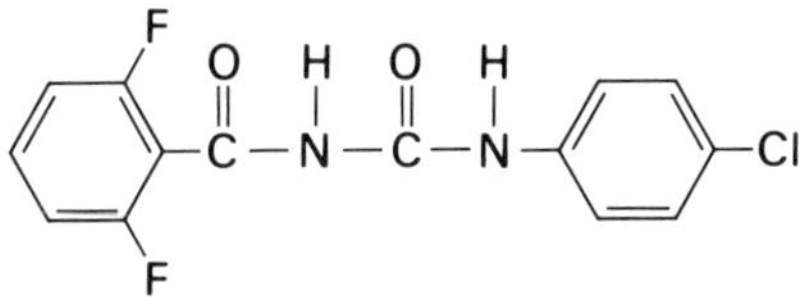
F
O
H
O
H
C
N
C
N
Cl
F

(**2**)

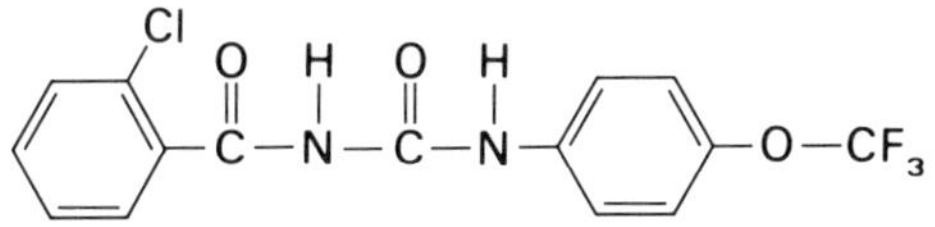
Cl
O
H
O
H
C
N
C
N
O
CF3

(**3**)

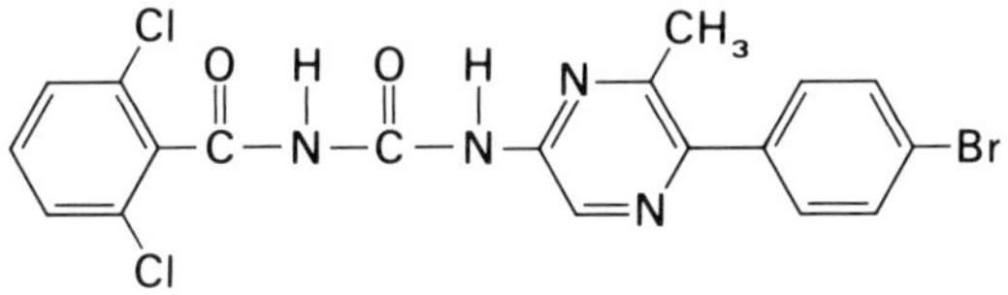
Cl
O
H
O
H
CH3
N
C
N
C
N
Br
N
Cl

(**4**)

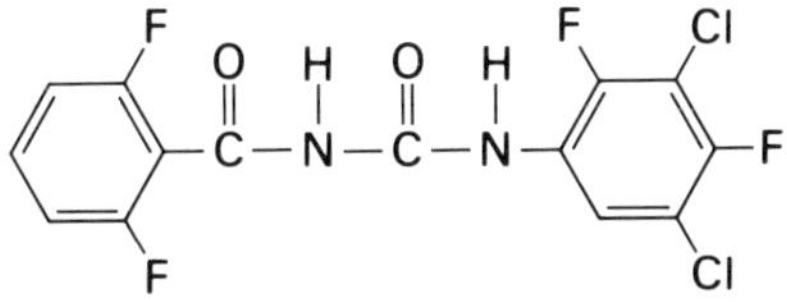
F
O
H
O
H
F
Cl
C
N
C
N
F
F
Cl

(**5**)

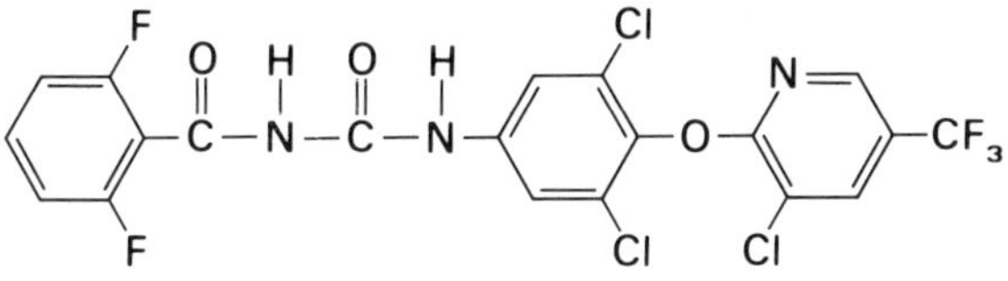
F
O
H
O
H
Cl
N
C
N
C
N
O
CF3
F
Cl
Cl

(**6**)

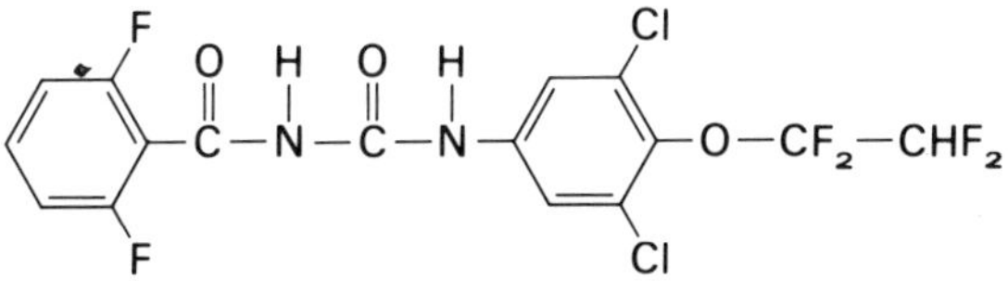
F
O
H
O
H
Cl
C
N
C
N
O
CF2
CHF2
F
Cl

(**7**)

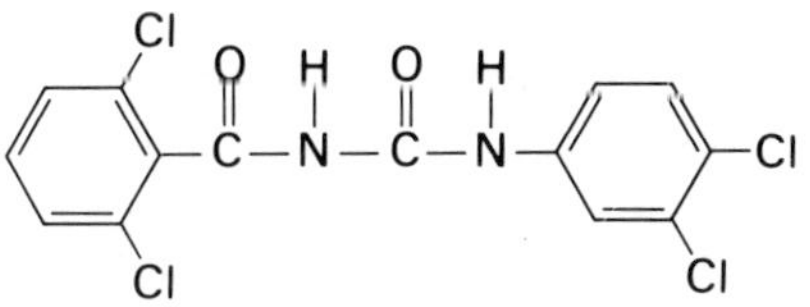
Cl
O
H
O
H
C
N
C
N
Cl
Cl
Cl

(**8**)

reduced by DFB treatment, but inhibition was not dose-dependent, and histological changes were more prominent under *in vitro* conditions (Becker, 1978). However, Cohen and Casida (1980a,b, Table 2) found that DFB had no effect on a cell-free CSase isolated from gut tissue of *T. castaneum.*

The biochemical effects of BPUs on *in vitro* chitin synthesis were first studied in the abdomen systems and tissue cultures described (Table 1). Inhibitory effects were compared to polyoxin D, the known competitive substrate inhibitor of CSase in fungi (Endo and Misato, 1969; Endo *et al.,* 1970). Inhibition of chitin biosynthesis in these systems varies with the precursor used (Hajjar and Casida, 1979; Mayer *et al.,* 1980b). However, the BPUs were found to be potent inhibitors of chitin synthesis, with I_{50} values as low as 2.1×10^{-11} M, in general, 1–2 orders of magnitude more potent than polyoxin D. DFB inhibits chitin synthesis in all these *in vitro* systems except the cuticle-producing system from *Plodia interpunctella* (Oberlander and Leach, 1974).

Hajjar and Casida (1978, 1979) studied the inhibition of [^{14}C]chitin biosynthesis from [^{14}C]glucose in the isolated milkweed bug abdomen by 24 BPUs in relation to their toxicity in fifth instar nymphs. A good correlation was found between toxicity to nymphs, expressed as the LD_{50} values, and inhibition of chitin synthesis. Inhibition by DFB was rapid and accumulation of UDP-NAGA, the presumed precursor to chitin, occurred within 20 minutes when NAGA was used as the substrate. The authors concluded that the primary action of BPUs appears to involve direct and rapid inhibition of a terminal step in chitin synthesis within the integument. Recently, Nakagawa *et al.* (1984) found a similar correlation between larvicidal activity and chitin biosynthesis inhibition in integuments from the larvae of the rice stem borer *Chilo suppressalis,* Walker.

Accumulation of UDP-NAGA was also observed in the cockroach leg regenerate (Marks and Sowa, 1976) and *M. domestica* systems (van Eck, 1979), but not in the *L. cuprina* abdomen (Turnbull and Howells, 1982). Mitsui *et al.* (1981) reported the accumulation of UDP-NAGA from *in vivo* experiments using *Mamestra brassicae* larvae treated with DFB. The accumulation of NAGA has been reported from *in vivo* experiments using *P. brassicae* larvae treated with DU 19111 but not with DFB (Post *et al.,* 1974; Deul *et al.,* 1978).

Although the above findings strongly indicate inhibition of the final polymerization step during chitin synthesis, a definitive conclusion on CSase inhibition can only be demonstrated with cell-free enzyme preparations. The findings by Turnbull and Howells (1983) are particularly interesting in that they provide the first report of a cell-free chitin synthesizing preparation from an insect tissue that is sensitive to inhibition by DFB. The time course incorporation of NAGA into chitin by the crude enzyme from *L. cuprina* is presented in Figure 1. The hyperbolic curve was reported to be due to the rapid utilization and/or degradation of UDP-NAGA (about 60 nmol/h) and increased inhibition by UDP. Both DFB and polyoxin D inhibit the enzyme preparation from *L.*

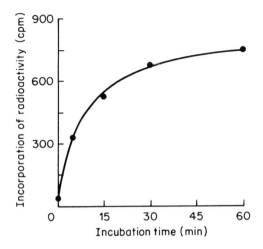

Figure 1 Incorporation of radioactivity from [14C]UDP-NAGA-insoluble material by homogenates of larval integuments from *L. cuprina*. The data are from a single experiment and each point is the mean of duplicate determinations (Reproduced by permission of I. F. Turnbull and A. J. Howells)

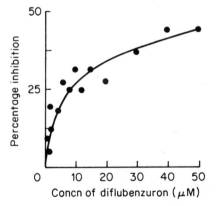

Figure 2 Inhibition of chitin synthesis in homogenates of larval integuments from *L. cuprina* by DFB. Mean values from seven experiments with different enzyme preparations, calculated as percentages of control (Reproduced by permission of I. F. Turnbull and A. J. Howells)

cuprina at low concentrations (Figure 2 and Table 2). However, inhibition is not complete, affecting only a maximum of about 50% of the total incorporation even at relatively high concentrations. Turnbull and Howells (1983) concluded that more than one type of labelled product was being formed in *L. cuprina* enzyme preparation, one being sensitive and the other(s) being insensitive to polyoxin D and DFB. The sensitive product is chitin and the insensitive product(s) are possibly mucopolysaccharides or glycoproteins.

Although both DFB and polyoxin D inhibit the enzyme preparation from *L. cuprina*, these chemicals apparently do not inhibit the same metabolic step since their inhibitory effects are not additive (Turnbull and Howells, 1983). This difference in the modes of action of polyoxin D and DFB has been widely accepted and is supported by similar findings with cell-free enzyme preparation from various insects, crustaceans and fungi. The CSase from *S. calcitrans* was insensitive to BPUs (**2,3**) at 20 μM concentrations, and high concentrations of polyoxin D were required for inhibition to occur (Mayer *et al.*, 1980a, 1981). Decreased sensitivity to polyoxin D and nikkomycin has also been reported for *H. cecropia* enzyme preparations (Cohen and Casida, 1982a). The enzymes from other insects (Table 2) were all very sensitive to the nucleoside antibiotics but not to the BPUs (**2,3**) (Cohen and Casida, 1980b, 1982a; Mitsui *et al.*, 1981). Similarly, an enzyme preparation from larvae of the silkworm *Bombyx mori* was not inhibited by DFB and its 2,6-dichlorobenzoyl analogue, but was inhibited by polyoxin D. About 48% inhibition was reported at 50 μM concentrations of polyoxin D (Kitahara *et al.*, 1983). In accordance with these findings, DFB was reported to stimulate *T. brevicornis* gut CSase activity *in vivo* (Cohen and Casida, 1980b). In contrast, the activity of enzymes from *M. brassicae* larvae treated with DFB *in vivo* was only one-tenth that of enzymes from untreated control larvae (Mitsui *et al.*, 1981).

Although DFB was found to inhibit CSase activity of only one insect species, inhibition has also been reported for a microsomal enzyme from 72-hour-old larvae of the brine shrimp *A. salina* (Horst, 1981). Inhibition was observed when the enzyme was preincubated for 1 hour with DFB at concentrations of 1 μg/ml (3.2 μM). Approximately 76–92% inhibition was noted with various enzyme fractions. This inhibition is consistent with the high *in vivo* toxicity of DFB to newly hatched larvae. At concentrations of about 10 ppb of DFB, all larvae died within 3 days (Cunningham, 1976). In comparison to DFB, polyoxin D had little, if any, effect on the enzyme (Horst, 1981).

DFB did not inhibit CSase from the fungi *Coprinus cinereus* (Brillinger, 1979) and *P. blakesleeanus* (Leighton *et al.*, 1981; Marks *et al.*, 1982), or the yeast *S. cerevisiae* (Cohen and Casida, 1980b).

Polyoxins and nikkomycins

The polyoxins and nikkomycins are two groups of pyrimidine nucleoside antibiotics isolated from *Streptomyces cacaoi* var. *asoensis* (Endo *et al.*, 1970) and *S.*

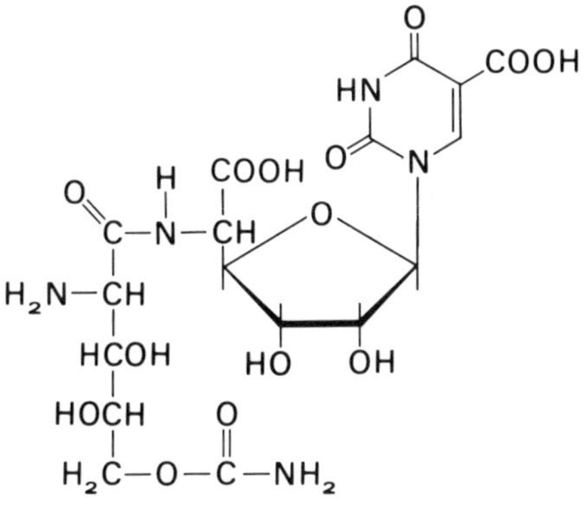

(9)

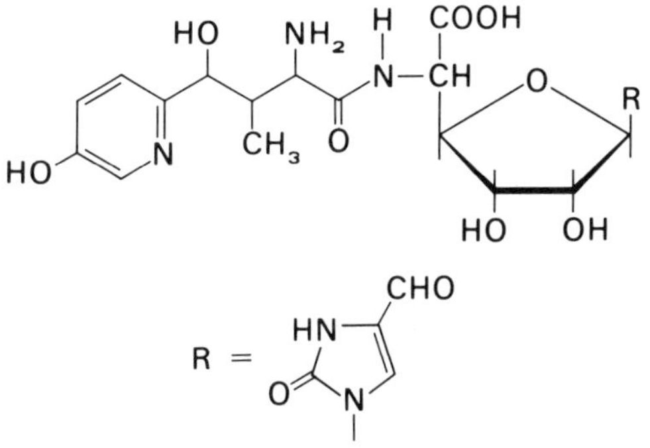

(10a)

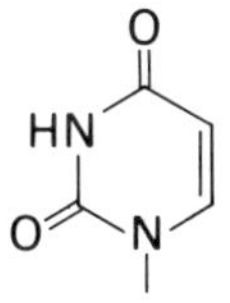

(10b)

tendae, Tü 901 (Dähn *et al.*, 1976), respectively. The polyoxins consist of at least 12 active components, with polyoxin D (9) being the major and most potent fungicidal component (Endo *et al.*, 1970). Similarly, several nikkomycins have been identified and (10a) and (10b) were reported to be the most potent components (Mothes and Seitz, 1982).

Both chemicals are structural analogues of UDP-NAGA and are competitive substrate inhibitors of CSase from several fungi (Dähn *et al.*, 1976; Brillinger 1979; Leighton *et al.*, 1981) and insects (Cohen and Casida, 1980b, 1982a,b). Consequently, they have been used extensively in comparative *in vitro* studies with insecticidal chitin synthesis inhibitors. The inhibitory properties of these antibiotics to *in vitro* systems, tissue cultures and cell-free preparation from various insects have already been discussed (see also Tables 1 and 2).

The inactivity of polyoxin D to CSase of the brine shrimp (Horst, 1981) is of interest, primarily because it has been recently reported that polyoxin D inhibits *in vivo* chitin synthesis in brine shrimp (Calcott and Fatig, 1984). About 70% inhibition was observed in 48-hour-old larvae at 100 µg/ml concentrations. The LD_{50} value of polyoxin D to *A. salina* larvae is 12 ppm (Calcott, 1984, personal communication). These findings indicate that chitin biosynthesis is inhibited *in vivo* by polyoxin D, but inhibition is apparently not associated with CSase. It is possible that *A. salina* CSase differs in many ways from those of the insect and fungal enzymes, including its susceptibility to both polyoxin D and DFB, particularly in view of recent findings that avermectin, a γ-aminobutyric acid receptor inhibitor, was also reported to inhibit chitin biosynthesis in this species (Calcott and Fatig, 1984).

Captan and sulphenimide fungicides

The inhibition of chitin synthesis by captan (**11**) was first reported in the cockroach leg regenerate system by Marks and Sowa (1976) (Table 4). Captan was also found to be a potent inhibitor of CSase activity from integuments of *T. ni*, and wing tissues of *H. cecropia* and gut of *T. castaneum* (Cohen and Casida, 1980b, 1982a; Table 2). It also inhibits the *in vitro* synthesis of peritrophic membrane tissues from adult *C. erythrocephala* (Becker, 1978).

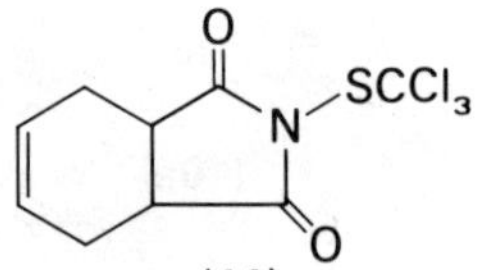

$$\text{(11)}$$

Table 4 Inhibitors of *Phycomyces blakesleeanus* chitin synthetase and cockroach chitin synthesis system[a]

Compound	I_{50} (M)	
	P. blakesleeanus	*L. maderae*
Chlordane	3.1×10^{-5}	7.3×10^{-6}
Endosulfan	3.0×10^{-5}	—
DDT	3.1×10^{-5}	3.9×10^{-8}
Dieldrin	1.6×10^{-4}	—
Captan	3.9×10^{-4}	10^{-4} M
Azidotriazine (CGA 19255)	1.5×10^{-3}	$<4.0 \times 10^{-9}$
Kitazin P	4.7×10^{-4}	$< 10^{-4}$ M
Dinocap	5.0×10^{-5}	—
Isoprothiolane	6.6×10^{-5}	1.3×10^{-7}
Tunicamycin	No effect	7.0×10^{-9}

[a] Marks and Sowa, 1976; Leighton *et al.*, 1981; Marks *et al.*, 1982.

In fungi, captan inhibited CSase activity from *P. blakesleeanus* (Leighton *et al.*, 1981; Table 4), but not enzyme activity from *C. cinereus* (Brillinger, 1979). The I_{50} value of captan for CSase of the yeast *S. cerevisiae* was reported to be 240 μM (Cohen and Casida, 1980b). However, inhibition by captan is non-competitive (Marks *et al.*, 1982), and may result from direct reaction with sulphydryl groups or from breakdown to thiophosgene, which reacts with the enzyme binding sites (Cohen and Casida, 1980b).

Triazine herbicides

The triazine insect growth inhibitor CGA 19255 (**12**) and other triazine herbicides did not inhibit CSase from the gut of *T. castaneum* (Cohen and Casida, 1980b) or from whole *S. calcitrans* (Mayer *et al.*, 1981). CGA 19255 was found

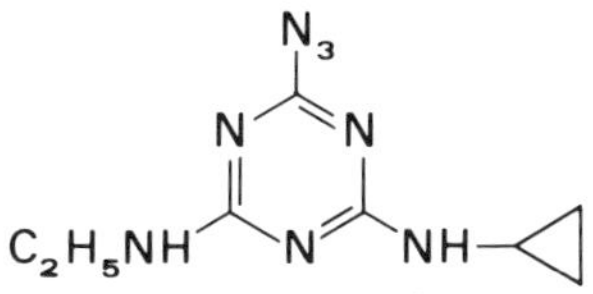

(**12**)

to be a weak inhibitor of CSase from *P. blakesleeanus* but was a potent inhibitor in the cockroach leg regenerate system (Leighton *et al.*, 1981; Marks *et al.*, 1982; Table 4). Miller, R. W. *et al.* (1981) reported that inhibition by CGA 19255 and its structural analogue CGA 72662 was less pronounced than DFB in the cockroach leg regenerate system. The exact mode of action of triazine insect growth regulators is unclear, but they apparently do not manifest their lethal action via effects on chitin synthesis.

Chlorinated hydrocarbons

Several chlorinated insecticides, e.g. chlordane and DDT, have been found to inhibit CSase from *P. blakesleeanus* and chitin synthesis in the cockroach leg regenerate system (Leighton *et al.*, 1981; Marks *et al.*, 1982; Table 4). Inhibition of the fungal enzyme by these insecticides is non-specific in all cases. Since the insecticidal actions of these neurotoxins is very rapid, inhibition of chitin biosynthesis, if any, is probably insignificant.

Phenylcarbamates

The herbicide Barban (**13**) and certain other phenylcarbamates (H-24108, H-22949) were reported to be moderate to weak inhibitors of CSase from *T. castaneum* and yeast (Cohen and Casida, 1980b). About 77% and 79% inhibition was observed for Barban in yeast at 240 μM and *Tribolium* at 400 μM,

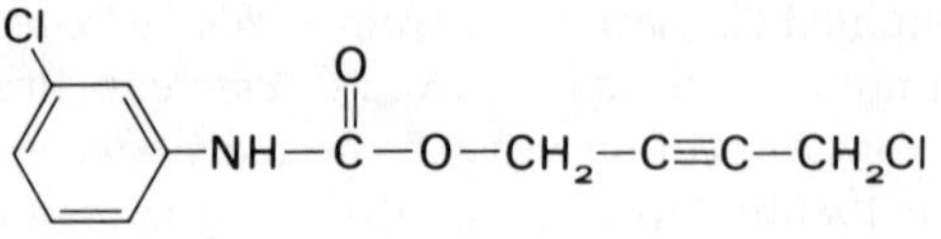

$$\text{Cl} \quad \overset{O}{\underset{\|}{C}}$$

Structure with Cl-substituted benzene ring: $-NH-\overset{O}{\underset{\|}{C}}-O-CH_2-C{\equiv}C-CH_2Cl$

(13)

respectively. Cohen and Casida (1980b) suggested that inhibition may be due to direct action of Barban with sulphydryl groups. In contrast, the phenylcarbamate herbicides propham, chlorpropham and phenmedipham and several other phenylcarbamate insecticides and acaracides were inactive (Cohen and Casida, 1980b).

Kitazin P

Kitazin P, S-benzyl-O,O-diisopropyl phosphorothiolate (14), is an organophosphate fungicide used against *Pyricularia oryzae*. Maeda *et al.* (1970) found that treatment of mycelia with Kitazin P resulted in the inhibition of cell wall synthesis and the accumulation of UDP-NAGA. The authors suggested that this organophosphate inhibited chitin synthesis or possibly the permeation of chitin precursors through cell membranes.

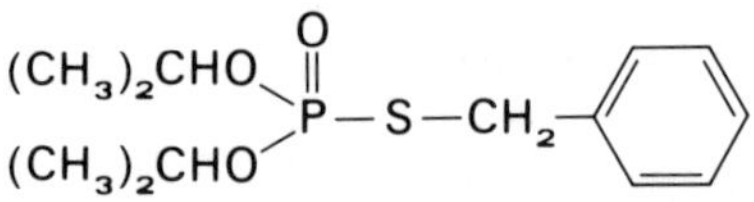

$$(CH_3)_2CHO \underset{(CH_3)_2CHO}{\overset{O}{\underset{}{\diagdown}}} \overset{\|}{P} - S - CH_2 -\bigcirc$$

(14)

Kitazin P also was found to inhibit CSase activity from *P. blakesleeanus* but not from the cockroach leg regenerate system (Marks *et al.*, 1982; Table 4). In contrast, Kitazin, the diethyl analogue, did not inhibit CSase activity from *T. castaneum* gut (Cohen and Casida, 1980b) but did inhibit chitin synthesis in the milkweed *in vitro* abdomen system (Hajjar and Casida, 1979). About 31% inhibition was found at 3×10^{-6} M. However, it has been determined that Kitazin P inhibits specifically the sequential transmethylation of phosphatidylethanolamine into phosphatidylcholine (Kodama and Akatsuka, 1982).

Tunicamycin

Tunicamycin (**15**) is a complex of closely related antibiotics isolated from *Streptomyces lysosuperficus* (Takatsuki *et al.*, 1971). This antibiotic is a

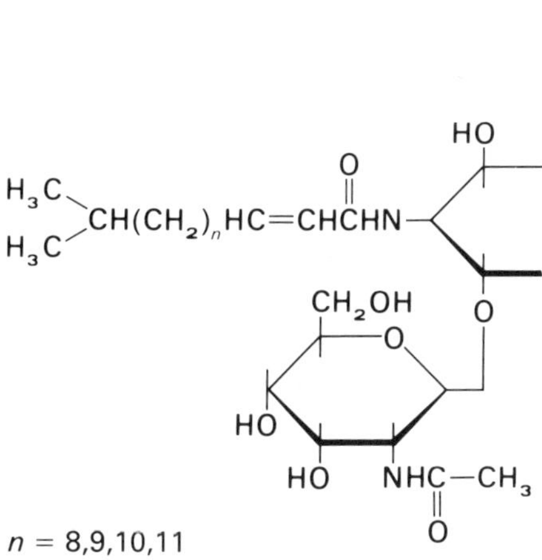

$n = 8,9,10,11$

(**15**)

non-competitive inhibitor of UDP-NAGA : dolichyl-P-NAGA-1-P transferase and prevents the transfer of NAGA to dolichyl phosphate in certain yeasts and mammals (Lehle and Tanner, 1976; Elbein *et al.*, 1979; Heifetz *et al.*, 1979). Tunicamycin also inhibits the incorporation of glucose into glycosyl-phosphoryl-dolichol and into lipid-linked oligosaccharides, but does not inhibit the synthesis of mannosyl-phosphoryl-dolichol or the addition of the second NAGA to NAGA-pyrophosphoryl-dolichol to form *N,N′*-diacetylchitobiosyl-pyrophosphoryl-dolichol (Elbein *et al.*, 1979).

The antibiotic inhibits chitin synthesis in the cockroach leg regenerate system (Marks *et al.*, 1982; Table 4) and in isolated abdomens of two insects (Quesada Allue, 1982) but not CSase activity from *T. castaneum* (Cohen and Casida, 1980b) and *S. calcitrans* (Mayer *et al.*, 1981). Cohen and Casida, (1980b, 1982b) suggested that chitin biosynthesis is not mediated by a lipid carrier. However, the role of these carriers in insects is not yet clear. The formation of NAGA lipid-linked oligosaccharides in insects has been investigated by Quesada Allue and his colleagues (Quesada Allue *et al.*, 1976a,b; Quesada Allue and Belocopitow, 1978). When either [^{14}C]-labelled UDP-NAGA, UDP-glucose, UDP-galactose or GDP-mannose was incubated with microsomal enzymes obtained from whole *Ceratitis capitata* pupae, the corresponding [^{14}C]dolichol-pyrophosphate was synthesized (Quesada Allue *et al.*, 1976a). Formation of the lipid-linked NAGA oligosaccharide occurred concomitantly with chitin synthesis *in vitro*, with [^{14}C]NAGA incorporation accounting for 0.71% and 2.62%, respectively (Quesada Allue *et al.*, 1976b)

Under other experimental conditions, dolichol-pyrophosphate-N,N'-diacetylchitobiose was formed when microsomal enzymes from *C. capitata* pupae were incubated with UDP-[^{14}C]NAGA (Quesada Allue and Belocopitow, 1978). The enzymes also catalyse the transfer of mannose from GDP-mannose into lipid-linked oligosaccharides of 8–9 maltose units in size. The authors suggested that these glycolipids may participate in glycoprotein biosynthesis but their role in chitin biosynthesis could not be defined.

Quesada Allue (1982) found that tunicamycin caused 30–33% inhibition of the incorporation of [^{14}C]GA into [^{14}C]chitin in an *in vitro* abdomen system from newly moulted imagos of *Triatoma infestans* (Hemiptera). Moreover, the amount of radioactivity released by chitinase digestion of *T. infestans* [^{14}C]chitin, synthesized in the presence of added tunicamycin, also decreased by about 35%. The author suggested that [^{14}C]GA was incorporated into two different pools of chitin, with the tunicamycin-susceptible pool being associated with the synthesis of a lipid-linked saccharide intermediate involved in the formation of chitin–protein complexes. However, a more definitive conclusion could be made if inhibition were demonstrated with a microsomal enzyme preparation instead of with cultured abdominal sections.

Studies conducted by Horst (1983) indicated the formation of similar lipid-linked NAGA and mannose oligosaccharides by brine shrimp microsomes. Incubations with [^{3}H] NAGA yielded a series of oligosaccharides ranging from 2 to 8 glycosyl units in size, with the longer components being sensitive to chitinase digestion. It was suggested that these lipid-linked chitin oligosaccharides may participate in the formation of an endogenous primer for chitin synthesis after their transfer to a protein acceptor, whereas mannose lipid-linked oligosaccharides may be involved with glycoprotein biosynthesis (Horst, 1983).

Tunicamycin was found to be a weak, competitive substrate inhibitor of CSase from *N. crassa* with a K_i of 480 $\pm$ 60 µM (Selitrennikoff, 1979). Based on the structural similarity of the antibiotic and the substrate, the linear competitive inhibition and the fact that the enzyme is not known to possess a lipid intermediate, Selitrennikoff (1979) suggested that tunicamycin inhibits the enzyme by directly competing with the substrate. In contrast, Marks *et al.* (1982) found that tunicamycin stimulates CSase from *P. blakesleeanus* by inactivating lipid-carrier reactions competing with chitin synthesis for UDP-NAGA. The authors suggested that the inhibition observed in *N. crassa* may be due to the inside-out nature of the vesicles involved in chitin synthesis, which necessitates the mediation of substrate entry into the vesicle by a non-physiological lipid-carrier process.

Avermectins

The avermectins are a mixture of antibiotics produced by *Streptomyces avermitilis* MA-4680 upon fermentation and have potent insecticidal, acaricidal and nematocidal properties (Burg *et al.*, 1979; Miller, *et al.*, 1979; Putter *et al.*, 1981). Four closely related major components and four homologous minor components of these macrocyclic lactone antibiotics were separated, characterized and identified (Miller, *et al.*, 1979; Albers-Schönberg *et al.*, 1981). Studies with avermectin B_{1a} (**16**) have shown that it interferes with γ-aminobutyric binding to synaptic receptors, thus disturbing Cl^- passage

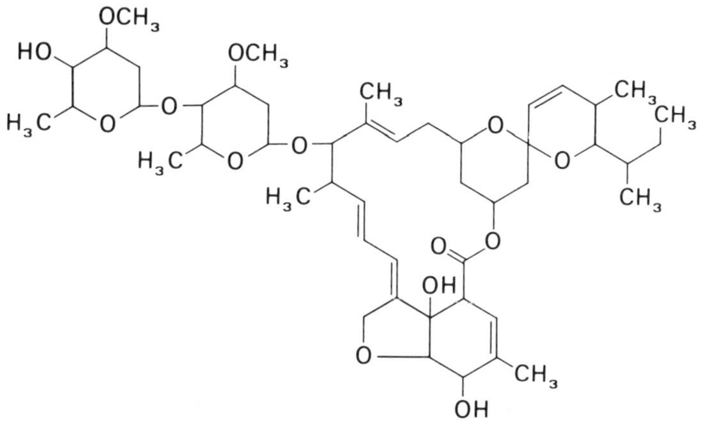

(**16**)

through the end-plate, resulting in inhibitory post-synaptic potentials and eventual paralysis. These findings were reported in studies with muscle membranes from lobsters (Fritz *et al.*, 1979; Mellin *et al.*, 1983) and the ventral nerve cord from *Ascaris* (Kass *et al.*, 1980), both of which are avermectin-susceptible organisms, and in studies with crude brain–membrane systems from rat, which is intrinsically resistant to avermectins (Pong and Wang, 1982; Pong *et al.*, 1982).

Although avermectins are neurotoxins and kill susceptible organisms by causing loss of motor function and paralysis, Calcott and Fatig (1984) reported that avermectins may also inhibit chitin turnover and synthesis in other organisms at low concentrations, therefore inhibiting the moulting process. These conclusions were based on studies with the fungus *Mucor miehei*, which lacks a nervous system associated with GABA receptors, and the brine shrimp *A. salina*. Both organisms were found to be susceptible to avermectin. The minimal inhibitory concentration to *M. miehei* was 2.1 µg/ml and the LC_{90} value for the brine shrimp was 0.8 ng/ml. Approximately 50% inhibition of chitin biosynthesis in *M. miehei* and *A. salina* was observed at concentrations of 4.5 µg/ml and 10 ng/ml, respectively. The organisms were preincubated with the test material for 20–30 minutes prior to the addition of [^{3}H]NAGA. In addition, avermectin also was found to decrease chitin turnover in the brine shrimp at 10 ng/ml, and to inhibit *Streptomyces antibioticus* chitinase activity at an apparent K_i of approximately 2–3 µg/ml (Calcott and Fatig, 1984). However, the possibility of chitin synthesis inhibition in the brine shrimp being associated with toxicity cannot be excluded.

The toxicity of avermectins to numerous insects, mites and ticks has been reported recently, but no studies were found on the inhibition of chitin synthesis. Avermectins, especially B_{1a}, are highly toxic to several tetranychid mites (Putter *et al.*, 1981; Grafton-Cardwell and Hoy, 1983) and to fire ants upon feeding (Lofgren and Williams, 1982). When administered orally to cattle and livestock at very low doses (20–200 µg/kg per day), avermectins effectively control numerous tick ectoparasites (Drummond *et al.*, 1981) and manure-breeding dipterous larvae (Miller, J. A. *et al.*, 1981; Schmidt, 1983). It was observed that mites, lepidopterous larvae (Putter *et al.*, 1981), ticks and fire ants treated with avermectin were unable to move and feed, which led to death in a few days. Mortality was apparently not associated with a moult. These observations and the relatively low concentrations needed to produce these effects strongly suggest that avermectins act as neurotoxins rather than as inhibitors of chitin synthesis or metabolism.

Plumbagin

Plumbagin is a naturally occurring product isolated from the root of a tropical medicinal shrub *Plumbago capensis* and found to inhibit CSase isolated from *T. ni* integuments (Kubo *et al.*, 1983). The isolated product was characterized and identified as plumbagin (**17**). Plumbagin was reported to inhibit the incorporation of NAGA from UDP-[^{3}H]NAGA into chitin by about 29% at a relatively high concentration of 0.3 mM. This activity was comparable to that of polyoxin D, which caused 31% inhibition at 0.3 mM. In feeding experiments, plumbagin inhibited the ecdysis of the first or second instar of *Pectinophora gossypiella*, *Heliothis zea*, *H. virescens* and *T. ni*. The effective doses for 90% inhibition ranged from 400 to 1400 ppm.

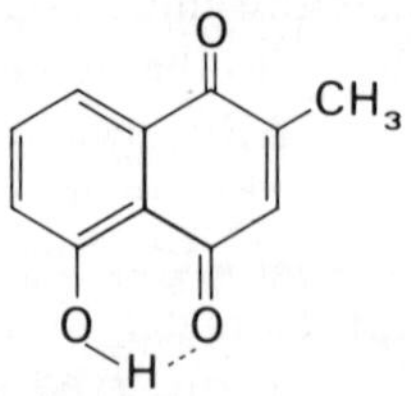

(**17**)

PROPOSED MODE OF ACTION OF CHITIN SYNTHESIS INHIBITORS

Although several groups of chemicals, including known pesticides, inhibit chitin synthesis in tissue cultures or cell-free preparations, inhibition was non-specific and/or reversible, and insecticidal action was not associated with chitin biosynthesis. The polyoxin and nikkomycin fungicides have been found to be competitive substrate inhibitors of CSase from several fungi (Endo *et al.*, 1970; Dähn *et al.*, 1976; Brillinger, 1979; Leighton *et al.*, 1981) and are thought to have similar inhibitory properties on insect enzymes (Cohen and Casida, 1980b, 1982a; Turnbull and Howells, 1983). However, their potential use as insecticides is apparently limited. Consequently, these chemicals will not be discussed further, and this section will be limited to BPUs.

Since the initial discovery of the unusual insecticidal action of BPUs, much research has been conducted to define the molecular basis of chitin synthesis inhibition. Although chitin synthesis inhibition in insects is the primary target of BPUs in most cases, several other biochemical and histological effects were determined. It has been suggested by various researchers that these effects are either primary, and associated directly with the action of BPUs, or secondary, resulting from overall physiological or biochemical changes in the treated insect. Several mechanisms of action were proposed. These mechanisms are identified for the purpose of this discussion as follows: hormonal effects, inhibition of DNA synthesis, active metabolite formation, inhibition of lipid-linked oligosaccharide intermediate, zymogen activation, effects on regulatory mechanisms and inhibition of UDP-NAGA transport.

Hormonal effects

Yu and Terriere (1975, 1977) reported that DFB interfered with ecdysone and juvenile hormone titres and these changes would lead to the various physiological and biochemical changes observed in BPU-treated insects. They found that DFB inhibited a β-ecdysone metabolizing enzyme in three species of flies, *M. domestica, S. bullata* and *Phormia regina*. Increased β-ecdysone levels would stimulate chitinase and phenoloxidase activities as reported by Ishaaya and Casida (1974) and interfere with normal chitin synthesis and enhance breakdown of chitin. β-Ecdysone was also suggested to stimulate microsomal oxidase activity which, in turn, would stimulate juvenile hormone metabolism and result in juvenile hormone deficiency (Yu and Terriere, 1975). Lung (1980) found that larvae of the wasp *Athalia rosae* were sensitive to DFB only during the first two-thirds of any instar. DFB treatments apparently did not impair cuticle formation but resulted in a premature moult, possibly associated with premature activity of the endocrine glands, corpora allata and ventral head gland observed in histological preparations. Lung (1980) concluded that DFB does not inhibit chitin synthesis, but rather affects the endocrine system.

Although these findings indicate possible involvement of hormone mediation in the primary mode of action of DFB, this mechanism is ruled out by several other findings. O'Neill *et al.* (1977) did not find significant differences in the endogenous β-ecdysone titres from DFB-treated and control pharate pupae of the stable fly *S. calcitrans*. Hajjar and Casida (1979) also found that DFB did not alter *in vivo* metabolism of α- or β-[³H]ecdysone by fifth instar milkweed nymphs. Moreover, the data on chitin synthesis inhibition in *in vitro* abdomen systems, tissue cultures and cell-free preparations, provide more direct evidence on the primary effects of BPUs.

Inhibition of DNA synthesis

DFB has also been found to inhibit DNA synthesis in several species as demonstrated by reduced radiolabelled thymidine incorporation following topical treatment. Mitlin *et al.* (1977) studied the effect of DFB on DNA synthesis in an attempt to determine the mechanism of DFB sterilization in the boll weevil, *A. grandis*. They found that DFB treatment inhibited the biosynthesis of DNA in 1- to 2-day-old adults. DFB also inhibited DNA biosynthesis and mitoses in the epidermis from abdominal sternites of *T. molitor* pupae (Soltani *et al.*, 1984). However, DNA inhibition was observed only at relatively high DFB concentrations associated with prematurely blocked pupal–adult development. In a histological study, Meola and Mayer (1980) found that DFB also inhibited the formation of the adult imaginal epidermis in pharate pupae of *S. calcitrans*. Inhibition was due to reduction of DNA synthesis specifically in the imaginal epidermal histoblasts. DeLoach *et al.* (1981) suggested that DFB may alter cell membrane properties in a manner similar to that observed for certain dithiocarbonilates or Kitazin P, and prevent active transport of nucleosides and amino acids.

Although inhibition of DNA synthesis would explain most of the observed effects of DFB on insects, it may be limited to species forming imaginal epidermis, such as the cyclorrhaphous Diptera and Hymenoptera. This specific action apparently cannot be recognized during the mitotic cycle of *Tenebrio* pupae. Inhibition of DNA synthesis in the abdominal epidermis was also observed at low DFB concentrations, which did not block pupal–adult development but led to an abnormal moult (Soltani *et al.*, 1984). Obviously, in the absence of the epidermal cells in *S. calcitrans*, an adult cuticle cannot be formed; but DFB also inhibits chitin synthesis by established epidermal cells (Mayer *et al.*. 1980b).

Active metabolite formation

Cohen and Casida (1980b) suggested that inhibition of gut CSase from *T. castaneum* may be due to an active and extremely potent metabolite of DFB.

However, they ruled out this hypothesis when preliminary findings showed that potential metabolites did not inhibit the enzyme, and that DFB was not readily metabolized by *T. castaneum* (Cohen and Casida, 1980b) or by other insects (Ivie and Wright, 1978). The rapid *in vitro* inhibition of chitin synthesis (Hajjar and Casida, 1979), and the absence of a common nucleus among the different BPUs for an active metabolite that might act on CSase further exclude this mechanism of action (Chen and Mayer, 1984).

Inhibition of lipid-linked oligosaccharide intermediates

Several researchers suggested that BPUs inhibit the transfer of NAGA into NAGA-pyrophosphoryl-dolichol in a manner similar to tunicamycin (Marks and Sowa, 1976; Brillinger, 1979; Mayer *et al.*, 1981). This would suggest that the transfer of NAGA to the growing chain of chitin by CSase is not the last step in the biosynthesis of chitin in insects. If such inhibition occurs, it would also explain the accumulation of UDP-NAGA found in certain *in vitro* systems (Marks and Sowa, 1976; Hajjar and Casida, 1979; van Eck, 1979). Inhibition of chitin synthesis by tunicamycin has been observed in certain *in vitro* systems (Marks *et al.*, 1982; Quesada Allue, 1982) but none has been demonstrated with cell-free preparations from insects (Cohen and Casida, 1980b; Mayer *et al.*, 1981). Although the existence of these intermediates in insects has been demonstrated (Quesada Allue *et al.*, 1976a,b; Quesada Allue and Belocopitow, 1978), their role in chitin biosynthesis has not yet been defined.

Inhibition of zymogen activation

It has been suggested that BPUs may prevent insect chitin synthesis by interfering with the proteolytic activation of the CSase zymogen (Leighton *et al.*, 1981; Marks *et al.*, 1982; Turnbull and Howells, 1983). Marks *et al.* (1982) reported that CSase activation in insects is controlled by ecdysone and catalysed by a protease. The action of BPUs may be explained by assuming that these compounds inhibit the proteases involved in enzyme activation. Inhibition would lead to a cascade effect similar to that observed in fungal preparations.

In support of this hypothesis Marks *et al.* (1982) demonstrated that known chymotrypsin inhibitors, such as chymostatin, 2-nitro-4-carboxyphenyl-*N,N*-diphenylcarbamate and lima bean and soya bean trypsin–chymotrypsin inhibitors, did inhibit chitin synthesis in the cockroach leg regenerate. Turnbull and Howells (1983) speculated that CSase present in the crude homogenates of *L. cuprina* is initially in an inactive form and is rapidly activated by endogenous proteases upon enzyme incubation; addition of DFB inhibits enzyme activation. This proposal would then be consistent with their findings that inhibition by polyoxin D and DFB is not additive, and with the results of

Horst (1981) who found that microsomal preparations from larvae of the brine shrimp show reduced capacity for chitin synthesis following preincubation with DFB. However, DFB does not inhibit integumental proteases from wing buds of the migratory locust nymphs (Cohen and Casida, 1982b) and insect enzyme preparations are active without the need for exogenous activation by proteases (Cohen and Casida, 1980a; Mayer *et al.*, 1980a; Horst, 1981). More importantly DFB inhibits chitin synthesis rapidly in *in vitro* abdomen systems even when CSase is in an active form (Hajjar and Casida, 1978, 1979).

Effects on regulatory mechanisms

Cohen and Casida (1980b, 1982a,b) have suggested yet another mechanism that may be involved in chitin synthesis inhibition. This involves a regulatory mechanism associated with the polymerization step in chitin synthesis that is inhibited by DFB *in vivo*, but not *in vitro*. Disruption of this mechanism may lead to DFB insensitivity and continuous *in vitro* chitin synthesis and fibrillo-geneses. DFB was postulated to affect such a mechanism by stabilizing certain proteins or altering a critical spatial organization of the enzyme within the cell membrane. However, this hypothesis would require a better understanding of the organization and function of the insect enzyme complex *in situ*, and there is yet no evidence for such a mechanism.

CONCLUSIONS

Although several groups of chemicals have been found to inhibit *in vivo* and/or *in vitro* chitin synthesis in various insects, crustaceans or fungi, only the BPUs have potent insecticidal properties associated with chitin biosynthesis. The primary effects of BPUs in most insects involve inhibition of chitin synthesis during the moulting process of immature stages. Their potent ovicidal activity is also associated with the disruption of cuticle formation in the developing embryos. Additional effects have been observed and are apparently associated with other primary or possibly secondary effects resulting from overall physio-logical or biochemical changes in treated insects. Most notable is the inhibition by DFB of DNA synthesis in the imaginal epidermal histoblasts of *S. calcitrans* pharate pupae; yet inhibition of chitin synthesis by established epidermal cells has also been demonstrated.

During active *in vitro* chitin biosynthesis in isolated abdomens from *O. fasciatus*, the action of BPUs (**2,3**) involves direct and rapid inhibition of the transfer of NAGA to the growing chitin chain. Inhibition of chitin synthesis in cell-free preparation by DFB has been demonstrated in the blowfly *L. cuprina* and the brine shrimp *A. salina*. Inhibition apparently involves a biochemical step or site that is different from that of polyoxin D. In *L. cuprina* homogenates, DFB and polyoxin D do not have additive effects, whereas the CSase from *A.*

salina is not susceptible to inhibition by polyoxin D. In addition, DFB does not inhibit CSase from several other insect species, yeasts or fungi, whereas polyoxin D and nikkomycin are good inhibitors. Both fungicides are structural analogues of UDP-NAGA and are competitive inhibitors of CSase at relatively low concentrations.

In an attempt to explain chitin synthesis inhibition by the BPUs, several mechanisms have been proposed. These include:

(1) interference with hormone metabolism, which in turn disrupts chitin synthesis;
(2) inhibition of DNA synthesis;
(3) inhibition of CSase by an active metabolite;
(4) interference with NAGA lipid-linked oligosaccharide synthesis, similar to tunicamycin;
(5) inhibition of the proteolytic activation of the CSase zymogen; and
(6) inhibition of an *in vivo* regulatory mechanism for CSase or alteration of a critical spatial organization of the enzyme.
(7) inhibition of UDP-NAGA transport across biomembranes

Although these mechanisms provide possible explanations, and some evidence was demonstrated for mechanisms 1,2,4,5 and 7, none of these provides a satisfactory answer. Another plausible explanation could well be that inhibition in insects is not associated with metabolic steps leading to chitin synthesis or CSase but rather with the disruption of events occurring during or following chitin synthesis, such as transport of macromolecules or spatial organization and interactions of chitin–protein components, which could cause feedback inhibition of CSase.

ACKNOWLEDGEMENTS

I gratefully acknowledge the assistance of my wife Sophie and Ms. Anne Gardner in the preparation of the manuscript, and the valuable criticism of Dr S. Zakhari.

REFERENCES

Agui, N., Yagi, S., and Fukaya, M. (1969). 'Induction of moulting of cultivated integument taken from a diapausing rice stem borer larva in the presence of ecdysterone (Lepidoptera: Pyralidae)', *Appl. Ent. Zool.*, **4**, 156–157.

Albers-Schönberg, G., Arison, B. H., Chabala, J. C., Douglas, A. W., Eskola, P., Fisher, M. H., Lusi, A., Mrozik, H., Smith, J. L., and Tolman, R. L. (1981). 'Avermectins. Structure determination', *J. Am. Chem. Soc.*, **103**, 4216–4221.

Anderson, S. O. (1974). 'Evidence for two mechanisms of sclerotisation in insect cuticle', *Nature*, **251**, 507–508.

Anderson, S. O. (1976). 'Cuticular enzymes and sclerotization in insects', in *The Insect Integument* (ed. H. R. Hepburn), pp. 121–144, Elsevier, Amsterdam.

Anderson, S. O. (1979). 'Biochemistry of insect cuticle', *Ann. Rev. Entomol.*, **24**, 29–61.

Becher, H-M., Becker, P., Prokic-Immel, R., and Wirtz, W. (1983). 'CME 134, a new chitin synthesis inhibiting insecticide', 10th International Congress of Plant Protection. *Plant Protection for Human Welfare*, 1, 408–415.

Becker, B. (1978). 'Effects of 20-hydroxy-ecdysone, juvenile hormone, dimilin and captan on *in-vitro* synthesis of peritrophic membranes in *Calliphora erythrocephala*', *J. Insect Physiol.*, 24, 699–705.

Brillinger, G. U. (1979). 'Metabolic products of microorganisms 181. Chitin synthase from fungi, a test model for substances with insecticidal properties', *Archs Microbiol.*, 121, 71-74.

Burg, R. W., Miller, B. M., Baker, E. E., Birnbaum, J., Currie, S. L., Hartman, R., Kong, Y-L., Monaghan, R. L., Olson, G., Putter, I., Tunac, J. B., Wallick, H., Stapley, E. O., Oiwa, R., and Omura, S. (1979). 'Avermectins, new family of potent anthelmintic agents: producing organism and fermentation', *Antimicrob. Agents Chemother.*, 15, 361–367.

Cabib, E. (1972). 'Chitin synthetase system from yeast', *Methods in Enzymology*, 28, 572–580.

Calcott, P. H., and Fatig, R. O. (1984). 'Inhibition of chitin metabolism by avermectin in susceptible organisms', *J. Antibiot.*, 37, 253–259.

Candy, D. J., and Kilby, B. A. (1962). 'Studies on chitin synthesis in the desert locust', *J. exp. Biol.* 39, 129–140.

Chen, A. C., and Mayer, R. T. (1984). 'Insecticides: effects on the cuticle', in *Comprehensive Insect Physiology, Biochemistry and Pharmacology* (eds G. A. Kerkut and L. I. Gilbert), Chapter 3, Volume 12, In Press. Pergamon Press, Oxford, England.

Chihara, C. J., Silvert, D. J., and Fristrom, J. W. (1982). 'The cuticle proteins of *Drosophila melanogaster*: stage specificity', *Devl. Biol.*, 89, 379–388.

Clarke, L., Temple, G. H. R., and Vincent, J. F. V. (1977). 'The effects of a chitin inhibitor—Dimilin—on the production of peritrophic membrane in the locust, *Locusta migratoria*', *J. Insect Physiol.*, 23, 241–246.

Cohen, E., and Casida, J. E. (1980a). 'Properties of *Tribolium* gut chitin synthetase', *Pestic. Biochem. Physiol.*, 13, 121–128.

Cohen, E., and Casida, J. E. (1980b). 'Inhibition of *Tribolium* gut chitin synthetase', *Pestic. Biochem. Physiol.*, 13, 129–136.

Cohen, E., and Casida, J. E. (1982a). 'Properties and inhibition of insect integumental chitin synthetase', *Pestic. Biochem. Physiol.*, 17, 301–306.

Cohen, E., and Casida, J. E. (1982b). 'Insect chitin synthetase as a biochemical probe for insecticidal compounds', in *Pesticide Chemistry, Human Welfare and the Environment, Volume 3. Mode of Action, Metabolism and Toxicology* (eds S. Matsunaka, D. H. Hutson and S. D. Murphy), pp. 25–32, Pergamon Press, New York.

Cunningham, P. A. (1976). 'Effects of Dimilin (TH 6040) on reproduction in the brine shrimp *Artemia salina*', *Environ. Entomol.*, 5, 701–706.

van Daalen, J. J., Meltzer, J., Mulder, R., and Wellinga, K. (1972). 'A new insecticide with a novel mode of action', *Naturwissenschaften*, 59, 312–313.

Dähn, U., Hagenmaier, H., Höhne, H., König, W. A., Wolf, G., and Zähner, H. (1976). 'Stoffwechselprodukte von Mikroorganismen. 154. Nikkomycin, ein neuer Hemmstoff der Chitinsynthese bei Pilzen', *Archs Microbiol.*, 107, 143–160.

Dean, R. L., Bollenbacher, W. E., Locke, M., Smith, S. L., and Gilbert, L. I. (1980). 'Haemolymph ecdysteroid levels and cellular events in the intermoult/moult sequence of *Calpodes ethlius*', *J. Insect Physiol.*, 26, 267–280.

DeLoach, J. R., Meola, S. M., Mayer, R. T., and Thompson, J. M. (1981). 'Inhibition of DNA synthesis by diflubenzuron in pupae of the stable fly *Stomoxys calcitrans* (L.)', *Pestic. Biochem. Physiol.*, 15, 172–180.

Deul, D. H., DeJong, B. J., and Kortenbach, J. A. M. (1978). 'Inhibition of chitin synthesis by two 1-(2,6 disubstituted benzoyl)-3-phenylurea insecticides II', *Pestic. Biochem. Physiol.*, **8**, 98–105.

Drummond, R. O., Whetstone, T. M., and Miller, J. A. (1981). 'Control of ticks systemically with Merck MK-933, an avermectin', *J. Econ. Entomol.*, **74**, 432–436.

Duran, A., Bowers, B., and Cabib, E. (1975). 'Chitin synthetase zymogen is attached to the yeast plasma membrane', *Proc. Natl Acad. Sci. USA*, **72**, 3952–3955.

van Eck, W. H. (1979). 'Mode of action of two benzoylphenyl ureas as inhibitors of chitin synthesis in insects', *Insect Biochem.*, **9**, 295–300.

Elbein, A. D., Gafford, J., and Kang, M. S. (1979). 'Inhibition of lipid-linked saccharide synthesis: comparison of tunicamycin, streptovirudin, and antibiotic 24010', *Archs Biochem. Biophys.*, **196**, 311-318.

Eli Lilly (1977), Technical Report on EL-494. Lilly Research Laboratories, Indianapolis, Indiana.

Endo, A., Kakiki, K., and Misato, T. (1970). 'Mechanism of action of the antifungal agent polyoxin D', *J. Bact.*, **104**, 189–196.

Endo, A., and Misato, T. (1969). 'Polyoxin D, a competitive inhibitor of UDP-*N*-acetylglucosamine: chitin *N*-acetyl-glucosaminyltransferase in *Neurospora crassa*', *Biochem. Biophys. Res. Commun.*, **37**, 718–722.

Fristrom, J. W. (1968). 'Hexosamine metabolism in imaginal disks of *Drosophila melanogaster*', *J. Insect Physiol.*, **14**, 729–740.

Fritz, L. C., Wang, C. C., and Gorio, A. (1979). 'Avermectin B_{1a} irreversibly blocks postsynaptic potentials at the lobster neuromuscular junction by reducing muscle membrane resistance', *Proc. Natl Acad. Sci., USA*, **76**, 2062–2066.

Gijswijt, M. J., Deul, D. H., and DeJong, B. J. (1979). 'Inhibition of chitin synthesis by benzoyl-phenylurea insecticides, III. Similarity in action in *Pieris brassicae* (L.) with polyoxin D', *Pestic. Biochem. Physiol.*, **12**, 87–94.

Glaser, L., and Brown, D. H. (1957). 'The synthesis of chitin in cell-free extracts of *Neurospora crassa*', *J. Biol. Chem.*, **228**, 729–742.

Grafton-Cardwell, E. E., and Hoy, M. A. (1983). 'Comparative toxicity of Avermectin B_1 to the predator *Metaseiulus occidentalis* (Nesbitt) (Acari: Phytoseiidae) and the spider mites *Tetranychus urticae* Koch and *Panonychus ulmi* (Koch) (Acari: Tetranychidae)', *J. Econ. Entomol.*, **76**, 1216–1220.

Grosscurt, A. C. (1978). 'Diflubenzuron: some aspects of its ovicidal and larvicidal mode of action and an evaluation of its practical possibilities', *Pestic. Sci.*, **9**, 373–386.

Hackman, R. H. (1974). 'Chemistry of the insect cuticle', in *The Physiology of Insecta* (ed. M. Rockstein), pp. 215–270, Academic Press, New York.

Hackman, R. H. (1976). 'The interactions of cuticular proteins and some comments on their adaptation to function', in *The Insect Integument* (ed. H. R. Hepburn), pp. 107–120, Elsevier, Amsterdam.

Hackman, R. H., and Goldberg, M. (1977). 'Molecular crosslinks in cuticles', *Insect Biochem.*, **7**, 175–184.

Hackman, R. H., and Goldberg, M. (1978). 'The non-covalent binding of two insect cuticular proteins by a chitin', *Insect Biochem.*, **8**, 353–357.

Hajjar, N. P. (1979). 'Diflubenzuron inhibits chitin synthesis in *Culex pipiens* L. larvae', *Mosquito News*, **39**, 381–384.

Hajjar, N. P., and Casida, J. E. (1978). 'Insecticidal benzoylphenyl ureas; structure-activity relationships as chitin synthesis inhibitors', *Science*, **200**, 1499–1500.

Hajjar, N. P., and Casida, J. E. (1979). 'Structure-activity relationships of benzoylphenyl ureas as toxicants and chitin synthesis inhibitors in *Oncopeltus fasciatus*', *Pestic. Biochem. Physiol.*, **11**, 33–45.

Hardy, J. C., and Gooday, G. W. (1983). 'Stability and zymogenic nature of chitin synthase from *Candida albicans*', *Current Microbiol.*, **9**, 51–54.

Heifetz, A., Keenan, R. W., and Elbein, A. D. (1979). 'Mechanism of action of tunicamycin on the UDP-GlcNAc: Dolichyl-phosphate GlcNAc-1-phosphate transferase', *Biochemistry*, **18**, 2186–2192.

Hepburn, H. R. (ed.) (1976). *The Insect Integument*, Elsevier, Amsterdam.

Hillerton, J. E., and Vincent, J. F. V. (1979). 'The stabilization of insect cuticles', *J. Insect Physiol.*, **25**, 957–963.

Horst, M. N. (1981). 'The biosynthesis of crustacean chitin by a microsomal enzyme from larval brine shrimp', *J. Biol. Chem.*, **256**, 1412–1419.

Horst, M. N. (1983). 'The biosynthesis of crustacean chitin. Isolation and characterization of polyprenol-linked intermediates from brine shrimp microsomes', *Archs Biochem. Biophys.*, **223**, 254–263.

Hunter, E., and Vincent, J. F. V. (1974). 'The effects of a novel insecticide on insect cuticle', *Experientia*, **30**, 1432.

Ishaaya, I., and Casida, J. E. (1974). 'Dietary TH 6040 alters composition and enzyme activity of housefly larval cuticle', *Pestic. Biochem. Physiol.*, **4**, 484–490.

Ivie, G. W., and Wright, J. E. (1978). 'Fate of diflubenzuron in the stable fly and house fly', *J. Agric. Food chem.*, **26**, 90–94.

Jan, Y. N. (1974). 'Properties and cellular localization of chitin synthetase in *Phycomyces blakesleeanus*', *J. Biol. Chem.*, **249**, 1973–1979.

Jeuniaux, C. (1963). *Chitine et Chitinolyse*, Masson, Paris.

Karlson, P., Koolman, J., and Hoffmann, J. A. (1975). 'Biochemistry of ecdysone', *Am. Zool.*, **15** (Suppl. 1), 49–59.

Karlson, P., and Sekeris, C. E. (1976). 'Control of tyrosine metabolism and cuticle sclerotization by ecdysone', in *The Insect Integument* (ed. H. R. Hepburn), pp. 145–156, Elsevier, Amsterdam.

Kaska, H. M., Mayer, R. T., Sowa, B. A., Meola, R. W., and Coppage, D. L. (1980). 'Chitin synthesis in *Heliothis zea* (Boddie) pupae and inhibition by chitin synthesis inhibitors', *Southwestern Entomol.*, **5**, 139–143.

Kass, I. S., Wang, C. C., Walrond, J. P., and Stretton, A. O. W. (1980). 'Avermectin B_{1a}, a paralyzing anthelmintic that affects interneurons and inhibitory motoneurons in *Ascaris*,' *Proc. Natl Acad. Sci., USA*, **77**, 6211–6215.

Ker, R. F. (1977). 'Investigation of locust cuticle using the insecticide diflubenzuron', *J. Insect Physiol.*, **23**, 39–48.

Kimura, S. (1973). 'The control of chitin deposition by ecdysterone in larvae of *Bombyx mori*', *J. Insect Physiol.*, **19**, 2177–2181.

Kitahara, K., Nakagawa, Y., Nishioka, T., and Fujita, T. (1983). 'Cultured integument of *Chilo suppressalis* as a bioassay system of insect growth regulators', *Agr. Biol. Chem.*, **47**, 1583–1589.

Kobinata, K., Uramoto, M., Nishii, M., Kusakabe, H., Nakamura, G., and Isono, K. (1980). 'Neopolyoxins A, B, and C, new chitin synthetase inhibitors', *Agric. Biol. Chem.*, **44**, 1709–1711.

Kodama, O., and Akatsuka, T. (1982). 'Kitazin P and edifenphos, possible inhibitors of phosphatidylcholine biosynthesis', in *Pesticide Chemistry, Human Welfare and the Environment, Volume 3, Mode of Action, Metabolism and Toxicology* (eds. S. Matsunaka, D. H. Hutson and S. D. Murphy), pp. 135–140, Pergamon Press, New York.

Kubo, I., Uchida, M., and Klocke, J. A. (1983). 'An insect ecdysis inhibitor from the African medicinal plant, *Plumbago capensis* (Plumbaginaceae); a naturally occurring chitin synthetase inhibitor', *Agr. Biol. Chem.*, **47**, 911–913.

Lehle, L., and Tanner, W. (1976). 'The specific site of tunicamycin inhibition in the formation of dolichol-bound *N*-acetylglucosamine derivatives', *FEBS Lett.*, **71**, 167–170.

Leighton, T., Marks, E., and Leighton, F. (1981). 'Pesticides: insecticides and fungicides are chitin synthesis inhibitors', *Science*, **213**, 905–907.

Lipke, H., Strout, K., Hensel, W., and Sugumaran, M. (1981). 'Structural proteins of sarcophagid larval exoskeleton', *J. Biol. Chem.*, **256**, 4241–4246.

Lofgren, C. S., and Williams, D. F. (1982). 'Avermectin B_{1a}: highly potent inhibitor of reproduction by queens of the red imported fire ant (Hymenoptera: Formicidae)', *J. Econ. Entomol.*, **75**, 798–803.

Lung, G. (1980). 'Untersuchungen zur Wirkung von Diflubenzuron (Dimilin^R) auf die Larven von *Athalia rosae* L.', *Z. Pflanzenkrankh. Pflanzensh.*, **87**, 13–26.

Maeda, T., Hiroshi, A., Kazuo, K., and Misato, T. (1970). 'Studies on the mode of action of the organophosphorus fungicide, Kitazin. Part II. Accumulation of an amino sugar derivative on Kitazin treated mycelia of *Pyricularia oryzae*', *Agr. Biol. Chem.*, **34**, 700–709.

Mandaron, P. (1976). 'Ultrastructure des disques de patte de drosophile cultives *in vitro*. Evagination, secretion de la cuticle nymphale et apolysis', *Wilhelm Roux' Arch. Dev. Biol.*, **179**, 185–196.

Marks, E. P., Leighton, T., and Leighton, F. (1982). 'Modes of action of chitin synthesis inhibitors', in *Insecticide Mode of Action* (ed. J. Coats), pp. 281–313. Academic Press, New York.

Marks, E. P., and Sowa, B. A. (1976). 'Cuticle formation *in vitro*', in *The Insect Integument* (ed. H. R. Hepburn), pp. 339–357, Elsevier, Amsterdam.

Mayer, R. T., Chen, A. C., and DeLoach, J. R. (1980a). 'Characterization of a chitin synthase from the stable fly *Stomoxys calcitrans* (L.)', *Insect Biochem.*, **10**, 549–556.

Mayer, R. T., Chen, A. C., and DeLoach, J. R. (1981). 'Chitin synthesis inhibiting insect growth regulators do not inhibit chitin synthase', *Experientia*, **37**, 337–338.

Mayer, R. T., Meola, S. M., Coppage, D. L., and DeLoach, J. R. (1980b). 'Utilization of imaginal tissues from pupae of the stable fly for the study of chitin synthesis and screening of chitin synthesis inhibitors', *J. Econ. Entomol.*, **73**, 76–80.

Mellin, T. N., Busch, R. D., and Wang, C. C. (1983). 'Postsynaptic inhibition of invertebrate neuromuscular transmission by Avermectin B_{1a}', *Neuropharmacology*, **22**, 89–96.

Meola, S. M., and Mayer, R. T. (1980). 'Inhibition of cellular proliferation of imaginal epidermal cells by diflubenzuron in pupae of the stable fly', *Science*, **207**, 985–987.

Miller, J. A., Kunz, S. E., Oehler, D. D., and Miller, R. W. (1981). 'Larvicidal activity of Merck MK-933, an avermectin, against the horn fly, stable fly, face fly, and house fly', *J. Econ. Entomol.*, **74**, 608–611.

Miller, R. W., Corley, C., Cohen, C. F., Robbins, W. E., and Marks, E. P. (1981). 'CGA-19255 and CGA-72662: Efficacy against flies and possible mode of action and metabolism', *Southwestern Entomol.*, **6**, 272–278.

Miller, T. W., Chaiet, L., Cole, D. J., Cole, L. J., Flor, J. E., Goegelman, R. T., Gullo, V. P., Joshua, H., Kempf, A. J., Krellwitz, W. R., Monaghan, R. L., Ormond, R. E., Wilson, K. E., Albers-Schönberg, G., and Putter, I. (1979). 'Avermectins, new family of potent anthelmintic agents: isolation and chromatographic properties', *Antimicrob. Agents Chemother.*, **15**, 368–371.

Mitlin, N., Wiygul, G., and Haynes, J. W. (1977). 'Inhibition of DNA synthesis in boll weevils (*Anthonomus grandis*, Boheman) sterilized by Dimilin', *Pestic. Biochem. Physiol.*, **7**, 559–563.

Mitsui, T., Nobusawa, C., and Fukami, J. (1981). 'Inhibition of chitin synthesis by diflubenzuron in *Mamestra brassicae* L.', *J. Pestic. Sci.*, **6**, 155–161.

Mitsui, T., Nobusawa, C., Fukami, J., Collins, J., and Riddiford, L. M. (1980). 'Inhibition of chitin synthesis by diflubenzuron in *Manduca* larvae', *J. Pestic. Sci.*, **5**, 335–341.

Mothes, U., and Seitz, K-A. (1982). 'Action of the microbial metabolite and chitin synthesis inhibitor nikkomycin on the mite *Tetranychus urticae*; an electron microscope study', *Pestic. Sci.*, **13**, 426–441.

Mulder, R., and Gijswijt, M. J. (1973). 'The laboratory evaluation of two promising new insecticides which interfere with cuticle deposition', *Pestic. Sci.*, **4**, 737–745.

Muzzarelli, R. A. A. (1977). *Chitin*, Pergamon Press, Oxford and New York.

Nakagawa, Y., Kitahara, K., Nishioka, T., Iwamura, H., and Fujita, T. (1984). 'Quantitative structure-activity studies of benzoylphenylurea larvicides. I. Effect of substituents at aniline moiety against *Chilo suppressalis* Walker', *Pestic. Biochem. Physiol.*, **21**, 309–325.

Neumann, R., and Guyer, W. (1983). 'A new chitin synthesis inhibitor CGA 112'913: its biochemical mode of action as compared to diflubenzuron.' 10th International Congress of Plant Protection. *Plant Protection for Human Welfare*, **1**, 445–451.

Neville, A. C. (1975). *Biology of the Arthropod Cuticle*, Springer-Verlag, New York.

Nishioka, T., Fujita, T., and Nakajima, M. (1979). 'Effect of chitin synthesis inhibitors on cuticle formation of the cultured integument of *Chilo suppressalis*', *J. Pestic. Sci.*, **4**, 367–374.

Oberlander, H., and Leach, C. E. (1974). 'Inhibition of chitin synthesis in *Plodia interpunctella*', *Proc. First. International Working Conference on Stored-Product Entomology*, Savannah, Georgia, pp. 651–655.

Ohta, N., Kakiki, K., and Misato, T. (1970). 'Studies on the mode of action of polyoxin D. Part II. Effect of polyoxin D on the synthesis of fungal cell wall chitin', *Agr. Biol. Chem.*, **34**, 1224–1234.

O'Neill, M. P., Holman, G. M., and Wright, J. E. (1977). 'β-Ecdysone levels in pharate pupae of the stable fly, *Stomoxys calcitrans* and interaction with the chitin inhibitor diflubenzuron', *J. Insect Physiol.*, **23**, 1243–1244.

Pong, S-S., DeHaven, R., and Wang, C. C. (1982). 'A comparative study of avermectin B_{1a} and other modulators of the γ-aminobutyric acid receptor chloride ion channel complex,' *J. Neurosci.*, **2**, 966–971.

Pong, S-S., and Wang, C. C. (1982). 'Avermectin B_{1a} modulation of γ-aminobutyric acid receptors in rat brain membranes', *J. Neurochem.*, **38**, 375–379.

Porter, C. A., and Jaworski, E. G. (1965). 'Biosynthesis of chitin during various stages in the metamorphosis of *Prodenia eridania*', *J. Insect Physiol.* **11**, 1151–1160.

Post, L. C., DeJong, B. J., and Vincent, W. R. (1974). '1-(2,6-Disubstituted benzoyl)-3-phenylurea insecticides: inhibitors of chitin synthesis', *Pestic. Biochem. Physiol.*, **4**, 473–483.

Post, L. C., and Vincent, W. R. (1973). 'A new insecticide inhibits chitin synthesis', *Naturwissenschaften*, **60**, 431–432.

Putter, I., MacConnell, J. G., Preiser, F. A., Haidri, A. A., Ristich, S. S., and Dybas, R. A. (1981). 'Avermectins: novel insecticides, acaricides and nematicides from a soil microorganism', *Experientia*, **37**, 963–964.

Quennedey, A., Quennedey, B., Delbecque, J-P., and Delachambre, J. (1983). 'The *in vitro* development of the pupal integument and the effects of ecdysteroids in *Tenebrio molitor* (Insecta, Coleoptera)', *Cell Tissue Res.*, **232**, 493–511.

Quesada Allue, L. A. (1982). 'The inhibition of insect chitin synthesis by tunicamycin', *Biochem. Biophys. Res. Commun.* **105**, 312–319.

Quesada Allue, L. A., and Belocopitow, E. (1978). 'Lipid-bound oligosaccharides in insects', *Eur. J. Biochem.*, **88**, 529–541.

Quesada Allue, L. A., Marechal, L. R., and Belocopitow, E. (1976a). 'Biosynthesis of polyprenol phosphate sugars by *Ceratitis capitata* extracts', *FEBS Lett.*, **67**, 243–247.

Quesada Allue, L. A., Marechal, L. R., and Belocopitow, E. (1976b). 'Chitin synthesis in *Triatoma infestans* and other insects', *Acta Physiol. Lat. Am.*, **26**, 349–363.

Roberts, P. E., and Willis, J. H. (1980). 'The cuticular proteins of *Tenebrio molitor*', *Devl. Biol.*, **75**, 59–69.

dc Rousset-Hall, A., and Gooday, G. W. (1975). 'A kinetic study of a solubilized chitin synthetase preparation from *Coprinus cinereus*', *J. gen. Microbiol.*, **89**, 146–154.

Rudall, K. M. (1963). 'The chitin/protein complexes of insect cuticles', *Adv. Insect Physiol.*, **1**, 257–313.

Rudall, K. M. (1965). 'Skeletal structure in insects', in *Aspects of Insect Biochemistry, Biochem. Soc. Symp.*, **25**, 83–92.

Rudall, K. M., and Kenchington, W. (1973). 'The chitin system', *Biol. Rev.*, **49**, 597–636.

Sbragia, R. J., Bisabri-Ershadi, B., Rigterink, R. H., Clifford, D. P., and Dutton, R. (1983). 'XRD-473, a new acylurea insecticide effective against *Heliothis*.' 10th International Congress of Plant Protection. *Plant Protection for Human Welfare*, **1**, 417–424.

Schmidt, C. D. (1983). 'Activity of an avermectin against selected insects in aging manure', *Environ. Entomol.*, **12**, 455–457.

Schneiderman, H. A., and Gilbert, L. I. (1964). 'Control of growth and development in insects', *Science*, **143**, 325–333.

Selitrennikoff, C. P. (1979). 'Competitive inhibition of *Neurospora crassa* chitin synthetase activity by tunicamycin', *Archs Biochem. Biophys.*, **195**, 243–244.

Soltani, N., Besson, M. T., and Delachambre, J. (1984). 'Effects of diflubenzuron on the pupal-adult development of *Tenebrio molitor* L. (Coleoptera, Tenebrionidae): growth and development, cuticle secretion, epidermal cell density, and DNA synthesis', *Pestic. Biochem. Physiol.*, **21**, 256–264.

Sowa, B. A., and Marks, E. P. (1975). 'An *in vitro* system for the quantitative measurement of chitin synthesis in the cockroach: inhibition by TH 6040 and polyoxin D', *Insect Biochem.*, **5**, 855–859.

Sridhara, S. (1983). 'Cuticular proteins of the silkmoth *Antheraea polyphemus*', *Insect Biochem.*, **13**, 665–675.

Surholt, B. (1975). 'Studies *in vivo* and *in vitro* on chitin synthesis during the larval-adult moulting cycle of the migratory locust *Locusta migratoria* L.', *J. Comp. Physiol.*, **102**, 135–147.

Takatsuki, A., Arima, K., and Tamura, G. (1971). 'Tunicamycin, a new antibiotic. I. Isolation and characterization of tunicamycin', *J. Antibiot.*, **24**, 215–223.

Turnbull, I. F., and Howells, A. J. (1982). 'Effects of several larvicidal compounds on chitin biosynthesis by isolated larval integuments of the sheep blowfly *Lucilia cuprina*', *Aust. J. Biol. Sci.*, **35**, 491–503.

Turnbull, I. F., and Howells, A. J. (1983). 'Integumental chitin synthase activity in cell-free extracts of larvae of the Australian sheep blowfly, *Lucilia cuprina*, and two other species of Diptera', *Aust. J. Biol. Sci.*, **36**, 251–262.

Vardanis, A. (1976). 'An *in vitro* assay system for chitin synthesis in insect tissue', *Life Sci.*, **19**, 1949–1956.

Verloop, A., and Ferrell, C. D. (1977). 'Benzoylphenyl ureas—a new group of larvicides interfering with chitin deposition', in *Pesticide Chemistry in the 20th Century* (ed. J. R. Plimmer), pp. 237–270, ACS Symposium Series No. 37.

Vincent, J. F. V., and Hillerton, J. E. (1979). 'The tanning of insect cuticle—a critical review and a revised mechanism', *J. Insect Physiol.*, **25**, 653–658.

Wellinga, K., Mulder, R., and van Daalen, J. J. (1973a). 'Synthesis and laboratory evaluation of 1-(2,6-disubstituted benzoyl)-3-phenylureas, a new class of insecticides. I. 1-(2,6-dichlorobenzoyl)-3-phenylureas', *J. Agric. Food Chem.*, **21**, 348–354.

Wellinga, K., Mulder, R., and van Daalen, J. J. (1973b). 'Synthesis and laboratory evaluation of 1-(2,6-disubstituted benzoyl)-3-phenylureas, a new class of insecticides. II. Influence of the acyl moiety on insecticidal activity', *J. Agric. Food Chem.*, **21**, 993–998.

Yu, S. J., and Terriere, L. C. (1975). 'Activities of hormone metabolizing enzymes in house flies treated with some substituted urea growth regulators', *Life Sciences*, **17**, 619–625.

Yu, S. J., and Terriere, L. C. (1977). 'Ecdysone metabolism by soluble enzymes from three species of Diptera and its inhibition by the insect growth regulator TH-6040', *Pestic. Biochem. Physiol.*, **7**, 48–55.

Zoebelein, G., Hammann, I., and Sirrenberg, W. (1980). 'BAY SIR 8514, a new chitin synthesis inhibitor', *J. appl. Entomol.*, **89**, 289–297.

Insecticides
Edited by D. H. Hutson and T. R. Roberts
© 1985 John Wiley & Sons Ltd

CHAPTER 8

Chemical mediation of insect behaviour

O. T. Jones

INTRODUCTION

The content of this chapter is restricted to naturally occurring chemicals which evoke behavioural responses in insects. In addition to pheromones which operate between individuals of the same species, some of the chemicals involved in interspecies interactions will also be considered, namely allomones and kairomones.

These terms are best defined as follows.

A *pheromone* is a chemical or mixture of chemicals released by one organism that induces a response in another individual of the same species (Karlson and Butenandt, 1959; Karlson and Lüscher, 1959; Kalmus, 1965).

311

An *allomone* is a chemical or mixture of chemicals released by one organism that induces a response in an individual of another species; the response is adaptively favourable to the emitter (Brown, 1968). Examples of allomones are defensive secretions which are released by many insects and which are poisonous or repugnant to attacking predators.

A *kairomone* is a chemical or a mixture of chemicals released from one organism which induces a response in an individual of another species; the response is adaptively favourable to the recipient (Brown *et al.*, 1970). Examples of kairomones are chemicals released by insects or from their food which are then used for host/prey location by other unrelated species that are parasitic/predatory on them.

Behaviour modifying chemicals have received considerable interest in recent years because of their potential use in future insect control strategies. Several recent reviews and publications discuss the practical application of pheromones, allomones and kairomones (Brand *et al.*, 1979; Minks, 1979; Ritter, 1979; Boness, 1980; Campion and Nesbitt, 1981; Mitchell, 1981; Roelofs, 1981; Kydonieus and Beroza, 1982; Ridgway *et al.*, 1982) and their uses can be summarized as follows.

Monitoring

By using pheromone-based trapping systems insect populations can be estimated and new areas of infestation can be detected at an early stage. Their use in timing and assessing the need for insecticide sprays can lead to a reduction in the use of toxic chemicals, and has done so.

Lure and kill

Pheromones can be used to attract insect pests to specific areas which are then treated with insecticides, hormone analogues or pathogens. Conversely a kairomone can be used to attract and/or encourage the activities of entomophagous insects in a pest infested area.

Mass trapping

Pheromones can be used for population suppression by having suitable densities of baited traps which reduce the pest population by male and/or female annihilation.

Mating disruption

By permeating the air over an infested area with the sex pheromone of a particular insect population suppression can be achieved by the process of mating disruption.

Repellents

By using a suitable repellent chemical, derived possibly from allomonal studies, crops and commodities can be protected from insect pests again without the use of toxic chemicals.

Rather than elaborating further, however, on the uses of behaviour modifying chemicals along the lines mentioned above, I hope to give a general review of their chemical structures, their origin and, where possible, their structure–activity relationships.

PHEROMONES

Sex pheromones of Lepidoptera

By far the largest number of pheromones elucidated to date have come from studies with moths and butterflies. Long-range sex attractants for males produced by females have received most attention, largely because of their potential use in pest management. It should be noted, however, that while some of these long-range sex attractants have been isolated chemically from the species concerned, a number have been arrived at by field or laboratory screening of candidate compounds. While such attractants may be chemically identical to the true pheromone, in the absence of confirmatory chemical evidence they are not dealt with under this section. It should be noted, however, that there is no sex attractant described for any species which has not been described previously for another as a true pheromone.

Several lists of both sex attractants and sex attractant pheromones have been produced in recent years and for detailed information of attractants for particular species the following references may be of use: Baker and Bradshaw, 1979, 1981, 1983; Baker and Evans, 1977, 1979; Baker and Herbert, 1984; Bestman and Vostrowsky, 1981; Kydonieus and Beroza, 1982; Roelofs and Brown, 1982; Steck *et al.*, 1982; Tamaki, 1984).

Even within those compounds confirmed as pheromones, the number of species is too great for listing in this chapter. Instead, the main features of the chemistry of these compounds will be described.

Structure

Most lepidopteran sex pheromones are derivatives of unsaturated, even-carbon (C_{10} to C_{18}) straight chain alcohols (Table 1). The commonest derivative is the acetate but the parent alcohols or the aldehydes are also frequently found. The main chain is sometimes saturated, but is more often monounsaturated, or less frequently, di- or triunsaturated.

Table 1 Sex pheromones isolated from female Lepidoptera

Nomenclature

Unsaturation: any unsaturation in the main chain is indicated by the geometry *Z* or *E* (*cis* or *trans* are used less frequently these days) followed by its position(s) and separated from the number of carbon atoms by a hyphen.

Chain length: the number of carbon atoms is designated by an arabic numeral.

Functional group: this is separated from the rest by a colon and is abbreviated Alc = alcohol Ac = acetate and Ald = aldehyde.

Examples:

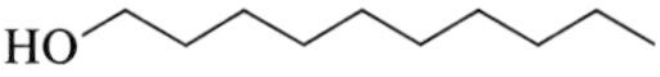

HO⌒⌒⌒⌒⌒⌒⌒   10:Alc

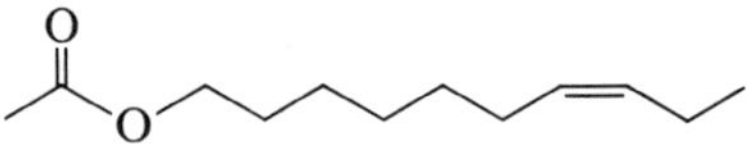

(acetate structure)   Z7–10:Ac

(aldehyde structure)   Z,E,9,11–14:Ald

10:Alc	E 9–12:Ac	14:Alc	Z 12–14:Ac	Z 10–16:Ac
E 5–10:Ac	Z 9–12:Ac	E 5–14:Alc	E,E 3,5–14:Ac	E 11–16:Ac
Z 5–10:Ac	E 10–12:Ac	Z 8–14:Alc	Z,Z 3,5–14:Ac	Z 11–16:Ac
Z 7–10:Ac	Z 10–12:Ac	E 9–14:Alc	E,E 9,11–14:Ac	E,Z 6,11–16:Ac
12:Alc	11–12:Ac	Z 9–14:Alc	Z,E 9,11–14:Ac	E,E 7,11–16: Ac
Z 5–12:Alc	E,E 7,9–12:Ac	E 11–14:Alc	Z,E 9,12–14:Ac	E,Z 7,11–16:Ac
Z 7–12:Alc	E,Z 7,9–12:Ac	Z 11–14: Alc	Z,Z 9,12–14:Ac	Z,Z 7,11–16: Ac
E 8–12:Alc	E,E 8,10–12:Ac	Z,E 9,11–14:Alc	14:Ald	Z,E 11,14–16:Ac
Z 8–12:Alc	Z,E 8,10–12:Ac	14:Ac	Z 7–14:Ald	16:Ald
E 9–12:Alc	E,E 9,11–12:Ac	E 5–14:Ac	Z 9–14: Ald	Z 7–16:Ald
E,E 5,7–12:Alc	Z 9,11–12:Ac	Z 5–14:Ac	E 11–14:Ald	Z 9–16:Ald
Z,E 5,7–12:Alc	10 Meth-12:Ac	E 6–14:Ac	Z 11–14:Ald	Z 11–16:Ald
E,E 8,10–12:Alc	Z,E 5,7–12:Ald	E 7–14:Ac	Z 9–15:Ac	E,Z 6,11–16:Ald
Z,E 8,10–12:Alc	E 4–13:Ac	Z 7–14:Ac	E 12–15:Ac	E,E 10,12–16:Ald
12:Ac	Z 7–13:Ac	Z 8–14:Ac	Z 12–15:Ac	E,Z 10,12–16:Ald
E 3–12:Ac	E 8–13:Ac	E 9–14:Ac	Z 7–16:Alc	Z,Z 11,13–16:Ald
E 5–12:Ac	Z 8–13:Ac	Z 9–14:Ac	E 11–16:Alc	Z 13–18: Alc
Z 5–12:Ac	Z 10–13:Ac	E 10–14:Ac	Z 11–16:Alc	E,Z 3,13–18:Alc
E 7–12:Ac	E 11–13:Ac	Z 10–14:Ac	E,E 10,12–16:Alc	Z,Z 3,13–18:Alc
Z 7–12:Ac	Z 11–13:Ac	E 11–14:Ac	E,Z 10,12–16:Alc	E,Z 3,13–18:Ac
E 8–12:Ac	E,Z 4,7–13:Ac	Z 11–14:Ac	Z 7–16:Ac	Z,Z 3,13–18: Ac
Z 8–12:Ac	E,Z,Z 4,7,10–13:Ac	E 12–14:Ac	Z 9–16:Ac	E,Z 6,11–18:Ald

Table 2 Examples of sex pheromones of female Lepidoptera where one or more components are not derivatives of straight chain primary alcohols

---

Lymantridae

Lymantria dispar

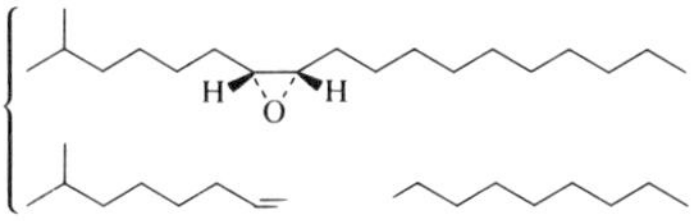

Bierl *et al.* (1970)
Carde *et al.* (1973)

Orgya pseudostugata

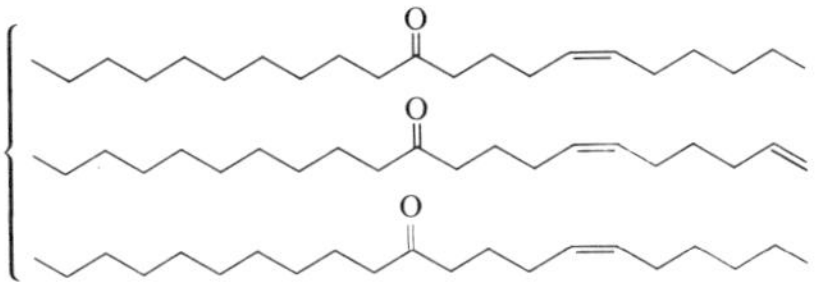

Smith *et al.* (1975)
Smith *et al.* (1978)

Carposinidae

Carposina nipponensis

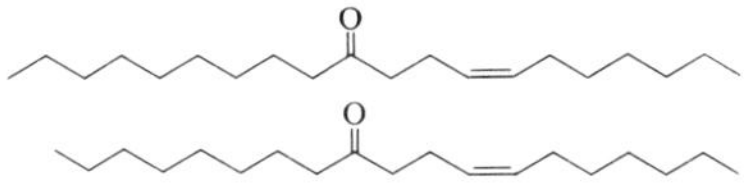

Tamaki *et al.* (1977)

Arctiidae

Estigmene acrea

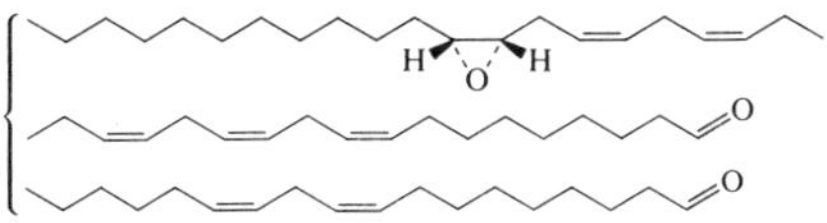

Hill and Roelofs (1981)

Holomelina sp.
Isia isabella

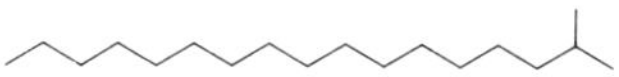

Roelofs and Carde (1971)

Utetheisa ornatrix

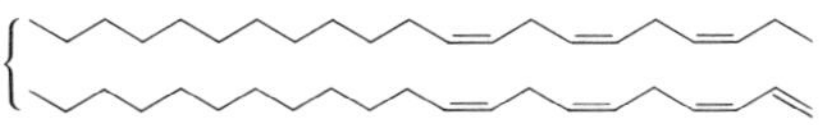

Conner *et al.* (1980)

Geometridae

Boarmia selenaria

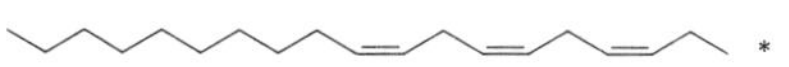

Becker *et al.* (1983)

Operophtera brumata

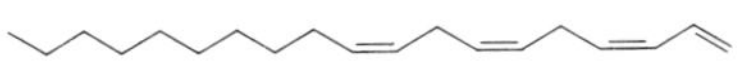

Bestmann *et al.* (1982)
Roelofs *et al.* (1982)

Table 2 (*continued*)

Alsophila pometaria	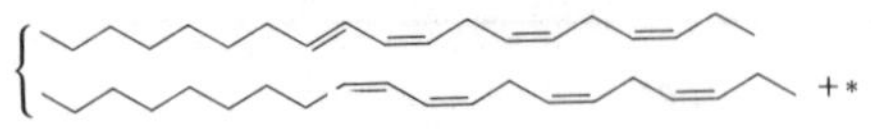 + *	Wong *et al.* (1984)
Caenurgina erechtea *Anticarsia gemmatalis*	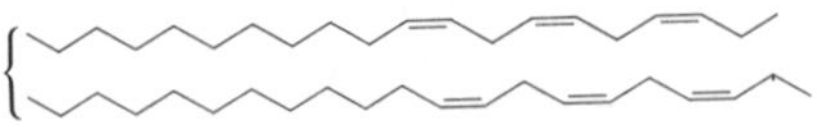	Underhill *et al.* (1983) Heath *et al.* (1983)
Notodontidae *Thaumetopoea* *pityocampa*		Guerrero *et al.* (1981)
Harrisina brillians	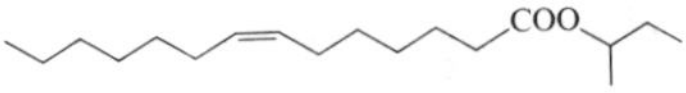	Myerson *et al.* (1982)
Thyridopteryx *ephemeraeformis*	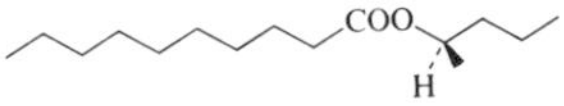	Leonhardt *et al.* (1983)

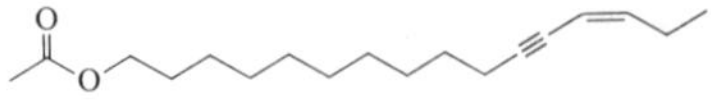

Ten carbon chains are comparatively uncommon but *Agrotis segetum* and *A. fucosa* have been shown to use (*Z*)-5-decenyl acetate (**1**) (Bestmann *et al.*, 1978; Wakamura, 1978). These short chain compounds are probably not frequently found in moths because of their reduced scope for structural diversity and their higher volatility. From the frequency with which they appear, the optimum chain length would appear to be 12, 14 and 16, these making up about 70% of the total compounds. Although odd-carbon chains are not generally found, the female potato tuberworm moth (*Phthorimaea operculella*) produces two 13-carbon compounds (**3**) and (**4**) (Persoons *et al.*, 1976b) and *Adoxophyes* sp. produces the chiral branched compound (**2**) (Tamaki *et al.*, 1979). Examples of other unusual lepidopteran sex pheromones are given in Table 2. Although in Lepidoptera it is usually the females that produce the long-range sex attractant pheromone, with the male then releasing the close-range aphrodisiac secretion, there are a few exceptions. In the pyralid moth *Eldana saccharina* the male produces both long- and short-range aphrodisiac components from glands in the wings and from abdominal hairpencils. Vanillin (**5**) and *p*-hydroxybenzaldehyde (**6**) have been identified from the hairpencils while the wing glands produce the γ-lactone (**7**) (Kunesch *et al.*, 1981; Zagatti, 1981; Zagatti *et al.*, 1981., Kunesch and Zagatti, 1982). Similarly in the wax moths, *Galleria mellonella* and *Achroia grisella*, the males produce aphrodisiac pheromones consisting of *n*-undecanal (**8**) and *n*-nonanal (**9**) in the case of *G. mellonella* (Röller *et al.*, 1968; Leyrer and Monroe, 1973) and *n*-undecanal and (*Z*)-11-octadecenal in *A. grisella* (Dahm *et al.*, 1971) but with the absence of any long-range attractants.

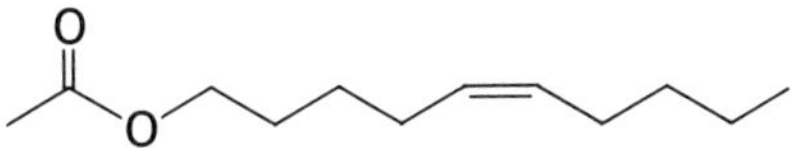

(1)

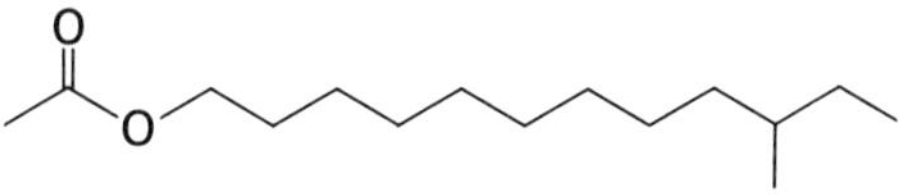

(2)

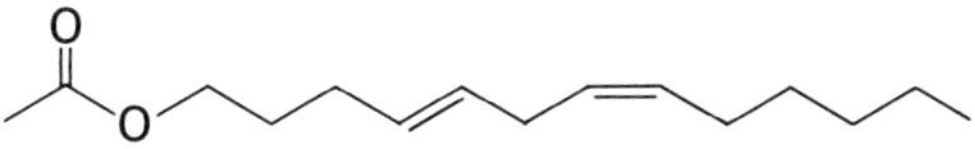

(3)

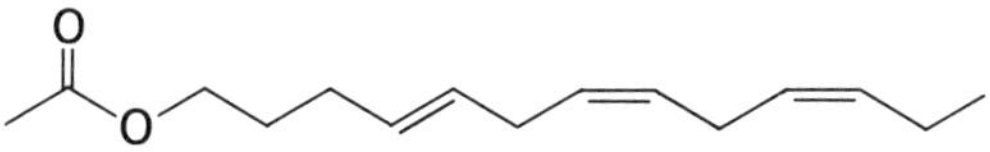

(4)

Eldana saccharina

wing glands

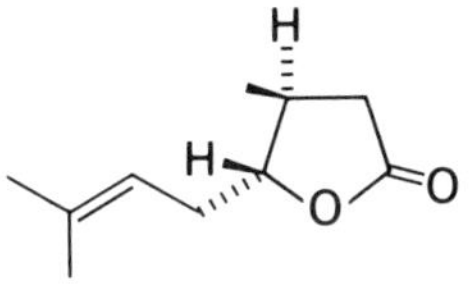

(7, (*E*)-3-methyl-4-dimethylallyl-γ-lactone)

hair pencils

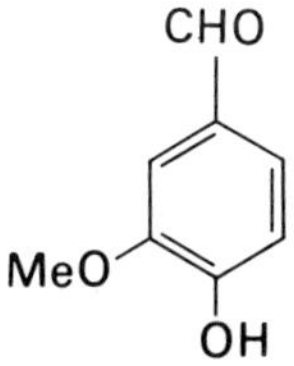

CHO

MeO

OH

(5)

CHO

OH

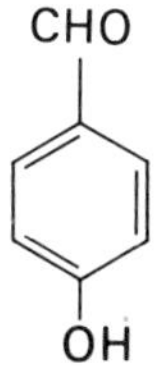

(6)

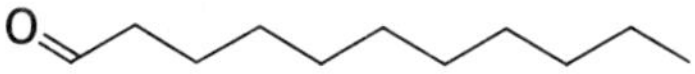

(8)

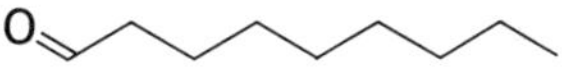

(9)

Table 3 Representative examples of sex pheromones of female Lepidoptera where all the components are derivatives of C_{12}, C_{14}, C_{16} or C_{18} straight chain primary alcohols. The ratios are generally those extracted from the females (After Bradshaw, 1984)

Species	Pheromone components	Ratio	Reference
Agrotis segetum	(Z)-5-Decenyl acetate	6	Lofstedt *et al.* (1982)
	Decyl acetate	8	
	(Z)-7-Dodecenyl acetate	100	
	(Z)-9-Tetradecenyl acetate	49	
Trichoplusia ni.	(Z)-7-Dodecenyl acetate	9	Bjostad *et al.* (1980b)
	Dodecanyl acetate	1	
Grapholitha molesta	(Z)-8-Dodecenyl acetate	100	Carde, *et al.* (1979)
	(E)-8-Dodecenyl acetate	7	
	(Z)-8-Dodecen-1-ol	30	
Rhyacionia frustrana	(E)-9,11-Dodecadienyl acetate	4	Hill *et al.* (1981)
	(E)-9-Dodecenyl acetate	96	
Laspeyresia nigricana	(E,E)-8,10-Dodecadienyl acetate	—	Wall *et al.* (1976)
Malacosoma disstria	(Z,E)-5,7-Dodecadien-1-ol	ca. 5	Chisholm *et al.* (1980b)
	(Z,E)-5,7-Dodecadienal	1	
Plusia chalcites	(Z)-7-Dodecenyl acetate	5	Dunkelblum *et al.* (1981)
	(Z)-9-Tetradecenyl acetate	1	
Spodoptera frugiperda	(Z)-9-Dodecenyl acetate	10	Jones and Sparks (1979)
	(Z)-9-Tetradecenyl acetate	1	
Sparganothis directana	(E)-9,11-Dodecadienyl acetate	35	Bjostad *et al.* (1980a)
	(E)-11-Tetradecenyl acetate	19	
	(Z)-11-Tetradecenyl acetate	28	
Archips argyrospilus	(Z)-11-Tetradecenyl acetate	15	Carde *et al.* (1977)
	(E)-11-Tetradecenyl acetate	10	
	(Z)-9-Tetradecenyl acetate	1	
	Dodecyl acetate	20	
Scotia exclamationis	(Z)-5-Tetradecenyl acetate	95	Bestmann *et al.* (1980a)
	(Z)-9-Tetradecenyl acetate	5	
Panolis flammea	(Z)-9-Tetradecenyl acetate	100	Baker *et al.* (1982a)
	(Z)-11-Tetradecenyl acetate	1	
	(Z)-11-Hexadecenyl acetate	5	
Heliothis virescens	(Z)-9-Tetradecenal	3	Klun *et al.* (1979)
	Tetradecanal	2	
	(Z)-7-Hexadecenal	1	
	(Z)-9-Hexadecenal	1	
	(Z)-11-Hexadecenal	81	
	Hexadecanal	9	
	(Z)-11-Hexadecen-1-ol	3	
Chilo sacchariphagus	(Z)-13-Octadecenyl acetate	7	Nesbitt *et al.* (1980)
	(Z)-13-Octadecen-1-ol	1	
Aegeria tibialis	(Z,Z)-3,13-Octadecadienyl acetate	4	
	(Z,Z)-3,13-Octadecadien-1-ol	1	Underhill *et al.* (1978)

The multicomponent nature of lepidopteran sex pheromones

It is now quite well established that the pheromones used by Lepidoptera are generally multicomponent and that many of the earlier reports of single-component pheromones presented an oversimplified picture (Roelofs and Carde, 1977). A few examples remain where a single biologically active compound will elicit a full sequence of attraction and precopulatory behaviour from a male, e.g. (*E,E*)-8,10-dodecadienyl acetate and the pea moth (*Laspeyresia nigricana*) (Wall *et al.*, 1976). In most cases, however, a precise blend of two or more compounds is required for the full response with the secondary components mediating several facets of the mating behavioural sequence. They no doubt also serve to maintain species specificity.

Examples of pheromone blends for a few representative species are given in Table 3. The ratios of compounds are usually critical in situations where closely related species use the same compounds but maintain specificity through having different ratios (Roelofs and Carde, 1974). Although individual females are often quite constant in the ratios of compounds which they emit and vary little also between females of the same species (Baker *et al.*, 1978) ratios of the compounds extracted from their glands may often differ greatly from that actually emitted (Tamaki, 1977).

From Table 3 it is seen that there are many structural similarities between the compounds that make up any particular blend including:

(1) esterification to acetate and/or oxidation to aldehyde of an unsaturated alcohol (e.g. *Chilo sacchariphagus, Malacosoma disstria* and *Aegeria tibialis*);
(2) movement of the double bond of a monounsaturated alcohol or its derivative, by two positions along the carbon chain (e.g. *Panolis flammea*);
(3) insertion or deletion of two methylene groups, either (a) between the functional group and the double bond (e.g. *Plusia chalcites*) or (b) between the double bond and the methyl end of the chain (e.g. *Spodoptera frugipera*);
(4) saturation of one of the double bonds either converting diene to monoene or monoene to saturated compound (e.g. *Trichoplusia ni*);
(5) alteration of the geometry of the one double bond from (*Z*) to (*E*) or vice versa (e.g. *Grapholitha molesta, Sparganothis directana*).

Steck *et al.*, (1982) have proposed for noctuid species of N. America a series of 'one-change' step rules for relating and even predicting the components of complex sex pheromone secretions. These rules have also been found to be applicable to multicomponent sex pheromone blends of a number of Tortricidae. The seven-component sex pheromone of *Heliothis virescens* (Table 3) can be used to illustrate this. The main components are (*Z*)-11-hexadecenal and (*E*)-9-tetradecenal which are related by (3a) above but the other given components are linked as shown in the scheme:

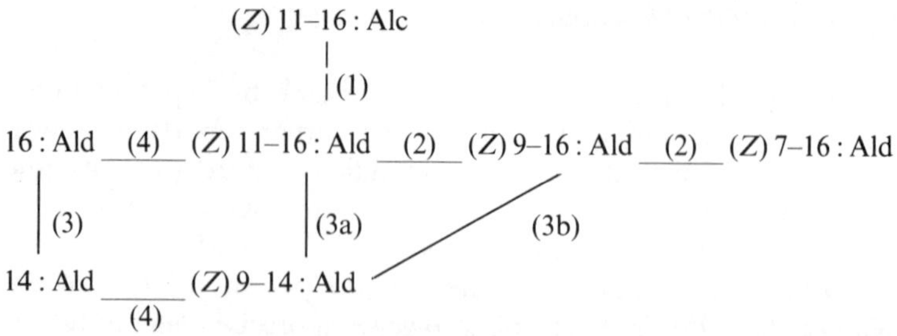

$(Z)\,11–16:Alc$
|
|(1)

$16:Ald$ ___(4)___ $(Z)\,11–16:Ald$ ___(2)___ $(Z)\,9–16:Ald$ ___(2)___ $(Z)\,7–16:Ald$

|(3) |(3a)   (3b)

$14:Ald$ _____ $(Z)\,9–14:Ald$
___(4)___

The biosynthetic and sensory mechanisms by which such changes occur or are perceived within moths have still to be unravelled in most species.

Structure–activity relationships

Chemoreception studies with moth antennae using electroantennogram (EAG) techniques have shown that the receptors are themselves chiral (Bestmann and Vostrowsky, 1981) and that natural chiral pheromone molecules must assume particular conformations before they are bound to the receptor (Chapman *et al.*, 1978; Bestmann *et al.*, 1980b). Similar studies based on amplitude measurements of EAG responses have shown that changes in the functional group reduce activity considerably whereas the geometry of the double bond appears to be less critical at the antennal level, though this may also be very significant in the field. The position of the double bond and the length of the chain are also both important: in general, any change in either will reduce activity compared with the optimum compound.

Field trapping experiments with blends of compounds and analogues which showed EAG activity have shown, however, that very precise combinations of compounds are required for behavioural responses and the analogues frequently fail to induce responses. This implies that there is some integration of antennal signals at a higher level in most insects.

Field trapping experiments have shown up a few features of structure–activity which are not observable by EAG techniques. Changes of one methylene group added or subtracted did not eliminate all activity from a pheromone (Steck *et al.*, 1982). Similarly the insertion of a second double bond between the original double bond and the end of the chain does not eliminate activity whereas the same change performed elsewhere on the chain is highly detrimental to trap catch (Chisholm *et al.*, 1980a).

Biosynthesis and biotransformation

Studies on the biosynthesis of the primary sex pheromone component of *Trichoplusia ni* have shown that the key steps involve Δ-11 desaturation and chain shortening:

$$16 \to (Z)\,11–16 \to (Z)\,9–14 \to (Z)\,7–12$$

The female pheromone gland contains (Z)-7-dodecenyl acetate, small amounts of (Z)-7-dodecenoate and large quantities of the precursor (Z)-11-hexadecenoate (Bjostad and Roelofs, 1983).

A similar pattern has also been observed in the pheromone biosynthesis of *Argyrotaenia citrana* except that the sequence of events is different:

$$16 \rightarrow 14 \rightarrow (Z) \, 11–14 \rightarrow (Z) \, 11–14 : \text{Alc} \rightarrow (Z) \, 11–14 : \text{Ac}.$$

A combination of chain shortening and desaturation steps could well prove to be a general mechanism in the biosynthesis of many lepidopteran sex pheromones. Studies to confirm the origin of lepidopteran sex pheromones in the fatty acid pathway have been reviewed by Roelofs and Brown (1982) and Bradshaw (1984).

Other behaviour-modifying chemicals involved in lepidopteran courtship

The importance of courtship stimulants released by male butterflies prior to mating is now firmly established. Volatile components of the hairpencil secretion of the butterfly *Amauris ochlea* include hex-3-enoic acid as a major component together with nonanal, octanal, methyl salicylate, eugenol, (Z)-jasmone and a pyrrolizidine alkaloid as minor components (Petty *et al.*, 1977). The lactone (10), structurally related to the aliphatic esterifying acids of the pyrrolizidine alkaloids, has been isolated from the costal fringes of male Ithomiinae butterflies (Edgar *et al.*, 1976). The scent scales on the wings of *Pieris* sp. butterflies produce mixtures of monoterpenes, the compositions of which differ between species (Hayashi *et al.*, 1978). Two species of Colias butterflies use esters and alkanes from the male wing glands both as aphrodisiacs and to maintain reproductive isolation (Grula *et al.*, 1980).

(10)   *Ithomiinae* butterflies

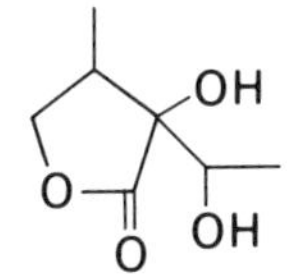

It has also been shown that male moths use courtship stimulants, but are released from genital scent brushes (or hairpencils). Males of *Grapholitha molesta* attract females at close range with hairpencil secretions containing (11–14) (Baker, T. C. *et al.*, 1981; Nishida *et al.*, 1982). Some noctuid moths employ aromatic compounds such as 2-phenylethanol (e.g. *Mamestra configurata*) (Bestmann *et al.*, 1977), while two species of Asian arctiid moths have been found to contain pyrrolizine-carboxaldehyde (15) in their coremata (Schneider *et al.*, 1982).

(11) 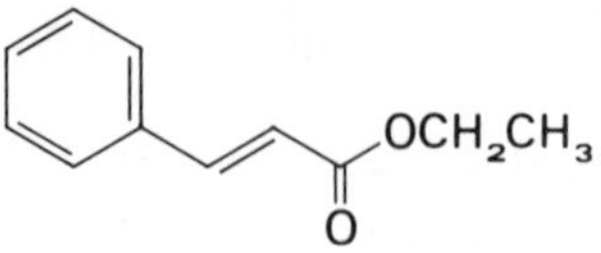 ethyl-(E)-cinnamate

(12) 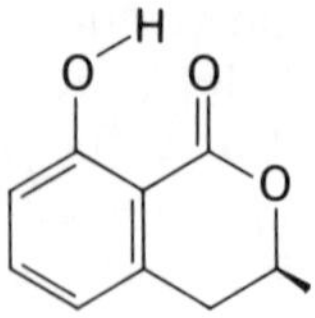 R-(−)-mellein

(13) 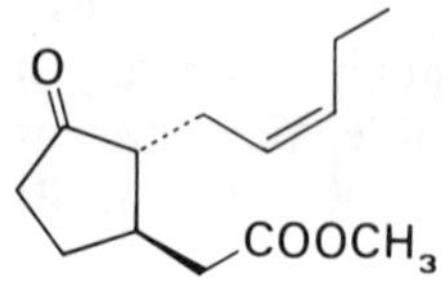 methyl jasmonate

(14) 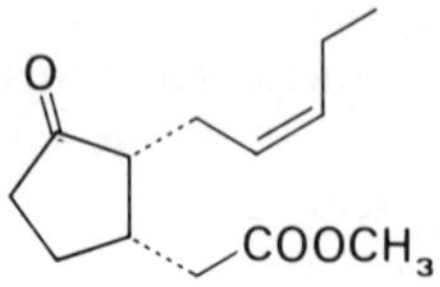 methyl 2-epijasmonate
Grapholitha molesta

(15)

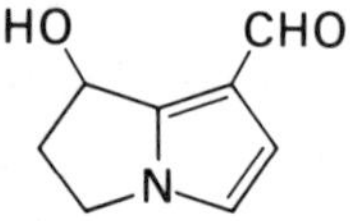

Pheromones produced by larval Lepidoptera

The larvae of a few lipidopteran species have also been shown to use pheromones: the eastern tent caterpillar *(Malacosoma americanum)* forages by perception of chemicals present in trails of silk leading from the communal tent of this species to its distant food sources. The distribution of *Ephestia kühniella* larvae in their food is also thought to be influenced by the β-triketones (16) produced by their mandibular glands (Mudd, 1978). This chemical secretion is therefore referred to as an epideictic pheromone.

(16)

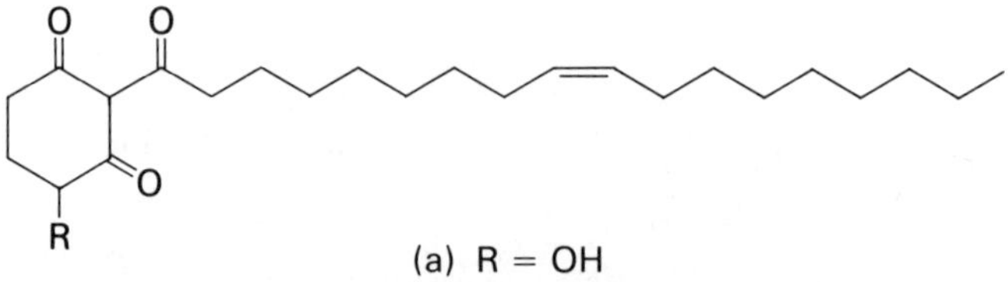

(a) R = OH
(b) R = H

Sex and oviposition pheromones of Diptera

In the Diptera (true flies) a wide variety of sexual aggregating systems is known. Some do not involve pheromones at all while others are partly or wholly influenced by behaviour-modifying chemicals. Several of the suborder Cyclo-rhapha use cuticular hydrocarbons for the last stages of courtship behaviour only. The common housefly *(Musca domestica)* female secretes a blend of (Z)-9-tricosene **(17)**, (Z)-9-heneicosene and some branched alkanes of 28–30 carbons and has them dispersed over the whole body cuticle (Carlson *et al.*, 1971; Richter *et al.*, 1976). These compounds cause the males to land on any fly-like object within 5–10 cm of the source (Mansingh *et al.*, 1972; Uebel *et al.*, 1976). The lesser housefly *(Fannia canicularis)* female on the other hand uses (Z)-9-pentacosene **(18)** to stimulate copulatory responses in males, but over a few centimetres only (Uebel *et al.*, 1978). It is interesting to note that newly emerged males and females have identical cuticular wax compositions but after 5 days very considerable differences are apparent. At this time (Z)-9-pentacosene constitutes approximately 66% of female wax but only 1% in the case of males. Furthermore, whereas the female wax consists entirely of hydrocarbons the male wax contains approximately 37% of heneicosan-8-ol acetate **(19)** (Uebel *et al.*, 1977).

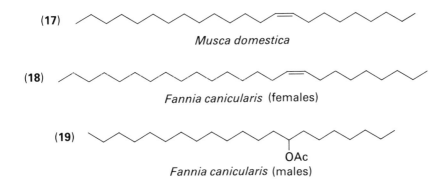

(17) *Musca domestica*

(18) *Fannia canicularis* (females)

(19) OAc
Fannia canicularis (males)

Cuticular pheromones are also known in the stable fly *(Stomoxys calcitrans)* (Sonnet *et al.*, 1979) and face fly *(Musca autumnalis)* (Uebel *et al.*, 1975). Female *S. calcitrans* produce a behaviourally active hydrocarbon fraction consisting of methyl-branched and 1,5-dimethyl-branched hydrocarbons (Sonnet *et al.*, 1977a) as well as the unsaturated compounds **(20–23)** (Sonnet *et al.*, 1979). The cuticular wax of the males on the other hand contains (Z,Z)-1,7,13-pectacosatriene **(24)** which is absent from that of the females (Sonnet *et al.*, 1977b). The three most active compounds isolated from the cuticle of tsetse flies, *Glossina morsitans*, are **(25)**, **(26)** and **(27)** (Carlson *et al.*, 1978) while those of *G. pallidipes* are **(28)** and **(29)** (McDowell *et al.*, 1981) and **(30)** (Carlson *et al.*, 1984). Contact pheromones have also been demonstrated in at least one primitive dipteran, the midge, *Culicoides melleus*, where the most active constituents were found to be 2-, 8- and 10-methyldocosanes (Linley and Carlson, 1978).

Because of their relatively low volatility, cuticular hydrocarbons tend to have only short-range activity and many of the compounds described above are perceived through chemoreceptors on the tarsi rather than the antennae. Lower

(20) 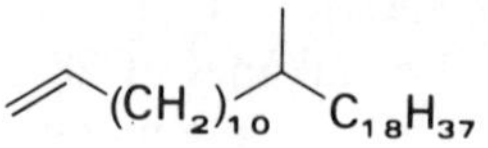 $\diagup\diagdown(CH_2)_{10}\diagup\diagdown C_{18}H_{37}$

(21) 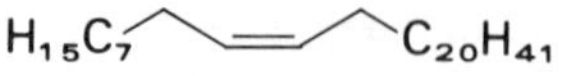 $H_{15}C_7\diagup\diagdown=\diagdown C_{20}H_{41}$

(22) 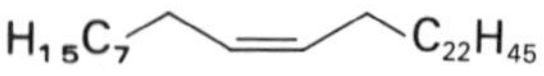 $H_{15}C_7\diagup\diagdown=\diagdown C_{22}H_{45}$

(23) 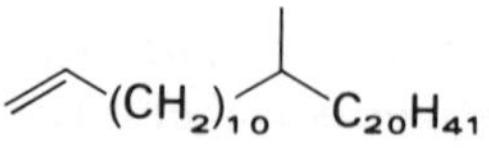 $\diagup\diagdown(CH_2)_{10}\diagup\diagdown C_{20}H_{41}$

Stomoxys calcitran
(females)

(24) 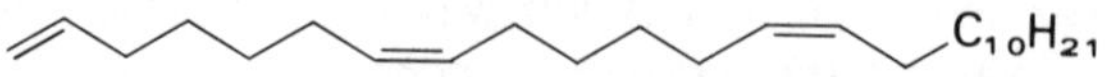 $\diagup\diagdown\diagup\diagdown\diagup=\diagdown\diagup\diagdown=\diagdown\diagup C_{10}H_{21}$

S. calcitrans (males)

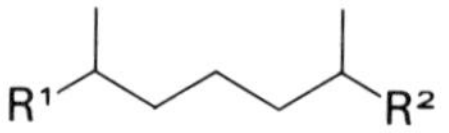

$R^1\diagup\diagdown\diagup\diagdown R^2$ *Glossina morsitans*

(25) $R^1 = C_{14}H_{29}, R^2 = C_{18}H_{37}$
(26) $R^1 = R^2 = C_{16}H_{33}$

(27) 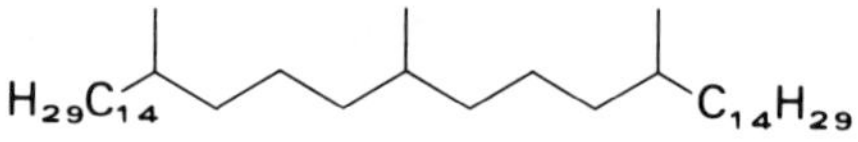 $H_{29}C_{14}\diagup\diagdown\diagup\diagdown\diagup\diagdown C_{14}H_{29}$

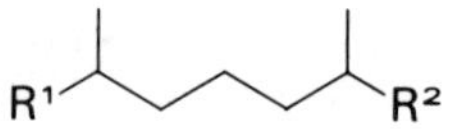

$R^1\diagup\diagdown\diagup\diagdown R^2$ *G. pallidipes*

(28) $R^1 = C_{12}H_{25}, R^2 = C_{18}H_{37}$
(29) $R^1 = C_{14}H_{29}, R^2 = C_{16}H_{33}$

(30) 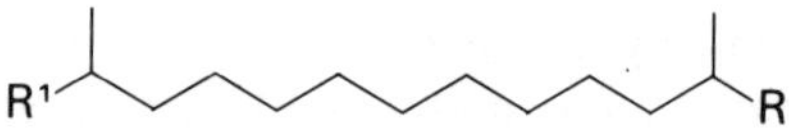 $R^1\diagup\diagdown\diagup\diagdown\diagup\diagdown R^2$ *G. pallidipes*

$R^1 = C_{12}H_{25}, R^2 = C_{12}H_{25}$

molecular weight compounds have been described for sciarid and phorid flies which act over greater distances. Females of the sciarid *Bradysia impatiens* attract males from at least one metre downwind using a pheromone released from the thorax and legs (Alberts *et al.*, 1981). The mushroom sciarid *(Lycoriella mali)* female releases *n*-heptadecane to attract the male (Kostelec *et al.*, 1980) while from the mushroom phorid *(Megaselia halterata)* 3,6-dimethyl heptan 2,4-dione (31) has been isolated (Baker *et al.*, 1982d).

(31) [structure] *Megaselia halterata*

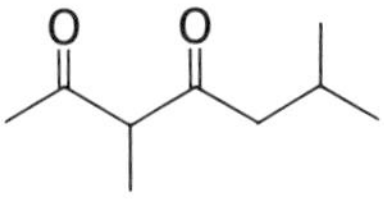

Of all dipteran families studied to date, perhaps the most advanced form of pheromone communication is that shown by fruit flies of the family Tephritidae. Here the pheromones are produced in special, sometimes visible, glands, and which can attract over long distances. Several compounds have been isolated from the males of Mediterranean fruit fly *(Ceratitis capitata)* including methyl-*(E)*-6-nonenoate, *(E)*-6-nonen-1-ol (Jacobson *et al.*, 1973) and β-fenchol (Jacobson and Ohinata, 1980). These compounds have only limited activity in the field and do not attract virgin females, suggesting that more compounds have still to be isolated from the males' rectal glands.

In the case of the olive fruit fly *(Dacus oleae)* it is the virgin females that attract the males and the major component of its sex pheromone was shown to be 1,7-dioxaspiro[5.5]undecane (32) (Baker *et al.*, 1980). Other minor components isolated from the female include the two hydroxy spiroacetals (33) and (34) (Baker *et al.*, 1982c). It has also been found that during non-reproductively active periods of the year, male *D. oleae* will also produce measurable quantities of (32) (Mazomenos and Pomonis, 1983).

(32) [structure] *Dacus oleae*

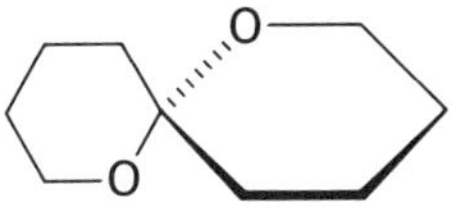

(33) [structure]

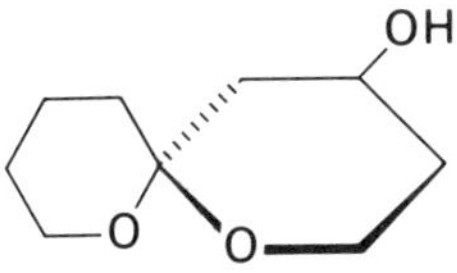

Dacus oleae

(34) [structure] OH

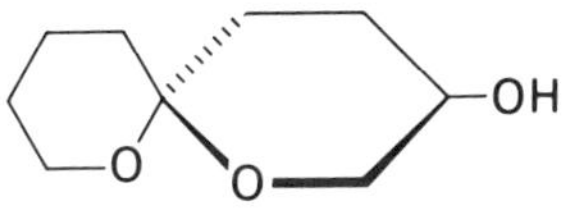

Dacus oleae

Several novel compounds have been identified from other Dacine species: a lactone (35) and *N*-(2-methyl butyl)-proprionamide have been isolated from airations of male melon flies *(Dacus curcurbitae)* and oriental fruit flies *(D. dorsalis)*, respectively (Ohinata *et al.*, 1982). In addition, from the rectal glands of *D. curcurbitae*, several pyrazines and amides have also been identified, including the novel amide (36) (Baker *et al.*, 1982b).

(35) [structure]

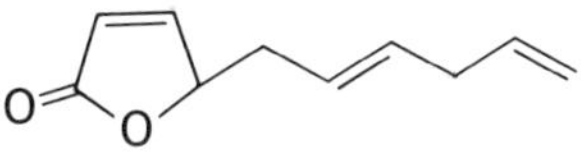

Dacus dorsalis

(36) [structure]

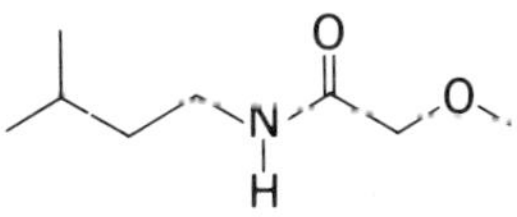

Dacus curcurbitae

Two new lactones, anastrephin (**37**) and epianastrephin (**38**) have been isolated from 16 sex pheromone components of *Anastrepha suspensa* and *A. ludens* as well as the alcohols (**39**) and (**40**) (Battiste *et al.*, 1983; Stokes *et al.*, 1983).

(**37**)

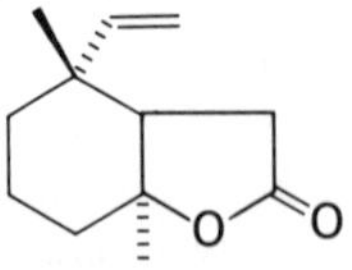

Anastrepha suspensa

(**38**)

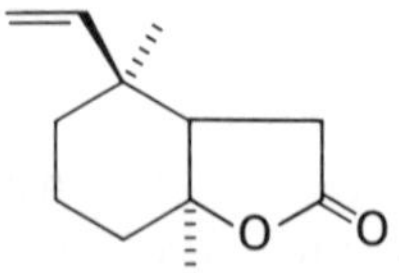

Anastrepha ludens

(**39**) HO

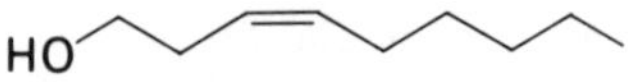

A. suspensa

(**40**) HO

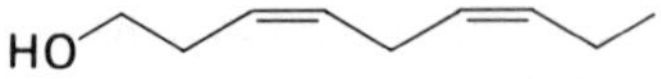

A. ludens

Recently, the first oviposition attractant pheromone from the apical droplet of eggs of the mosquito *Culex pipiens fatigans* was identified as *erythro*-6-acetoxy-5-hexadecanolide (**41**) (Laurence and Pickett, 1982).

(**41**)

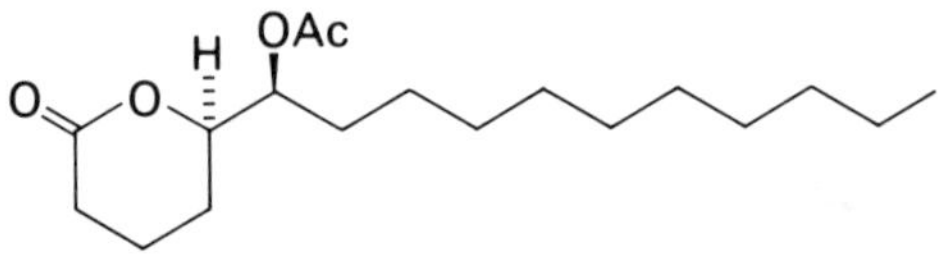

Culex pipiens fatigans

Aggregation and sex pheromones of Coleoptera

Shorey (1973) has postulated that aggregation pheromones in the Coleoptera arose as mechanisms for gathering beetles at a suitable food source, implying that a sex pheromone function arose secondarily. Aggregation of beetles may also serve to overcome and kill even apparently healthy trees through shear weight of numbers, a phenomenon known as 'mass attack'. Several chemicals are usually involved in the initiation, continuance and termination of such attacks.

Primary attraction to a potential host has been shown in some species to be mediated by host plant metabolites while in others the approach is random but suitability of the host for colonization is then determined by making test borings into the bark and evaluating its susceptibility to attack from the quantity and/or quality of resin produced in response.

Once the first, or first few, beetles have bored into a suitable host tree, they begin to release pheromone which attracts both sexes of the same species. The first bark beetle aggregation pheromone to be identified was that of *Ips paraconfusus*. The pheromone is produced by the males and released from the hind guts along with the faeces. It has been shown to be a three-component mixture of ipsdienol (42), ipsenol (43) and cis-verbenol (44) (Silverstein et al., 1966), with the combination of the three compounds working synergisticly in the field but little or no attraction by any of them alone. Since then, several other aggregation pheromones have been identified for a number of bark beetle species. The *Dendroctonus brevicomis* aggregation pheromone was shown to be a synergistic mixture of bicyclic ketals, *exo*-brevicomin (45) and frontalin (46) and the terpene alcohol myrcene (47) (Bedard et al., 1969, 1980). In this species both sexes produce components of the aggregation pheromone; *exo*-brevicomin is produced by the females while frontalin is produced by the males, but in addition, the host tree (ponderosa pine) also contributes to the blend since myrcene is released from it as a result of the beetle's attack. Inhibition of further attraction to the aggregation site is effected through the release of *trans*-verbenol (48) and verbenone (49).

(42) (+) ipsdienol

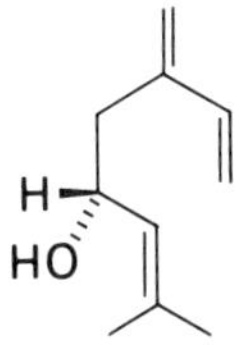

(43) (−) ipsenol

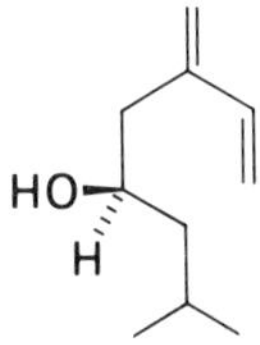

(44) cis-verbenol

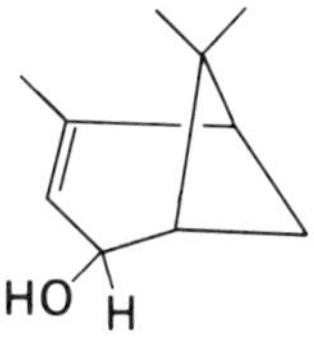

(45) exo-brevicomin

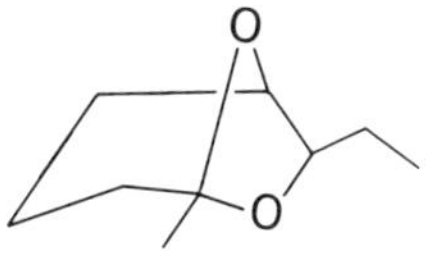

(46) frontalin

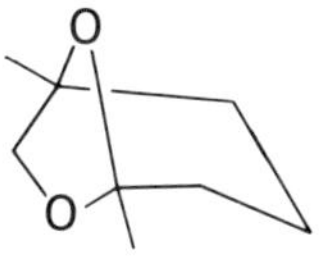

(47) myrcene

(48) trans-verbenol

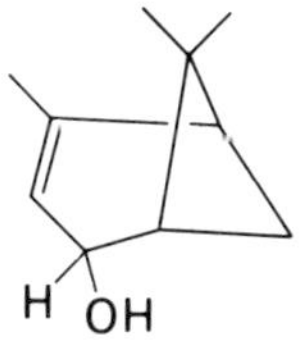

(49) verbenone

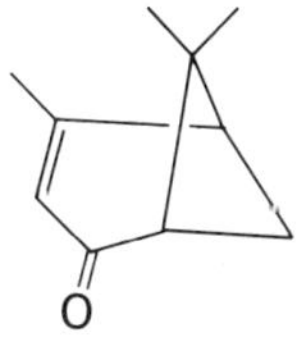

Ipsenol, originally described as a component of the pheromone of *I. paracon-fusus* also attracts the bark beetle *Pityokteines curvidens* (Harring *et al.*, 1975), whereas *Pityogenes chalcographus* a pest of Norway spruce is attracted by (50) (Francke *et al.*, 1977).

(50)

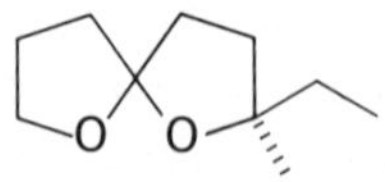

In contrast with the bark beetles described above, which construct their galleries in the phloem tissues, ambrosia beetles bore in the xylem. The male-produced aggregation pheromone of *Gnathotrichus sulcatus* was shown to be 6-methyl-5-hepten-2-ol (sulcatol) (51) and was found to be very effective in field tests (Byrne *et al.*, 1974; McLean and Borden, 1975).

(51)

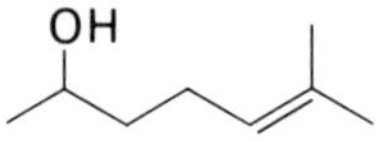

sulcatol

More recently the aggregation pheromone of the smaller European elm bark beetle, *Scolytus multistriatus,* the principal carrier of the Dutch elm disease pathogen, has also been isolated and identified. It consists of two female-produced components, 4-methyl-3-heptanol (52) and α-multistriatin (53) and one host-tree compound (α-cubebene) (54) (Pearce *et al.*, 1975). These compounds were also later found to be used by the related species, the larger European elm bark beetle, *S. scolytus* (Blight *et al.*, 1977).

(52)

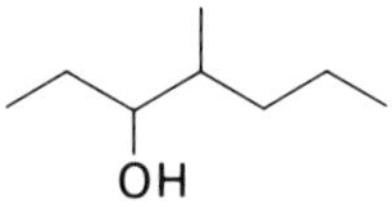

OH

(53)

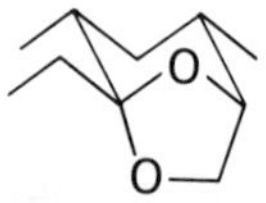

α-multistriatin

(54)

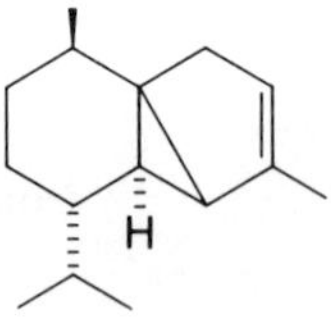

α-cubebene

Many of the pheromone components of bark beetles are chiral compounds and different enantiomers have very different biological activities (Silverstein, 1979). Thus it is (1*R*,5*S*,7*R*)-(+)-*exo*-brevicomin and (1*S*,5*R*)(−)-frontalin that synergize with myrcene in the pheromone of *D. brevicomis* (Wood *et al.*, 1976). The opposite enantiomers are inactive. Similarly, *Ips grandicollis* responds to

(S)(−)-ipsenol and not to (R)(+)-ipsenol (Vite *et al.*, 1976). The absolute configuration of the *Ips paraconfusus* pheromone was described by Silverstein *et al.*, (1966) as (S)(+)-*cis*-verbenol, predominantly the (S)(+) enantiomer of ipsdienol and (S)(+)-*cis*-verbenol while that of *I. pini* is (R)(−)-ipsdienol (Birch *et al.*, 1980). However, in *I. pini* the activity of (R)(−)-ipsdienol is completely negated by as little as 3% of (S)(+)-ipsdienol (Birch *et al.*, 1980). A similar interruption of response by *Gnathotrichus retusus* to its pheromone (S)(+)-sulcatol (**51**) is achieved by small percentages of (R)(−)-sulcatol (Borden *et al.*, 1980). A racemic mixture of (+) and (−) isomers on the other hand attracts *G. sulcatus* (Borden *et al.*, 1980). In the species *Ips pini* differences in the production of, and response to, optical isomers of ipsdienol have been found between populations from different geographical areas (Lanier *et al.*, 1980).

There is therefore a high behavioural resolution in response to enantiomeric pheromones and this implies that mechanism of synthesis and perception are also chiral in nature.

As to the biosynthesis of these pheromones in bark beetles, there has been much speculation as to whether beetles accumulate and release compounds from their host tree, whether they convert host compounds in their tissues or extracellularly in the hind gut to pheromones, whether all the components are produced by the insect *de novo* from non-host-derived compounds or even whether the pheromone components are actually generated by microorganisms in the beetle's gut.

The biosynthetic routes of the pheromone components of *I. paraconfusus* are now known (Figure 1). Both sexes produce *cis*-verbenol when exposed to (−)-α-pinene and *trans*-verbenol from (+)-α-pinene (Renwick *et al.*, 1976). When exposed to the vapour of myrcene, ipsdienol and ipsenol appeared in the hindguts of male beetles, but not in females (Hughes, 1974; Hughes and Renwick, 1977; Byers *et al.*, 1979), whereas no pheromone was detected in the guts of beetles not exposed to myrcene. Deuterium-labelling techniques have now confirmed that myrcene is converted in male *I. paraconfusus* to ipsdienol and ipsenol (Hendry *et al.*, 1980). All three components of the aggregation pheromone of *I. paraconfusus* are therefore produced by the insect from simple oxidation of host plant chemicals. Brand *et al.*, (1975), however, isolated a bacterium *Bacillus cerenus*, from the hindgut of *I. paraconfusus* which could convert α-pinene to *cis*- and *trans*-verbenol *in vitro*. It has also been shown that *I. paraconfusus* males fed on the antibiotic streptomycin no longer synthesize ipsdienol and ipsenol on exposure to myrcene implying that bacteria are involved in the sex-specific conversion of myrcene to the pheromone components, but the lack of effect of these antibiotics on *cis*-verbenol synthesis remains unexplained. It has been shown also in *D. frontalis* that the mycangial fungus carried by the female is capable of oxidizing *trans*-verbenol to verbenone (Brand *et al.*, 1976). It may be therefore that microorganisms have a much more important role in bark beetle pheromone production than has been thought to date.

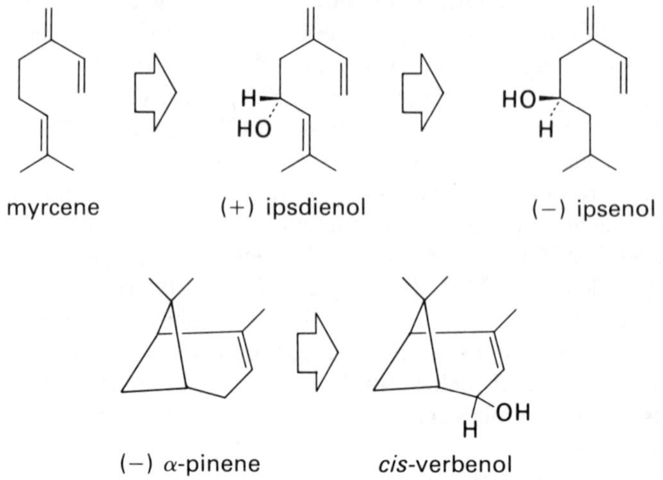

Figure 1 Biosynthesis of the components of the aggregation pheromone of *Ips paraconfusus* by simple conversion from the host-tree terpenes myrcene and α-pinene

Turning now to the sex pheromone of non-bark beetle species, perhaps the most complex pheromone described for a coleopteran species is that produced by the male boll weevil, *Anthonomus grandis* (Tumlinson *et al.*, 1969) (**55–58**). All of the compounds are required for effective attraction in the field. Grandisal and grandisol have also been shown to be present in males of *Pissodes stroloi* and *P. approximatus* (Booth *et al.*, 1983).

(**55**)
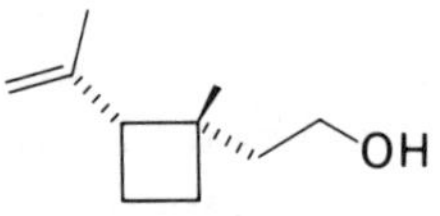

grandisol

(**56**)
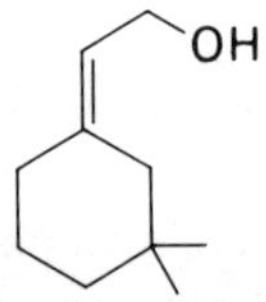

(**57**)
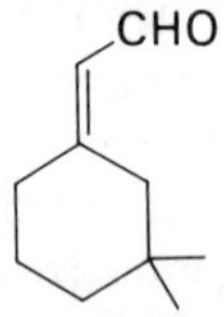

grandisal

(**58**)
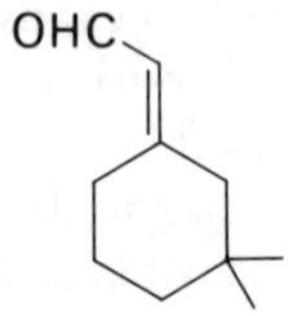

grandisal

In complete contrast, male black carpet beetles, *Attagenus megatoma,* can be attracted by a single compound, (*E,Z*)-3,5-tetradecadienoic acid (**59**), which is secreted by the female of the species. Other dermestid beetles, pests of stored products, however, produce multicomponent pheromones. *Trogoderma glabrum* has six active components (**60–65**) but only (**61**) and (**65**) are required for trapping. A high degree of cross-attraction occurs among a number of *Trogoderma* species (Vick *et al.,* 1970) (Table 4). The sex and aggregation pheromones of other non-dermestid stored product Coleoptera are also shown in Table 4.

Table 4   Major pheromone components of some stored product beetles

Species (sex producing pheromone)	Compound	References
Sex pheromones		
Acanthoscelides obtectus (♂)	(*E*)-(—)-Methyl-2,4,5-tetradecatrienoate	Hope *et al.* (1967); Horler (1970)
Anthrenus flavipes (♀)	(*Z*)-3-Decenoic acid	Burkholder *et al.* (1974); Fukui *et al.* (1974)
Attagenus megatoma (♀)	(*E,Z*)-3.5-Tetradecadienoic acid	Burkholder and Dicke (1966); Silverstein *et al.* (1967)
A. elongatulus (♀)	(*Z,Z*)-3,5-Tetradecadienoic acid	Barak and Burkholder (1977a) Fukui *et al.* (1977)
Lasioderma serricorne (♀)	4,6-Dimethyl-7-hydroxy nonan-3-one	Coffelt and Burkholder (1972) Chuman *et al.* (1979a,b)
Stegobium paniceum (♀)	2,3-Dihydro-2,3,5-trimethyl-6(methyl-2-oxo-butyl)-4*H*-pyran-4-one	Kuwahara *et al.* (1975, 1978)
Trogoderma inclusum and *T. variabile*	(*Z*)-14-Methyl-8-hexadecenol (*Z*)-14-Methyl-8-hexadecenal	Burkholder and Dicke (1966); Rodin *et al.* (1969); Cross *et al.* (1976)
Trogoderma glabrum (♀)	(*E*)-14-Methyl-8-hexadecenol (*E*)-14-Methyl-8-hexadecenal	Burkholder and Dicke (1966); Yarger *et al.* (1975); Cross *et al.* (1976)
Trogoderma granarium (♀)	92:8 (*Z:E*)-14-Methyl-8-hexadecenal	Levinson and Barilan (1967); Cross *et al.* (1976)
Aggregation pheromones		
Prostephanus truncatus	Isopropyl (*E*)-2-methyl-2-pentenoate	Hall *et al.* (1984)
Rhyzopertha dominica (♂)	1-Methylbutyl-(*E*)-2-methyl-pentenoate 1-Methylbutyl-(*E*)-2,4-di-methyl-2-pentenoate	Khorramshahi and Burkholder (1981); Williams *et al.* (1981)
Tribolium castaneum (♂) and *T. confusum* (♂)	4,8-Dimethyl decanal	Suzuki (1980) (1981)
Sitophilus oryzae and *S. zeamais*	(*R,S*)-4-Methyl-5-hydroxy-3-heptanone	Schmuff *et al.* (1984)

(59) ![structure 59]

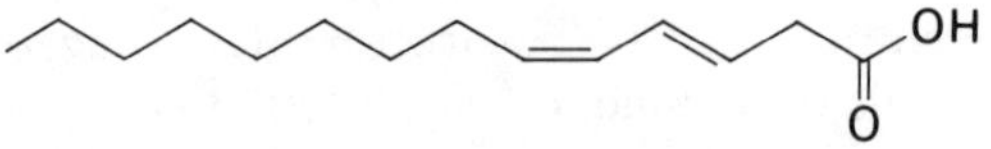

(60) 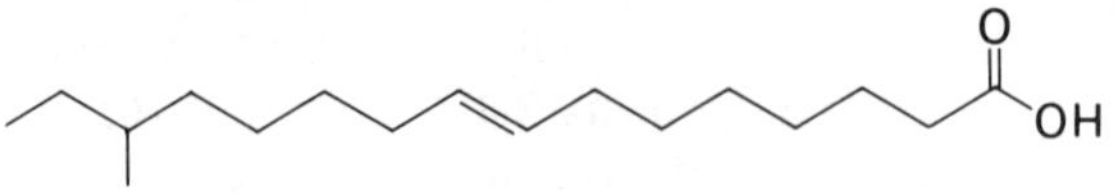 ![structure 60]

(61) ![structure 61]

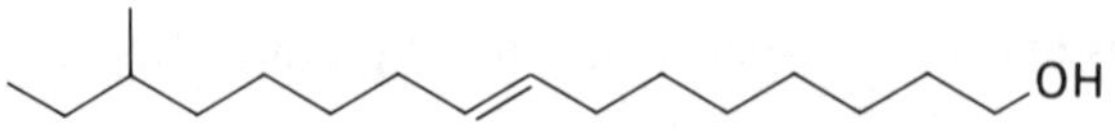

(62) 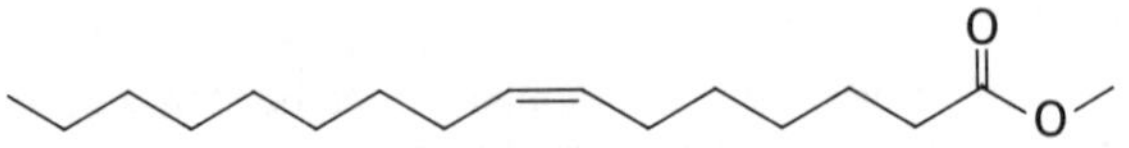 ![structure 62]

(63) ![structure 63] 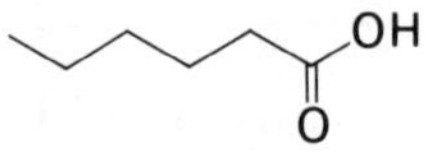   **(64)** 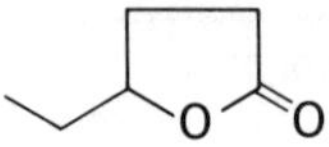 ![structure 64]

(65) ![structure 65]

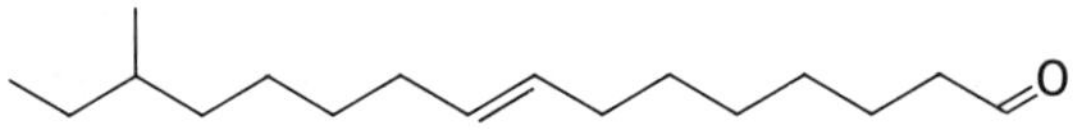

As in the case of bark beetle pheromones many of the stored product beetle pheromones are chiral. In *Trogoderma granarium* the $(+)(S)$ enantiomer of **(65)** in a $92:8$ mixture of the (Z) and (E) isomers is much more active than the $(-)(R)$ form (Rossi and Niccoli, 1978). *Tribolium castaneum* and *T. confusum* have both been found to use $(4R,8R)$-$(-)$-dimethyl decanal **(66)** as a sex and aggregation pheromone (Levinson and Mori, 1983; Suzuki and Mori, 1983). Similarly the major component of the sex pheromone of *Lasioderma serricorne* has been fully identified as $(4S,6S,7S)$-4,6-dimethyl-7-hydroxynonan-3-one **(67)** (Chuman *et al.*, 1982a,b).

(66) 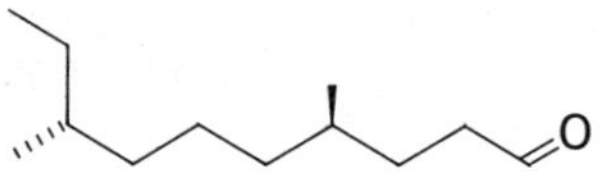 ![structure 66]   **(67)** ![structure 67]

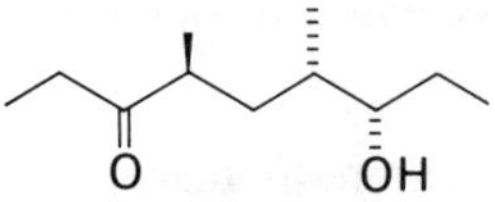

The sex pheromones of a few other important coleopteran pests have also been identified. Japanese beetle, *Papillia japonica,* has been shown to be a long chain lactone **(68)**. The absolute configuration was established by an enantiomer-specific synthesis from $(-)$-glutamic acid **(69)** (Tumlinson *et al.*, 1977). The sex pheromones of three *Diabrotica* spp. have also been identified: *D. cristata* produces $(2S,8R)$-8-methyl-2-decyl acetate **(70)**, *D. virgifera virgifera* $(2R,8R)$-8-methyl-2-decyl propionate **(71)** and *D. decimpunctata howardi* (R)-10-methyl-2-tridecanoate **(72)** (Guss *et al.*, 1982, 1983a,b).

(68)

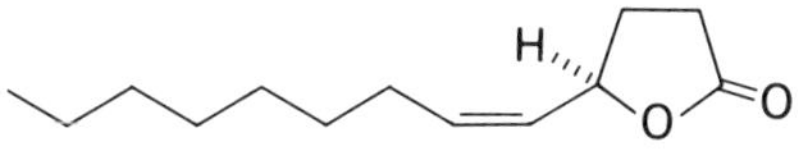

(69) 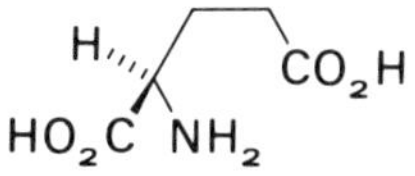 HO_2C NH_2 ... H ... CO_2H **(70)** 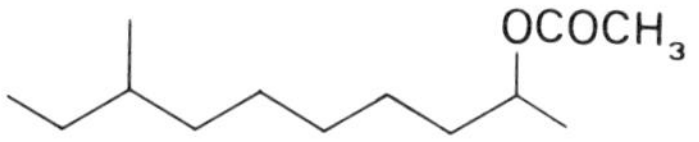 ... $OCOCH_3$

(71) 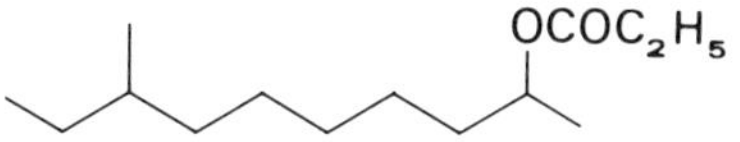 ... $OCOC_2H_5$

(72) 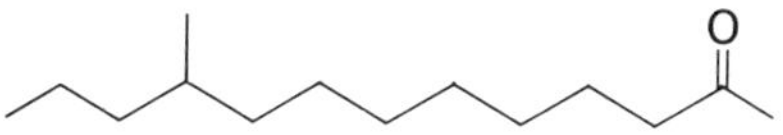 ... O

Compounds with pheromonal activity have also been found in the larvae of at least one beetle species; larvae of the house longhorn beetle, *Hylotrupes bajulus* produce a number of oxygenated monoterpenes in their faecal material and of these (–)-verbenone, synergized by *p*-cymen-8-ol, stimulates egg-laying by the adult female (Higgs and Evans, 1978).

Pheromones of social and non-social Hymenoptera

Ants

Of the social Hymenoptera, behaviour-modifying chemicals have been most widely studied in ants. These chemicals originate in a variety of exocrine glands dispersed over the body of the ant. Sex pheromones are known in ants but since sexual activity is normally restricted to a few days in the year they are not easily studied under laboratory conditions, and no sex attractant pheromones have been characterized to date. Perhaps the best studied pheromones of ants are those involved in recruitment behaviour and food retrieval. Scout ants recruit workers to a food source by means of chemical trails. Characterization of the chemicals involved has been difficult, however, because of the extremely low levels of compound present in trail secretions. Structures characterized to date are shown in Figure 2.

Atta cephalotes, a leaf cutting ant species, uses a pheromone secretion for marking leaves before and during transport to the nest. The main component which induces attraction of other workers to the marked leaf was found to be *n*-tridecane while (Z)-9-nonadecene releases pickup and remarking behaviour (Bradshaw *et al.*, 1983).

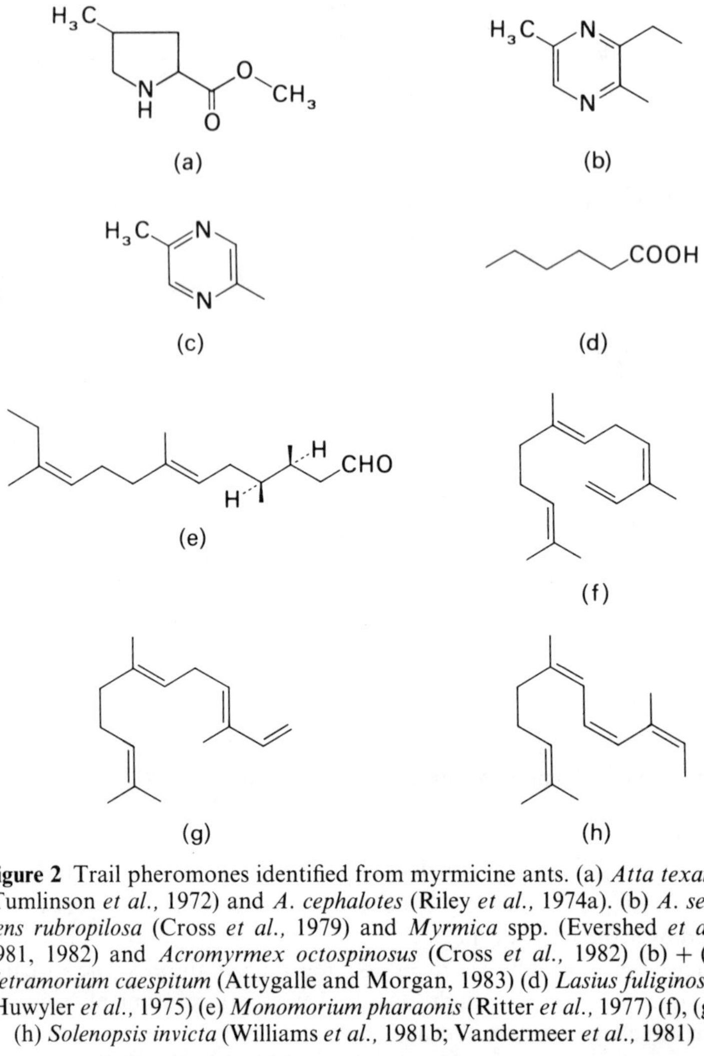

Figure 2 Trail pheromones identified from myrmicine ants. (a) *Atta texana* (Tumlinson *et al.*, 1972) and *A. cephalotes* (Riley *et al.*, 1974a). (b) *A. sexdens rubropilosa* (Cross *et al.*, 1979) and *Myrmica* spp. (Evershed *et al.*, 1981, 1982) and *Acromyrmex octospinosus* (Cross *et al.*, 1982) (b) + (c) *Tetramorium caespitum* (Attygalle and Morgan, 1983) (d) *Lasius fuliginosus* (Huwyler *et al.*, 1975) (e) *Monomorium pharaonis* (Ritter *et al.*, 1977) (f), (g), (h) *Solenopsis invicta* (Williams *et al.*, 1981b; Vandermeer *et al.*, 1981)

The vast majority of ants use chemical secretions for defensive purposes and a number of them will be discussed later in this chapter. However, in many species these secretions also have a communicative function as well as the defensive or offensive purpose. These so-called 'alarm pheromones' have been studied widely and have been characterized for many species. They are most easily divided into two categories: simple and multicomponent alarm pheromones. In the former, one or a small number of compounds release identical behaviour patterns in worker ants. Longhurst *et al.*, (1978) have shown that in *Odontomachus troglodytes* 2,6-dimethyl-3-butyl pyrazine and its 3-pentyl- and 3-hexyl-homologues all release the same alarm behaviour in the workers. The alarm pheromone in every species of *Atta* studied so far has been found to be 4-methyl-3-heptanone, and the (+) isomer of that compound in at least *Atta texana* and *A. cephalotes* (73) (Riley *et al.*, 1974b). In *A. texana* this compound releases attraction at a concentration of 3×10^9 molecules/cm^3 and 'alarm', fast running with open mandibles, at approximately 10 times that concentration (Moser *et al.*, 1968).

(73)

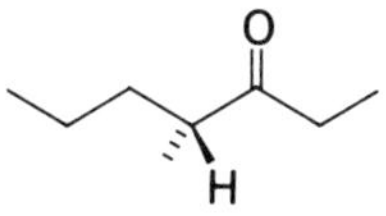

Atta spp.

Multicomponent alarm pheromones are those where two or more compounds are present, each of which releases different elements of behaviour. The poison glands of *Myrmicaria* spp. contain large quantities of monoterpene hydrocarbons such as β-myrcene, β-pinene and limonene (Brand *et al.*, 1974; Longhurst *et al.*, 1983). In *M. eumenoides* all three of these compounds will alert and attract workers, while limonene alone also causes the ants to circle 1–2 cm around the source. In some species where the alarm pheromone is multicomponent, the various components do not necessarily come from the same gland. The most complex alarm communication system to be investigated to date is that of the African weaver ant *Oecophylla longinoda* where no less than four glands are involved. 'Alerting' in this species is released by hexanal, attraction and probably circling by 1-hexanol and biting by the less volatile 3-undecanone and 2-butyl-2-octenal (Bradshaw *et al.*, 1979a,b,c).

The queen recognition pheromone of the red imported fire ant, *Solenopsis invicta* has also been characterized recently as a mixture of the three compounds (74), (75) and (76) (Rocca *et al.*, 1983a,b).

(74)

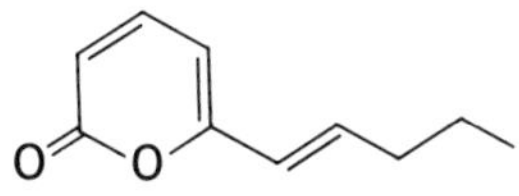

(75)

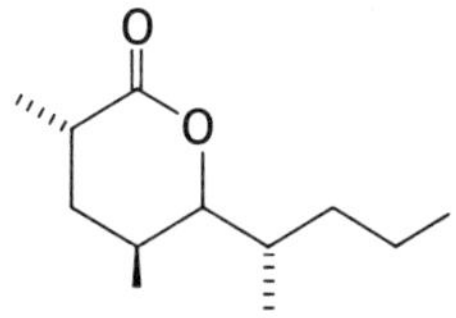

(76)

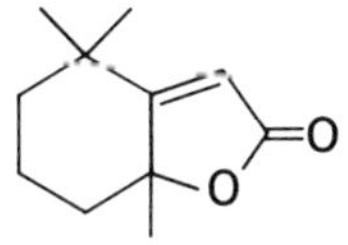

Solenopsis invicta

Several other compounds have been isolated from the various glands of a number of ant species but since they are so numerous and their behavioural significance largely unknown they have not been included here but the following references could be of interest in this context: Lloyd *et al.*, 1975; Duffield *et al.*, 1976, 1977; Parry and Morgan, 1979 (review); Bradshaw and Howse, 1984.

Bees

Of the bee species studied to date, the honeybee *(Apis mellifera)* has, because of its economic importance, received considerable attention as regards its pheromonal secretions. Chemical signals come from the secretions of the mandibular and Nasonov glands as well as the lining of the sting pouch. The major component of the queen's mandibular gland pheromone was shown to be (*E*)-9-*oxo*-2-decenoic acid (**77**) (Butler *et al.*, 1961; Callow *et al.*, 1964) and of the many other minor components, 9-hydroxy-(*E*)-2-decenoic acid was also thought by some to have a role in attracting worker bees (Butler and Simpson, 1967) but this was later disputed (Boch *et al.*, 1975). Such conflicting results are not unexpected, however, when the chirality of the compounds used in the test were unknown, but enantiomeric synthesis of these compounds is now possible (Kandil and Slessor, 1983).

(**77**)

Apis mellifera

The workers' Nasonov pheromone has also been shown to be multi-component, the total secretion being a mixture of the following seven compounds: (*Z*) and (*E*)-citral, nerol, geraniol, nerolic acid, geranic acid and (*E,E*)-farnesol (Shearer and Boch, 1966; Butler and Calam, 1969; Pickett *et al.*, 1980). The alarm pheromone derived from the sting pouch has been shown to have an even greater number of components. Of the following 13 compounds identified to date: 1-decanol, phenol, 1-butanol, isopentyl acetate, isopentyl alcohol, 1-hexanol, 2-heptyl acetate, 2-heptanol, 1-octanol, 1-acetoxy-2-octenol, 2-nonyl acetate, 1-acetoxy-2-nonene, (*Z*)-11-eicosen-1-ol, only the last 11 mentioned compounds produce alarm responses when tested individually on honeybee workers (Pickett *et al.*, 1982; Collins and Blum, 1983). A mixture of alkanes and esters has also been isolated from the sting of *Apis mellifera* queens (Blum *et al.*, 1983).

Bumble bees have also been studied quite extensively in terms of their pheromonal secretions: male bees use secretions from their labial glands to scent mark sites along their flight paths for territorial demarcation, sex attraction and copulatory excitation. Terrestrol, 2,6-dihydro-6-*E*-farnesol (**78**) was the first marking pheromone to be described from *Bombus terrestris* (Bergström *et al.*,

1968; Ställberg-Stenhagen, 1970). Fourteen *Bombus* and *Psithyrus* species were subsequently studied (Calam, 1969; Kullenberg *et al.*, 1970; Svensson and Bergström, 1977) and a large variety of compounds were identified including monoterpenes, sesquiterpenes, diterpenes, aliphatic alcohols, acetates, esters, aldehydes and hydrocarbons.

(78)

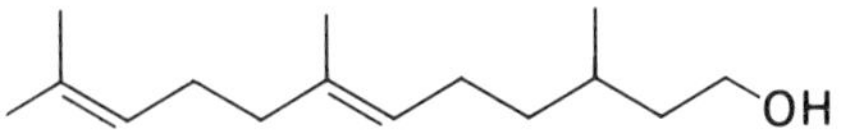

terrestrol

Bombus terrestris

Although bees of the genus *Andrena* have a solitary existence, the males produce an aggregation pheromone in their mandibular glands which attracts both males and females to communal mating areas (Tengö and Bergström, 1976, 1977). The secretion of these glands is highly complex and contains monoterpenes, straight chain fatty acid derivatives and a series of spiroketals including 1,6-dioxaspiro[4.4]nonanes, 1,6-dioxaspiro[4.5]decanes, 1,7-dioxaspiro[5.5]undecanes, a 1,6-dioxaspiro[4.6]undecane and a 1,7-dioxaspiro[5.6]dodecane (Franke *et al.*, 1980, 1981). The Dufour's gland of this genus has also been shown to contain a variety of terpenes including farnesene, farnesol, geraniol and a series of farnesyl and geranyl esters with either geranyl octanoate **(79)** or farnesyl hexanoate **(80)** dominating depending on the species (Bergström and Tengö, 1974). The same gland in the Colletidae and Halictidae families of solitary bees produces a series of macrocyclic lactones both saturated and unsaturated (Bergström, 1974; Hefetz *et al.*, 1978; Bergström and Tengö, 1979; Duffield *et al.*, 1981a).

(79)

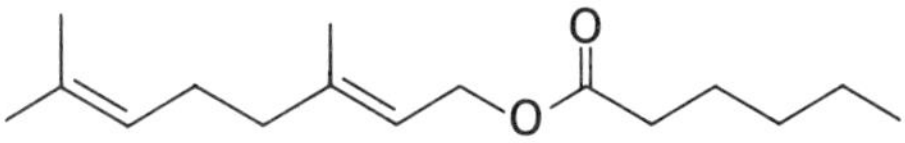

geranyl octanoate

(80)

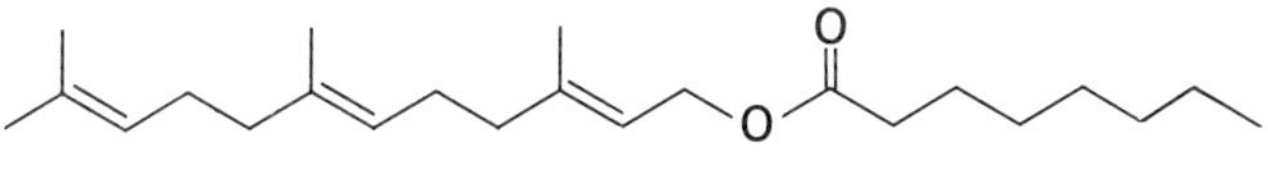

farnesyl hexanoate

A recent review of the sociochemicals of bees has been published which provides a very useful source of references (Duffield *et al.*, 1984).

Wasps

The mandibular glands of a number of species of solitary wasps from the family Sphecidae have also been analysed and found in all cases but one to contain 2,5- and 2,6-dimethylalkyl-pyrazines (Borg-Karlson and Tengö, 1980; Hefetz and Batra, 1980; Duffield *et al.*, 1981b).

Some pheromone structures have also been described for social wasps; both (Z) and (E) isomers of the spiroacetals (**81a** and **81b**) have been isolated from *Paravespula vulgaris* and it has been suggested that they serve to protect individuals from attack by fellow workers (Franke *et al.*, 1978). The compound (Z)-9-pentacosene, which initiates brood warming has also been identified from pupae in the brood cells of *Vespa crabro* (Veith and Koeniger, 1978).

(81) R¹—   *Paravespula vulgaris*

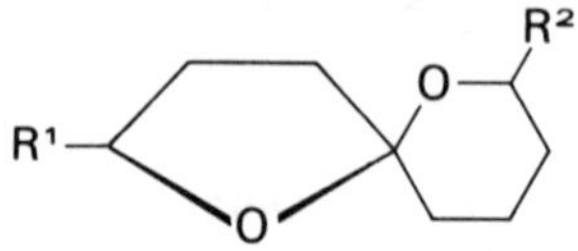

a; R¹ = H, R² = Me
b; R¹ = Me, R² = H

Sawflies

The only non-social group of Hymenoptera studied in terms of their pheromones are the sawflies (S.O. Symphyta) where a series of chiral compounds have been described. The sex pheromones of *Diprion* spp., *Neodiprion sertifer*, *N. lecontei* and *N. pinetum* have been shown to consist of (**82a,b** and **c**). The optimum pheromone blend of *Neodiprion sertifer* has shown to be (2S,3S,7S)-3,7-dimethyl-pentadecan-1-ol acetate with a trace of the (2S,3R,7R)-isomer as a synergist (Kikukawa *et al.*, 1983). In another sawfly species, *Pikonema alaskensis*, the yellowhead spruce sawfly, the primary sex pheromone has been reported to be a series of (Z,Z)-9,19 alkadienes ($C_{29}, C_{31}, C_{33}, C_{35}, C_{37}$) (Bartlet *et al.*, 1983) but later, the biological activity was shown to be due to (Z)-10-nonadecenal produced by atmospheric oxidation of these alkadienes (Bartlet and Jones, 1983).

(82)   *Diprion* and
                                       Neodiprion spp.

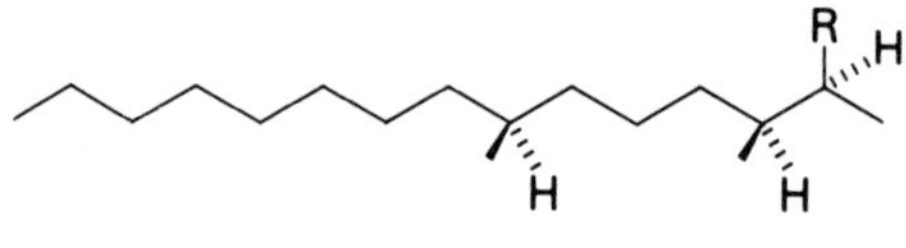

a; R = OH
b; R = Ac
c; R = Pr

Pheromones of other insect and acarine orders

Dictyoptera: cockroaches

Because of their economic importance, cockroaches have been widely studied in terms of their pheromone components. An oxygenated germacrene structure was first proposed for the American cockroach, *Periplaneta americana* (**83**)

(Nishida *et al.*, 1976; Persoons *et al.*, 1976a) followed by another oxygenated cyclic sesquiterpene structure (84) (Talman *et al.*, 1978; Adams *et al.*, 1979; Persoons *et al.*, 1979). The exact structural elucidation of periplanone A (83) was eventually achieved by comparison of its spectral data with that of its stable rearrangement product (85) (Persoons *et al.*, 1982).

(83)

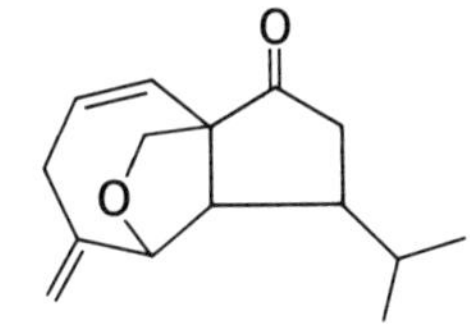

periplanone A

(84)

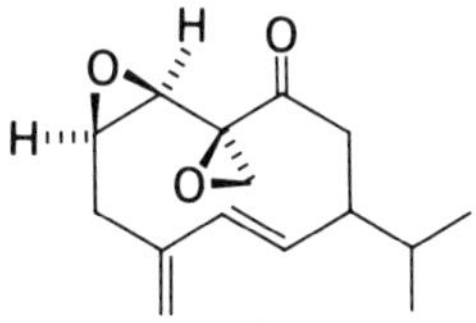

periplanone B
(1*R*,2*R*,7*S*,10*R*)

(85)

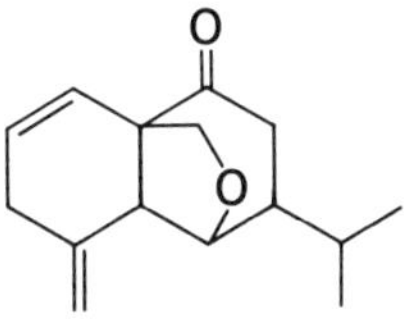

Periplaneta americana

For the German cockroach, *Blattella germanica*, two long chain ketones (86) and (87) have been described (Nishida *et al.*, 1976) with the (*S,S*) isomer of (86) being the one present in the insect but with the racemate working just as well (Nishida *et al.*, 1979). These compounds only produce wing-raising responses in the males and do not apparently produce any attractant responses.

(86)

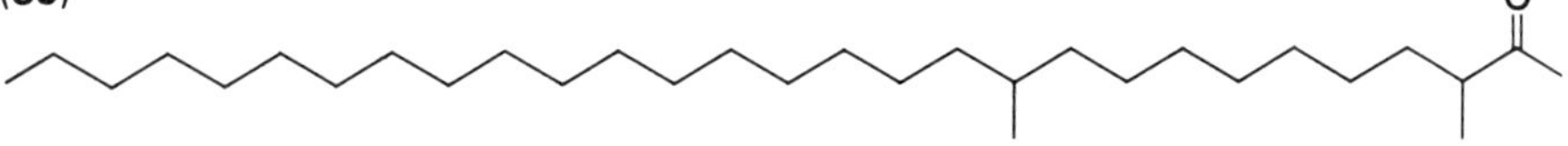

(*S,S*)

(87) $HO(CH_2)_{16}$

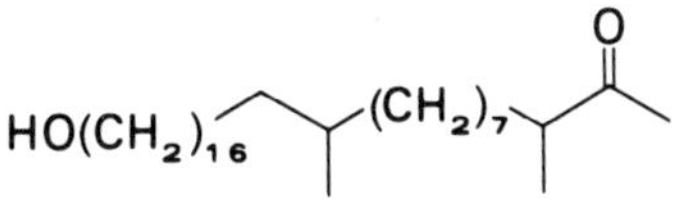

Blattella germanica

Hemiptera: aphids, bugs and scale insects

Since sexual reproduction in aphids tends to be restricted to certain times of the year and parthenogenetic reproduction predominates, the study of their sex pheromones has received little or no attention. Conversely, the pheromones which control their alarm responses have been studied extensively. Sesquiterpene hydrocarbons such as (E)-β-farnesene (88) and related homologues have been shown to be involved in eliciting alarm behaviour in aphids from 19 genera and three aphid subfamilies (Bowers et al., 1972; Edwards et al., 1973; Wientjens et al., 1973; Nishino et al., 1976a,b; Pickett and Griffiths, 1980). More recently germacrene A (89) has been shown to be the alarm pheromone of Thereoaphis spp. (Bowers et al., 1977) and β-selinine, one of its breakdown products, may also be used by aphids of the genera Eucallipterus and Calaphis (Nault and Montgomery, 1979).

(88)

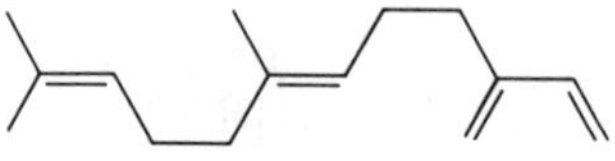

(E)-β-farnesene

(89)

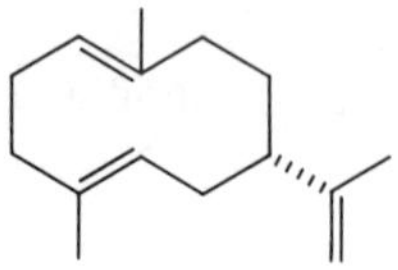

germacrene A

Levinson et al., (1974) have shown that the bed bug Cimex lectularius, uses a mixture of (E)-2-octenal and (E)-2-hexenal as an alarm pheromone while another pyrrhocorid bug, Dysdercus intermedius, uses similar straight chain alkene and alkane compounds as alarm pheromones including: dodecane, tridecane, pentadecane, hexanal, (E)-2-hexenal, 4-keto-2-hexenal, 2-octenal and 4-keto-2-octenal (Calam and Youdeowei, 1968). The alarm pheromone of a triatomine bug, Dipetalogaster anaximus, has on the other hand been shown to have the branched ketone, 3-methyl-2-hexanone (Rossiter and Staddon, 1983).

Scale insects from the genus Aonidiella have yielded pheromones with interesting structures. The sex pheromone of A. aurantii, the California red scale, is a mixture of (90) and (91) (Roelofs et al., 1978; Tashiro et al., 1979) and the related structure (92) serves the same function in A. citrina, the yellow scale (Gieselmann et al., 1979a).

(90)

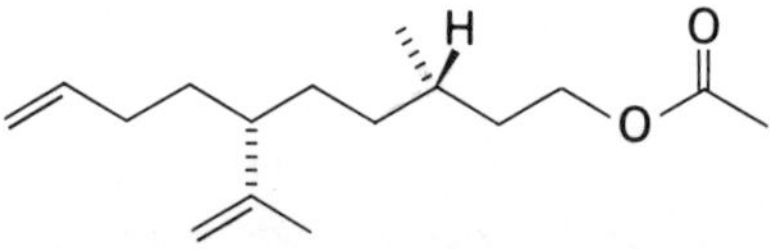

(3S,6R)

Aonidiella aurantii

(91)

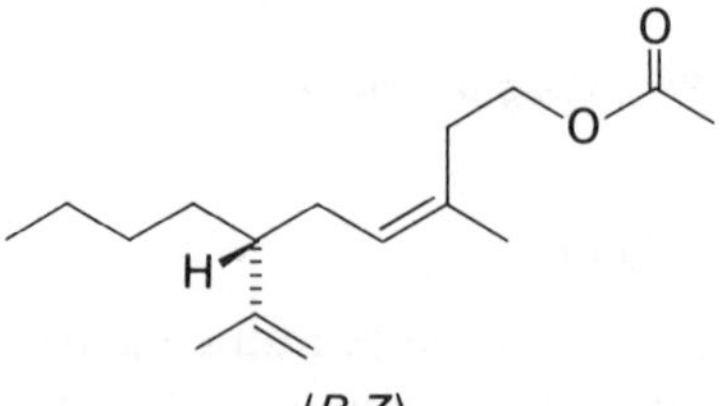

(R,Z)

Aonidiella aurantii

(92) *Aonidiella citrina*

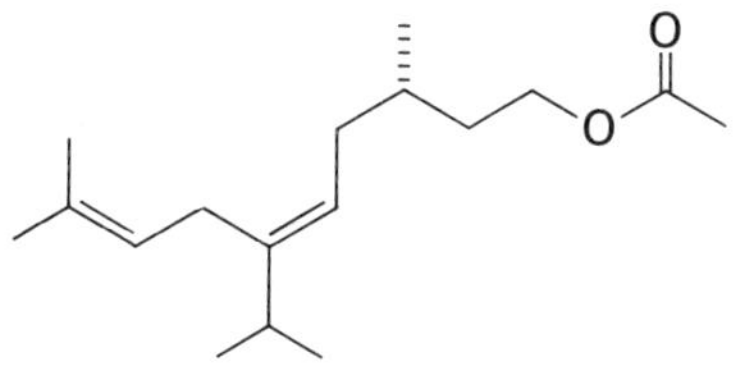

(S)

The female comstock mealybug, *Pseudococcus comstocki* attracts males with a norterpene, 2,6-dimethyl-3-acetoxy-1,5-heptadiene **(93)** (Bierl-Leonhardt *et al.*, 1980, Negishi *et al.*, 1980) and the citrus mealybug, *Planococcus citri*, uses the related compound **(94)** (Bierl-Leonhardt *et al.*, 1981). The sex pheromones of the San José scale *Quadraspidiotus perniciosus* **(95)** (Gieselmann *et al.*, 1979b) and the white peach scale *Pseudaulacaspis pentagona* **(96)** (Heath *et al.*, 1979) have also been described. All these compounds are chiral but great progress has been made in recent years on their stereoselective synthesis (Mori, 1982).

(93)

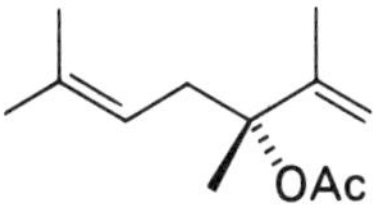

(R)-(+)-
Pseudococcus comstocki

(94)

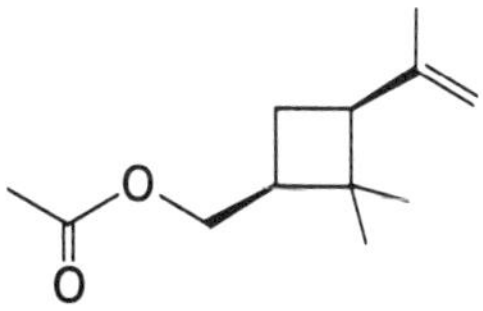

Planococcus citri

(95a)

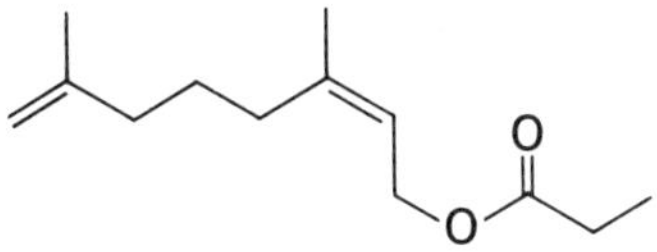

(95b)

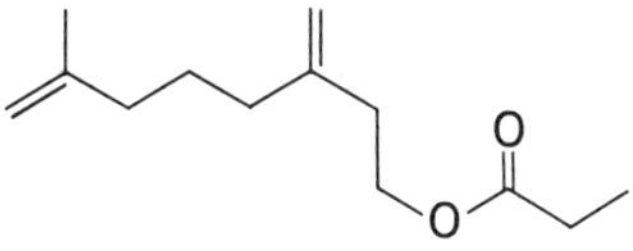

Quadraspidiotus perniciosus

(96)

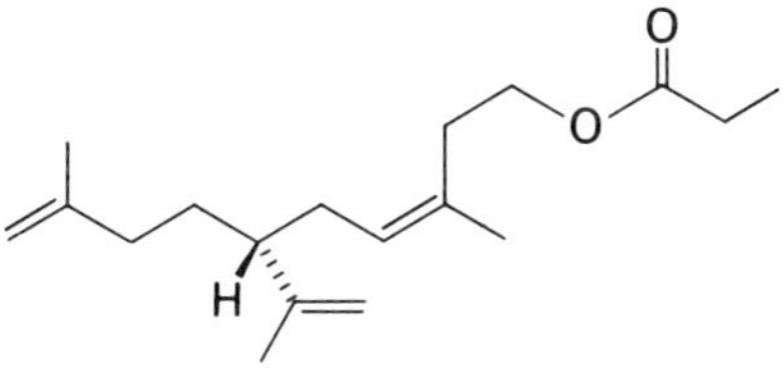

Pseudaulacaspis pentagona

Isoptera: termites

A number of compounds have been described as trail pheromones for termites; the diterpene neocembrene A **(97)** was characterized from *Nasutitermes* spp.

(Birch *et al.,* 1972) while (*Z,Z,E*)-3,6,8-dodecatrienene-1-ol (**98**) has been isolated from *Reticulitermes flavipes* and *R. virginicus* (Matsumura *et al.,* 1968) and produces a trail-following response in that species. It also induces trail following in a number of other *Reticulitermes* species as well as *Coptotermes formosanus* and *Leucotermes speratus* (Ritter and Coenen-Saraber, 1969; Honda *et al.,* 1975; Howard *et al.,* 1976).

(**97**)

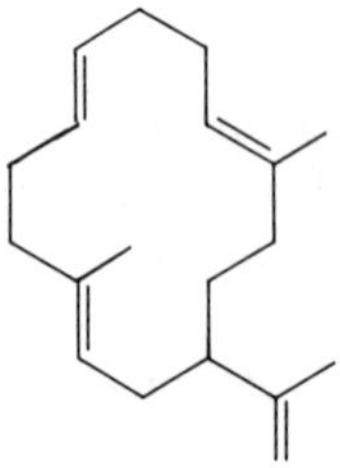

Nasutitermes spp.

(**98**) HO

Reticulitermes spp.

Acari: mites and ticks

The alerting (or alarm) pheromone of the mite *Tyrophagus putrescentiae* has been identified as (**99**) (Kuwahara, 1978) while that of four other mite species has been shown by the same author to be principally citral. The aggregation pheromone of the mite, *Lardoglyphus konoi* has been identified as 1,3,5,7-tetra methyldecyl formate (**100**) with the *R*-configuration at C_1 (Kuwahara *et al.,* 1982). Little else is known about the stereochemistry of these chiral compounds but studies with analogues have shown that the primary (*Z*)-2-alkenyl formate moiety of (**99**) is essential in producing alarm pheromone activity in *T. putrescentiae* (Kuwahara and Sakuma, 1982).

(**99**)

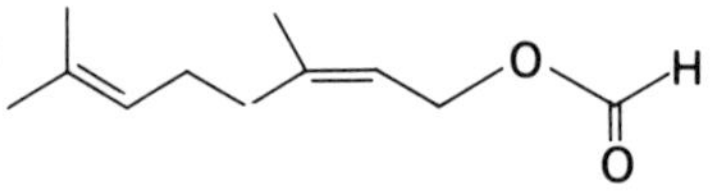

Tyrophagus putrescentiae

(**100**)

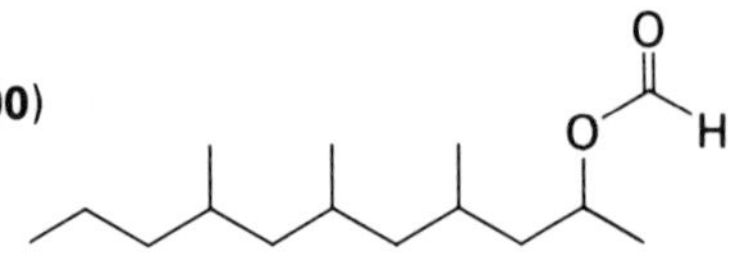

Lardoglyphus konoi

Some information is now available on the pheromones of ixodid ticks and is reviewed by Gothe (1983).

ALLOMONES

The defence chemistry of insects

Chemical defence is well known as a form of protection against predation in many animals. Arthropod defence chemicals have received particular attention because of the possibility of discovering novel insecticides. Since such chemicals are also present in relatively large amounts compared with pheromones, many valuable advances have been made in this field in recent years.

The subject of arthropod defence against predation through the use of chemicals has been well reviewed recently (Eisner, 1970; Weatherston and Persy, 1970; Schildknecht, 1971, 1976; Blum *et al.*, 1981; Pasteels *et al.*, 1983) and because of limitations of space, only a selection of chemical groups will be considered in this chapter.

Alkanes and esters

Simple alkanes and esters are used by a number of social insects for defence purposes but they may be harmless by themselves. Such compounds are used by *Formica rufa* and *F. sanguinaea* for instance in combination with formic acid as a defensive secretion. It is thought in fact that both alkanes and esters act as wetting agents for the formic acid, in the above examples (Lofquist, 1977).

Similar compounds have also been found in the defensive secretions of soldier termites. The alkane chains are made up of two-carbon acetate units linked together and are from 21 to 35 carbon atoms long depending on the species, the colony and nest location (Prestwich, 1983). Some shorter chained compounds have also been found; a mixture of hexadecanal and heptadecanal occurs in the defensive secretion of the termite *Coptotermes testacens* (Blum *et al.*, 1982).

The use of straight chain compounds in defence secretions is not confined to social insects: secretions from the tergal gland of the beetle *Aleochara curtala* have been shown to contain alkanes, alkenes, aldehydes and quinones (Peschke and Metzler, 1982). Three species of rove beetles, *Bledius spectabilis, Platystethus arenarius* and *Oxytelus piceus*, employ a similarly complex mixture of 1-alkenes, acetates, citral, quinones and lactones (Dettner and Schwinger, 1982).

Terpenoids and lactones

The lower terpenoids are relatively non-specific toxicants found in the defensive secretions of many insects (structures (**101**)–(**106**)). Because of their volatility and powerful smell, their odour may be sufficient to deter the attacker. The sawfly species *Neodiprion sertifer* for example discharges an oily effluent from diverticular pouches in the foregut when attacked. This secretion is identical to the terpenoid resin of its host plant, *Pinus sylvestris*, and is probably sequestered from its food while feeding (Eisner *et al.*, 1974). Compounds shown to be present in both host tree and insect include α- and β-pinene, pinifolic acid, pimeric acid, palustric acid, dihydroabietic acid, abietic acid, neoabietic acid and a (−)-pimaric acid. Since these compounds are taken by the sawfly larva directly from its host plant, it is probably more economical in terms of energetics than having to synthesize defensive toxins *de novo*. In a number of cases it has been shown that several insect species make their own mono-, sesqui- and diterpene defence chemicals from simpler starting materials. It has been shown for instance that the walking stick insect *Anisomorpha buprestoides* and the ant *Acanthomyops claviger* produce their terpenes from acetate through mevalonate according to the usual biosynthetic pathway. The former makes dolichodial (**104**), while the latter uses citronellal (**101**) and citral (**102**) as defence chemicals (Chadha *et al.*, 1962; Meinwald *et al.*, 1962; Regnier and Wilson, 1968).

(**101**) CHO

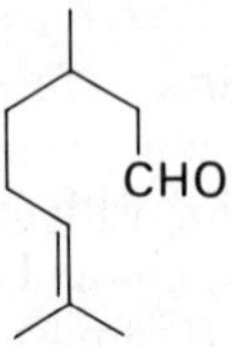

citronellal
Acanthomyops claviger

(**102**) CHO

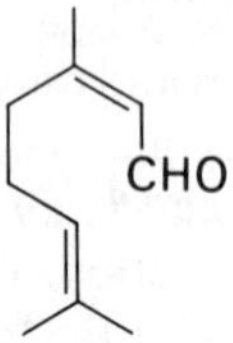

citral
A. claviger

(**103**)

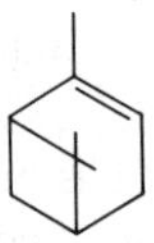

α-pinene
Neodiprion spp.

(**104**) CHO
CHO

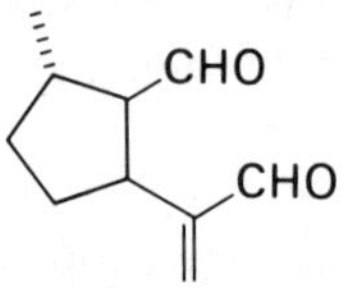

dolichodial
Anisomorpha buprestoides

Cantharidin (**106**), the terpenoid defence compound of the meloid beetle, *Lytta vesicatoria*, is in fact the basis of the well known 'aphrodisiac' Spanish fly and has vesicant properties in humans. It is also, however, quite toxic to man; the lethal dose being about 0.5 mg/kg body weight. Another beetle which has been shown to use monoterpenoid compounds amongst others is the devil's coach horse beetle *Staphylinus olens* which secretes a mixture of the terpenoid iridodial (**105**) together with 4-methylhexan-3-one when attacked (Fish and Pattenden, 1975).

(**105**) CHO
CHO

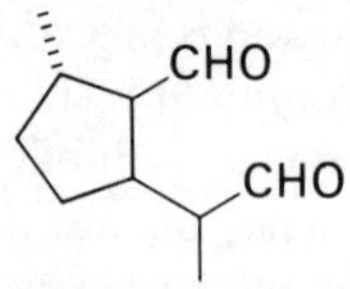

iridodial
Staphylinus olens

(**106**) O
O
O
O

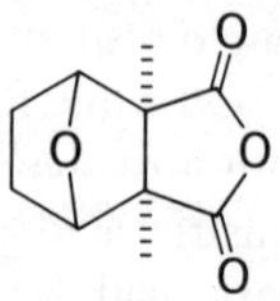

cantharidin
Lytta vesicatoria

It is, however, only in termite species where higher terpenoids are really used extensively for defensive purposes. Indeed, only a few of the highly evolved termite genera are capable of synthesizing diterpenes. Soldiers of the genus *Cubitermes* manufacture at least 16 diterpenes similar to (**107**) at least five of which are unique to termites. Perhaps the most complex diterpenoid compounds, however, are produced by members of the family Nasutitermitinae. Bicyclic (secotrinervitene) (**109**) and tetracyclic (kempene-like) (**108**) (**110**) diterpenes have been isolated from the soldier caste of species belonging to this family (Braekman *et al.*, 1980; Dupont *et al.*, 1981; Prestwich *et al.*, 1981). Many species such as *Trinervitermes gratiosus* use mixtures of diterpenes (**111**) and monoterpenes (**103** and similar) (Prestwich *et al.*, 1976; Prestwich, 1978).

(**107**)

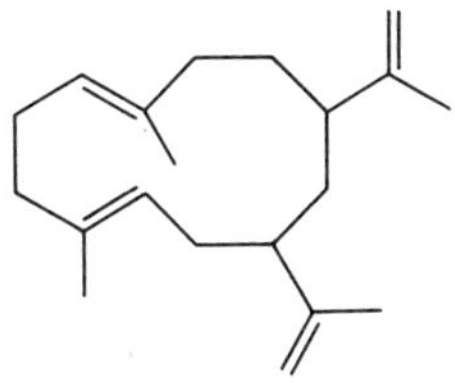

cubitene
Cubitermes

(**108**)

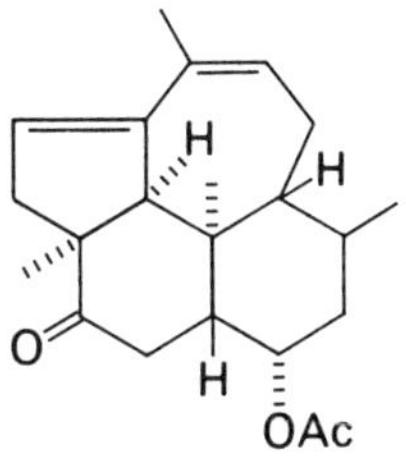

kempene-1
Nasutitermes kempae

(**109**)

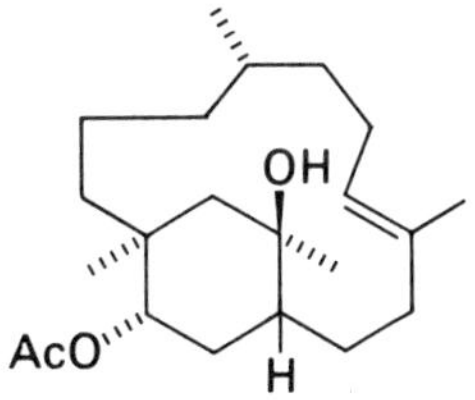

seco-trinervitene
N. princeps

(**110**)

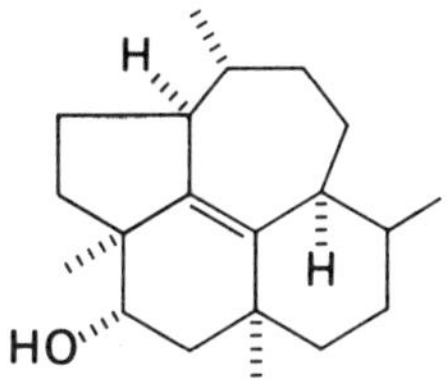

rippertene
N. rippertii

(**111**)

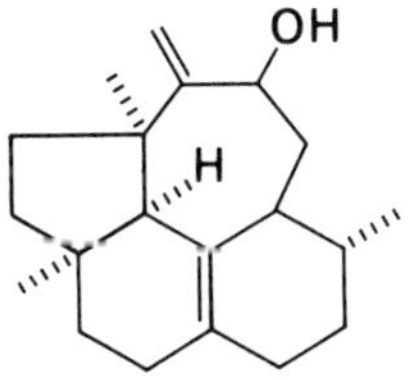

trinervitene
Trinervitermes gratiosus

Some West African species such as *Amitermes evuncifer* use a sesquiterpene ether, 4,11-epoxy-(*Z*)-eudesmane (>90%) (Wadhams *et al.*, 1974) with 10-epi-eudesma-3,11-diene, 8-epi-cararrapi oxide, cararrapi oxide, and (*Z*)-β-ocimene as minor components (**112–116**) (Baker *et al.*, 1978b). Similar mono- and sesquiterpenoid mixtures have been found in the defence secretions of *Syntermes* (**117–120**), and *Ancistrotermes* spp. (**121–124**) (Baker *et al.*, 1978a; Baker, R. *et al.*, 1981).

(**112**)

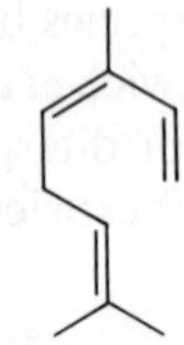

(*Z*)-β-ocimene

(**113**)

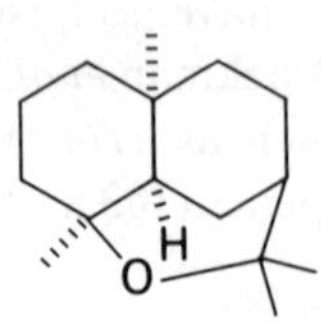

4,11-epoxy-(*Z*)-eudesmane

(**114**)

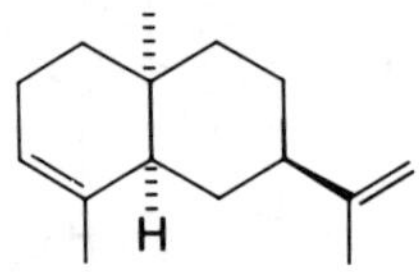

10-epi-eudesma-3,11-diene

(**115**)

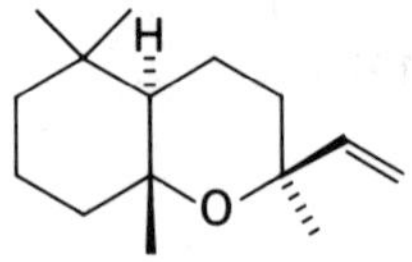

8-epi-cararrapi oxide

(**116**)

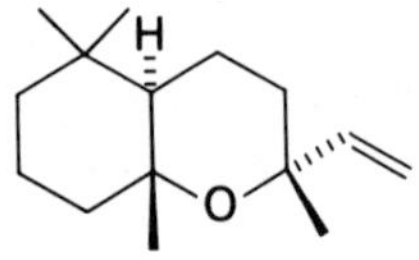

cararrapi oxide

(**117**)

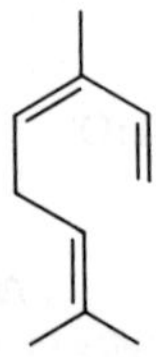

(*Z*)-β-ocimene

(**118**)

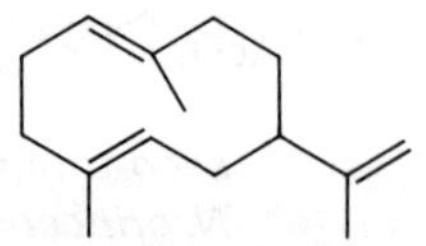

germacrene A

(**119**)

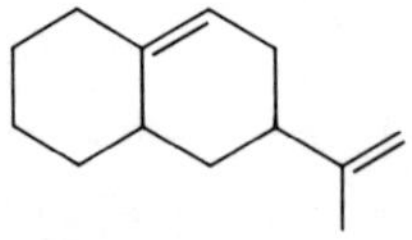

aristolochene

(**120**)

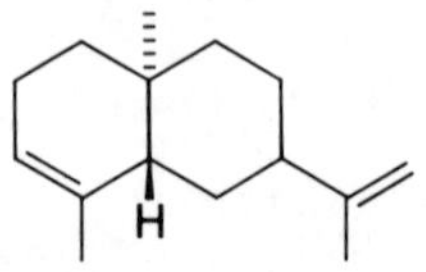

epi-α-selenine

(121) 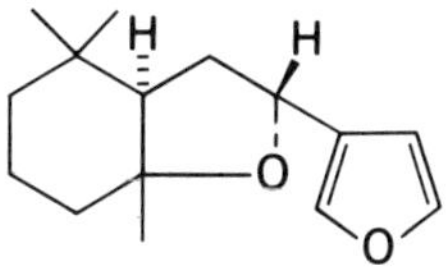 ancistrofuran

(122) 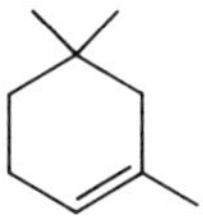 α-cyclogeraniolene

(123) 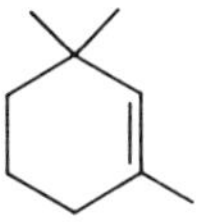 β-cyclogeraniolene

(124) 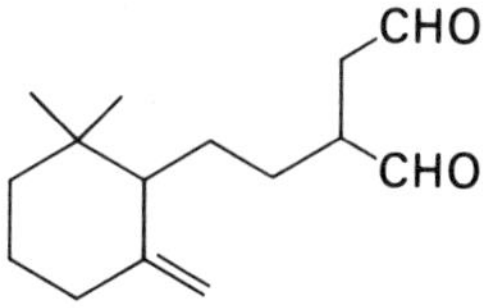 ancistrodial

Alkaloids

It has only recently been realized the extent to which alkaloids are used for defence by arthropods (Tursch et al., 1976). Many species of Lepidoptera such as the cinnabar and tiger moths (Arctiidae) feed as larvae on plants rich in alkaloids, and in so doing sequester them for protection during both the larval and the adult stages (Rothschild, 1973). Sequestration of alkaloids is not, however, confined to the Lepidoptera but has also been recorded in other orders such as the Hemiptera, Coloptera and Orthoptera (Duffey, 1980; Blum, 1981).

Many arthropods, however, are capable of synthesizing these compounds *de novo* and a number are shown in Figure 3. A number of these species can release the alkaloids with the haemolymph by the process of reflex bleeding.

Phenols, quinones and cyanogenic systems

Perhaps the best documented use of phenols as a defensive chemical is that of the bombardier beetle, *Brachymus crepitans*. When attacked, it discharges a hot explosive cloud of toxic chemicals at its attacker. This it achieves by reacting a phenol substrate, hydroquinone, with hydrogen peroxide through the use of an enzyme catalase in an explosion chamber (Figure 4). A highly exothermic reaction occurs with the phenol hydroquinone being oxidized to benzoquinone and being forced out in vapour form mixed with steam, the other product of the reaction (Schildknecht and Holoubek, 1961).

The millipedes *Apheloria corrugata* and *Pseudopolydesmus serratus* release hydrogen cyanide by a similar 'reaction chamber' system (Figure 5). In the upper chamber it stores a mandelonitrile (benzaldehyde and hydrogen cyanide) which can be released into the reaction chamber by a muscular valve. Again enzymic dissociation occurs but the products are not explosive—they are released over about 30 minutes (Eisner *et al.*, 1963a,b).

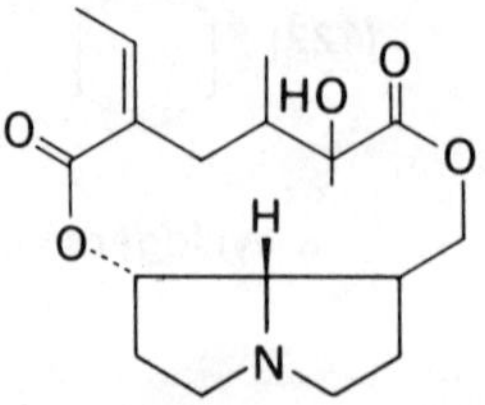

senecionine
Tyria jacobaeae
(Aplin & Rothschild, 1972)

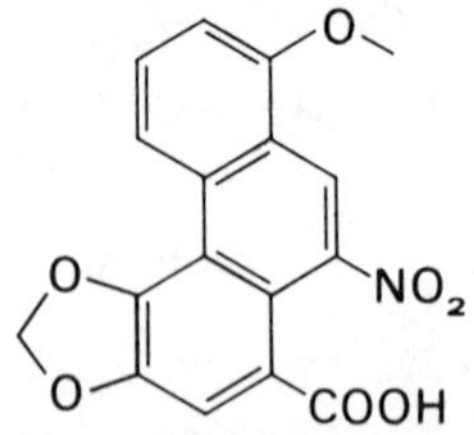

aristolochic acid
Pachlioptera aristolochiae
(Von Euw *et al.*, 1968)

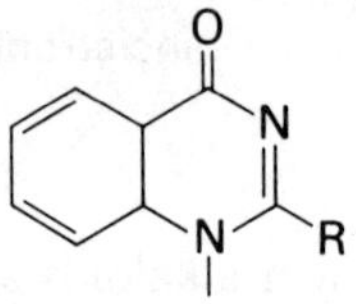

glomerin, R = Me
homoglomerin, R = Et
Glomeris marginata
(Meinwald *et al.*, 1966)

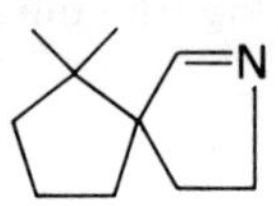

polyzonimine
Polyzonium rosalbum
(Smolanoff *et al.*, 1975)

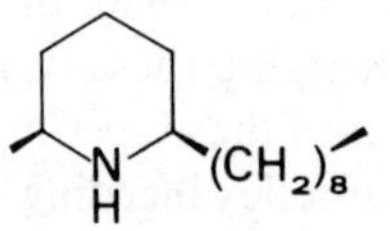

2-methyl-6-nonyl
piperidine
Solenopsis sp.
(Brand *et al.*, 1972)

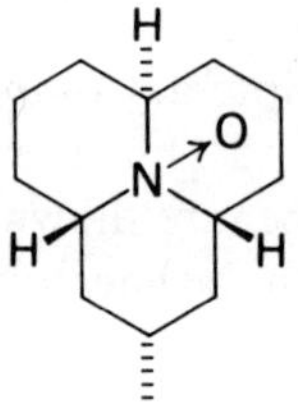

coccinelline
Coccinella sp.
(Tursch *et al.*, 1971)

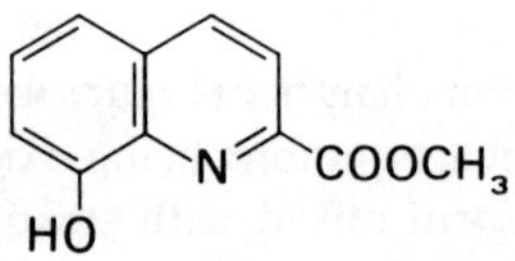

8-hydroxyquinoline
2-carboxylate
Ilybius sp. (Schildknecht *et al.*, 1969)

Figure 3 Alkaloid defences of arthropods

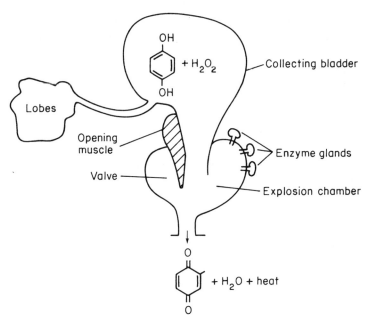

Figure 4 The defensive organ of the bombardier beetle (After Eisner, 1970 Reproduced by permission of Academic Press Inc.)

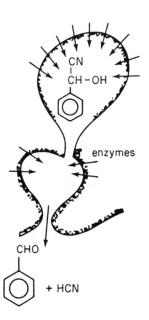

Figure 5 'Reaction chamber' of the millipede *Apheloria* (Reproduced by permission of Masson S. A., Paris, from Fig. 9, p. 28, Barbier: *Introduction à l'écologie chimique*, 1976, Masson S. A., Paris, France)

Quinones and phenols also occur in other arthropod species: arachnids, millipedes, earwigs and termites (Moore, 1968) have been shown to use these compounds as defence chemicals. Besides quinone itself, other compounds such as 2-methyl-2,3-dimethyl and 2,3,5-trimethyl benzoquinones have been found. Some simple phenols such as *m* and *p*-cresol (125) and salicylaldehyde (126) have been found in some beetles and bugs (Pattenden and Staddon, 1972) and in one grasshopper species it is claimed that its 2,5-dichlorophenol defensive compound was actually derived from ingested herbicide (Eisner *et al.*, 1971). Some water beetle species use the phenolic compounds (127 and 128) for bacterio- or fungistatic purposes in maintaining their body surface free of such microorganisms (Schildknecht, 1971).

(125)

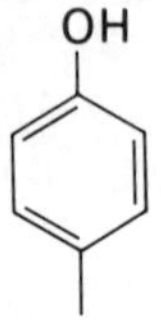

p-cresol
Calasoma sp.

(126)

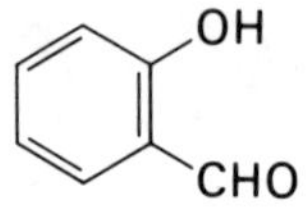

salicylaldehyde
Notonecta sp.

(127)

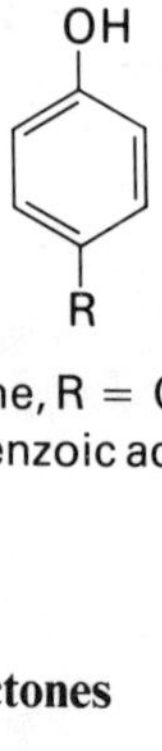

hydroquinone, R = OH
p-hydroxybenzoic acid, R = CO$_2$H
Dytiscus sp.

(128)

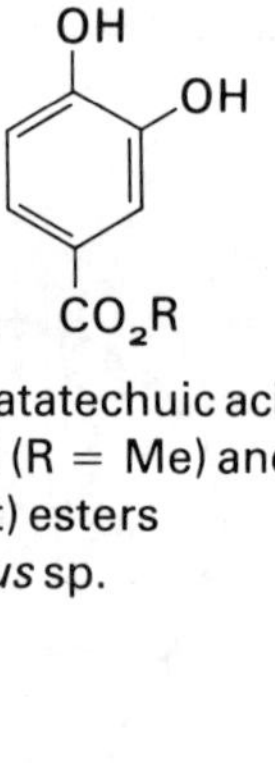

protocatatechuic acid
methyl (R = Me) and ethyl
(R = Et) esters
Dytiscus sp.

Macrocyclic lactones

One of the more primitive genera of termites, *Armitermes*, produces macrocyclic lactones, including the novel α- and β-hydroxy macrolides (129) and (130) (Prestwich and Collins, 1981; Prestwich, 1982). Such lactones have also been found in beetles (Moore and Brown, 1976) and bees (Hefetz *et al.*, 1978).

(129)

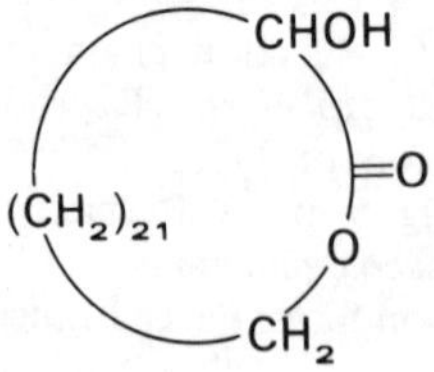

(CH$_2$)$_{21}$

CHOH

CH$_2$

=O

O

(130)

(CH$_2$)$_{20}$

CHOH

CH$_2$

CH$_2$

=O

O

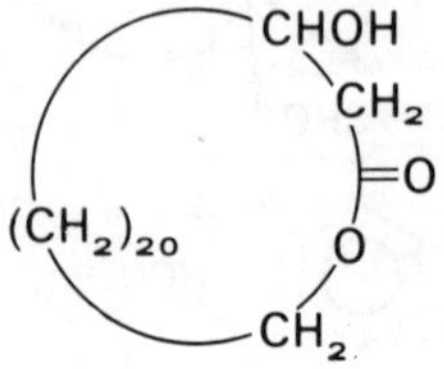

Armitermes

In this rather brief review of some of the compounds used by insects for defence, consideration has only been given to those compounds that are secreted or released externally as a defensive strategy. Compounds used for defence but which are not secreted have not been considered here but have been extensively studied. Cardiac glycosides (Reichstein *et al.*, 1968; Brower, 1969; Rothschild, 1973; Roeske *et al.*, 1976), pyrrolizidine alkaloids, (Rothschild and Aplin, 1971; Rothschild, 1973), cyanogenic glycosides (Jones *et al.*, 1962; Duffey *et al.*, 1974) and steroidal compounds (Schildknecht, 1970, 1971) are well documented examples of such 'passive' defence chemicals. Similarly, toxic venom injected by arthropods has not been considered in this chapter but the subject has been reviewed by Bettini (1978).

KAIROMONES

During the process of coevolution between plants and animals, many phytophagous insects, having once overcome a particular plant species defence compounds, attune their sensory receptors to some of these secondary metabolites and use the information for host plant finding and recognition. In this way, a compound which normally acts as an allomone protecting a plant species against attacks from herbivorous insects, may also serve as a kairomone for an insect species which uses that plant species as a food host. Several publications have appeared on this aspect of insect/host plant chemical ecology (Dethier, 1970; Schoonhoven, 1972; Harborne, 1977; Chapman and Bernays, 1978) and a number of important insect behaviour-modifying chemicals have been discovered from this field of study but the subject is too extensive to be dealt with in this chapter. Here we will confine ourselves to reports of animal–animal interactions, and in particular, host or prey selection behaviour of parasitoids and predators.

Insect parasitoids

In searching for an insect host, a parasitoid may first of all locate its host's habitat before it searches out the host itself. Plant volatiles emanating from the host's food or food plant have been shown to be important cues in host habitat location for a number of hymenopterous parasitoids (Arthur, 1962; Streams *et al.*, 1968; Read *et al.*, 1970; Camors and Payne, 1971; Vinson, 1976, 1980) as well as dipterous parasitoids (Dowden, 1934; Monteith, 1956, 1958, 1964; Herrebout, 1969a,b). Camors and Payne (1972) showed that α-pinene alone attracted *Heydenia unica*, a parasitoid of the bark beetle *Dendroctonus frontalis*, but the host-derived compounds also influenced this response. Similarly, both male and female *Diaeretiella rapae*, an aphid parasitoid, are attracted to fresh collard leaves and to very low concentrations of mustard oil (allyl isothiocyanate) which occurs in collard leaves (Read *et al.*, 1970).

Greany *et al.* (1977) reported that female *Biosteres longicaudatus*, a parasitoid of tephritid fruit fly larvae, are attracted to rotting fruit, a probable source of host larvae. These females were also attracted by acetaldehyde, ethanol and acetic acid, all of which are produced by the fungi responsible for the rotting of the fruit.

To locate the host, once in the host's location, a parasitoid may use a number of odour cues. Some insects for instance release mandibular gland secretions during feeding which may aid host location by a parasitoid. The mandibular gland secretion of larvae of the flour moth, *Ephestia kühniella*, for instance, contains an epideidic pheromone which elicits oviposition movements in the hymenopteran parasite *Venturia canescens* (Corbet, 1971). The unusual β-triketones and some 2-acyl-cyclohexane-1,3-diones characterized from this secretion have been shown to be involved in these kairomonal responses (Mudd, 1978, 1981). Other sources for compounds that release this behavioural change have been the host's frass or webbing (Kajita and Drake, 1969; Lewis and Jones, 1971; Kennedy and Golford, 1972; Hendry *et al.*, 1973). The braconid parasitoid, *Microplisis croceipes*, for instance, uses 13-methyl hentriacontane from the frass of its host, *Heliothis zea*, for host selection (Jones *et al.*, 1971). Another braconid parasitoid, *Cardiochiles nigriceps*, of the closely related species *H. virescens* uses several methyl hentria- dotria- and tritriacontanes from the mandibular glands and frass of the host larvae (Vinson *et al.*, 1975).

In some species the parasitoids employ the host insect's own sex pheromone as a kairomone for host location. This has been shown to be the case with *Trichopoda pennipes*, a parasitoid of stink bugs (Mitchell and Man, 1970); parasitoids of *Ips confusus* and other bark beetle species (Rice, 1969; Vite and Williamson, 1970; Dixon and Payne, 1979; Greany and Hagen, 1960) and scale insect parasitoids (Sternlicht, 1973).

Perhaps the most extensively studied kairomonal effects have been those involved in host location by *Trichogramma* spp. Chemicals released by the eggs of *Heliothis zea* have been shown to increase the rates at which *T. evanescens* found and parasitized them (Lewis *et al.*, 1971). Lewis *et al.* (1972) showed that the host-seeking response of *T. evanescens* was stimulated by compounds emanating from the scales of *H. zea* adults and Jones *et al.* (1973) identified tricosane as the most active of several chemicals. Detailed studies of the response of the parasitoid to tricosane have now shown conclusively that the kairomone functions primarily by releasing and maintaining host-seeking responses rather than by attracting and serving as a steering mechanism (Lewis *et al.*, 1975a,b).

Predators

Host selection behaviour of parasitoids and prey selection behaviour of predators are generally very similar with both habitat location and host location being exhibited by both entomophagous groups. The coccinellid beetle, *Anatis ocellata*, for instance, a predator of aphids found on pine needles, is attracted to infested trees by chemicals found in pine needles (Kesten, 1969).

Predaceous *Chrysopa carnea* larvae have also been shown to respond to kairomones from the scales of *H. zea* adults by initiating prey-seeking behaviour (Lewis *et al.*, 1977) but acceptance of *H. zea* eggs is stimulated by an additional kairomone probably originating in the accessory gland secretion which is associated with egg deposition (Nordlund *et al.*, 1977). Some predators respond to chemicals from both the prey and its habitat: *Enoclerus lecontei* and *Temnochila chlorodia*, two beetle predators of the bark beetle pest, *Dendroctonus brevicomis*, have been shown to arrive during and very shortly after the mass arrival of their prey. *T. chlorodia* responds very specifically to *exo*-brevicomin (prey derived, **45**) while its response decreases when *trans*-verbenol and verbenone (host tree compounds modified by the beetles) are also present (Bedard *et al.*, 1969, 1980). *E. lecontei*, on the other hand, which is a significant predator of *D. brevicomis*, responds to the host tree volatiles, but not significantly to the pheromone of *D. brevicomis* although it does respond to the pheromone of *Ips typographus*, another of its prey species (Lanier *et al.*, 1972; Wood, 1972).

CONCLUSIONS

A wide variety of behaviour-modifying chemicals has therefore been characterized from arthropods but, to date, their use in insect control strategies has still to be developed in a majority of cases. Pheromones have been used quite extensively over the last decade for monitoring pest populations and this has undoubtedly led to a reduced dependence on, and more effective use of, insecticides. On the other hand, the use of pheromones for direct insect population suppression by mass trapping or mating disruption has still to be established as a generally acceptable means of insect control. Although encouraging results have been obtained in a number of cases, and with a number of species, our lack of understanding of the basic behavioural mechanisms of response by the insects to pheromones, plus our failure to control several biological, physical and human factors, has prevented the establishment of the fundamental principles which are essential prerequisites for the successful wide-scale exploitation of these techniques. Attempts at developing effective pest management tools from allomones or kairomones are still in their infancy but given their high level of biological activity, as in the case of pheromones, and given sufficient understanding of their role in the interactions between insect species then they too will find their niche in future integrated pest management strategies.

REFERENCES

Adams, M. A., Nakanishi, K., Still, W. C., Arnold, E. V., Clardy, J., and Persoons, C. J. (1979). 'Sex pheromone of the American cockroach: absolute configuration of Periplanone-B', *J. Am. Chem. Soc.*, **101**, 2495–2498.

Alberts, S. A., Kennedy, M. K., and Carde, R. T. (1981). 'Pheromone-mediated anemotactic flight and mating behaviour of the sciarid fly *Bradysia impatients'*, *Environ. Entomol.*, **10**, 10–15.

Aplin, R. T., and Rothschild, M. (1972). 'Poisonous alkaloids in the body tissues of the garden tiger moth (*Arctia caja*) and the cinnabar moth (*Tyria* (= *Callimorpha*) *jacobaeae*) (Lepidoptera)', in *Toxins of Animal and Plant Origin* (*Proc. Int. Symp. Anim. Plant Toxins*) (eds A. de Vries and K. Kochva) Gordon and Breach, New York, pp. 579–595.

Arthur, A. P. (1962). 'Influence of host tree on abundance of *Itoplectis conquisitor* (Say), a polyphagous parasite of the European Shoot Moth *Rhyacionia buoliana* (Schiff.)', *Can. Entomol.*, **94**, 337–347.

Attygalle, A. B., and Morgan, E. D. (1983). 'Trail pheromone of the ant *Tetramorium caespitum* L.', *Naturwissenschaften*, **70**, 364–365.

Baker, J. L., Hill, A. S., and Roelofs, W. L. (1978). 'Seasonal variations in pheromone catches of male omnivorous leafroller moths *Platynota stultana'*, *Environ. Entomol.*, **7**, 399–401.

Baker, R., and Bradshaw, J. W. S. (1979). 'Insect chemistry', *Ann. Rep. Chem. Soc. B.*, **76**, 404–432.

Baker, R., and Bradshaw, J. W. S. (1981). 'Insect pheromones and related behaviour-modifying chemicals,' *Aliphatic & Relat. Nat. Prod. Chem.*, **2**, 46–75.

Baker, R., and Bradshaw, J. W. S. (1983). 'Insect pheromones and related natural products,' *Aliphatic & Relat. Nat. Prod. Chem.*, **3**, 66–106.

Baker, R., Bradshaw, J. W. S., and Speed, W. (1982a). 'Methoxymercuration-demercuration and mass spectrometry in the identification of sex pheromones of *Panolis flamea*, the pine beauty moth', *Experientia*, **38**, 233–234.

Baker, R., Briner, P. H. and Evans, D. A. (1978a). 'Chemical defence in the termite *Ancistrotermes cavithorax*: ancistrodial and ancistrofuran', *J. Chem. Soc., Chem. Commun.*, 410–411.

Baker, R., Coles, H. R., Edwards, M., Evans, D. A., Howse, P. E., and Walmsley, S. (1981). 'Chemical composition of the frontal gland secretion of *Syntermes* soldiers (Isoptera, Termitidae)', *J. Chem. Ecol.*, **7**, 135–145.

Baker, R., and Evans, D. A. (1977). 'Insect chemistry', *Ann. Rep. Chem. Soc. B.*, **74**, 367–391.

Baker, R., and Evans, D. A. (1979), 'Insect pheromones and related behaviour-modifying chemicals', *Aliphatic & Relat. Nat. Prod. Chem.*, **1**, 102–127.

Baker, R., Evans, D. A., and McDowell, P. G. (1978b). 'Mono- and sesquiterpenoid constituents of the defence secretion of the termite *Amitermes evuncifer'*, *Tetrahedron Lett.*, **42**, 4073–4076.

Baker, R., and Herbert, R. H. (1984). 'Insect pheromones and related natural products', *Nat. Prod. Reports*, **1**, 299–318.

Baker, R., Herbert, R. H., Howse, P. E., Jones, O. T., Francke, W., and Reith, W. (1980). 'Identification and synthesis of the major sex pheromone of the olive fly (*Dacus oleae)'*, *J. Chem. Soc., Chem. Commun.*, **1106**, 52–53.

Baker, R., Herbert, R. H., and Lomer, R. A. (1982b). 'Chemical components of the rectal gland secretions of male *Dacus cucurbitae*, the melon fly', *Experientia*, **38**, 232–233.

Baker, R., Herbert, R. H., and Parton, A. H. (1982c). 'Isolation and synthesis of 3- and 4-hydroxy-1,7-dioxaspiro[5.5]undecanes', *J. Chem. Soc., Chem. Commun.*, 601–603.

Baker, R., Parton, A. H., Bhaskar Rao, V., and Jyothi Rao, V. (1982d). 'The isolation, identification and synthesis of 3,6-dimethylheptan-2,4-dione, a pheromone of the mushroom fly, *Megaselia halterata* (Diptera: Phoridae)', *Tetrahedron Lett.*, **23**, 3103–3104.

Baker, T. C., Nishida, R., and Roelofs, W. L. (1981). 'Close-range attraction of female oriental fruit moths to herbal scent of male hairpencils', *Science*, **214**, 1359–1361.

Barak, A. V., and Burkholder, W. E. (1977a). 'Behaviour and pheromone studies with *Attagenus elongatulus* Casey (Coleoptera: Dermestidae)', *J. Chem. Ecol.*, **3**, 219–237.

Barak, A. V., and Burkholder, W. E. (1977b). 'Studies on the biology of *Attagenus elongatulus* Casey (Coleoptera: Dermestidae) and the effects of larval crowding on pupation and life cycle', *J. Stored Prod. Res.*, **13**, 169–175.

Bartlet, R. J., and Jones, R. L. (1983). '(Z)-10-nonadecanal: a pheromonally active air oxidation product of (Z,Z,)-9.19-dienes in Yellowheaded Spruce Sawfly,' *J. Chem. Ecol.*, **9**, 1333–1341.

Bartlet, R. J., Jones, R. L., and Krick, T. P. (1983). '(Z)-5-tetradecen-1-ol: a secondary pheromone of the Yellowheaded Spruce Sawfly and its relationship to (Z)-10-nonadecenal', *J. Chem. Ecol.*, **9**, 1343–1352.

Battiste, M. A., Strekowski, L., Vanderbilt, D. P., Visnick, M., King, R. W., and Nation, J. L. (1983). 'Anastrephin and Epianastrephin, novel lactone components isolated from the sex pheromone blend of male Caribbean and Mexican fruit flies', *Tetrahedron Lett.*, **24**, 2611–2614.

Becker, D., Kimmel, T., Cyjon, R., Moore, I., Wysoki, M., Bestmann, H. J. Platz, H., Roth, K., and Vostrowsky, O. (1983). '(3Z,6Z,9Z)-3,6,9-nonadecatriene—a component of the sex pheromonal system of the giant looper, *Boarmia* (*Ascotis*) *selenaria* Schiffermüller (Lepidoptera: Geometridae)', *Tetrahedron Lett.*, **24**, 5505–5508.

Bedard, W. D., Tilden, P. E., Wood, D. L., Silverstein, R. M., Brownlee, R. G., and Rodin, J. O. (1969). 'Western pine beetle: Field response to its sex pheromone and a synergistic host terpene myrcene', *Science*, **164**, 1284–1285.

Bedard, W. D., Wood, D. L., Tilden, P. E., Lindahl, K. Q., Jr, Silverstein, R. M., and Rodin, J. O. (1980). 'Field responses of the western pine beetle and one of its predators to host- and beetle-produced compounds,' *J. Chem. Ecol.*, **6**, 625–641.

Bergström, G. (1974). 'Studies on natural odoriferous compounds X. macrocyclic lactones in the Dufour gland secretion of the solitary bees *Colletes cunicularius* L. and *Halictus calceatus* Scop (Hymenoptera, Apidae)', *Chem. Scr.*, **5**, 39–46.

Bergström, G., and Tengö, J. (1974). 'Studies on natural odoriferous compounds IX. Farnesyl- and geranyl-esters as main volatile constituents of the secretion from Dufour's gland in 6 species of *Andrena* (Hymenoptera, Apidae)', *Chem. Scr.*, **5**, 28–34.

Bergström, G., and Tengö, J. (1979). 'C_{24}-,C_{22}-,C_{20}- and C_{18}-macrocyclic lactones in halictide bees', *Acta Chem. Scand. Ser. B.*, **33**, 390.

Bergström, G., Kullenberg, B., Ställberg-Stenhagen, S., and Stenhagen, E. (1968). 'Studies on natural odoriferous compounds. II. Identification of 2,3-dihydrofarnesol as the main component of the marking perfume of male bumble-bees of the species *Bombus terrestris* L.', *Arkiv, Kemi.*, **28**, 453–469.

Bestmann, H. J., Brosche, T., Koschatzky, K. H., Michaelis, K., Platz, H., Roth, K., Sub, J., Vostrowsky, O., and Knauf, W. (1982). 'Pheromone XLII. 1,3,6,9-nonadecatetraen, das Sexualpheromon des Frostspanners *Operophtera brumata* (Geometridae)', *Tetrahedron Lett.*, **23**, 4007–4010.

Bestmann, H. J., Brosche, T., Koschatzky, K. H., Michaelis, K., Platz, H., Vostrowsky, O., and Knauf, W. (1980a). 'Pheromon XXX. Identifizierung eines neuartigen pheromonkomplexes aus der graseule *Scotia exclamationis* (Lepidoptera)', *Tetrahedron Lett.*, **21**, 747–750.

Bestmann, H. J., Hirsch, H. L., Platz, H., Rheinwald, M., and Vostrowsky, O. (1980b). 'Differentiation between chiral pheromone analogues by chemoreceptors', *Angew. Chem. Internat. Ed.*, **19**, 475–477.

Bestmann, H. J., and Vostrowsky, O. (1981). 'Chemistry of insect pheromones,' In *Chemie der Pflanzenschutz- und Schadlingsbekämpfungsmittel* (ed. R. Wegler), Vol. 6, pp. 29–164, Springer Verlag, Berlin.

Bestmann, H. J., and Vostrowsky, O. (1982). 'Insektenpheromone. Strukture und vorkommen Isolierung und Strukturaufklärung- Synthese-Biologische Aktivität und Verhaltensauslösung-Anwendung in Pflanzenschuz', *Naturwissenschaften*, **69**, 457–471.

Bestmann, H. J., and Vostrowsky, O., Koschatzky, K. H., Platz, H., Brosche, T., Kantardjiew, I., Rheinwald, M., and Knauf, W. (1978). (Z)-5-decenyl acetate, a sex attractant for the male turnip moth *Agrotis segetum* (Lepidoptera)', *Angew. Chem.*, **90**, 815–816.

Bestmann, H. J., Vostrowsky, O., and Platz, H. (1977). 'Pheromone XII. Mannchenduftstoffe von Noctuiden (Lepidoptera)', *Experientia*, **33**, 874–875.

Bettini, S. (ed.) (1978). *'Arthropod Venoms', Handb. Exp. Pharmacol.* (N.S.) Vol. 48, Springer Verlag, Berlin.

Bierl, B. A., Beroza, M., and Collier, C. W. (1970). 'Potent sex attractant of the gypsy moth: its isolation, identification and synthesis', *Science*, **170**, 87–89.

Bierl-Leonhardt, B. A., Moreno, D. S., Schwarz, M., Fargerlund, J., and Plimmer, J. R. (1981). 'Isolation, identification and synthesis of the sex pheromone of the citrus mealybug, *Planococcus citri* (Risso)', *Tetrahedron Lett.*, **22**, 389–392.

Bierl-Leonhardt, B. A., Moreno, D. S., Schwarz, M., Forster, H. S., Plimmer, J. R., and DeVilbiss, E. D. (1980). 'Identification of the pheromone of the comstock mealybug', *Life Sci.*, **27**, 399–402.

Birch, A. J., Brown, W. V., Corrie, J. E. T., and Moore, B. P. (1972). 'Neocembrene-A, a termite trail pheromone', *J. Chem. Soc. Perkin, Trans.*, I, pp. 2653–2658.

Birch, M. C., Light, D. M., Wood, D. L., Browne, L. E., Silverstein, R. M., Bergot, B. J., Ohloff, G., West, J. R., and Young, J. C. (1980). 'Pheromonal attraction and allomonal interruption of *Ips pini* in California by the two enantiomers of ipsdienol', *J. Chem. Ecol.*, **6**, 703–717.

Bjostad, L. B., Gaston, L. K., Noble, L. L., Moyer, J. H., and Shorey, H. H. (1980a). 'Dodecyl acetate, a second pheromone component of the cabbage looper moth, *Trichoplusia ni'*, *J. Chem. Ecol.*, **6**, 727–734.

Bjostad, L. B., and Roelofs, W. L. (1983). 'Sex pheromone biosynthesis in *Trichoplusia ni*: key steps involve delta-11 desaturation and chain shortening', *Science*, **220**, 1387–1389.

Bjostad, L. B., Taschenberg, E. F., and Roelofs, W. (1980b). 'Sex pheromone of the woodbine leafroller moth, *Sparganothis* sp.,' *J. Chem. Ecol.*, **6**, 797–804.

Blight, M. M., Mellon, F. A., Wadhams, L. J., and Wenham, M. J. (1977). 'Volatiles associated with *Scolytus scolytus* beetles on English elm,' *Experientia*, **33**, 845–847.

Blum, M. S. (1981). *Chemical Defenses of Arthropods*, Academic Press, New York.

Blum, M. S., Fales, H. M., Jones, T. H., Rinderer, T. E., and Tucker, K. W. (1983). 'Caste-specific esters derived from the queen honey bee sting apparatus', *Comp. Biochem. Physiol. B.*, **75B**, 237–238.

Blum, M. S., Jones, T. H., Howard, D. F., and Overal, W. L. (1982). 'Biochemistry of termite defenses: *Coptotermes, Rhinotermes* and *Cornitermes* species', *Comp. Biochem. Physiol. B.,* **71B**, 731–733.

Boch, R., Shearer, D. A., and Young, J. C. (1975). 'Honey bee pheromones: field tests of natural and artificial queen substance', *J. Chem. Ecol.,* **1**, 133–148.

Boness, M. (1980). In *Chemie der Pflanzenschutz und Schädlingsbekämpfungsmittel* (ed. R. Wegler), Vol. 6, pp. 165–185, Springer Verlag, Berlin.

Booth, D. C., Phillips, T. W., Claesson, A., Silverstein, R. M., Lanier, G. N., and West, J. R. (1983). 'Aggregation pheromone components of two species of *Pissodes* weevils (Coleoptera: curculionidae): isolation, identification and field activity', *J. Chem. Ecol.,* **9**, 1–12.

Borden, J. H., Handley, J. R., McLean, J. A., Silverstein, R. M., Chong, L., Slessor, K. N., Johnston, B. D., and Schuler, H. R. (1980). 'Enantiomer-based specificity in pheromone communication by two sympatric *Gnathotrichus* species (Coleoptera: Scolytidae)', *J. Chem. Ecol.,* **6**, 445–456.

Borg-Karlson, A. K., and Tengö, J. (1980). 'Pyrazines as marking volatiles in philanthine and nyssonine wasps (Hymenoptera: Sphecidae)', *J. Chem. Ecol.,* **6**, 827–835.

Bowers, W. S., Nault, L. R., Webb, R. E., and Dutky, S. R. (1972). 'Aphid alarm pheromone: Isolation, identification, synthesis', *Science,* **177**, 1121–1122.

Bowers, W. S., Nishino, C., Montgomery, M. E., Nault, L. R., and Nielsen, M. W. (1977). 'Sesquiterpene progenitor, Germacrene A: an alarm pheromone in aphids', *Science,* **196**, 680–681.

Bradshaw, J. W. S. (1984). 'Insect natural products—compounds derived from acetate, shikimate and amino acids', In *Comprehensive Insect Physiology, Biochemistry and Pharmacology* (eds G. A. Kerkut and L. I. Gilbert), Vol. 11, Chapter 17, Pergamon Press, Oxford.

Bradshaw, J. W. S., Baker, R., and Howse, P. E. (1979a). 'Multicomponent alarm pheromones in the mandibular glands of major workers of the African weaver ant, *Oecophylla longinoda*', *Physiol. Entomol.,* **4**, 15–25.

Bradshaw, J. W. S., Baker, R., and Howse, P. E. (1979b). 'Chemical composition of the poison apparatus secretions of the African weaver ant, *Oecophylla longinoda,* and their role in behaviour', *Physiol. Entomol.,* **4**, 39–46.

Bradshaw, J. W. S., Baker, R., Howse, P. E., and Higgs, M. D. (1979c). 'Caste and colony variations in the chemical composition of the cephalic secretions of the African weaver ant, *Oecophylla longinoda*', *Physiol. Entomol.,* **4**, 27–38.

Bradshaw, J. W. S., and Howse, P. E. (1984). 'Sociochemicals of ants', In *Chemical Ecology of Insects* (eds W. J. Bell and R. T. Cardé), pp. 429–473, Chapman and Hall, London.

Bradshaw, J. W. S., Howse, P. E., and Baker, R. (1983). 'A novel pheromone regulating chain transport of leaves in *Atta cephalotes*', *(In Preparation)*.

Braekman, J. C., Daloze, D., Dupont, A., Pasteels, J., Tursch, B., Declerq, J. P., Germain, G., and Van Meerssche, M. (1980). 'Secotrinervitane, a novel bicyclic diterpene skeleton from a termite soldier', *Tetrahedron Lett.,* **21**, 2761–2762.

Brand, J. M., Blum, M. S., Fales, H. M., and MacConnell, J. G. (1972). 'Fire ant venoms. Comparative analyses of alkaloidal components', *Toxicon,* **10**, 259–271.

Brand, J. M., Blum, M. S., Lloyd, H. A., and Fletcher, D. J. C. (1974). 'Monoterpene hydrocarbons in the poison gland secretion of the ant *Myrmicaria natalensis* (Hymenoptera: Formicidae)', *Ann. entomol. Soc. Am.,* **67**, 525–526.

Brand, J. M., Brack, J. W., Britton, L. N., Markovetz, A. J., and Barras, S. J. (1976). 'Bark beetle pheromones: production of verbenone by a mycangial fungus of *Dendroctonus frontalis*', *J. Chem. Ecol.,* **2**, 195–199.

Brand, J. M., Bracke, J. W., Markovetz, A. J., Wood, D. L., and Browne, L. E. (1975). 'Production of verbenol pheromone by bacterium isolated from bark beetles', *Nature*, **254**, 136–137.

Brand, J. M., Young, J. C., and Silverstein, R. M. (1979). 'Insect pheromones: a critical review of recent advances in their chemistry, biology and application', *Prog. Chem. Org. Nat. Prod.*, **37**, 1–190, Springer Verlag, New York.

Brower, L. P. (1969). 'Ecological chemistry', *Sci. Am.*, **220**, 22–29.

Brown, W. L. Jr (1968). 'An hypothesis concerning the function of the metapleural glands in ants', *Am. Nat.*, **102**, 188–191.

Brown, W. L. Jr, Eisner, T., and Whittaker, R. H. (1970). 'Allomones and kairomones: transpecific chemical messengers', *Bio Science* **20**, 21–22.

Burkholder, W. E., and Dicke, R. J. (1966). 'Evidence of sex pheromones in females of several species of Dermestidae', *J. Econ. Entomol.*, **59**, 540–543.

Burkholder, W. E., Ma, M., Kuwahara, Y., and Matsumura, F. (1974). 'Sex pheromone of the furniture carpet beetle, *Anthrenus flavipes* (Coleoptera: Dermestidae)', *Can. Entomol.*, **106**, 835–839.

Butler, C. G., and Calam, D. H. (1969). 'Pheromones of the honey bee: the secretion of the Nassonoff pheromones of the worker', *J. Insect Physiol.*, **15**, 237–244.

Butler, C. G., Callow, R. K., and Johnston, N. C. (1961). 'The isolation and synthesis of queen substance, 9-oxo-trans-2-enoic acid, a honeybee pheromone', *Proc. R. Soc. (B)*, **155**, 417–432.

Butler, C. G., and Simpson, J. (1967). 'Pheromones of the queen honeybee *(Apis mellifera L.)* which enable her workers to follow her when swarming', *Proc. R. Entomol. Soc. (A)* **42**, 149–154.

Byers, J. A., Wood, D. L., Browne, L. E., Fish, R. H., Piatek, B., and Hendry, L. B. (1979). 'Relationship between a plant compound myrcene, and pheromone production in the bark beetle *Ips paraconfusus'*, *J. Insect Physiol.*, **25**, 477–482.

Byrne, K. J., Swigar, A. A., Silverstein, R. M., Borden, J. H., and Stokkink, E. (1974). 'Sulcatol: population aggregation pheromone in the scolytid beetle, *Gnathotrichus sulcatus'*, *J. Insect Physiol.*, **20**, 1895–1900.

Calam, D. H. (1969). 'Species and sex-specific compounds from the heads of male bumble bees *(Bombus* spp.)', *Nature*, **221**, 856–857.

Calam, D. H., and Youdeowei, A. (1968). 'Identification and functions of secretion from the posterior scent gland of the fifth-instar larva of the bug, Dysdercus intermedius,' *J. Insect Physiol.*, **14**, 1147–1158.

Callow, R. K., Chapman, J. R., and Paton, P. N. (1964). 'Pheromones of the honeybee: chemical studies of the mandibular gland secretion of the queen', *J. apic. Res.*, **3**, 77–89.

Camors, F. B., and Payne, T. L. (1972). 'Resistance of *Heydenia unica* (Hymenoptera: Pteromalidae) to *Dendroctonus frontalis* (Coleoptera: Scolytidae) pheromones and a host tree terpene', *Ann. Entomol. Soc. Am.*, **65**, 31–33.

Campion, D. G., and Nesbitt, B. F. (1981). 'Lepidopteran sex pheromones and pest management in developing countries', *Trop. Pest. Manag.*, **27**, 53–61.

Carde, A. M., Baker, T. C., and Carde, R. T. (1979). 'Identification of a four-component sex pheromone of the oriental fruit moth, *Grapholitha molesta* (Lepidoptera: Tortricidae)', *J. Chem. Ecol.*, **5**, 423–427.

Carde, R. T., Roelofs, W. L., and Doane, C. C. (1973). 'Natural inhibitor of the gypsy moth sex attractant', *Nature*, **241**, 474–475.

Carde, R. T., Carde, A. M., Hill, A. S., and Roelofs, W. L. (1977). 'Sex pheromone specificity as a reproductive isolating mechanism among the sibling species *Archips argyrospilus* and *A. mortuanus* and other sympatric tortricine moths (Lepidoptera: Tortricidae)', *J. Chem. Ecol.*, **3**, 71–84.

Carlson, D. A., Langley, P. A., and Huyton, P. (1978). 'Sex pheromone of the tsetse fly: isolation, identification and synthesis,' *Science*, **201**, 750–753.

Carlson, D. A., Mayer, M. S., Silhacek, D. L., James, J. P., Beroza, M., and Bierl, B. (1971). 'Sex attractant pheromone of the house fly: isolation, identification and synthesis', *Science*, **174**, 76–78.

Carlson, D. A., Nelson, D. R., Langley, P. A., Coates, T. W., Davis, T. L., and Leegwater-vanderlinden, M. E. (1984). 'Contact sex pheromone in the tsetse fly *Glossina pallidipes* (Austen). Identification and synthesis', *J. Chem. Ecol.*, **10**, 429–450.

Chadha, M. S., Eisher, T., Monro, A., and Meinwald, J. (1962). 'Defence mechanisms of Arthropods-VII Citronellal and citral in the mandibular gland secretion of the ant *Acanthomyops claviger* (Roger)', *J. Insect Physiol.*, **8**, 175–179.

Chapman, R. F., and Bernays, E. A. (eds) (1978). *Insect and Host Plant, Ent. Exp. Appl.*, **24**, 200–766 (Proceedings of the 4th International Symposium, Nederland. Entomol. Vereniging, Amsterdam.

Chapman, O. L., Mattes, K. C., Sheridan, R. S., and Klun, J. A. (1978). 'Stereochemical evidence of dual chemoreceptors for an achiral sex pheromone in Lepidoptera', *J. Am. Chem. Soc.*, **100**, 4878–4884.

Chisholm, M. D., Steck, W., and Underhill, E. W. (1980a). 'Effects of additional double-bonds on some olefinic moth sex attractants', *J. Chem. Ecol.*, **6**, 203–212.

Chisholm, M. D., Underhill, E. W., Steck, W., Slessor, K. N., and Grant, G. G. (1980b). '(Z)-, (E)-7-dodecadienal and (Z)-5,(E)-7-dodecadien-1-ol, sex pheromone components of the forest tent caterpillar, *Malacosoma disstria'*, *Environ. Entomol.*, **9**, 278–282.

Chuman, T., Kato, K., and Noguchi, M. (1979a). 'Synthesis of (+)-serricornin, 4,6-dimethyl-7-hydroxy-nonan-3-one, a sex pheromone of cigarette beetle (*Lasioderma serricorne* F.)', *Agric. Biol. Chem.*, **43**, 2005.

Chuman, T., Kohno, M., Kato., and Noguchi, M. (1979b). '4,6-dimethyl-7-hydroxynonan-3-one, a sex pheromone of the cigarette beetle (*Lasioderma serricorne* F.)', *Tetrahedron Lett.*, **25**, 2361–2364.

Chuman, T., Mochizuki, K., Mori, M., Kohno, M., Kato, K., Nomi, H., and Mori, K. (1982a). 'Behavioural and electroantennogram responses of male cigarette beetle (*Lasioderma serricorne* F.) to optically active serricornins', *Agric. Biol. Chem.*, **46**, 3109—3112.

Chuman, T., Mochizuki, K., Mori, M., Kohno, M., Ono, M., Onishi, I., and Kato, K. (1982b). 'The pheromone activity of (±)-serricornins for male cigarette beetle (*Lasioderma serricorne* F.)', *Agric. Biol. Chem.*, **46**, 593–595.

Coffelt, J. A., and Burkholder, W. E. (1972). 'Reproductive biology of the cigarette beetle, *Lasioderma serricorne*. I. Quantitative laboratory bioassay of the female sex pheromone from females of different ages', *Ann. Entomol. Soc. Am.*, **65**, 447–450.

Collins, A. M., and Blum, M. S. (1983). 'Alarm responses caused by newly identified compounds derived from honeybee sting', *J. Chem. Ecol.*, **9**, 57–65.

Conner, W. E., Eisner, T., Vandermeer, R. K., Guerrero, A., Ghiringelli, D., and Meinwald, J. (1980). 'Sex attractant of an arctiid moth (*Utetheisa ornatrix*): role of the pheromone derived from dietary alkaloids,' *Behav. Ecol. Sociobiol.*, **9**, 227–235.

Corbet, S. A. (1971). 'Mandibular gland secretion of larvae of the flour moth, *Anagasta kuehniella*, contains an epideictic pheromone and elicits oviposition movement in a hymenopteran parasite', *Nature*, **232**, 481–484

Cross, J. H., Byler, R. C., Cassidy, R. F. Jr, Silverstein, R. M., Greenblatt, R. E., Burkholder, W. E., Levinson, A. R., and Levinson, H. Z. (1976). 'Porapak-O collection of pheromone components and isolation of (Z)- and (E)-14-methyl-8-hexadecenal, potent sex attracting components from females of four species of *Trogoderma* (Coleopetera: Dermestidae)', *J. Chem. Ecol.*, **2**, 457–468.

Cross, J. H., Byler, R. C., Ravid, U., Silverstein, R. M., Robinson, S. W., Baker, P. M., Sabino de Oliveira, J., Jutsum, A. R., and Cherrett, J. M. (1979). 'The major component of the trail pheromone of the leaf cutting ant, *Atta sexdens rubropilosa* Forel: 3-ethyl-2,5-dimethylpyrazine', *J. Chem. Ecol.,* **5**, 187–203.

Cross, J. H., West, J. R., Silverstein, R. M., Jutsum, A. R., and Cherrett, J. M. (1982). 'Trail pheromone of the leaf-cutting ant, *Acromyrmex octospinosus* (Reich) (Formicidae: Myrmicinae)', *J. Chem. Ecol.,* **8**, 1119–1124.

Dahm, K. H., Meyer, D., Finn, W. E., Reinhold, V., and Roller, H. (1971). 'The olfactory and auditory mediated sex attraction in *Achroia grisella* (Fabr.)', *Naturwissenschaften,* **58**, 265–266.

Dethier, V. G. (1970). 'Chemical interactions between plants and insects', in *Chemical Ecology* (eds E. Sondheimer and J. B. Simeone), pp. 83–102, Academic Press, New York.

Dettner, K., and Schwinger, G. (1982). 'Defensive secretions of three oxytelinae rove beetles (Coleoptera: Staphylinidae)', *J. Chem. Ecol.,* **8**, 1411–1420.

Dixon, W. N., and Payne, T. L. (1979). 'Aggregation of *Thanasimus dubius* on trees under mass-attack by the southern pine beetle', *Environ. Entomol.,* **8**, 178–181.

Dowden, P. B. (1934). '*Zenillia lebatrix* Panzer, a tachinid parasite of the gypsy moth and brown-tail moth', *J. Agric. Res.,* **48**, 97–114.

Duffey, S. S. (1980). 'Sequestration of plant natural products by insects', *Ann. Rev. Entomol.,* **23**, 447–477.

Duffey, S. S., Underhill, E. W., and Towers, G. H. N. (1974). Intermediates in the biosynthesis of HCN and benzaldehyde by a polydesmid millipede *Harpaphe haydeniana* (Wood)', *Comp. Biochem. Physiol.,* **47B**, 753–766.

Duffield, R. M., Blum, M. S., and Wheeler, J. W. (1976). 'Alkylpyrazine alarm pheromones in primitive ants with small colonial units', *Comp. Biochem. Physiol.,* **54B**, 439–440.

Duffield, R. M., Brand, J. M., and Blum, M. S. (1977). '6-Methyl-5-hepten-2-one in *Formica* species: identification and function as an alarm pheromone (Hymenoptera: Formicidae)', *Ann. Entomol. Soc. Am.,* **70**, 309–310.

Duffield, R. M., Fernandes, A., Lamb, C., Wheeler, J. W., and Eickwort, G. C. (1981a). 'Macrocyclic lactones and isopentenyl esters in the Dufour's gland secretion of halictine bees (Hymenoptera: Halicitidae)', *J. Chem. Ecol.,* **7**, 319–331.

Duffield, R. M., Shamin, M., Wheeler, J. W., and Menke, A. S. (1981b). 'Alkylpyrazines in the mandibular gland secretions of *Aminophila* wasps (Hymenoptera: Sphecidae)', *Comp. Biochem. Physiol., B.,* **70**, 317–318.

Duffield, R. M., Wheeler, J. W., and Eickwort, G. C. (1984). 'Sociochemicals of bees', in *Chemical Ecology of Insects* (eds W. J. Bell and R. T. Cardé), pp. 387–428, Chapman and Hall, London.

Dunkelblum, E., Gothilf, S., and Kehat, M. (1981). 'Sex pheromone of the tomato looper, *Plusia chalcites (esp.),* *J. Chem. Ecol.,* **7**, 1081–1088.

Dupont, A., Braekman, J. C., Daloze, D., Pasteels, J. M., and Tursch, B. (1981). 'Chemical composition of the frontal gland secretions from Neo-Guinean nasute termite soldiers', *Bull. Soc. Chim. Belg.,* **90**, 485–499.

Edgar, J. A., Culvenor, C. C. J., and Pliske, T. E. (1976). 'Isolation of a lactone structurally related to the esterifying acids of pyrrolizidine alkaloids from the costal fringes of male Ithomiinae', *J. Chem. Ecol.,* **2**, 263–270.

Edwards, L. J., Siddall, J. B., Dunham, L. L., Uden, P., and Kislow, C. J. (1973). 'Trans-β-farnesene, alarm pheromone of the green peach aphid, *Myzus persicae* (Sulzer), *Nature,* **241**, 126–127.

Eisner, T. (1970). 'Chemical defense against predation in arthropods', in *Chemical Ecology* (eds E. Sondheimer and J. B. Simeone) pp. 157–218, Academic Press, New York.

Eisner, T., Eisner, H. E., Hurst, J. J., Kafatos, F. C., and Meinwald, J. (1963a). 'Cyanogenic glandular apparatus of a millipede', *Science*, **139**, 1218–1220.

Eisner, T., Hendry, L. B., Peakall, D. B., and Meinwald, J. (1971). '2,5-Dichlorophenol (from ingested herbicide?) in defensive secretions of grasshopper', *Science*, **172**, 277–278.

Eisner, T., Hurst, J. J., and Meinwald, J, (1963b). 'Defense mechanisms of arthropods. XI. The structure, function and phenolic secretions of the glands of a chordeumoid millipede and a carabid beetle', *Psyche*, **70**, 95–116.

Eisner, T., Johanessee, J. S., Carrel, J. Hendry, L. B., and Meinwald, J. (1974). 'Defensive use by an insect of a plant resin', *Science*, **184**, 996–999.

Evershed, R. P., Morgan, E. D., and Cammaerts, M. C. (1981). 'Identification of the trail pheromone of the ant *Myrmica rubra* L. and related species,' *Naturwissenschaften*, **68**, 374–375.

Evershed, R. P., Morgan, E. D., and Cammaerts, M. C. (1982). '3-Ethyl-2,5-dimethyl pyrazine, the trail pheromone from the venom gland of eight species of *Myrmica* ants', *Insect Biochem.*, **12**, 383–391.

Fish, L. J., and Pattenden, G. (1975). 'Iridodial, and a new alkanone, 4-methylhexan-3-one, in the defensive secretion of the beetle, *Straphylinus olens'*, *J. Insect Physiol.*, **21**, 741–744.

Francke, W., Heeman, V., Gerken, B., Renwick, J. A. A., and Vite, J. P. (1977). '2-Ethyl-1,6-dioxaspiro[4.4]nonane, principal aggregation pheromone of *Pityogenes chalcographus* (L.)', *Naturwissenschaften*, **64**, 590–591.

Francke, W., Hindorf, G., and Reith, W. (1978). 'Methyl-1,6-dioxaspiro[4.5]decanes as odours of *Paravespula vulgaris* (L.)', *Angew. Chem. Int. Ed. Engl.*, **17**, 862.

Francke, W., Reith, W., Bergström, G., and Tengö, J. (1980). 'Spiroketals in the mandibular glands of *Andrena* bees', *Naturwissenschaften*, **67**, 149–150.

Francke, W., Reith, W., Bergström, G., and Tengö, J. (1981). 'Pheromone bouquet of the mandibular glands in *Andrena haemorrhoa* (Hym., Apoidea)., *Z. Naturforsch. Teil C.*, **36**, 928–932.

Fukui, H., Matsumura, F., Ma, M. C., and Burkholder, W. E. (1974). 'Identification of the sex pheromone of the furniture carpet beetle *Anthrenus flavipes* Le Conté, *Tetrahedron Lett.*, **40**, 3563–3566.

Fukui, H., Matsumura, F., Barak, A. V., and Burkholder, W. E. (1977). 'Isolation and identification of a major sex-attractant component of *Attagenus elongatulus* (Casey) (Coleoptera: Dermestidae)', *J. Chem. Ecol.*, **3**, 539–548.

Gieselmann, M. J., Moreno, D. S., Fargerlund, J., Tashiro, H., and Roelofs, W. L (1979a). 'Identification of the sex pheromone of the yellow scale', *J. Chem. Ecol.*, **5**, 27–33.

Gieselmann, M. J., Rice, R. E., Jones, R. A., and Roelofs, W. L. (1979b). 'Sex pheromone of the San José scale', *J. Chem. Ecol.*, **5**, 891–927.

Gothe, R. (1983). 'Pheromones in ixodid and argasid ticks. Part I: ixodid icks', *V.M.R.; Vet. Med. Rev.*, **1983**, 16–37.

Greany, P. D., and Hagen, K. S. (1980). 'Prey selection', in *Semiochemicals: Their Role in Pest Control* (eds D. A. Nordlund, R. L. Jones and W. J. Lewis), Wiley, New York.

Greany, P. D., Tumlinson, J. H., Chambers, D. L., and Boush, G. M. (1977). 'Chemically mediated host finding by *Biosteres (Opius) longicaudatus*, a parasitoid of tephritid fruit fly larvae', *J. Chem. Ecol.*, **3**, 189–195.

Grula, J. W., McChesney, J. D., and Taylor, O. R. Jr. (1980). 'Aphrodisiac pheromones of the sulphur butterflies *Colias eurytheme* and *C. philodice* (Lepidoptera: Pieridae)', *J. Chem. Ecol.*, **6**, 241–256.

Guerrero, A., Camps, F., Coll, J., Riba, M., Einhorn, J., Descoins, C., and Lallemand, J. Y. (1981). 'Identification of a potential sex pheromone of the processionary moth *Thaumetopoea pityocampa* (Lepidoptera, Notodontidae)', *Tetrahedron Lett.*, **22**, 2013–2016.

Guss, P. L., Carney, R. L., Sonnet, P. E., and Tumlinson, J. H. (1983a). 'Stereospecific sex attractant for *Diabrotica cristata* (Harris) (Coleoptera: Chrysomelidae)', *Environ. Ent.*, **12**, 1296–1297.

Guss, P. L., Tumlinson, J. H., Sonnet, P. E., and McLaughlin, J. R. (1983b). 'Identification of a female-produced sex pheromone from the southern corn rootworm, *Diabrotica undecimpunctata howardi*', *J. Chem. Ecol.*, **9**, 1363–1375.

Guss, P. L., Tumlinson, J. H., Sonnet, P. E., and Proveaux, A. T. (1982). 'Identification of a female-produced sex pheromone of the western corn rootworm', *J. Chem. Ecol.*, **8**, 545–556.

Hall, D. R., Cork, A., and Hodges, R. J. (1984). 'Chemical and biological studies of the aggregation pheromone of the greater grain borer, *Prostephanus truncatus* (Coleoptera: Bostrichidae)', Poster presented at the XVII International Congress of Entomology, 20–26 Aug., 1984, Hamburg, F.R.G.

Harborne, J. B. (1977). *Introduction to Ecological Biochemistry*, pp. 103–129, Academic Press, London.

Harring, C. M., Vite, J. P., and Hughes, P. R. (1975) 'Ipsenol'der Populationslockstoff des krummzähningen Tannenborkenkäfers', *Naturwissenschaften*, **62**, 488.

Hayashi, N., Kuwahara, Y., and Komae, H. (1978). 'The scent scale substance of male Pieris butterflies (*Pieris meleta* and *Pieris napi*)', *Experientia*, **34**, 684–685.

Heath, R. R., McLaughlin, J. R., Tumlinson, J. H., Ashley, T. R., and Doolittle, R. E. (1979). 'Identification of the white peach scale sex pheromone. An illustration of micro techniques', *J. Chem. Ecol.*, **5**, 941–953.

Heath, R. R., Tumlinson, J. H., Leppla, N. C., McLaughlin, J. R., Dueben, B., Dundulis, E., and Guy, R. H. (1983). 'Identification of a sex pheromone produced by female velvetbean caterpillar moth', *J. Chem. Ecol.*, **9**, 645–656.

Hefetz, A., and Batra, S. W. T. (1980). 'Chemistry of the cephalic secrtetions of eumenid wasps', *Comp. Biochem. Physiol.* **65B**, 455–456.

Hefetz, A., Blum, M. S., Eickwort, G. C., and Wheeler, J. W. (1978). 'Chemistry of the Dufour's gland secretion of halictine bees', *Comp. Biochem. Physiol.* **61B**, 129–132.

Hendry, L. B., Greany, P. D., and Gill, R. J. (1973). 'Kairomone mediated host-finding behaviour in the parasitic wasp *Orgilus lepidus*', *Entomol. Exp. Appl.*, **16**, 471–477.

Hendry, L. B., Piatek, B., Browne, L. E., Wood, D. L., Byers, J. A., Fish, R. H., and Hicks, R. A. (1980). '*In vivo* conversion of a labelled host plant chemical to pheromones of the bark beetle *Ips paraconfusus*', *Nature*, **284**, 485.

Herrebout, W. M. (1969a). 'Some aspects of host selection in *Eucarcelia rutilla* Vill', *Neth. J. Zool.*, **19**, 1–104.

Herrebout, W. M. (1969b). 'Habitat selection in *Eucarcelia rutilla* Vill. II. Experiments with female of known age', *Z. Angew. Entomol.*, **63**, 336–349.

Higgs, M. D., and Evans, D. A. (1978). 'Chemical mediators in the oviposition behaviour of the house longhorn beetle, *Hylotrupes bajulus*', *Experientia*, **34**, 46–47.

Hill, A. S., Berisford, C. W., Brady, U. E., and Roelofs, W. L. (1981). 'Nantucket pine tip moth, *Rhyacionia frustrana*: identification of two sex pheromone components', *J. Chem. Ecol.*, **7**, 517–528.

Hill, A. S., and Roelofs, W. L. (1981). 'Sex pheromone of the saltmarsh caterpillar moth, *Estigmene acrea.*', *J. Chem. Ecol.*, 7, 655–667.

Honda, H., Oshima, K., and Yamamoto, I. (1975). 'Chemical nature of trail following substances from a Japanese termite, *Reticulitermes speratus* Kolbe', *J. Agric. Sci. Tokyo Univ. Agric.*, 20, 121–128.

Hope, J. A., Horler, D. F., and Rowlands, D. G. (1967). 'A possible pheromone of the bruchid, *Acanthoscelides obtectus* (Say)', *J. Stored Prod. Res.*, 3, 387–388.

Horler, D. F. (1970). '(−)-Methyl-*n*-tetradeca-trans-2,4,5-trienoate, an allenic ester produced by the male dried bean beetle, *Acanthoscelides obtectus* (Say)', *J. Chem. Soc. (C)*, 859–862.

Howard, R., Matsumura, F., and Coppel, H. C. (1976). 'Trail-following pheromones of Rhinotermitidae: approaches to their authentication and specificity', *J. Chem. Ecol.*, 2, 147–166.

Hughes, P. R. (1974). 'Myrcene: a precursor of pheromones in *Ips* beetles', *J. Insect Physiol.*, 20, 1271–1275.

Hughes, P. R., and Renwick, J. A. A. (1977). 'Neural and hormonal control of pheromone biosynthesis in the bark beetle *Ips paraconfusus*', *Physiol. Entomol.*, 2, 117–123.

Huwyler, S., Grob, K., and Visconti, M. (1975). ' The trail pheromone of the ant *Lasius fuliginosus*: identification of six components,' *J. Insect Physiol.*, 21, 299–304.

Jacobson, M., and Ohinata, K. (1980). 'Unique occurrence of fenchol in the animal kingdom', *Experientia*, 36, 629–630.

Jacobson, M., Ohinata, K., Chambers, D. L., Jones, W. A., and Fujimoto, M. S. (1973). 'Insect sex attractants. 13. Isolation, identification and synthesis of sex pheromone of the male Mediterranean fruit fly', *J. Med. Chem.*, 16, 248–251.

Jones, D. A., Parsons, J., and Rothschild, M. (1962). 'Release of hydrocyanic acid from crushed tissues of all stages in the life cycle of species of the Zygaeninae (Lipidoptera)', *Nature*, 193, 52–53.

Jones, R. L., Lewis, W. J., Bowman, M. C., Beroza, M., and Bierl, B. A. (1971). 'Host seeking stimulant for the parasite of corn earworm: isolation, identification and synthesis', *Science*, 173, 842–843.

Jones, R. L., Lewis, W. J., Beroza, M., Bierl, B. A., and Sparks, A. N. (1973). 'Host-seeking stimulants (Kairomones) for the egg-parasite *Trichogramma evanescens*', *Environ. Entomol.*, 2, 593–596.

Jones, R. L., and Sparks, A. N. (1979). '(*Z*)-9-Tetradecen-1-ol acetate. A secondary sex pheromone of the fall armyworm, *Spodoptera frugiperda* (J. E. Smith)', *J. Chem. Ecol.*, 5, 721–725.

Kajita, H., and Drake, E. F. (1969). Biology of *Apanteles chilonis* and *Apanteles flavipes* parasites of *Chilo suppressalis*', *Mushi*, 42, 163–174.

Kalmus, H. (1965). 'Possibilities and constraints of chemical telecommunication', *Proc. 2nd Int. Congr. Endocrinol. Lond.* pp. 188–192.

Kandil, A. A., and Slessor, K. (1983). 'Enantiomeric synthesis of 9-hydroxy-(*E*)-2-decenoic acid, a queen bee pheromone', *Can. J. Chem.*, 61, 1166–1168.

Karlson, P., and Butenandt, A. (1959). 'Pheromones (ectohormones) in insects', *Ann. Rev. Entomol.*, 4, 39–58.

Karlson, P,, and Lüscher, M. (1959). 'Pheromones: a new term for a class of biologically active substances', *Nature*, 183, 55–56.

Kennedy, B. H., and Golford, J. R. (1972). 'Development of *Dendrosoter protuberans* on larvae of the smaller European elm bark beetle being reared on an artificial medium', *Ann. Entomol. Soc. Am.*, 65, 757–759.

Kesten, U. (1969). 'Zur Morfologie und Biologie von *Anatis ocellata* (L.) (Coleoptera, Coccinellidae)', *Z. Angew. Entomol.*, **63**, 412–445.

Khorramshahi, A., and Burkholder, W. E. (1981). 'Behaviour of the lesser grain borer *Rhyzopertha dominica* (Coleoptera: Bostrichidae): Male-produced aggregation pheromone attracts both sexes', *J. Chem. Ecol.*, **7**, 33–38.

Kikukawa, T., Matsumura, F., Olaifa, J., Kraemer, M., Coppel, H. C., and Tai, A. (1983). 'Field evaluation of chiral isomers of the sex pheromone of the European pine sawfly, *Neodiprion sertifer*', *J. Chem. Ecol.*, **9**, 673–693.

Klun, J. A., Plimmer, J. R., Bierl-Leonhardt, B. A., Sparks, A. N., and Chapman, O. L. (1979). 'Trace chemicals: the essence of sexual communication systems in *Heliothis* species', *Science*, **204**, 1328–1330.

Kostelec, J. G., Girard, J. E., and Hendry, L. B. (1980). 'Isolation and identification of a sex attractant of a mushroom-infesting sciarid fly', *J. Chem. Ecol.*, **6**, 1–11.

Kullenberg, B., Bergström, G., and Ställberg-Stenhagen, S. (1970). 'Volatile components of the marking secretion of the male bumblebees', *Acta Chem. Scand.*, **24**, 1481–1483.

Kunesch, G., and Zagatti, P. (1982). 'Male sex pheromones of the African sugar cane borer: *Eldana saccharina* Wlk.: identification and behaviour', *Les Med. Chim.* Versailles, 16–20 Nov., 1981, Ed INRA Publ., 1982 (Les Colloques de l'INRA, 7) pp 281–287.

Kunesch, G., Zagatti, P., Lallemand, J. Y., Debal, A., and Vigneron, J. P. (1981). 'Structure and synthesis of the wing gland pheromone of the male African sugar cane borer: *Eldana saccharina* (Wlk.) (Lepidoptera, Pyralidae)', *Tetrahedron Lett.*, **22**, 5271–5274.

Kuwahara, Y. (1978). 'Alerting pheromone produced by acaridae', *Shokubutsu Boeki*, **32**, 62–68.

Kuwahara, Y., Fukami, H., Ishii, S., Matsumura, F., and Burkholder, W. E. (1975). 'Studies on the isolation and bioassay of the sex pheromone of the drugstore beetle *Stegobium paniceum* (Coleoptera: Anobiidae)', *J. Chem. Ecol.*, **1**, 413–422.

Kuwahara, Y., Fukami, H., Howard, R., Ishii, S., Matsumura, F., and Burkholder, W. E. (1978). 'Chemical studies on the Anobiidae: sex pheromone of the drugstore beetle, *Stegobium paniceum* (L.) (Coleoptera)', *Tetrahedron*, **34**, 1769–1774.

Kuwahara, Y., and Sakuma, L. (1982). Pheromone study on acarid mites. Part IX. Syntheses of alarm pheromone analogs of the mold mite, *Tyrophagus putrescentiae* and their biological activities', *Agric. Biol. Chem.*, **46**, 1855–1860.

Kuwahara, Y., Yen, L. T. M., Tominaga, Y., Matsumoto, K., and Wada, Y. (1982). 'Pheromone study on acarid mite. Part X. 1,3,5,7-tetramethyldecyl formate, lardolure: aggregation pheromone of the acarid mite, *Lardoglyphus konoi* (Sasa and Asanuma) (Acarina: Acaridae)', *Agric. Biol. Chem.*, **46**, 2283–2291.

Kydonieus, A. F., and Beroza, M. (eds) (1982). *Insect Suppression with Controlled Release Pheromone Systems*, Vols 1 and 2, CRC Press, Boca Raton, Fla.

Lanier, G. N., Birch, M. C., Schmitz, R. F. and Furniss, M. M. (1972). 'Pheromones of *Ips pini* (Coleoptera: Scolytidae): variation in response among three populations', *Can. Ent.*, **104**, 1917–1923.

Lanier, G. N., Classon, A., Stewart, T., Piston, J. J., and Silverstein, R. M. (1980). '*Ips pini*: the basis for interpopulational differences in pheromone biology', *J. Chem. Ecol.*, **6**, 677–687.

Laurence, B. R., and Pickett, J. A. (1982). Erythro-6-acetoxy-5-hexadecanolide, the major component of a mosquito oviposition attractant pheromone', *J. Chem. Soc., Chem. Com.*, **1982**, 59–60.

Leonhardt, B. A., Neal, J. W. Jr, Klun, J. A., Schwartz, M., and Plimmer, J. R. (1983). 'An unusual lepidopteran sex pheromone system in the bagworm moth', *Science*, **219**, 314–316.

Levinson, H. Z., and Barilan, A. R. (1967). 'Function and properties of an assembling scent in the Khapra beetle', *Trogoderma granarium. Riv. Parassitol.*, **28**, 27–42.

Levinson, H. Z., and Levinson, A. R. (1979). 'Trapping of storage insects by sex and food attractants as a tool of integrated control., in *Chemical Ecology: Odour Communication in Animals* (ed. F. J. Ritter), pp. 327–341, Elsevier North Holland Biomedical Press, Amsterdam.

Levinson, H. Z., Levinson, A. R., and Maschwitz, U. (1974). 'Action and composition of the alarm pheromone of the bedbug, *Cimex lectularius* L.', *Naturwissenschaften*, **12**, 684–685.

Levinson, H. Z., and Mori, K. (1983). 'Chirality determines pheromone activity for flour beetles', *Naturwissenschaften*, **70**, 190–192.

Lewis, W. J., and Jones, R. L. (1971). 'Substance that stimulates host-seeking by *Microplitis croceipes*, a parasite of *Heliothis* species', *Ann. Ent. Soc. Am.*, **64**, 471–473.

Lewis, W. J., Jones, R. L., Nordlund, D. A., and Gross, H. R. Jr (1975a). 'Kairomones and their use for management of entomophagous insects II. Mechanisms causing increase in the rate of parasitization by *Trichogramma* spp., *J. Chem. Ecol.*, **1**, 349–360.

Lewis, W. J., Jones, R. L., Nordlund, D. A., and Sparks, A. N. (1975b). 'Kairomones and their use for management of entomophagous insects I. Evaluation for increasing rates of parasitization by *Trichogramma* spp. in the field', *J. Chem. Ecol.*, **1**, 343–348.

Lewis, W. J., Nordlund, D. A., Gross, H. R. Jr, Jones, R. L., and Jones, S. L. (1977). 'Kairomones and their use for management of entomophagous insects: V. Moth scales as a stimulus for predation of *Heliothis zea* (Boddie) eggs by *Chrysopa carnea* Stephens larvae', *J. Chem. Ecol.*, **3**, 483–487.

Lewis, W. J., Sparks, A. N., Jones, R. L., and Barras, D. J. (1972). Efficiency of *Cardiochiles nigriceps* as a parasite of *Heliothis virescens* on cotton', *Environ. Entomol.*, **1**, 468–471.

Lewis, W. J., Sparks, A. N., and Redlinger, L. M. (1971). 'Moth odor: a method of host finding by *Trichogramma evanescens*', *J. Econ. Entomol.*, **64**, 557–558.

Leyrer, R. L., and Monroe, R. E. (1973). 'Isolation and identification of the scent of the moth, *Galleria mellonella* and a re-evaluation of its sex pheromone', *J. Insect Physiol.*, **19**, 2267–2271.

Linley, J. R., and Carlson, D. A. (1978). 'A contact mating pheromone in the biting midge, *Culicoides melleus*', *J. Insect Physiol.*, **24**, 423–427.

Lloyd, H. A., Blum, M. S., and Duffield, R. M. (1975). 'Chemistry of the male mandibular gland secretion of the ant *Camponotus clarithorax*', *Insect Biochem.*, **5**, 489–494.

Lofquist, J. (1977). 'Toxic properties of the chemical defence systems in the competitive ants *Formica rufa* and *F. sanguinea*', *Oikos*, **28**, 137–151.

Lofstedt, C., Vanderpers, J., Lofquist, J., Lanne, B. S., Appelgren, M., Bergström, G., and Thelin, B. (1982). 'Sex pheromone components of the turnip moth, *Agrotis segetum*: chemical identification, electrophysiological evaluation and behavioural activity', *J. Chem. Ecol.*, **8**, 1305–1321.

Longhurst, C., Baker, R., Howse, P. E., and Speed, W. S. (1978). 'Alkyl pyrazines in ants: their presence in three genera, and caste-specific behavioural responses to them in *Odontomachus troglodytes*', *J. Insect Physiol.*, **24**, 833–837.

Longhurst, C., Bolwell, S., Bradshaw, J. W. S., Howse, P. E., and Evans, D. A. (1983). 'Multicomponent alarm pheromones from poison gland secretion of *Myrmicaria eumenoides* and *M. striata* (Hymenoptera: Formicidae)', in Preparation.

Mansingh, A., Steele, R. W., Smallman, B. N., Meresz, O., and Mozogai, C. (1972). 'Pheromone effects of *cis*-9 long alkenes on the common housefly. An improved sex attractant combination', *Can. Entomol.*, **104**, 1963–1965.

Matsumura, F., Coppel, H. C., and Tai, A. (1968). 'Isolation and identification of termite trail-following pheromone', *Nature*, **219**, 963–964.

Mazomenos, B. E., and Pomonis, J. G. (1983). 'Male olive fruit fly pheromone: Isolation identification and laboratory bioassays', in *Fruit Flies of Economic Importance* (ed. R. Cavalloro), Proc. CEC/IOBC Int. Symp., Athens, Greece, 16–19 Nov, 1982, pp. 96–103.

McDowell, P. G., Whitehead, D. L., Chaudhury, M. F. B., and Snow, W. F. (1981). 'The isolation and identification of the cuticular sex-stimulant pheromone of the tsetse, *Glossina pallidipes* Austen (Diptera: Glossinidae)', *Insect Sci. Appl.*, **2**, 181–187.

McLean, J. A., and Borden, J. H. (1975). 'Survey for *Gnathotrichus sulcatus* (Coleoptera: Scolytidae) in a commercial sawmill with the pheromone, sulcatol', *Can. J. Forest Res.*, **5**, 586–591.

Meinwald, J., Chadha, M. S., Hurst, J. J., and Eisner, T. (1962). 'Defense mechanisms of arthropods IX. Anisomorphal, the secretion of a phasmid insect', *Tetrahedron Lett.*, 29–33.

Meinwald, Y. C., Meinwald, J., and Eisner, T. (1966). '1,2-Dialkyl-4(3H)-quinazolines in the defensive secretion of a millipede *(Glomeris marginata)*', *Science*, **154**, 390–391.

Minks, A. K. (1979). 'Present status of insect pheromones in agriculture and forestry', in *Proc. Int. Symp. IOBC/WPRS on Integrated Control in Agriculture and Forestry* (eds K. Russ and H. Berger), Vienna, Oct, 1979, pp. 127–136.

Mitchell, E. R. (ed.) (1981). *Management of Insect Pests with Semiochemicals: Concepts and Practice,* Plenum, New York.

Mitchell, W. C., and Man, R. F. L. (1970). 'Response of the female southern green stink bug and its parasite *Trichopoda pennipes* to male stink bug pheromones', *J. Econ. Entomol.*, **64**, 856–859.

Monteith, L. G. (1956). 'Influence of host movement on selection of hosts by *Drino bohemica* Mesn. as determined in an olfactor', *Can. Entomol.*, **88**, 583–586.

Monteith, L. G. (1958). 'Influence of host and its food plant on host finding by *Drino bohemica* Mesn. and interaction of other factors', *Proc. Int. Cong. Entomol. 10th, 1956*, **2**, 603–606.

Monteith, L. G. (1964). 'Influence of health of the food plant on the host and host-finding by tachinid parasites, *Can. Entomol.*, **96**, 1477–1482.

Moore, B. P. (1968). 'Studies on the chemical composition and function of the cephalic gland secretion in Australian termites', *J. Insect Physiol.*, **14**, 33–39.

Moore, B. P., and Brown, W. V. (1976). 'The chemistry of the metasternal gland secretion of the Eucalypt Longicorn *Phoracantha synonyma* (Coleoptera: Cerambycidae)', *Aust, J. Chem.*, **29**, 1365–1374.

Mori, K. (1982). 'Recent progress in the synthesis of optically active pheromones, In *Les Mediateurs Chimiques*. INRA Publ., Ed) Les Colloques de l'INRA No. 7, Proc. Int. Symp., Versailles, 16–20 Nov., 1981, pp. 41–53.

Moser, J. C., Brownlee, R. G., and Silverstein, R. M. (1968). 'The alarm pheromones of the ant *Atta texana'*, *J. Insect Physiol.*, **14**, 529–535.

Mudd, A. (1978). 'Novel β-triketones from Lepidoptera', *J. Chem. Soc., Chem. Comm.*, 1075–1076.

Mudd, A. (1981). 'Novel 2-acylcyclohexane-1,3-diones in the mandibular glands of lepidopteran larvae. Kairomones of *Ephestia kuehniella* Zeller', *J. Chem. Soc., Perkin Trans.*, **1**, 2357–2362.

Myerson, J., Haddon, W. F., and Soderstrom, E. L. (1982). '*sec*-Butyl (*Z*)-7-tetradecenoate. A novel sex pheromone component from the western grapeleaf skeletonizer, *Harrisina brillians'*, *Tetrahedron Lett.*, **23**, 2757–2760.

Nault, L. R., and Montgomery, M. E. (1979). 'Aphid alarm pheromones', *Misc. Pub. Entomol. Soc. Am.*, **11**, 23–31.

Negishi, T., Uchida, M., Tamaki, Y., Mori, K., Ishiwatari, T., Asano, S., and Nakagawa, K. (1980). 'Sex pheromone of the comstock mealybug, *Pseudococcus comstocki* Kuwana: isolation and identification', *Appl. Entomol. Zool.*, **15**, 328–333.

Nesbitt, B. F., Beevor, P. S., Hall, D. R., Lester, R., and Willians, J. R. (1980). 'Components of the sex pheromone of the female sugar cane borer, *Chilo sacchariphagus* (Bojer) (Lepidoptera: Pyralidae)—identification and field trials', *J. Chem. Ecol.*, **6**, 385–394.

Nishida, R., Baker, T. C., and Roelofs, W. L. (1982). 'Hairpencil pheromone components of the male oriental fruit moths, *Grapholitha molesta*', *J. Chem. Ecol.*, **8**, 947–959.

Nishida, R., Kuwahara, Y., Fukami, H., and Ishii, S. (1979). 'Female sex pheromone of the German cockroach *Blattella germanica* (L.) (Orthoptera: Blattellidae) responsible for male wing-raising IV. The absolute configuration of the pheromone, 3,11-dimethyl-2-nonacosanone', *J. Chem. Ecol.*, **5**, 289–297.

Nishida, R., Sato, T., Kuwahara, Y., Fukami, H., and Ishii, S. (1976). 'Female sex pheromone of the German cockroach *Blattella germanica* (L.) (Orthoptera: Blattellidae), responsible for male wing-raising II. 29-hydroxy-3,11-dimethyl-2-nonacosanone', *J. Chem. Ecol.*, **2**, 449–455.

Nishino, C., Bowers, W. S., Montgomery, M. E., and Nault, L. R. (1976a). 'Aphid alarm pheromone mimics: sesquiterpene hydrocarbons', *Agric. Biol. Chem.*, **40**, 2303–2304.

Nishino, C., Bowers, W. S., Montgomery, M. E., and Nault, L. R. (1976b). 'Aphid alarm pheromone mimics: the *nor*-farnesenes', *Appl. Entomol. Zool.*, **11**, 340-343.

Nordlund, D. A., Lewis, W. J., Jones, R. L., Gross, H. R. Jr, and Hagen, K. S. (1977). 'Kairomones and their use for management of entomophagous insects: VI. An examination of the kairomones for the predator *Chrysopa carnea* Stephens at the oviposition sites of *Heliothis zea* (Boddie)', *J. Chem. Ecol.*, **3**, 507–511.

Ohinata, K., Jacobson, M., Kobayashi, R. M., Chambers, D. L., Fujimoto, M. S., and Higa, H. H. (1982). 'Oriental fruit fly and melon fly: biological and chemical studies of smoke produced by males', *J. Environ. Sci. Hlth*, **A17**, 197–216.

Parry, K., and Morgan, E. D. (1979). 'Pheromones of ants: a review', *Physiol. Entomol.*, **4**, 161–189.

Pasteels, J. M., Gregoire, J. C., and Rowell-Rahier, M. (1983). 'The chemical ecology of defense in arthropods', *Ann Rev. Entomol.*, **28**, 263–289.

Pattenden, G., and Staddon, B. W. (1972). 'Identification of *iso*-butyric acid in secretion from Brindley's scent glands in *Rodnius prolixus* (Heteroptera: Reduviidae)', *Ann. Entomol. Soc. Am.*, **65**, 1240–1241.

Pearce, G. T., Gore, W. E., Silverstein, R. M., Peacock, J. W., Cuthbert, R. A., Lanier, G. N., and Simeone, J. B. (1975). 'Chemical attractants for the smaller European elm bark beetle *Scolytus multistriatus* (Coleoptera: Scolytidae)', *J. Chem. Ecol.*, **1**, 115–124.

Persoons, C. J., Verwiel, P. E. J., Ritter, F. J., and Nooijen, W. J. (1982). 'Studies on sex pheromones of American cockroach, with emphasis on structure elucidation of periplanone-A', *J. Chem. Ecol.*, **8**, 439–451.

Persoons, C. J., Verwiel, P. E. J., Ritter, F. J., Talman, E., Nooijen, P. J. F., and Nooijen, W. J. (1976a). 'Sex pheromones of the American cockroach, *Periplaneta americana*: a tentative structure of Periplanone-B', *Tetrahedron Lett.*, **24**, 2055–2058.

Persoons, C. J., Verwiel, P. E. J., Talman, E., and Ritter, F. J. (1979). 'Sex pheromone of the American cockroach, *Periplaneta americana*. Isolation and structure elucidation of Periplanone-B', *J. Chem. Ecol.*, **5**, 221–236.

Persoons, C. J., Voerman, S., Verwiel, P. E. J., Ritter, F. J., Nooijen, W. J., and Minks, A. K. (1976b). 'Sex pheromone of the potato tuberworm moth *Phthorimaea operculella* (Zeller) (Lepidoptera: Gelichiidae) and field experiments with them', *Entomol. Exp. Appl.*, **20**, 289–300.

Peschke, K., and Metzler, M. (1982). 'Defensive and pheromonal secretion of the tergal gland of *Aleochara curtula* I. The chemical composition', *J. Chem. Ecol.*, **8**, 773–783.

Petty, R. L., Boppre, M., Schneider, D., and Meinwald, J. (1977). 'Identification and localization of volatile hairpencil components in male *Amauris ochlea* butterflies (Danaidae)', *Experientia, 33*, 1324–1326.

Pickett, J. A., and Griffiths, D. C. (1980). 'Composition of aphid alarm pheromones', *J. Chem. Ecol., 6*, 349–360.

Pickett, J. A., Williams, I. H., and Martin, A. P. (1982). '(*Z*)-11-eicosen-1-ol, an important new pheromonal component from the sting of the honey bee, *Apis mellifera* L. (Hymenoptera, Apidae)', *J. Chem. Ecol., 8*, 163–175.

Pickett, J. A., Williams, I. H., Martin, A. P., and Smith, M. C. (1980). 'Nasonov pheromone of the honey bee, *Apis mellifera* L. (Hymenoptera: Apidae) Part I. Chemical characterization', *J. Chem. Ecol., 6*, 425–434.

Prestwich, G. D. (1978). 'Isotrinervi-2 β-ol. Structural isomers in the defense secretions of allopatric populations of the termite *Trinervitermes gratiosus*', *Experientia, 34*, 682–683.

Prestwich, G. D. (1982). 'From tetracycles to macrocycles. Chemical diversity in the defense secretions of nasute termites', *Tetrahedron, 38*, 1911–1919.

Prestwich, G. D. (1983). 'The chemical defenses of termites', *Sci. Am., 249*, 78–81, 84–87.

Prestwich, G. D., and Collins, M. S. (1981). 'Macrocyclic lactones as the defense substances of the termite genus *Armitermes*', *Tetrahedron Lett., 22*, 4587–4590.

Prestwich, G. D., Tanis, S. P., Pilkiewicz, F., Miura, I., and Nakanishi, K. (1976). 'Nasute termite frontal gland secretions II. Structures of trinervitane congeners from *Trinervitermes* soldiers', *J. Am. Chem. Soc., 98*, 6062–6064.

Prestwich, G. D., Jones, R. W., and Collins, M. S. (1981). 'Terpene biosynthesis by nasute termite soldiers (Isoptera: Nasutitermitinae)', *Insect Biochem, 11*, 331–336.

Read, D. P., Feeny, P. P., and Root, R. B. (1970). 'Habitat selection by the aphid parasite *Diaeretiella rapae*', *Can. Entomol., 102*, 1567–1578.

Regnier, F. E., and Wilson, E. O. (1968). 'The alarm-defense system of the ant *Acanthomyops claviger*', *J. Insect Physiol., 14*, 955–970.

Reichstein, T., von Euw, J., Parsons, J. A., and Rothschild, M. (1968). 'Heart poisons in the monarch butterfly', *Science, 161*, 861–866.

Renwick, J. A. A., Hughes, P. R., and Krull, I. S. (1976). 'Selective production of *cis*- and *trans*-verbenol from (−)- and (+)-α-pinene by a bark beetle', *Science, 191*, 199–201.

Rice, R. E. (1969). 'Response of some predators and parasites of *Ips confusus* (Le C.) to olfactory attractants', *Contrib. Boyce Thompson Inst., 24*, 189–194.

Richter, I., Krain, H., and Mangold, H. K. (1976). 'Long-chain (*Z*)-9-alkenes are 'psychedelics' to houseflies with regard to visually stimulated sex attraction and aggregation', *Experientia, 32*, 186–188.

Ridgway, R. L., Lloyd, E. P., and Pross, W. H. (eds) (1983). *Cotton Insect Management with Special Emphasis on the Boll Weevil*. USDA Handbook, Govt Printing Office, Washington, DC.

Riley, R. G., Silverstein, R. M., Carroll, B., and Carrol, R. (1974a). 'Methyl 4-methylpyrrole-2-carboxylate; a volatile trail pheromone from the leaf cutting ant, *Atta cephalotes*', *J. Insect Physiol., 20*, 651–654.

Riley, R. G., Silverstein, R. M., and Moser, J. C. (1974b). 'Isolation, identification, synthesis and biological activity of volatile compounds from the heads of *Atta* ants', *J. Insect Physiol., 20*, 1629–1637.

Ritter, F. J. (ed.) (1979). *Chemical Ecology: Odour Communication in Animals*, pp. 249–402, Elsevier/North Holland, Amsterdam.

Ritter, F. J., Bruggeman-Rotgans, I. E. M., Verwiel, P. E. J., Persoons, C. J., and Talman, E. (1977). 'Trail pheromones of the Pharaoh's ant, *Monomorium pharaonis*: isolation and identification of faranal, a terpenoid related to juvenile hormone II', *Tetrahedron Lett.*, 2617–2618.

Ritter, F. J., and Coenen-Saraber, C. M. A. (1969). 'Food attractants and a pheromone as trail-following substances for the Saintonge termite', *Entomol. Exp. Appl.*, **12**, 611–622.

Rocca, J. R., Tumlinson, J. H., Glancey, B. M., and Lofgren, C. S. (1983a). 'The queen recognition pheromone of *Solenopsis invicta*, preparation of (*E*)-6-(1-pentenyl)-2*H*-pyran-2-one.' *Tetrahedron Lett.*, **24**, 1889–1892.

Rocca, J. R., Tumlinson, J. H., Glancey, B. M., and Lofgren, C. S. (1983b). 'Synthesis and stereochemistry of tetrahydro-3,5-dimethyl-6-(1-methylbutyl)-2*H*-pyran-2-one, a component of the queen recognition pheromone of *Solenopsis invicta*.' *Tetrahedron Lett.*, **24**, 1893–1896.

Rodin, J. O., Silverstein, R. M., Burkholder, W. E., and Gorman, J. E. (1969). 'Sex attractant of female dermestid beetle *Trogoderma inclusum* Le Conte', *Science*, **165**, 904–906.

Roelofs, W. L. (1981). 'Attractive and aggregating pheromones', In *Semiochemicals: Their role in Pest Control* (eds D. A. Nordlund, R. L. Jones and W. J. Lewis), pp. 215–235, Wiley, New York.

Roelofs, W. L., and Brown, R. L. (1982). 'Pheromones and evolutionary relationships of Tortricidae', *Ann. Rev. Ecol. Syst.*, **13**, 395–422.

Roelofs, W. L., and Carde, R. T. (1971). 'Hydrocarbon sex pheromones in tiger moths (Arctiidae)', *Science*, **171**, 684–686.

Roelofs, W. L., and Carde, R. T. (1974). 'Sex pheromones in the reproductive isolation of lepidopterous species', In *Insect Pheromones* (ed. M. C. Birch), pp. 96–114, Elsevier, New York.

Roelofs, W. L., and Carde, R. T. (1977). 'Responses of lepidoptera to synthetic sex pheromone chemicals and their analogues', *Ann. Rev. Entomol.*, **22**, 377–405.

Roelofs, W. L., Gieselmann, M., Carde, A., Tashiro, H., Moreno, D. S., Henrick, C. A., and Anderson, R. J. (1978). 'Identification of the California red scale sex pheromone', *J. Chem. Ecol.*, **4**, 211–224.

Roelofs, W. L., Hill, A. S., Linn, C. E., Meinwald, J., Jain, S. C., Herbert, H. J., and Smith, R. F. (1982). 'Sex pheromone of the winter moth, a geometrid with unusually low temperature precopulatory responses', *Science*, **217**, 657–659.

Roeske, C. N., Seiber, J. N., Brower, L. P., and Moffitt, C. M. (1976). 'Milkweed cardenolides and their comparative processing by monarch butterflies (*Danaus plexippus* L.)', in *Biochemical Interactions between Plants and Insects* (Recent Adv. Phytochem.) (eds J. M. Wallace and R. L. Mansell), pp. 93–167, Plenum, New York.

Röller, H., Biermann, K., Bjerke, J., Norgard, D., and McShan, W. H. (1968). 'Sex pheromones of pyralid moths. I. Isolation and identification of the sex attractant of *Galleria mellonella* (greater wax moth)', *Acta Entomol. Bohemoslov.*, **65**, 208–211.

Rossi, R., and Niccoli, A. (1978). 'Relationship between chirality and biological activity', *Naturwissenschaften*, **65**, 259.

Rossiter, M., and Staddon, B. W. (1983). '3-Methyl-2-hexanone from the triatomine bug *Dipetalogaster maximus* (Uhler) (Heteroptera; Reduviidae)', *Experientia*, **39**, 380–381.

Rothschild, M. (1973). 'Secondary plant substances and warning coloration in insects', in *Insect–Plant Relationships. Symp. R. Ent. Soc. Lond. 6th*, (ed. H. F. van Emden), pp. 59–83, Blackwell, Oxford.

Rothschild, M., and Aplin, R. T. (1971). 'Toxins in tiger moths (Arctiidae: Lepidoptera)', in *Chemical Releasers in Insects* (*Pesticide Chemistry III*) (ed. A. S. Tahori), pp. 177–182, Gordon and Breach, New York.

Schildknecht, H. (1970). 'The defensive chemistry of land and water beetles (Dytiscidae)', *Angew. Chem. Int. Ed.*, **9**, 1–9.

Schildknecht, H. (1971). 'Evolutionary peaks in the defensive chemistry of insects', *Endeavour*, **30**, 136–141.

Schildknecht, H. (1976). 'Chemical ecology—a chapter of modern natural products chemistry', *Angew. Chem. Int. Ed.*, **15**, 214–222.

Schildknecht, H., Birringer, H., and Krauss, D. (1969). 'Aufklärung des gelben Prothorakalwehrdrüsen-Farbstoffes von *Ilybius fenestratus*', *Z. Naturforsch.*, **24B**, 38–47.

Schildknecht, H., and Holoubek, K. (1961). 'Die Bombardierkäfer und ihre Explosionschemie. V. Mitteilung über Insekten-Abwehrstoffe', *Angew. Chem.*, **73**, 1–7.

Schmuff, N. R., Phillips, J. K., Burkholder, W. E., Fales, H. M., Chen, C. W., Roller, P. P., and Ma, M. (1984). 'The chemical identification of the rice weevil and maize weevil aggregation pheromone', *Tetrahedron Lett.*, **25**, 1533–1534.

Schneider, D., Boppre, M., Zweig, J., Horsley, S. B., Bell, T. W., Meinwald, J., Hansen, K., and Diehl, E. W. (1982). 'Scent organ development in *Creatonotes* moths: regulation by pyrrolizidine alkaloids', *Science*, **215**, 1264–1265.

Schoonhoven. L. M. (1972). 'Secondary plant substances and insects', *Recent Adv. Phytochem.*, **5**, 197–224.

Shearer, D. A., and Boch, R. (1966). 'Citral in the Nassonoff pheromone of the honey bee', *J. Insect Physiol.*, **12**, 1513–1521.

Shorey, H. H. (1973). 'Behavioural responses to insect pheromones', *Ann. Rev. Entomol.*, **18**, 349–380.

Silverstein, R. M. (1979). 'Enantiomeric composition and bioactivity of chiral semiochemicals in insects', In *Chemical Ecology: Odour Communication in Animals* (ed. F. J. Ritter) pp. 133–146, Elsevier/N. Holland, Amsterdam.

Silverstein, R. M., Rodin, J. O., Burkholder, W. E., and Gorman, J. E. (1967). 'Sex attractant of the black carpet beetle', *Science*, **157**, 85–87.

Silverstein, R. M., Rodin, J. O., and Wood, D. L. (1966). 'Sex attractants in frass produced by male *Ips confusus* in ponderosa pine', *Science*, **154**, 509–510.

Smith, L. M., Smith, R. G., Loehr, T. M., Daves, G. D., Daterman, G. E., and Wohleb, R. H. (1978). 'Douglas-fir tussock moth pheromone: identification of a diene analogue of the principal attractant and synthesis of stereochemically defined 1,6-, 2,6- and 3,6-heneicosadien-11-ones', *J. Org. Chem.*, **43**, 2361–2366.

Smith, R. G., Daterman, G. E., and Daves, G. D. (1975). 'Douglas-fir tussock moth: sex pheromone identification and synthesis', *Science*, **188**, 63–64.

Smolanoff, J., Kluge, A. F., Meinwald, J., McPhail, A., Miller, R. W., Hicks, K., and Eisner, T. (1975). 'Polyzonimine: A novel terpenoid insect repellent', *Science*, **188**, 734–736.

Sonnet, P. E., Uebel, E. C., Harris, R. L., and Miller, R. W. (1977a). 'Sex pheromone of the stable fly: evaluation of methyl- and 1,5-dimethyl alkanes as mating stimulants', *J. Chem. Ecol.*, **3**, 245–249.

Sonnet, P. E., Uebel, E. C., Lusby, W. R., Schwarz, M., and Miller, R. W. (1979). 'Sex pheromone of the stable fly. Identification, synthesis and evaluation of alkenes from female stable flies', *J. Chem. Ecol.*, **5**, 353–361.

Sonnet, P. E., Uebel, E. C., and Miller, R. W. (1977b). 'An unusual polyene from male stable flies', *J. Chem. Ecol.*, **3**, 251–255.

Ställberg-Stenhagen, S. (1970). 'The absolute configuration of terrestrol', *Acta Chem. Scand.*, **24**, 348–360.

Steck, W., Underhill, E. W., and Chisholm, M. D. (1982). 'Structure–activity relationships in sex attractants for North American noctuid moths', *J. Chem. Ecol.*, **8**, 731–754.

Sternlicht, M. (1973). 'Parasitic wasps attracted by the sex pheromone of the coccid host', *Entomophaga*, **18**, 339–342.

Stokes, J. B., Uebel, E. C., Warthen, J. D. Jr, Jacobson, M., Flippen-Anderson, J. L. Gilardi, R., Spishakoff, L. M., and Wilzer, K. R. (1983). 'Isolation and identification of novel lactones from male Mexican fruit flies', *J. Agric. Food Chem.*, **31**, 1162–1167.

Streams, F. A., Shahjahan, M., and LeMasurier, H. G. (1968). 'Influence of plants on the parasitization of the tarnished plant bug by *Leiophron pallipes*', *J. Econ. Entomol.*, **61**, 996–999.

Suzuki, T. (1980). '4,8-Dimethyl decanal: the aggregation pheromone of the flour beetles, *Tribolium castaneum* and *T. confusum* (Coleoptera: Tenebrionidae)', *Agric. Biol. Chem.*, **44**, 2519–2520.

Suzuki, T. (1981). 'Identification of the aggregation pheromone of the flour beetles *Tribolium castaneum* and *T. confusum* (Coleoptera: Tenebrionidae)', *Agric. Biol. Chem.*, **45**, 1357–1363.

Suzuki, T., and Mori, K. (1983). '(4R, 8R)-(-)-4,8-Dimethyl decanal: the natural aggregation pheromone of the red flour beetle, *Tribolium castaneum* (Coleoptera: Tenebrionidae)', *Appl. Entomol. Zool.*, **18**, 134–136.

Svensson, B. G., and Bergström, G. (1977). 'Volatile marking secretions from the labial gland of north European *Pyrobombus* D. T. males (Hymenoptera, Apidae)', *Insectes Sociaux*, **24**, 213–224.

Talman, E., Verwiel, P. E. J., Ritter, F. J., and Persoons, C. J. (1978). 'Sex pheromone of the American cockroach, *Periplaneta americana*', *Israel J. Chem.*, **17**, 227–235.

Tamaki, Y. (1977). 'Complexity, diversity and specificity of behaviour modifying chemicals in Lepidoptera and Diptera', In *Chemical Control of Insect Behaviour: Theory and Application* (eds H. H. Shorey and J. J. McKelvey Jr), pp. 253–285, Wiley, New York.

Tamaki, Y. (1984). 'Sex pheromones', In *Comprehensive Insect Physiology, Biochemistry and Pharmacology* (eds G. A. Kerkut and L. I. Gilbert), Vol. 9, pp. 145–191, Pergamon Press, Oxford.

Tamaki, Y., Honma, K., and Kawasaki, K. (1977). 'Sex pheromone of the peach fruit moth *Carposina nipponensis* Walsingham (Lepidoptera: Carposinidae); isolation, identification and synthesis.' *Appl. Ent. Zool.*, **12**, 60–68.

Tamaki, Y., Noguchi, H., Sugie, H., Sato, R., and Kariya, A. (1979). 'Minor components of the female sex-attractant pheromone of the smaller tea tortrix moth (Lepidoptera: Tortricidae): isolation and identification', *Appl. Entomol. Zool.*, **14**, 101–113.

Tashiro, H., Gieselmann, M. J., and Roelofs, W. L. (1979). 'Residual activity of a California red scale synthetic pheromone component', *Environ. Entomol.*, **8**, 931–934.

Tengö, J., and Bergström, G. (1976). 'Comparative analysis of lemon-smelling secretions from heads of *Andrena* F. (Hymenoptera: Apoidea) bees', *Comp. Biochem. Physiol.*, **55B**, 179–188.

Tengö, J., and Bergström, G. (1977). 'Cleptoparasitism and odor mimetism in bees. Do *Nomada* males imitate the odor of *Andrena* females?', *Science*, **196**, 1117–1119.

Tumlinson, J. H., Hardee, D. D., Gueldner, R. C. M., Thompson, A. C., Hedin, P. A., and Minyard, J. P. (1969). 'Sex pheromone produced by male boll weevil: isolation, identification and synthesis', *Science*, **166**, 1010–1012.

Tumlinson, J. H., Klein, M. G., Doolittle, R. E., Ladd, T. L., and Proveaux, A. T. (1977). 'Identification of the female Japanese beetle sex pheromone: inhibition of male response by an enantiomer', *Science*, **197**, 789–792.

Tumlinson, J. H., Moser, J. C., Silverstein, R. M., Brownlee, R. G., and Ruth, J. M. (1972). 'A volatile pheromone of the leaf-cutting ant, *Atta texana*', *J. Insect Physiol.*, **18**, 809–814.

Tursch, B., Braekman, J. C., and Daloze, D. (1976). 'Arthropod alkaloids', *Experientia,* **32**, 401–407.

Tursch, B., Daloze, D., Dupont, M., Pasteels, J. M., and Tricot, M-C (1971). 'A defense alkaloid in a carnivorous beetle', *Experientia,* **27**, 1380–1381.

Uebel, E. C., Schwarz, M., Miller, R. W., and Menzer, R. E. (1978). 'Mating stimulant pheromone and cuticular lipid constituents of *Fannia femoralis* (Stein) (Dipetra: Muscidae)', *J. Chem. Ecol.,* **4**, 83–93.

Uebel, E. C., Sonnet, P. E., Menzer, R. E., Miller, R. W., and Lusby, W. R. (1977). 'Mating stimulant pheromone and cuticular lipid constituents of the little house fly, *Fannia canicularis* (L.)', *J. Chem. Ecol.,* **3**, 269–278.

Uebel, E. C., Sonnet, P. E., and Miller, R. W. (1976). 'House fly sex pheromone: Enhancement of mating strike activity by combination of (Z)-9-tricosene with branched saturated hydrocarbons', *Environ. Entomol.,* **5**, 905–908.

Uebel, E. C., Sonnet, P. E., Miller, R. W., and Beroza, M. (1975). 'Sex pheromone of the face fly, *Musca autumnalis* De Geer (Diptera: Muscidae)', *J. Chem. Ecol.,* **1**, 195–202.

Underhill, E. W., Palaniswamy, P., Abrams, S. R., Bailey, B. K., Steck, W. F., and Chisholm, M. D. (1983). 'Triunsaturated hydrocarbons, sex pheromone components of *Caenurgina erechtea*', *J. Chem. Ecol.,* **9**, 1413–1422.

Underhill, E. W., Steck, W., Chisholm, M. D., Worden, H. A., and Howe, J. A. G. (1978). 'A sex attractant for the cottonwood crown borer, *Aegeria tibialis*', *Can. Entomol.,* **110**, 495–498.

Veith, H. J., and Koeniger, N. (1978). 'Identifizierung von cis-9-pentacosen als Ausloser für das Warmen der Brut bei der Hornisse', *Naturwissenschaften,* **65**, 263.

Vick, K. W., Burkholder, W. E., and Gorman, J. E. (1970). 'Interspecific response to sex pheromones of *Trogoderma species* (Coleoptera: Dermestidae)', *Ann. Entomol. Soc. Am.,* **63**, 379–381.

Vinson, S. R., Jones, R. L., Sonnet, P., Bierl, B. A., and Beroza, M. (1975). 'Isolation, identification and synthesis of host-seeking stimulants for *Cardiochiles nigriceps,* a parasitoid of the tobacco budworm', *Entomol. Exp. Appl.* **18**, 443–450.

Vite, J. P., and Williamson, D. L. (1970). '*Thanasimus dubius*: prey perception', *J. Insect Physiol.,* **16**, 233–239.

Vite, J. P., Heddon, R., and Mori, K. (1976). '*Ips grandicollis*: field responses to the optically pure pheromone', *Naturwissenschaften,* **63**, 43–44.

Von Euw, J., Reichstein, T., and Rothschild, M. (1968). 'Aristolochic acid – I in the swallowtail butterfly *Pachiloptera aristolochiae*', *Israel J. Chem.,* **6**, 659–670.

Wadhams, L. J., Baker, R., and Howse, P. E. (1974). '4,11-Epoxy-*cis*-eudesmane, a novel oxygenated sesquiterpene in the frontal gland secretion of the termite *Amitermes evuncifer*', *Tetrahedron Lett.,* 1697–1700.

Wakamura, S. (1978). 'Sex attractant pheromone of the common cutworm moth, *Agrotis fucosa* Butler (Lepidoptera: Noctuidae): isolation and identification', *Appl. Entomol. Zool.,* **13**, 290–295.

Wall, C., Greenway, A. R., and Burt, P. E. (1976). 'Electroantenographic and field responses of the pea moth *Cydia nigricana* to sex attractants and related compounds', *Physiol. Entomol.,* **1**, 151–157.

Weatherston, J., and Percy, J. E. (1970). 'Arthropod defensive secretions, in *Chemicals Controlling Insect Behaviour* (ed. M. Beroza), pp. 95–144, Academic Press, New York.

Wientjens, W. H. J. M., Lakwijk, A. C., and Van Der Marel, T. (1973). 'Alarm pheromone of grain aphids', *Experientia,* **29**, 658–660.

Williams, H. J., Silverstein, R. M., Burkholder, W. E., and Khorramshahi, A. (1981a). 'Dominicalure 1 and 2, the components of the aggregation pheromone from the male lesser grain borer, *Rhyzopertha dominica* (F.) (Coleoptera: Bostrichidae)', *J. Chem. Ecol.*, **7**, 759–780.

Williams, H. J., Strand, M. R., and Vinson, S. B. (1981b). 'The trail pheromone of the imported fire ant *Solenopsis invicta* (Buren)', *Experientia*, **37**, 1159–1160.

Wong, J. W., Palaniswamy, P., Underhill, E. W., Steck, W. F., and Chisholm, M. D. (1984). 'Novel sex pheromone components from the fall cankerworm moth, *Alsophila pometaria*', *J. Chem. Ecol.*, **10**, 463–473.

Wood, D. L. (1972). 'Selection and colonization of ponderosa pine by bark beetles', in *Insect/Plant Relationships* (ed. H. F. Van Emden) *Symp. R. Ent. Soc. Lond.*, **6**, 101–117, Blackwell, Oxford.

Wood, D. L. (1980). *Environmental Protection and Biological Forms of Control of Pest Organisms* (eds B. Lundhold and M. Stackerud), Ecol. Bull. No. 31, Stockholm, p. 41.

Wood, D. L., Browne, L. E., Ewing, B., Lindahl, K., Bedard, W. D., Tilden, P. E., Mori, K., Pitman, G. B., and Hughes, P. R. (1976). 'Western pine beetle: specificity among enantiomers of male and female components of an attractive pheromone', *Science*, **192**, 896–898.

Yarger, R. G., Silverstein, R. M., amd Burkholder, W. E. (1975). 'Sex pheromone of the female dermestid beetle *Trogoderma glabrum* (Herbst)', *J. Chem, Ecol.*, **1**, 323–334.

Zagatti, P. (1981). 'Comportement sexuel de la pyrale de la canne à sucre *Eldana saccharina* (Wlk.) lié à deux phéromones émises par le mâle', *Behaviour*, **78**, 81–98.

Zagatti, P., Kunesch, G., and Morin, N. (1981). 'La vanilline, constituant de la secretion aphrodisiaque emise par les androconis du mâle de la pyral de la canne à sucre: *Eldana saccharina* (Wlk.) (Lepidoptera, Pyralidae, Galleriinae)', *C.R. Acad. Sci. Paris*, **292**, 633–635.

Index

Contents—Volume 1

Contents—Volume 2

Contents—Volume 3

Contents—Volume 4

Contents—Volume 5

Insecticides